*Proceedings of the Conference*

# Applying Research To Hydraulic Practice

*Sponsored by the*
*Hydraulics Division*
*of the*
*American Society of Civil Engineers*

*Jackson, Mississippi*
*August 17-20, 1982*

**Peter E. Smith, Editor**

Published by the
American Society of Civil Engineers
345 East 47th Street
New York, New York 10017

The Society is not responsible for any statements made or opinions expressed in its publications.

FOREWORD

The theme of the 1982 ASCE Hydraulics Division Special Conference is "Applying Research to Hydraulic Practice" -- a very timely theme considering the state of affairs in our profession today. Results of research are suggesting many new approaches to solving hydraulic problems and it is evident that there is a need to see where these approaches can be applied in hydraulic practice. It is hoped that this proceedings will help to partially fulfill this need.

A special historical note marks this years conference. The holding of the conference in Jackson, Mississippi is a return to the site of the very first Hydraulics Division Specialty Conference, and for that matter the first Specialty Conference in all of ASCE. The first was held in 1950 and its purpose was tQ provide a forum for the serious consideration and discussion of hydraulic problems.

The approximately 75 papers in this volume cover a wide range of topics and are arranged into 30 sessions in the order of presentation. There are a number of papers that were not received at the time of printing and are omitted from the proceedings, as noted in the Table of Contents. Copies of these papers are available from the authors.

This years conference was jointly sponsored by the Hydraulics Division and the Mississippi Section. Organizing and planning for the conference has been the responsibility of the following committee members:

Co-Chairmen:
   Marshall E. Jennings, Conference Chairman
   Lex P. Collins, Program Chairman

Executive Committee Representation:
   Patrick J. Ryan

Program:
   Danny L. King
   David L. Schreiber
   H. T. Shen

Local Arrangements:
   Marden B. Boyd
   Billy E. Colson
   Mark S. Dortch
   James E. Foster
   John L. Grace, Jr.
   Gerald G. Parker, Jr.
   Verne R. Schneider
   Peter E. Smith
   Kenneth V. Wilson, deceased

# CONTENTS

## DROP STRUCTURES
### (Session 3A)

## TURBULENCE MODELS IN HYDRAULICS COMPUTATIONS I
### (Session 3B)

## SMALL HYDRO DESIGN INNOVATIONS AND APPLICATIONS I
### (Session 3C)

## NEW SPILLWAY CONCEPTS
### (Session 4A)

*Papers not available at time of printing

*Papers not available at time of printing

*Papers not available at time of printing

## METRICATION IN WATER RESOURCES
### (Session 9 A,B,C)

## MODELING OF HYDRAULIC SYSTEMS II
### (Session 10A)

## TIDAL HYDRAULICS
### (Session 10B)

## SEDIMENT TRANSPORT I
### (Session 11A)

## VERIFICATION OF MODELS OF HYDROLOGIC TRANSPORT
## AND DISPERSION I
### (Session 11B)

*Papers not available at time of printing

*Papers not available at time of printing

## CONTROL OF WATER QUALITY IN RESERVOIRS AND RELEASES
### (Session 14A)

## HYDRAULICS OF TIDAL INLETS
### (Session 14B)

*Papers not available at time of printing

MODEL-PROTOTYPE SCOUR AT YOCONA DROP STRUCTURE

By Fred W. Blaisdell,[1] F.ASCE, Kenneth M. Hayward,[2] M.ASCE
and Clayton L. Anderson,[3] M.ASCE

INTRODUCTION

    The scour measured at the exit of a straight drop spillway
stilling basin model and its prototype are compared. There are three
major parts to the paper:  1) We begin with a description of those
features of the stilling basin design and its performance that are
pertinent to our presentation.  2) Following this brief description
of the stilling basin, its performance, and the dimensionless laboratory
model tests, we describe the Yocona River prototype structure.  3) We
conclude our presentation with a comparison of the model and prototype
scour patterns.

STRAIGHT DROP SPILLWAY STILLING BASIN

    Basin Geometry.--As shown in Fig. 1, the straight drop spillway is
a rectangular overfall weir. Its primary use is as a grade control
structure in waterways.  In Soil Conservation Service (SCS) terminology,
this spillway with its stilling basin is called the Type C structure.
The stilling basin contains floor blocks and an end sill. The sloping
wingwalls prevent scour caused by recirculating flow along the sides of
the main stream leaving the basin.

    The total length of the stilling basin $L_B$ is made up of the three
sub-lengths shown in Fig. 1. First, there is the distance from the weir
to the point where the upper surface of the nappe strikes the basin
floor $x_a$. Second, there is the distance from the nappe to the floor
blocks $x_b$. Finally, there is the distance from the floor blocks to the
end of the basin $x_c$. These latter two lengths depend only on the
critical depth $d_c$. ($x_b=0.8d_c$ and $x_c \geqslant 1.75d_c$) The critical depth is
computed from the discharge per unit width of the basin.

[1]Research Hydr. Engr., U.S. Dept. of Agr., Agricultural Research
Service, St. Anthony Falls Hydr. Lab., Third Avenue SE at Mississippi
River, Minneapolis, Minnesota  55414.

[2]Head, State Design Unit, U.S. Dept. of Agr., Soil Conservation
Service, P.O. Box 157, University, Mississippi  38677.

[3]Research Hydr. Engr., U.S. Dept. of Agr., Agricultural Research
Service, St. Anthony Falls Hydr. Lab., Third Avenue SE at Mississippi
River, Minneapolis, Minnesota  55414.

1

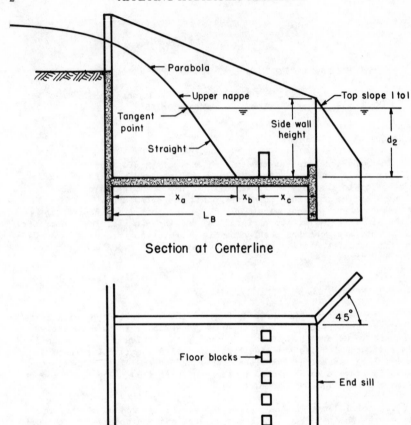

Fig. 1.-Straight Drop Spillway Stilling Basin

Nappe Distance.--The nappe distance $x_a$ depends on the tailwater depth. The nappe is free falling until it enters the tailwater surface, and its trajectory approximates that of a projectile. After entering the tailwater surface the nappe continues at a slope that is tangent to the free-falling nappe at the tailwater surface. If the drop is high and the tailwater is relatively low, the nappe slope is steep and the nappe impingement length is relatively short. However, as the tailwater rises, the point of tangency also rises and the nappe slope at the point of tangency flattens. The trajectory of the nappe below the tailwater level also flattens and the nappe impingement

length increases. If the higher tailwater causes the nappe to impinge
on or downstream of the floor blocks, the floor blocks become
ineffective.

For tailwater levels that are higher than the spillway crest--as is
the case for the Yocona spillway--, the submerged nappe trajectory is so
flat the jet may impinge on the stream bed beyond the basin exit. If
this occurs, the jet will scour the downstream channel. The solution is
to lengthen the distance to the floor blocks $x_a$ to ensure that the jet
strikes the basin floor upstream of the floor blocks.

Although high tailwater levels require long stilling basins, there
is a maximum basin length. If the tailwater level is more than
two-thirds the critical depth above the spillway crest, the jet floats
on the tailwater surface. The jet then will not attack the stream bed.
As a result, the maximum length of basin is determined by that tailwater
level that is above the spillway crest by two-thirds of the critical
depth. This condition determined the length of the Yocona basin.

Research Philosophy.--An additional point regarding a philosophy
employed during the laboratory tests needs to be mentioned. The
stilling basin could have been made long enough to completely dissipate
the destructive energy within the basin. This was felt to be
uneconomical. As a result, the research philosophy employed was to
initiate the energy dissipation within the basin, and complete the
energy dissipation in a stream enlargement between the basin exit and
the downstream channel. The laboratory development made the enlargement
a required component of the stilling basin. Unfortunately, this
requirement was not sufficiently emphasized when the test results were
published.[4] Although a paragraph and field photograph were included in
the closing discussion[5] to correct this omission, the requirement of a
stream enlargement downstream of the basin exit frequently has not been
followed in practice.

The assumption made during the generalized laboratory studies was
that in the field the enlargement would be self-formed by erosion. This
erosion would take place rapidly during flows approaching design
capacity. But, by properly proportioning the stilling basin components
to control the flow pattern beyond the basin exit, the scour would be
self-limiting. All the evidence we have obtained from the field
indicates this assumption was valid.

Performance Evaluation.--In the laboratory, the stilling basin
performance was evaluated by the scour that developed in concrete sand

---

[4]Donnelly, C. A. and Blaisdell, F. W., "Straight Drop Spillway
Stilling Basin," Journal of the Hydraulics Division, ASCE, Vol. 91, No.
HY3, Proc. Paper 4328, May, 1965, pp. 101-131. (a) Table 6.

[5]Donnelly, C. A. and Blaisdell, F. W., closing discussion of
"Straight Drop Spillway Stilling Basin," Journal of the Hydraulics
Division, ASCE, Vol. 92, No. HY4, Proc. Paper 4859, July, 1966,
pp. 140-145.

placed in the model downstream channel. This is illustrated in Fig. 2 where the scour holes developed at two half-models are compared: One half-model has floor blocks and an end sill; the other half-model has a plain apron. It was recognized that quantitative scour data was not being obtained from the model tests and that scaled prototype scour dimensions would differ from those obtained during the model studies. However, the sand was adequate to qualitatively measure the effects of changes in the arrangements and proportions of the basin elements and to establish the form of the scour patterns to be expected at prototype structures.

Fig. 2.-Comparison of Scour at Straight Drop Spillway Stilling Basins Left, half-model with floor blocks and end sill; right, half-model with plain apron.

A further consideration regarding the use of non-similar bed materials for dimensionless model studies is that, because the flow velocities decrease rapidly with lateral distance from the edge of the stream leaving the basin, the velocity gradient is steep and the distance between the velocities that scour and velocities that do not scour is small. As a result, the change in scour dimensions with change in the bed material size is relatively small and the dimensions of the scour are more dependent, within limits, on the flow pattern than they are on the size of the bed material. Therefore, prototype scour dimensions predicted from the model studies, while admittedly qualitative, are, nevertheless, roughly quantitative.

Application History.--The generalized criteria for the design of
the straight drop spillway stilling basin--the SCS Type C
structure--were developed from laboratory studies during the early
1950's.[4] Originally intended for use on farm-size watersheds, in the
years since then the stilling basin has been successfully used with
increasingly larger structures. Also, originally intended for
infrequent full-capacity operation only during 50- to 100-year frequency
storms, large structures are now being designed for 1- to 5-year
frequency storms. Although these changes in intended size and frequency
of full-capacity operation have severely tested the original design
criteria, the prototype installations have performed satisfactorily.
The Yocona River straight drop spillway built by SCS is a good example
of a large structure designed for frequent, full-capacity flows.


YOCONA RIVER GRADE CONTROL STRUCTURE

The Yocona River grade control structure is located near Oxford,
Mississippi. It is upstream of the Corps of Engineers' Enid Reservoir.
The lower portion of the Yocona River was excavated in the early 1900's.
This excavation was terminated about one mile downstream of the drop
structure. After the channel was excavated, overfalls developed and
migrated upstream. By the 1970's, the overfall at the structure site
was about 19 feet high--the upstream channel was 14 feet deep and the
downstream channel was 33 feet deep. The structure was designed to
control the 19-foot overfall.

Structure Description.--The structure is near the middle of the
Yocona River watershed. The drainage area at the structure site is 158
square miles and the 100-year frequency storm discharge is 39,000 cfs.
However, the capacity of the drop structure is 11,200 cfs. This is
approximately a 5-year frequency flow. This is also the flow required
to fill bank-full the meandering and unimproved downstream channel.
Since the controlled drop is 19 feet and the channel is 33 feet deep,
for bank-full flow the tailwater submerges the crest by 14 feet.
Flows in excess of 11,200 cfs cover the flood plain, pass over an
emergency spillway 1,500 feet long, and reenter the river at a planned
downstream point.

The Yocona River straight drop spillway was completed in 1976 at a
cost of $925,061. The spillway crest is 100 feet long and 19 feet deep.
The drop to the stilling basin floor is 21.5 feet. The end sill is 2.5
feet high, making the controlled drop 19 feet. The stilling basin is
116 feet long.

Riprap was placed on the downstream dam slopes outside of the
wingwalls. More pertinent to our subsequent discussion is that riprap
placed level with the top of the end sill extends 48 feet downstream.

Flow History.--Within less than a year of its completion, the
Yocona drop structure experienced two near full-capacity flows. These
occurred in March and April 1977 and gave the structure and the design
concepts a thorough test. These two storms did not create any emergency
spillway flow, but they did verify the SCS design concept that the

downstream channel had sufficient meander and roughness to insure
bank-full flow prior to flow over the emergency spillway.  In fact, both
inlet and outlet channels flowed bank-full.  Photographs show that the
drop in the water surface through the structure was completely "swamped
out."

A U.S. Geological Survey stream recorder on the Yocona River
approximately 7.5 miles downstream from the Yocona structure and a
rainfall recorder at Oxford, Mississippi, about 10 miles from the
structure, will be used to project a qualitative history of large flows
experienced at the Yocona structure between 1977 and 1982.  The March
and April 1977 storms, whose flows were observed and photographed, will
be the reference storms.

One additional storm of slightly less magnitude occurred in 1977.
In 1978, one storm approximately equal to the reference storms and one
slightly smaller storm occurred.  In 1979, the largest storm during the
1977-1982 period exceeded the reference storms by a considerable amount.
Also during 1979, four storms of significant magnitude occurred, but
were considerably smaller than the reference storms.  During 1980, four
large storms occurred but all were smaller than the reference storms.
In 1981 and 1982 this area experienced a marked reduction in large
storms and the flows have not been nearly as large as the 1977 storms.

Scour History.--The Soil Conservation Service surveyed cross
sections of the exit channel in 1976 after completion of the
construction.  A cross section range survey was made after the 1977
storms.  This survey showed that there had been no movement of the
riprap, but 100 feet downstream from the basin exit the scour depth had
reached 3.0 feet.  Also, sand deposits of as much as 2.5 feet had formed
above the excavated ditch bottom on both sides of the channel.  Although
some of this sand deposit may have come from downstream scour, probably
most of it originated upstream of the structure.

Naturally, there was concern that the scour might endanger the
structure, and plans were made to repair the scour hole.  However, a
study of the scour pattern showed a similarity to that observed during
the original laboratory studies.  In other words, the prototype scour
and deposition apparently was self-adjusting itself to the enlargement
anticipated during the laboratory tests.

As a result of these considerations, the intended repair was not
made.  Instead, a program of monitoring the progress of the scour and
deposition was initiated.  Annual surveys have shown no movement of the
riprap which extends 48 feet from the basin exit.  Scour begins at the
end of the riprap and reaches depths at distances from the stilling
basin exit listed in Table 1.

The thalweg depths of scour are shown in Fig. 3 for the six surveys
made since completion of construction.  There has been alternate scour
and deposition, both in the scour hole downstream of the riprap and in
the downstream channel.  The maximum depth of scour occurs about 28 feet
downstream of the riprap or 75 feet downstream of the stilling basin
exit.  Unless there is a change in the hydrologic or geomorphologic

TABLE 1.--Maximum depths and locations of scour

| Survey date | Maximum scour depth below end sill elevation, feet | Distance from end of basin, feet |
|---|---|---|
| July 1977 | 3.0 | 100 |
| June 1978 | 8.7 | 65 |
| October 1979 | 6.6 | 75 |
| September 1980 | 10.1 | 75 |
| March 1982 | 6.1 | 75 |
| Model ck. test 20 | 0.9 | 14.6 |
| " 18 | 2.5 | 21.6 |
| " 9 | 4.1 | 84.6 |

conditions, continued scour and deposition, probably within the approximate limits shown in Fig. 3, can be expected.

A computer-generated contour map for one of the six surveys is shown in Fig. 4. Although the scour holes are located further downstream in the prototype because of the riprap, the shape and location of the prototype scour holes are similar to those observed during the laboratory tests.

The flow leaving the stilling basin generates back eddies along the sides of the channel. Deposition of sediment occurs in the center of the eddies. The deposits show well in Fig. 4, the 1982 survey map. The height of the deposit has increased between each survey and the deposit is now about 14 feet deep. For the last four surveys, the increase in the height of the deposit has been about 2 feet between each survey.

SCOUR COMPARISONS

Included in Table 1 are scaled up measurements of the scour observed during laboratory check tests of the stilling basin criteria.[4(a)] The data have been scaled to prototype dimensions on the basis of critical depths in the model and prototype. The linear scale ratios represented in the table are in the range of 1 to 18 to 1 to 24.

All of the depths of scour predicted from the laboratory tests are less than those observed in the prototype. One reason for this may be ascribed to the relative sizes of the bed material. The coarse sand used in the model had a Unified Soil Classification SW bed material and a mean size of 0.70 mm, whereas the prototype SM and CL bed material had a mean size of 0.18 mm--only 26 percent of the actual model bed material size and only 1.4 to 1.1 percent of the scaled-up model bed material size. Therefore, more relative scour should be expected in the prototype because of the relatively much finer bed material. Also the deepest prototype scour occurs further downstream than the model scour. However, if the prototype scour is measured from the end of the riprap, the distances to the deepest scour are comparable to those observed in the model.

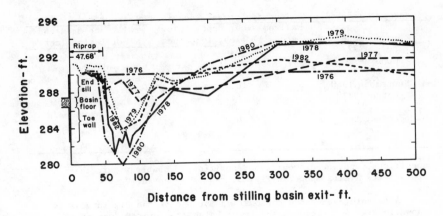

Fig. 3.-Thalweg Scour Depth

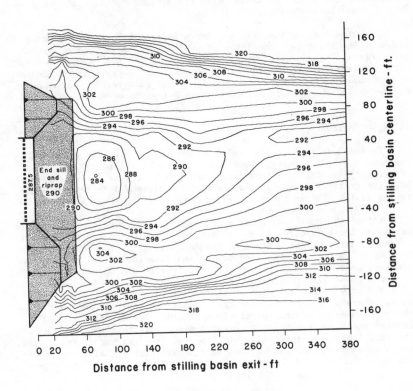

Fig. 4.-Scour Contours on March 11, 1982.

SUMMARY

The laboratory and field observations of scour are in general agreement, the prototype scour pattern is qualitatively similar to the model scour pattern, and the prototype scour depth exceeds the model scour depth due, in part at least, to the relatively finer prototype bed material. In any case, the scour that has been observed in the prototype does not endanger the structure. The monitoring of the scour and deposition indicates that the "watchful, do nothing" recommendation regarding repair of the initial scour that developed the predicted self-formed scour pattern has been proved valid.

# ROCK RIPRAP GRADIENT CONTROL STRUCTURES

Chodie T. Myers, Jr.,[1] M. ASCE

ABSTRACT: In some instances a rock riprap gradient control structure can be used economically to dissipate excess energy and establish a stable gradient in a channel where the gradient without some such control would be too steep and would cause erosive velocities. The structure consists of a riprap prismatic channel with a riprap transition at each end designed to flow within the subcritical range. Its essential design features are that the specific energy of the flow at design discharge is constant throughout the structure and is equal to the specific energy of the flow in the channel immediately upstream and downstream from the structure. Thus, for the design discharge the dissipation of hydraulic energy in the structure is at the same rate as the energy gain due to the gradient.

The Soil Conservation Service in Mississippi has constructed 14 rock riprap gradient control structures. In Tippah County, 12 are located on Muddy Creek and one on Tippah River. One is located on Running Slough Ditch in Panola County. The Design capacity of the structures range from 622 cfs (17.4 m$^3$/s) to 20,600 cfs (576.8 m$^3$/s). Since these type structures are designed based on sound theoretical hydraulics, they have not been model tested, but they have been inspected in the field periodically. All the structures have performed as designed with regard to establishing a stable gradient in the channel in which they are constructed.

In May 1980, field surveys revealed a deep scour hole forming at the downstream end of the exit transition on the three structures farthest downstream on Muddy Creek. However, no damages have occurred to the structures themselves. It was determined that a prototype study should be made on one of the structures on which the scour hole was developing. The Prototype Evaluation Branch of the U. S. Army Corps of Engineers, Waterways Experiment Station in Vicksburg entered into an agreement with the Soil Conservation Service and a prototype evaluation study was initiated. Measurement of point velocity and depth at various cross sections within the structure for the determination of velocity distribution and a roughness factor were performed. A comparison of assumed designed parameters versus actual measured parameters was made to verify the design assumptions and procedures.

---

[1]Civil Engineer, USDA, SCS, State Design Unit, University, MS.

## HYDRAULIC AND STABILITY DESIGN

The detailed design procedure for a rock riprap gradient control structure is published in Soil Conservation Service Technical Release No. 59[1]. The discussion of design procedures in this paper is limited to general principles necessary for understanding the operation of this type structure. The structure consists of a downstream transition, a prismatic channel section and an upstream transition. The general layout is shown in Fig. 1.

The design procedures for stability have been confined to clear water flows. The conveyance of bedloads presents a much more complex set of considerations. The design requires that the banks will not slough, erode or slide laterally, and that the bed will not aggrade or degrade beyond the design limits.

The design discharge is assumed to be wholly within the banks of the structure. The design discharge used for proportioning the structure is the same as the discharge used for proportioning of the earth channel. Usually structures that are stable for the design discharge will also be stable for all discharges less than the design discharge; however, if the structure tailwater decreases very rapidly with small decrease in discharges, discharges less than the design discharge may cause higher tractive stresses than the design discharge. Thus when tailwater changes rapidly with discharge, the design should be studied for the design discharge and lesser discharges. Also the design of the structure should be checked for stability for flows greater than design discharge. When out-of-bank flow occurs, the rapid rise in tailwater that usually occurs decreases the energy slope, thereby decreasing the tractive stresses on the rock riprap; however, this condition only exists when the out-of-bank flow has easy access to return to the channel flow downstream of the structure.

The structure is designed to maintain a constant specific energy head within the subcritical flow range. In a channel system having mild slopes, flows are subcritical. Thus the depth of flow for a given discharge is physically fixed by the downstream channel characteristics of the channel system and upstream characteristics have no effect on the depth. Water surface profiles are usually required for the channel downstream of the structure to evaluate the starting depth tailwater for the structure. For the purpose of stability analysis, the lowest probable starting depth corresponding to the design discharge should be used.

The basic criteria governing the design of the rock riprap structure are:

---

[1]Hydraulic Design of Riprap Gradient Control Structures, U. S. Department of Agriculture, Soil Conservation Service, Engineering Division

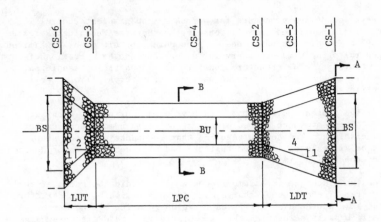

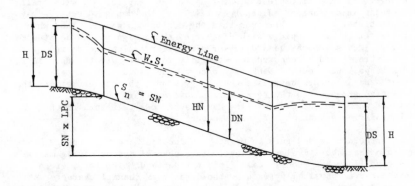

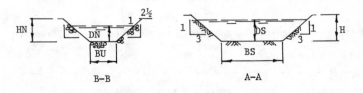

FIGURE 1

Structure Number 3, Muddy Creek

1.  The specific energy head, H, at every section of the riprap structure is set equal to the specific energy head at the junction of the downstream transition of the downstream channel, Section A-A of Figure 1. Specific energy head is given by the following equation (see Appendix for explanation of symbols).

$$H = d + \frac{v^2}{2g} = d + \frac{Q^2}{2ga^2} \tag{1}$$

2.  The prismatic channel bottom slope, $S_n$, is set equal to or less than 0.7 of the critical slope, $S_c$. The bottom slope, $S_n$, is expressed as a fraction of the critical slope, i. e.,

$$S_n = CS(s_c) \tag{2}$$

    Where          $0 \quad CS \quad 0.7$

3.  Manning's coefficient of roughness, n, is a function of the $D_{50}$ size of the riprap and has been evaluated to be

$$n = 0.0395 \ (D_{50})^{1/6} \tag{3}$$

4.  The critical tractive stress is a linear function of the $D_{50}$ size of the riprap, i.e.,

$$\tau_{bc} = 4.0 \ D_{50} \tag{4}$$

$$\tau_{sc} = K(4.0 \ D_{50}) \tag{5}$$

5.  The riprap size and structure dimensions are selected so that for the design discharge the maximum tractive stress on the riprap does not exceed the allowable tractive stress. Either side or bottom tractive stress may control.

For a given design discharge, Q, specific energy head, H, and side slope, z, the variables that must be adjusted to meet these conditions are bottom width, b, bottom slope, $S_n$, and riprap size, $D_{50}$.

The length of the prismatic channel, LPC, is equal to the vertical drop of the prismatic channel divided by the bottom slope, $S_n$. The vertical drop of the prismatic channel depends on the amount of gradient control required.

## PERFORMANCE OF CONSTRUCTED STRUCTURES

The Soil Conservation Service in Mississippi has constructed 14 rock riprap gradient control structures. One structure is located on Tippah River in Tippah County and became operational in October 1979. One was constructed on Running Slough Ditch in Panola County and became operational in November 1981. Both structures have performed as designed. The remaining 12 structures were constructed on Muddy Creek in Tippah County. Structures on Muddy Channel are numbered consecutively with structure No. 1 being located on the downstream end of this series of structures. Structures No. 1, 2 and 3 became operational during the fall of 1979. Structure No. 4 became operational during the summer of 1981. Structure No. 5 became operational during the winter of 1981 with the remaining structures becoming operational in the summer of 1982. Structures No. 1 through 4 have been operational a sufficient time to observe their performances. They have generally performed as designed with the following exceptions.

Surveys were obtained in May 1980 on structures No. 1, 2 and 3 on Muddy Creek. The design discharges range from 3,000 to 4,600 cfs. The design depths range from 8.5 to 9.8 feet. The surveys revealed that structures No. 1 and 2 had very deep scour holes immediately downstream from the end of the rock riprap. Surveys indicated that structure No. 1 had a scour hole 13 feet (4.0 m) below the constructed ditch bottom grade and structure No. 2 had a scour hole that had developed to a depth of 15 feet (4.6 m) below the constructed ditch bottom grade. The length of each of the scour holes below structures No. 1 and 2 was approximately 200 feet (61 m). Structure No. 3 had a scour hole that had developed approximately 4 feet (1.2 m) below the constructed ditch bottom grade and no pattern could be observed on the formation of the scour hole at this structure.

A series of hydraulic analyses were performed in an effort to determine the effects of ranges of Manning's "n" values on the hydraulics of the structure for the design depths of flow on structures No. 1-4. The results could be compared with field observations of the structures which have been subject to flows of these depths. The analyses were made assuming the earth channel "n" = 0.022 and 0.026 in combination with the riprap structure "n" = 0.035 and 0.04.

The design for capacity of the earth channel was based on aged condition "n" = 0.03 and the as built stability check was made for "n" = 0.025. The size of the rock riprap structure was based on an aged condition and for the "n" value for the riprap obtained from the National Cooperative Highway Research Program Report 108[1] (See Equation 3). The earth channel was stable for the as-built conditions but the effect of lowered tailwater for the as-built hydraulics on the structures was not checked in the original design.

---

[1]Tentative Design Procedure for Riprap Lined Channels, National Cooperative Highway Research Program Report 108.

The hydraulic analyses employed standard accepted computation procedures for computation of water surface profiles. Computation of the
rock riprap factor of safety against movement was based on the tractive
stress procedure with the critical tractive stress on the bed of the
channel assumed to be 4 x $D_{50}$.

Water surface profiles were run for the channel with a Manning's "n" of
0.022 and structures with a Manning's "n" of 0.035; for the channel
with a Manning's "n" of 0.022 and structures with a Manning's "n" of
0.04; for the channel with a Manning's "n" of 0.026 and structures with
a Manning's "n" of 0.035; and for the channel with a Manning's "n" of
0.026 and structures with a Manning's "n" of 0.04. The study indicates
that for most cases critical depth was obtained at the downstream end
of the prismatic channel and factors of safety for rock riprap movement
were less than one at this location (Fig. 2).

If it is assumed that the theoretical structural hydraulics conformed
to the actual field performance, then the results of these hydraulic
analyses did not definitively confirm the field observations regarding
riprap movement. The lowest riprap factors of safety computed were
for the side slopes at the downstream end of the prismatic section.
The only riprap movement observed in the field was in the bottom of
the downstream transition. The results of the analyses do, however,
tend to focus attention to the downstream end of the prismatic section
when the tailwater is lower than anticipated.

It seems probable that flow from the prismatic section did not
diverge at the assumed rate. This could have allowed the higher
velocity flow from the prismatic section to "jet" through the diverging transition. This could then permit upstream circulation to occur
along the outer portions of the diverging transition. This flow could
merge with the central flow from the prismatic section, thus increasing
the unit discharge and velocity in the center portion of the transition
and downstream channel. These factors would significantly increase
the tractive stresses on the bottom of both of these areas.

The lack of observable riprap movement on the side slopes could be due
to:

a.   Conservatism in the riprap design procedure.

b.   Larger riprap provided than specified.

c.   Resistance to movement created by interlocking of the angular
     rock.

## PROTOTYPE EVALUATION

Because of the development of the scour holes below structure Nos. 1, 2
and 3 on Muddy Creek channel and the determination during the hydraulic
and stability analysis of these sites that a range of "n" values could

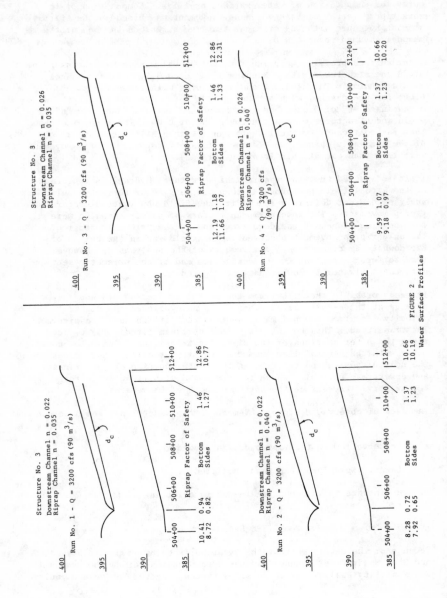

FIGURE 2
Water Surface Profiles

create flow conditions with a variety of factors of safety against
rock movement, it was determined that a prototype evaluation of
structure No. 3, Muddy channel should be performed.  The Prototype
Evaluation Branch of the U. S. Army Corps of Engineers, Waterways
Experiment Station at Vicksburg, Mississippi entered into an agreement
with the Soil Conservation Service and a prototype evaluation study was
initiated.  The purpose of the study was the measurement of point
velocities and depths at various cross sections within the structure
for determination of velocity distribution and roughness factor "n".

At the time of writing, only two measurements had been performed.  One
measurement was made on March 31, 1981 and an additional measurement on
January 4, 1982.  The cross sections on which measurements were taken
are shown in Fig. 1.  The large storms producing sufficient runoff for
measurement have been infrequent during the period the agreement has
been in effect.  The measurements taken have been when the flow ranged
from 2-4 feet (.6-1.2 m) with the design flow depth being 8.5 feet (2.6
m).  The velocity measurements were taken with a current meter suspended
from a boat.  Measurements were taken when adequate depth existed at a
ratio of 0.2 and 0.8 of the total depth  for cross sections 2, 3, 4 and
6.  At cross sections 1 and 5, the velocities were taken at 0.2, 0.4,
0.6 and 0.8 of the total depth for obtaining the velocity profile at
each measurement location of the cross section.  Figure 3 shows the
cross sections at locations 1, 5 and 2 with velocities for 0.2 depth
and the 0.8 depth for the measurement taken on January 4, 1982.  As can
be seen from studying these cross sections, the velocities at sections
1 and 5 indicate that there is an unequal velocity distribution or a
jet effect through the center of the section.  Cross section 2 showed
that there was an almost equal distribution of velocity throughout the
section.  Cross sections 4, 3 and 6, which are not shown, indicate an
almost equal velocity distribution throughout the cross section.  The
measurements made on March 31, 1981 showed similar velocity distribution for the same cross sections.

The discharge computed for both measurements indicate a considerable
change in quantity for the measurements at cross section 1 to the
measurement at cross section 6.  The measurements taken on March 31,
1981 had a range of discharge from 492 cfs (13.8 m$^3$/s) at cross
section 1 to 360 cfs (10.1 m$^3$/s) at cross section 6.  The measurements
taken on January 4, 1982 had a range of discharge from 906 cfs (25.4
m$^3$/s) at cross section 1 to 474 cfs (13.3 m) at cross section 6.  A
time span of approximately 4 hours existed for the measurements to be
made from cross section 1 to the measurements made at cross section 6.
The change in discharge values indicates that the measurements were
taken on the receding side of the hydrograph.  One of the purposes for
the prototype study was to determine the Manning "n" value for the
structure.  The varying discharges that occurred over the four hour
time span has made computation of Manning's "n" difficult and the
degree of accuracy is uncertain.  Manning's "n" values can be computed
by assuming the water surface profile to be parallel to the bottom
slope of the structure.  The "n" values computed by this method vary

Cross-section No. 1

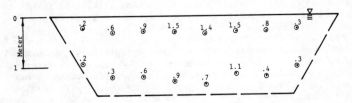

Cross-section No. 5

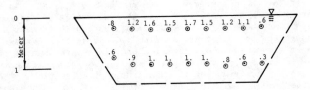

Cross-section No. 2

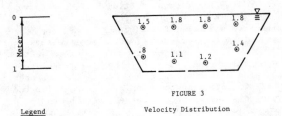

FIGURE 3

Legend

Velocity Distribution

⊙- Velocity in meters per second

from .036 to .055. The "n" values for design were approximately .04. Because of the small depth of flow during the measurements and the varying discharge, this method for calculation of "n" values is questionable.

## RECOMMENDATIONS

It is recommended that the prototype study be continued so that measurements are taken when the flow is near the design depth of 8.5 feet (2.6 m), and Manning "n" values can be accurately determined.

It is also recommended that model studies be performed to determine necessary changes in design of the structure components necessary to prevent the unequal velocity distribution (jet effect) that is occurring at the downstream exit of the structure as shown by the prototype study.

## APPENDIX I - NOTATION

$a$ = Flow area

$BS$ = Bottom width at the ends of the riprap structure

$BU$ = Bottom width of the prismatic channel

$C$ = Constant

$d$ = Depth of flow

$D_{50}$ = Size of rock in riprap of which 50% by weight is finer

$DN$ = Normal depth corresponding to the design

$DS$ = Depth of flow corresponding to the design discharge at the ends of the structure.

$g$ = Accelerations of gravity

$H$ = Specific energy head

$HN$ = Normal specific energy head corresponding to the design discharge in the prismatic channel

$K$ = Ratio of critical tractive stress on side slopes to critical tractive stress on the bottom of the trapezoidal channel

$LPC$ = Length of prismatic channel

$n$ = Manning's coefficient of roughness

$Q$ = Design discharge

Sc = SC = critical slope corresponding to the design discharge in the prismatic channel

$S_n$ = SN = Bottom slope of the prismatic channel

V = Velocity

z = Side slope of trapezoidal section expressed as a ratio of horizontal to vertical

$\tau_{bc}$ = The critical tractive stress for the riprap lining on the bottom of the trapezoidal channel

$\tau_{sc}$ = The critical tractive stress for the riprap lining on the side slopes of the trapezoidal channel

Design and Construction of Low Drop Structures

W. C. Little[1], M. ASCE and Robert C. Daniel[2]

## Abstract

Hydraulic design criteria for low drop channel grade control structures are reviewed. Guidelines and experiences in structural design, layout and construction are given. The hydraulic performance of a field structure is discussed. These low drop structures overcome the problems associated with low drop structures at high discharges where undulating waves are normally generated. A rock lined stilling basin with either a baffle pier or plate is provided to dissipate the energy through the drop.

## Introduction

Streams in northwestern Mississippi are incising into flood plains which are composed of several strata of fine grained alluvium of redeposited loess. These fine materials overlie unconsolidated sands and gravels. Vertical overfalls or knickpoints form and work upstream when a reach of channel erodes or is dug through the fine grained material, exposing the sands and gravels. As the headcut moves upstream, the height of overfall increases, leaving the downstream channel at a lower gradient. These headcuts vary in height from a fraction of a meter to several meters and move sporadically upstream as large runoff events occur. As they move farther upstream and become higher, bank instability problems become acute and large volumes of bank material are lost during extreme runoff events from mass bank failures.

These vertical headcuts present a great need for in-channel gradient control structures to halt the upstream migration of the headcuts and consequent loss of cropland from widening channels.

## Review of Hydraulic Design Procedures

Prior to 1980, hydraulic design procedures were not available for low drop, channel grade control structures. However, the need was so great that many action agencies designed and constructed structures which functioned as low drop structures. The authors observed the early attempts to provide bed control through low drop structures in Mississippi, Georgia, and South Carolina. The majority

---

[1]  Research Hydraulic Engineer, Erosion and Channels Research Unit, USDA Sedimentation Laboratory, Oxford, MS.

[2]  Civil Engineer, State Design Unit, Soil Conservation Service, University, MS.

of these structures were constructed by simply digging a trench
laterally across the channel and filling the trench with rock riprap
to form a horizontal approach apron, a short sloping chute and then a
horizontal apron at the downstream elevation. However, most of these
structures failed, generally from the water scouring a hole
immediately downstream. As the hole deepened the rocks were
undermined, fell into the scour hole causing complete failure,
permitting a new wave of erosion to continue upstream. Some of the
structures failed because no filter was provided so that the water
flowing through the rock voids eroded the bed material away until the
rocks caved into the scour hole.

Two major facts became apparent from observing these structures.
Firstly, a scour hole was eroded immediately below every structure.
Secondly, since no water cutoff was provided, the structures were
easily "undermined" by erosion of the bed through the voids in the
rock.

Little and Murphey (1982) incorporated both of these features
into the specific case model studies which began in 1975. They
provided a positive water cutoff with a sheet pile laterally across
the stream to serve as a weir and cutoff wall. They then dug a
stilling basin, similar in shape to those observed below the rock
structures, and provided a filter underneath the rock lined stilling
basin. During these experiments, another most important phenomenon
was observed. When the depth of flow over the weir (critical depth)
became approximately equal to the height of physical drop, undulating
waves (undular hydraulic jump) began to develop. As the flow depth
at the weir became greater, the undulating waves became higher and
persisted farther downstream. These undulating waves cause excessive
streambed erosion at the troughs where the velocity was highest.
Chow (1959, Fig. 15-4, page 397) shows that the undular hydraulic
jump exists from a Froude number of 1.0 up to approximately 1.73 and
that less than 5 percent of the energy in the flow is dissipated.

Observations by Little and Murphey (1981, 1982) that undulating
waves (undular hydraulic jump) are characteristic of low drop
structures led to an engineering definition of low drops. They
defined a low drop as "an hydraulic drop with a difference in
elevation between the upstream and downstream channel beds, H, a
discharge, Q with a corresponding critical depth, Yc, such that the
relative drop height, H/Yc, is equal to or less than 1.0."
Conversely, they defined a high drop as one with a relative drop
height, H/Yc, greater than 1.0.

Little and Murphey (1981, 1982) developed designs for a baffle
to disorganize and break up the undulating waves in a stilling basin
so the waves would not enter the downstream channel. Two types of
baffles were used, the baffle pier and baffle plate. The baffle pier
is a rectangular pier extending into the bed of the channel and the
baffle plate is a flat rectangular plate mounted vertically with the
flat side perpendicular to the flow. Little and Murphey (1982)
presented the following design criteria for the size and shape of
stilling basin and the size and placement of the baffle.

TENTATIVE DESIGN CRITERIA FOR LOW DROP STRUCTURES

Refer to Fig. 1 for the geometric layout.

Stilling Basin Dimensions (Fig. 1)

Stilling Basin Width, $W_{SB}$:

$$W_{SB} = 2B \tag{1}$$

where $W_{SB}$ is the maximum width of stilling basin at the elevation of the weir crest and B is length of weir crest.

Stilling Basin Length, $L_{SB}$:

$$L_{SB} = 2 \ X_b \tag{2}$$

where $L_{SB}$ is the length of stilling basin, measured from the weir to the beginning of the downstream channel. $X_b$ is defined as the distance between the weir and the baffle pier or plate and is obtained by

$$\frac{X_b}{Y_c} = 3.54 + 4.26 \ (\frac{H}{Y_c}) \tag{3}$$

where H is absolute drop height and $Y_c$ is critical depth for the design discharge.

Stilling Basin Depth, $Y_{SB}$:

$$Y_{SB} = Y_c + H \tag{4}$$

where $Y_{SB}$ is the depth of stilling basin measured from the crest of the weir.

Stilling Basin Side Slopes, $S_B$:

$$1.0:2.0 < S_B \leqq 1.0:2.5 \tag{5}$$

where $S_B$ is the side slope (vertical to horizontal) of the stilling basin.

Baffle Dimensions (Fig. 1)

Baffle Width, $W_b$

$$W_b = B/2 \tag{6}$$

where $W_b$ is the width of baffle pier or plate. The measurement and placement of the baffle is perpendicular to the flow and centered in the basin.

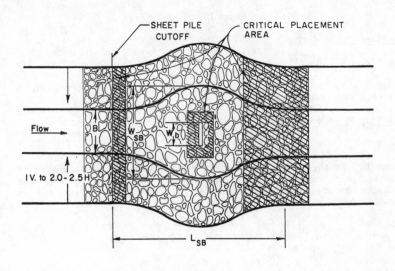

PLAN

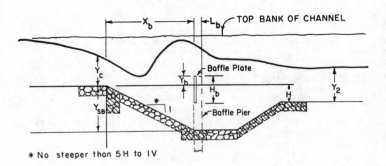

PROFILE

Figure 1.   Basin Dimensions and Riprap Placement.

Baffle Plate Height, $H_b$

The baffle height is pertinent only to the baffle plate since the baffle pier extends into the bottom of the basin. The baffle plate height is given by $H_b = Y_c$.

Height of Baffle Pier or Plate Above Weir Crest, $Y_b$:

$$Y_c/4 \leq Y_b \leq Y_c/3 \tag{7}$$

## DESIGN, LAYOUT, AND CONSTRUCTION

Several types of structures are commonly used for gradient control structures. In its search for an economical structure that would perform satisfactorily, the U. S. Department of Agriculture, Soil Conservation Service (SCS) in Mississippi has constructed six low drop grade control structures from the criteria given by Little and Murphey (1982) and have developed guidelines for their structural design.

### Hydraulic Design

Before hydraulic design computations can be made, a design discharge, weir cross section and absolute drop height must be determined. Various factors should be considered in selection of these parameters.

A detailed survey of a proposed specific site is required to adequately describe the channel geometry both upstream and downstream and to obtain a good estimate of the channel slope. The sheet pile weir crest length should be selected such that the flow is neither expanded nor constricted through the inlet section from the upstream channel.

The weir crest elevation should be determined from both engineering considerations and channel bank composition, especially in physiographic regions where the streambanks are highly stratified alluvium. If an extremely erodible stratum is present near the proposed weir crest elevation, severe bank instability problems would be expected. For example, Grissinger, Murphey and Little (1982) found that on unstable streams in the Yazoo basin, a layer of unconsolidated sand is generally found at or slightly above thalweg elevation which often extends several feet below the channel bottom. Above this sand layer, cohesive materials are found. Thus, by placing the weir crest elevation above this zone, the channel upstream will aggrade above the extremely erodible material. The upstream bank height is also reduced, thereby increasing the mass stability of the channel banks.

Before any hydraulic design can proceed, a design discharge must be selected. Design discharge for grade control structures naturally depends upon the usual hydrologic parameters, but also on specific site conditions. At present, there are no rigid criteria. However, the design discharge should be either bank-full capacity of the upstream channel or the 100-year return event, whichever is less.

This guide has been used by SCS and has not presented any evident problems. Tailwater depth for the stilling basin is assumed to be normal flow depth in the downstream channel.

After selection of design discharge, the criteria given by Eqs. (1) through (7) should be used to determine the geometry of the stilling basin and the size and placement of the baffle plate or pier. Riprap size and placement will be covered separately.

Structural Design

The low drop grade control structure requires less structural design than conventional drop structures. In most cases, the hydraulic design, structural design, and layout can be performed within a few days.

Undisturbed soil cores should be obtained at the location of the baffle pier or plate. Standard penetration tests should be run throughout the depth of boring, which should extend into firm foundation material. Unconfined compressive strength measurements of the foundation material must be made from the core samples. The composition of the soil material should be used to determine dewatering requirements during construction. The unconfined compressive strength of the weakest soil should be used in the pile design and slope stability analysis of the excavated slopes.

The design force for the moving water striking the baffle pier or plate can be computed from the momentum equation,

$$F = \rho\, Q\, V_c \qquad\qquad (8)$$

where $F$ = force on baffle plate in pounds (Newton)
   $\rho$ = mass density of water in lb-sec$^2$/ft$^4$ (Kg/m$^3$)
   $Q$ = design discharge in cfs (m$^3$/s)
and $V_c$ = critical velocity in fps (m/s).

For design, all of the flow should be used as striking the baffle. Since the baffle plate must be supported entirely by beam piles, the pile supports should be designed as a cantilever beam. The bending moment and shear on the pile and required depth of pile embedment may be determined by use of appropriate formulae. Knee braces should be added to each pile for reinforcement, anticipating floating debris increasing the force on the baffle plate. If bedrock is encountered at shallow depths below the bottom of the stilling basin, holes should be drilled for the piles and filled with concrete after placement of the pile.

Sheet piling for the baffle pier is driven to firm foundation material. The rectangular piling enclosure is then filled with rock riprap and capped with concrete. The width of the baffle pier, $W_b$, is determined by Eq. 6 and the length, $L_b$, from structural strength requirements.

Structural design is required for the sheet pile weir and cutoff wall only when the piling is driven at a natural vertical overfall or the soil in the stilling basin must be removed during construction for some reason. If soil material is removed from the downstream side, the sheet pile should be designed for the differential load.

## Layout

Layout of the circular type scour basin recommended in Fig. 1 is a somewhat difficult problem for field technicians. On some of the structures designed by the Soil Conservation Service, the basin was designed with straight sides (in plan view) rather than the circular shape. This layout also increases the reliability of quantity computations for contract payment. However, on some sites with 2 to 1 side slopes and a 2 to 1 chute slope, the SCS has experienced very steep (1.4 to 1) transition slopes where the chute between the weir and baffle intersect with the side slope of the stilling basin. If the flat (lateral to flow) bottom chute section between the weir and baffle was on a lower gradient, this would not occur. On those structures that were constructed with these steep slopes, the rocks were grouted with concrete to prevent movement. From these experiences and from previous model studies, the authors now recommend the chute gradient to be a minimum of 1 vertical to 5 horizontal.

Consideration must be given to the upstream out-of-bank flow around the structure. This flow will re-enter the downstream channel wherever it is allowed to do so. If the water surface elevation in the downstream channel is lower than top bank, the out of bank flow from upstream will cause channel bank erosion where it re-enters. Provisions can be made to minimize this by (1) constructing a dike on each side of the downstream channel bank a sufficient distance downstream so that the downstream water surface elevation in the channel is near top bank elevation before re-entry of the out of bank flow occurs, (2) limiting the drop through the structure to a small amount, four feet (1.22 m) or less, or (3) a combination of these.

Upstream approach channel alignment is extremely important for good hydraulic operation of the structure. Curved approach channels cause more of the flow around the baffle pier or baffle plate on one side setting up a singular vortex in the stilling basin. Sand and gravel will then be deposited on one side, preventing optimum energy dissipation within the stilling basin. On the outside of the approach bend, a large percentage of the flow goes around the side of the baffle and directly into the downstream channel. Thus, straight alignment of the approach channel is necessary for good energy dissipation. The approach channel should be straight for at least 10 times the channel bottom width.

## Riprap Size and Placement

Rock riprap should be used to line the scour basin to top bank of the channel. The rock size may be computed from criteria developed by Anderson and Paintal (1970) and later modified by Blaisdell (1973). An adequate safety factor (applied to mean rock size) should be used.

A filter underneath the riprap should be used and may be either well graded gravel material or a permeable blanket of plastic cloth. Anderson and Paintal (1970) recommend 1.5 to 2.0 times the $D_{50}$ size for the thickness of the riprap. The SCS in Mississippi uses 2.0 times the $D_{50}$ size for the thickness of riprap and has found this to be adequate.

In addition to the riprap lined stilling basin, riprap should also be placed upstream from the sheet pile weir for a minimum distance of 3 $Y_c$ and should also extend at least 10 feet (3.1 m) into the exit channel.

The construction specifications should require that the larger rocks (> $D_{85}$) be placed in the critical areas delineated on Fig. 1. If adequate size rock are not available, the riprap may be grouted. A larger riprap thickness should also be specified in the chute section between the weir cutoff wall and baffle.

## Construction

Nine low drop grade control structures as described in Table 1 have been built to date. The U. S. Army Corps of Engineers, Vicksburg District has built structures on Perry Creek and North Fork Tillatoba Creek in Mississippi. A photograph of the structure on North Fork Tillatoba Creek is shown in Fig. 2. The Soil Conservation Service has built structures on Indian Creek, Butterbowl Creek, Big Creek, Hurricane Creek, West Prong Muddy Creek and Muddy Creek in Mississippi. The Indian Creek structure is shown in Fig. 3. The Soil Conservation Service in Arkansas built a structure on Kerch Canal in central Arkansas. The Indian Creek, North Fork Tillatoba Creek and Kerch Canal structures were modeled by Little and Murphey (1981) as specific case studies. The other structures were designed based on preliminary data from a generalized model study of low drop structures by Little and Murphey (1982).

## Construction Costs

Even though some of these structures are very similar, their costs differed considerably (See Table 1). Variations in costs of materials and labor, construction of channel cutoffs, dewatering, channel realignment, downstream bank protection and provisions for out-of-bank flow may increase the cost significantly.

## Performance

Except for one structure, all installed structures have performed satisfactorily. The Butterbowl Creek structure has experienced movement of the rock immediately downstream from the cutoff wall. A study has not been performed to determine the exact cause.

The North Fork Tillatoba Creek and Perry Creek structures listed in Table 1 have been monitored for performance since February, 1979. A recording depth gage and precipitation gages were installed on each watershed. On June 24, 1980 the North Fork Tillatoba Creek structure

TABLE 1. SUMMARY OF INSTALLED DROP STRUCTURE CHARACTERISTICS

| SITE | DESIGN DISCHARGE CFS ($m^3$/s) | DROP HEIGHT FT. (m) | CHANNEL BOTTOM WIDTH FT. (m) | DISTANCE TO BAFFLE FT. (m) | CONSTRUCTION COST (YEAR) | TYPE OF BAFFLE |
|---|---|---|---|---|---|---|
| Indian Creek | 1800 (51.0) | 4.0 (1.22) | 18 (5.5) | 42 (12.8) | $83,560 (1975) | Pier |
| Kerch Canal | 1720 (48.7) | 4.0 (1.22) | 30 (9.1) | 55 (16.8) | $140,000 (1975) | Plate |
| North Fork Tillatoba Creek | 8000 (226.6) | 4.0 (1.22) | 70 (21.3) | 45 (13.7) | $214,000 (1977) | Pier |
| Perry Creek | 4100 (116.1) | 2.0 (0.61) | 45 (13.7) | 30 (9.1) | $105,000 (1978) | Plate |
| Butterbowl Creek | 5800 (164.3) | 5.0 (1.52) | 35 (10.7) | 48 (14.6) | $305,000 (1978) | Plate |
| Big Creek | 624 (17.7) | 4.5 (1.37) | 16 (4.9) | 25 (7.6) | $60,000 (1981) | Plate |
| Hurricané Creek | 1000 (28.3) | 3.4 (1.04) | 15 (4.6) | 21 (6.4) | $48,000 (1981) | Plate |
| West Prong Muddy Creek | 1015 (28.7) | 1.9 (0.58) | 16 (4.9) | 25 (7.6) | $49,892 (1981) | Plate |
| Muddy Creek | 740 (21.0) | 3.1 (0.95) | 11 (3.4) | 16 (4.9) | $41,214 (1981) | Plate |

FIGURE 2. Low Drop Grade Control Structure with Baffle Pier on North
Fork Tillatoba Creek.

FIGURE 3. Low Drop Grade Control Structure with Baffle Pier on Indian
Creek.

experienced a peak flow depth of 12.7 feet. The calculated discharge was 14500 cfs (410.8 $m^3/s$), 1.8 times the design discharge of 8000 cfs (226.6 $m^3/s$). Eq. 9 was used to calculate the riprap size for the design discharge. The $D_{50}$ calculated size was 20 inches (0.51 m). This median size was used on the structure. Despite the extreme event of June 24, 1980 and other subsequent extreme events equal to or greater than design discharge, the structure has maintained its integrity with no significant movement of riprap.

Summary

Procedures for the hydraulic design, structural design, layout, and construction of low drop channel grade control structures have been given. These structures have shown excellent hydraulic performance as channel grade control structures.

They have distinct advantages over other types of drop structures which are:

1.  More economical.
2.  Design and layout time is minimal.
3.  Only a short construction period is required preventing need for streamflow diversion.
4.  Dewatering can normally be accomplished with sump pump.

REFERENCES

1.  Anderson, Alvin G., Paintal, Amreek S., and John T. Davenport, "Tentative Design Procedure for Riprap-Lined Channels, National Cooperative Highway Research Program, Highway Research Board, National Academy of Sciences, Report 108, 1970.
2.  Blaisdell, Fred W., "Model Test of Box Inlet Drop Spillway and Stilling Basin," ARS-NC-3, Jan. 1973.
3.  Chow, Ven Te, "Open-Channel Hydraulics," McGraw-Hill Book Company, Inc., New York, 1959.
4.  Grissinger, E. H., Murphey, J. B., and Little, W. C., Late Quaternary valley-fill deposits in North-Central Mississippi. Southeastern Geology. Vol. 23 No. 3, 1982.
5.  Little, W. C. and Murphey, J. B., Model Study of Low Drop Grade Control Structures. Appendix B. Report prepared for U. S. Army Corps of Engineers, Vicksburg District, on Stream Channel Stability. 44 pages. April, 1981.
6.  Little, W. C., and Murphey, J. B., A Model Study of Low Drop Grade Control Structures, Journal of the Hydraulics Division, Proceedings, ASCE, 1982. (In Press)

# APPLYING LES TURBULENCE MODELING TO OPEN CHANNEL FLOW

by

Keith W. Bedford[1], AM ASCE; and

Youssef M. Dakhoul[2]

## ABSTRACT

A brief review of the different methods for turbulence modeling is given with a special concentration on the LES filtering method and its advantages over the traditional methods. A new spatial-temporal filtering procedure is also suggested. This is followed by a brief review of the existing river flow and transport models. The turbulence modeling aspects of these models are stressed and critiqued. Finally, certain techniques are suggested for applying the new filtering procedure in deriving a higher order river flow and transport model.

## INTRODUCTION

The physical process of river flow includes turbulence as an aspect of vital importance and large impact on the other constituents of the process. Therefore, a proportional segment of the modeling efforts should be devoted to the presentation of turbulence in the governing equations. Since Saint Venant (1871) published his famous equations, efforts for improving the river flow and transport models have been continuous and persistent, e.g. references [4,6,10,11,14,18,19,20]. As a result of these efforts, numerous forms of the one-dimensional river flow and transport equations were developed. It appears, although not to many researchers, that these equations possess all the turbulent energy generation, trasnport, and dissipation characteristics displayed by other three-dimensional turbulent models. See, for example, the work of Love (1976) on the one-dimensional Burgers' equation. The mathematical representation of these turbulent processes (turbulence modeling) is being heavily investigated; impressive developments are achieved and applied to all sorts of hydraulic problems, see Rodi (1980). Unfortunately however, from the viewpoint of turbulence modeling, the persistent efforts of the river flow modelers have not gone any farther than using the traditional model formulation methods.

The available river flow models include a number of coefficients that have to be determined empirically, or tuned until the model reproduces the field data. These two methods do not consistently give the

---

[1]Associate Professor, Civil Engineering Department, The Ohio State University, Columbus, Ohio 43210

[2]Research Associate, Civil Engineering Department, The Ohio State University, Columbus, Ohio 43210

same values for the coefficients which reflects the ambiguity and the difficulty involved in defining and understanding these coefficients. It is reasonable to believe that the degree of ambiguity associated with the coefficients is a function of the method used for modeling turbulence in the governing equations. The large-eddy simulation (LES) technique has recently emerged as a new turbulence modeling procedure that introduces a higher order definition of the large-scale turbulent motions, thereby attaching less importance to the small-scale motions and the coefficients describing them. See for example references [1,2, 3,5,7,8,9,12,13]. This work is intended as a review of the LES method as used in three-dimensional simulation of turbulence. The performance and advantages of the method are explicitly explained. Application of the method to the formulation of a new higher order river transport model is discussed. The discussion is intended to demonstrate how the various aspects of the LES method, as developed for three-dimensional problems can contribute necessary improvements in the existing one-dimensional river transport models.

## THE LES METHOD

### Averaging

Whenever a system of equations containing turbulent variables is to be solved numerically, the equations must be "prepared" commensurate with the grid spacings. That is because any numerical grid is capable of resolving only a certain portion of a rapidly fluctuating variable. This portion, being a function of the grid spacings, is called the large scale component. The remaining portion is called the subgrid-scale (SGS) component. Although not calculable, the SGS components have an important impact on the calculated large-scales. The term "preparing" the equations then means replacing the total variables by their large-scale components and, some way or another, modeling the SGS effects in terms of the large-scale components.

The large-scale component of a three dimensional, time dependent turbulent quantity, $\alpha(\underline{x},t)$, can be defined by the Fourier convolution integral

$$\overline{\alpha}(\underline{x},t) = \int_{-\infty}^{\infty} G(\underline{x}-\underline{x}',t-t')\alpha(\underline{x}',t')d\underline{x}'dt'. \tag{1}$$

Here $\underline{x}$ is the position vector, t is the time coordinate, and $G(\underline{x},t)$ is a weight or filter function defined as

$$G(\underline{x},t) = G_t(t) \prod_{i=1}^{n} G_i(x_i) \tag{2}$$

where $G_i(x_i)$ is the component of $G(\underline{x},t)$ in the $x_i$ direction; $G_t(t)$ is the temporal component; and n in the total number of spatial directions in which averaging is desired. The filter function must satisfy the condition

$$\int_{-\infty}^{\infty} G(\underline{x}-\underline{x}',t-t')dx'dt' = 1 \tag{3}$$

so that the averaged or large-scale component of a constant becomes the same constant. If $\overline{\alpha}(\underline{x},t)$ is really the large-scale component of $\alpha(\underline{x},t)$, its Fourier transform should vanish for wave lengths and periods equal to or smaller than the corresponding grid spacings. The Fourier transform of $\alpha(\underline{x},t)$ is given by the convolution (Faltung) theorem which states that if (1) is true then

$$F\{\overline{\alpha}(\underline{x},t)\} = F\{G(\underline{x},t)\} \cdot F\{\alpha(\underline{x},t)\} . \tag{4}$$

It is seen from (4) that the Fourier transform of $\overline{\alpha}(\underline{x},t)$ is directly proportional to the Fourier transform of the filter function. Therefore, for $\overline{\alpha}(\underline{x},t)$ to satisfy the requirement for being the large-scale component of $\alpha(\underline{x},t)$, the Fourier transform of $G(\underline{x},t)$ must vanish for wave lengths and periods equal to or smaller than the corresponding grid spacings. Special cases of the general averaging procedure, equations (1) and (2), are given in figure 1. Column (a) shows the traditional Reynolds temporal averaging. Column (b) shows the spatial-temporal averaging, with constant filter components, used mainly in the three-dimensional transient weather models. Column (c) gives the Leonard's n-dimensional spatial averaging with Gaussian filter components, Leonard (1974). Finally, a new spatial-temporal averaging, with Gaussian filter components, is suggested in column (d). Figure 1 shows that only the Gaussian filter components possess the required property if $\Delta_i$ or $\Delta_t$ are chosen equal to twice the corresponding grid spacing. The advantage of the suggested operation (column d), over the purely spatial Leonard's method, is its capability of removing temporal (as well as spatial) high frequency or SGS components of $\alpha(\underline{x},t)$.

## The LES Method Applied to Three-dimensional Problems

In references [1,2,3,5,7,8,9,12,13], the LES modelers use Leonard's version of equation (1) to define the large scale component. This version, given in figure 1(c), is transformed into an explicit form by expressing $\alpha(\underline{x}',t)$ in terms of $\alpha(\underline{x},t)$ through a Taylor series expansion

$$\overline{\alpha}(\underline{x},t) = \int_{x_1'=-\infty}^{\infty} \cdots \int_{x_n'=-\infty}^{\infty} G(\underline{x}-\underline{x}') \{ \alpha(\underline{x},t) + \sum_{i=1}^{n} (x_i'-x_i)\partial\alpha(\underline{x},t)/\partial x_i$$

$$+ \sum_{i=1}^{n} 0.5(x_i'-x_i)^2 \partial^2\alpha(\underline{x},t)/\partial^2 x_i + H.O.T. \} dx_1' \ldots dx_n' . \tag{5}$$

Using the Leonard's definition for $G(\underline{x},t)$, and neglecting the higher order terms (H.O.T.), the result of the above integral is

$$\overline{\alpha}(\underline{x},t) = \alpha(\underline{x},t) + \sum_{i=1}^{n} (\Delta_i^2/4\gamma)\partial^2\alpha(\underline{x},t)/\partial x_i^2. \tag{6}$$

Another important property of the Leonard's procedure is that the order of differentiation and averaging (filtering) can be interchanged. That is

$$\overline{\partial\alpha(\underline{x},t)/\partial x_i} = \partial\overline{\alpha}(\underline{x},t)/\partial x_i, \tag{7}$$

$$\overline{\partial\alpha(x,t)/\partial t} = \partial\overline{\alpha}(\underline{x},t)/\partial t. \tag{8}$$

The proof can be found in Leonard (1974) and Kwak et al. (1975).

The above properties of the Leonard's LES method are used, in references [1,2,3,5,7,8,9,12,13], to prepare the three-dimensional

| | a Reynolds Temporal Averaging | b Spatial-Temporal Averaging With Constant Filter | c Leonard's Averaging | d Suggested by the authors |
|---|---|---|---|---|
| | 1882 to Present | 1950 to Present | 1974 to Present | 1982 |
| Definitions of the large scale component, equation (1). | $\bar{a}(\underline{x},t)= \int_{-\infty}^{\infty} G_t(t-t')a(\underline{x},t')dt'$ | $\bar{a}(\underline{x},t) =$ $\int_{-\infty}^{\infty} G(\underline{x}-\underline{x}',t-t')a(\underline{x}',t')d\underline{x}'dt'$ | $\bar{a}(\underline{x},t)= \int_{-\infty}^{\infty} G(\underline{x}-\underline{x}')a(\underline{x}',t)d\underline{x}'$ | $\bar{a}(\underline{x},t) =$ $\int_{-\infty}^{\infty} G(\underline{x}-\underline{x}',t-t')a(\underline{x}',t')d\underline{x}'dt'$ |
| Definition of the filter function, equation (2). | $G(\underline{x},t) = G_t(t)$ $= \frac{1}{\Delta_t}$ for $\frac{-\Delta_t}{2} \le t \le \frac{\Delta_t}{2}$ $= 0$ otherwise | $G(\underline{x},t) = G_t(t) \prod_{i=1}^{n} G_i(x_i)$ where: $G_t(t) = \frac{1}{\Delta_t}$ for $\frac{-\Delta_t}{2} \le t \le \frac{\Delta_t}{2}$ $= 0$ otherwise $G_i(x_i) = \frac{1}{\Delta_i}$ for $\frac{-\Delta_i}{2} \le x_i \le \frac{\Delta_i}{2}$ $= 0$ otherwise | $G(\underline{x},t) = G(\underline{x}) = \prod_{i=1}^{n} G_i(x_i)$ where $G_i(x_i) = \sqrt{\frac{\gamma}{\pi}}\frac{1}{\Delta_i} e^{-\gamma x_i^2/\Delta_i^2}$ for all $x_i$ | $G(\underline{x},t) = G_t(t) \prod_{i=1}^{n} G_i(x_i)$ where $G_t(t) = \sqrt{\frac{\gamma}{\pi}}\frac{1}{\Delta_t} e^{-\gamma t^2/\Delta_t^2}$ for all $t$ $G_i(x_i) = \sqrt{\frac{\gamma}{\pi}}\frac{1}{\Delta_i} e^{-\gamma x_i/\Delta_i^2}$ for all $x_i$ |
| | | | | |
| Fourier Transform of the filter function | $F\{G(\underline{x},t)\} = F\{G_t(t)\}$ $= \frac{\sin(f_t\Delta_t/2)}{(f_t\Delta_t/2)}$ | $F\{G(\underline{x},t)\} = F\{G(t)\} \prod_{i=1}^{n} F\{G_i(x_i)\}$ $= \frac{\sin(f_t\Delta_t/2)}{(f_t\Delta_t/2)} \prod_{i=1}^{n} \frac{\sin(f_i\Delta_i/2)}{(f_i\Delta_i/2)}$ | $F\{G(\underline{x},t)\}=F\{G(\underline{x})\}=\prod_{i=1}^{n}F\{G_i(x_i)\}$ $= \prod_{i=1}^{n} \exp[-f_i^2\Delta_i^2/4\gamma]$ | $F\{G(\underline{x},t)\} = F\{G_t(t)\} \prod_{i=1}^{n} F\{G_i(x_i)\}$ $= \exp[-f_t^2\Delta_t^2/4\gamma] \cdot$ $\prod_{i=1}^{n} \exp[-f_i^2\Delta_i^2/4\gamma].$ |
| | | | | |

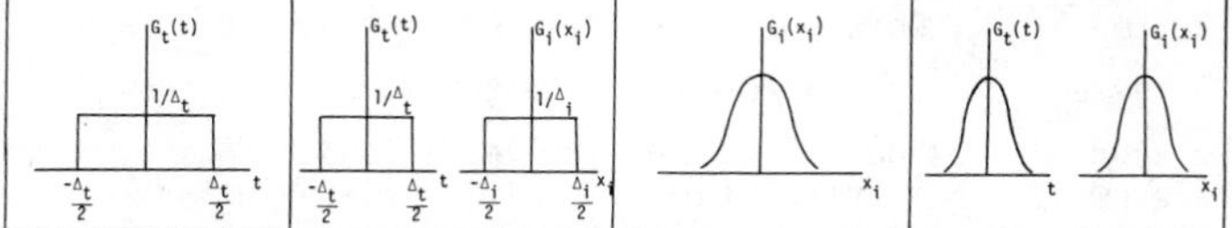

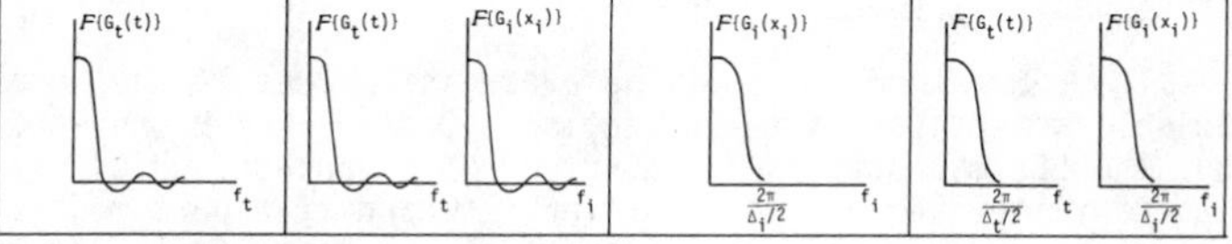

Legend:    $\Delta_t$ = temporal averaging scale

$\Delta_i$ = spatial averaging scale in the $x_i$ direction

$\gamma$ = dimensionless constant with optimum value of 6

$f_t$ = frequency = $2\pi/T$   where $T$ = wave period

$f_i$ = wave number in the $x_i$ direction = $2\pi/L_i$   where $L_i$ = wave length in the $x_i$ direction

Figure 1. History of the Averaging Procedures

Navier-Stokes equation for numerical solution on spatial grids with widely spaced nodes. The "prepared" or filtered equation is

$$\partial \bar{u}_i/\partial t + \partial[\bar{u}_i\bar{u}_j+(\Delta_\ell^2/4\gamma)\partial^2\bar{u}_i\bar{u}_j/\partial x_\ell^2]/\partial x_j = (-1/\rho)(\partial\bar{p}/\partial x_i) - \partial R_{ij}/\partial x_i.$$

(9)

Here $\bar{u}_i$ is the large-scale component of $u_i$, the total velocity in the $x_i$ direction; $\rho$ is the fluid's density; $p$ is the pressure intensity; $R_{ij}$ is the residual stress term with respect to which the molecular viscosity is negligible; and the repetition of a subscript in a single term implies summation over the three spacial directions. $R_{ij}$ is defined as

$$R_{ij} = \overline{\bar{u}_i u'_j} + \overline{u'_i \bar{u}_j} + \overline{u'_i u'_j},$$

(10)

where the prime denotes SGS components. Note that the second term in the left hand side of equation (9) is obtained by applying (6), with n=3, to $\bar{u}_i\bar{u}_j$. On the other hand, the Reynolds method approximates $\overline{u_i u_j}$ by $\bar{u}_i\bar{u}_j$. The implications of both methods are shown in figure 2. The Reynolds method (figure 2a) implies a constant $\bar{u}_i$ over the averaging scale, while the variation of $\bar{u}_i$ over the averaging scale is explicitly accounted for by the LES method (figure 2b). As a result, the Reynolds averaging (or filtering with constant filter) method attaches more importance to the SGS components and consequently to the residual stress term which must be modeled in terms of the large-scales through the ambiguous eddy coefficients (residual stress modeling).

Advantages of the LES Method

     The LES method is simply an averaging procedure that employes a variable weight (or filter) function. Due to this higher order averaging, the LES method posseses a number of advantages which may contribute important improvements to the river flow and transport models. First, the frequency or wave number content of the large-scale components, as defined by the LES method, is consistent with the grid spacings. On the other hand, the uniform weight function averaging leaves high frequency noise attached to the large scale components (see figure 1). Therefore, the governing equations would be much better prepared for numerical solution if the LES method is used. Second, the minimal importance attached by the LES method to the SGS components may very well reduce the gap between the tuned and the observed values of the coefficients. The eddy coefficients; dispersion coefficients; bottom friction coefficients and the like reflect our ignorance concerning the behavior of the SGS components. Logically, this ignorance should become more tolerable by reducing the magnitude of the SGS components. Finally, the LES method was proved successful in preserving the expected statistical features in the predicted turbulent fields. Since turbulence is stochastic in nature, the statistics of the predicted variables may be the only logical tools for judging the performance of a given turbulent flow model. See Bedford and Babajimopoulos (1980). Among these statistics are the mean, variance, standard deviation, skewness, and the spectral distribution. The one-dimensional application by Love (1976) (figure 3) shows a tremendous improvement in the velocity spectra when the LES method is used. Note how little the importance of the residual stress model is, when compared to the importance of the averaging procedure, in preserving the spectral distribution of the calculated turbulent velocity field. Other similar results are found in the two and three-dimensional applications in references [1,2,3,5,7,8,9,12,13] and are not presented here for space limitations. The reason for the LES method's superiority in reproducing the expected spectral features is its higher order definition of the large-scale components. These large-scale components are responsible for the most part of the turbulent energy transport process which is directly described by the spectral distribution.

## EXISTING MODELS

### Brief Review

Most river flow modelers use the one-dimensional shallow water equations derived under the following assumptions: 1) the slope angle $\theta$, figure 4, is so small that the coordinate system is considered orthogonal, and $\sin \theta = \tan \theta = S_0$. 2) The fluid is incompressible and thermally homogeneous. 3) The flow is one dimensional such that $U_y$ and $U_z$, as defined later in equation (16), are both nil. 4) The channel is prismatic such that $\xi(z)$ is a function of z only. 5) the cross-sectional area A, its perimeter $\sigma$, the depth h, and the hydraulic radius R are non-fluctuating quantities. 6) The molecular viscosity and diffusivity are negligible. 7) Flood plains, lateral discharges, source-sink terms are expressed in many different mathematical forms. They are omitted herein for simplicity. 8) The pressure is hydrostatically distributed.

As in Yen (1973, 1979), Holley & Harleman (1965), Liggett (1975), and many other works, the model can be derived by integrating the basic three-dimensional continuity, momentum, and transport equations over the cross-sectional area of the channel:

$$\int_A (\partial u_i / \partial x_i) dydz = 0, \tag{11}$$

$$\int_A \{\partial u_i / \partial t + \partial u_i u_j / \partial x_j + (1/\rho)\partial p / \partial x_i - F_i\} dydz = 0, \tag{12}$$

$$\int_A \{\partial c / \partial t + \partial u_i c / \partial x_i\} dydz = 0 . \tag{13}$$

Here $F_i$ is the external force per unit mass, and c is the concentration of containment. Yen (1973) defines the Leibnitz rule, and the boundary condition for the cross-section in the following form

$$\int_A (\partial \alpha / \partial x_i) dydz = \partial [\int_A \alpha dydz] / \partial x_i - [\alpha \, \partial A / \partial x_i] , \tag{14}$$

$$[\partial A / \partial t + u_i \partial A / \partial x_i]_\sigma = 0, \tag{15}$$

in which $[\ ]_\sigma$ denotes integration along the boundary of A. Using (14), (15), and the definitions

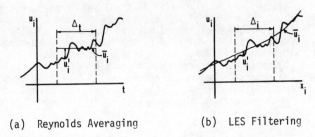

(a)  Reynolds Averaging          (b)  LES Filtering

Figure 2.   Definition of the Large-Scale and the
            SGS Components, Leonard (1974)

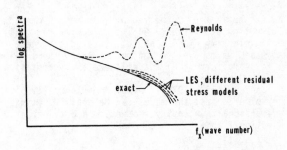

Figure 3.   One-Dimensional Velocity Spectra, Love (1976)

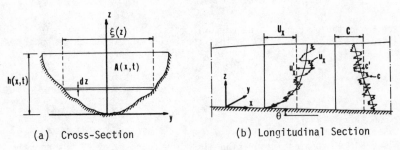

(a)  Cross-Section              (b)  Longitudinal Section

Figure 4.   Coordinate System and Notation

$$U_i = (1/A) \int_A u_i dy dz, \quad C = (1/A) \int_A c \, dy dz, \tag{16}$$

$$\beta U_i U_j = (1/A) \int_A u_i u_j dy dz, \quad \beta' U_i C = (1/A) \int_A u_i c \, dy dz, \tag{17}$$

$$F_x = g(S_o - S_f), \quad F_y = 0, \quad F_z = -g, \tag{18}$$

$$p = \rho g(h-z), \tag{19}$$

equations (11) through (13) yield the following system of one-dimensional equations

$$\partial A/\partial t + \partial U_x A/\partial x = 0, \tag{20}$$

$$\partial U_x A/\partial t + \partial \beta U_x^2 A/\partial x + gA\partial h/\partial x = gA(S_o - \frac{|U_x|U_x}{\varepsilon^2 R}), \tag{21}$$

$$\partial CA/\partial t + \partial \beta' U_x CA/\partial x = 0. \tag{22}$$

Here $\beta$ and $\beta'$ are momentum and flux correction factors; g is the gravitational acceleration; and $S_f$ is the friction slope. The friction slope in equation (21) is approximated by the Chezy's formula in which $\varepsilon$ is a bottom friction coefficient.

Note that the y and z components of the turbulent fluctuations in $u_i$ and c are removed by the areal integration procedure. The effects of these fluctuation components, $u_i'$ and c' in figure 4b, are modeled by the momentum and flux correction factors. An equivalent method for modeling these effects is outlined in Holley & Harleman (1965) as follows. The integration of (13), using (14) and (15), gives

$$\partial CA/\partial t + \partial[\int_A u_i c \, dy dz]/\partial x_i = 0. \tag{23}$$

Now, instead of using definition (17), $u_i$ is replaced by $(U_i + u_i')$ and c by $(C + c')$. The term is then expanded to obtain

$$\partial CA/\partial t + \partial U_i CA/\partial x_i + \partial[\int_A (U_i c' + u_i' C + u_i' c')dy dz]/\partial x_i = 0. \tag{24}$$

Using the definition

$$(1/A) \int_A (U_i c' + u_i' C + u_i' c')dy dz = - e_i(\partial C/\partial x_i), \tag{25}$$

and recalling that $U_y = U_z = \partial C/\partial y = \partial C/\partial z = 0$, equation (24) becomes

$$\partial CA/\partial t + \partial U_x CA/\partial x = \partial[A e_x (\partial C/\partial x)]/\partial x. \tag{26}$$

Both the flux correction factor, $\beta'$, and the dispersion coefficient, $e_x$, model the lateral fluctuations $u_i'$ and c' in terms of the areal means $U_i$ and C. We realize, however, that $U_i$ and C still possess high wave number and frequency components in the logitudinal direction and the time domain. Very rarely are equations (20,21,22 or 26) prepared, as explained in the preceding section, for numerical solution on the x-t grid. This fact has been generally overlooked except by very few modelers such as Holley and Harleman (1965) who applied the Reynolds procedure to average equation (26). This yields

$$\partial \overline{C} A/\partial t + \partial \overline{U}_x \overline{C} A/\partial x = \partial [A(e_x + D_x) \ \partial \overline{C}/\partial x]/\partial x, \qquad (27)$$

where $D_x$ is the turbulent diffusivity coefficient, and the sum $(e_x+D_x)$ is called the "longitudinal dispersion coefficient."

Critique

In deriving equations (20,21,22), which are extensively used for simulating river flows, turbulence modeling is done solely by integrating the original three-dimensional equations over the cross-sectional area. This replaces the total variables $u_i$ and c by their respective areal means $U_i$ and C. The diffusive effects of the differences $u_i'$ and c' are reflected solely in the correction factors $\beta$ and $\beta'$. In most practical river flow solutions of equation (21), the momentum correction factor, $\beta$, is systematically assumed equal to unity for turbulent flow. We believe that this assumption is conceptually erroneous since, according to definition (17), it denies the existance of $u_i'$. True that a certain portion of $u_i'$ (the difference between the smooth curve and the dashed vertical line in figure 4b) becomes smaller as the turbulence level increases. But the remaining portion represents rapid fluctuations with a diffusive effect that should not be ignored especially as the turbulence level increases. Note that no other turbulence modeling (or smoothing of the $u_i$ distribution) has preceded the areal integration. This conceptual difficulty is properly alleviated in deriving equation (26). Here the fluctuations $u_i'$ and c' are modeled by a mathematically diffusive term containing the dispersion coefficient. Although this has not been done before, the same notion can be easily applied to the momentum equation.

Whether the correction factors or the dispersion coefficients are used in the areal integration, the resulting equations (20,21,22 or 26) still contain the turbulent fluctuating components in the longitudinal direction and time. Therefore, the equations are not properly prepared for numerical solution on the widely spaced x-t grids. Not only conceptually, but also numerically these equations generate turbulence because of the non-linear partial differential terms. Equations (20,21, 22), however, do not contain dissipative terms and therefore, without proper care, can be difficult to solve numerically, Bedford and Sykes (1980). The bottom friction term in equation (21) will then have to assume all the responsibility for dissipating the generated turbulent energy. This, along with the observation in the preceding paragraph, may explain why a friction coefficient, $\varepsilon$, much different from its observed values is needed for the numerical solution to match field measurements. To alleviate this problem, the Reynolds turbulence modeling scheme is used to derive equation (27) from equation (26). Besides its inability to simultaneously prepare the equations with respect to the longitundial direction and temporal grid spacings, the Reynolds method attaches unnecessarily excessive importance to the eddy coefficients and fails to preserve the correct statistical properties in the predicted fields.

## SUGGESTIONS FOR A NEW MODEL

Based on the above review and critique, we seek a new model that possesses the following minimum requirements: 1) The model should recognize the existence of rapid turbulent fluctuations in all the three spatial directions and in time. 2) The model should include a minimum number of coefficients left to be determined empirically. The importance of these coefficients should be minimized so as to reduce the effect of their ambiguous values on the model's predictions. 3) As well as the correct mean flow features, the model should produce the correct statistical structures in the predicted fields. According to the observations in the preceding sections, it is desirable to achieve the three requirements by applying the averaging procedure suggested in column (d) of figure 1 to the original three-dimensional equations. To demonstrate, the averaged (or filtered) contaminant transport equation is

$$\partial \bar{c}/\partial t + \partial \overline{u_i c}/\partial x_i = 0 . \tag{28}$$

$u_i$ and c are decomposed to $(\bar{u}_i + u_i')$ and $(\bar{c} + c')$ respectively. The equation is expanded to yield

$$\partial \bar{c}/\partial t + \partial \overline{\bar{u}_i \bar{c}}/\partial x_i = -\partial [\overline{\bar{u}_i c'} + \overline{u_i' \bar{c}} + \overline{u_i' c'}]/\partial x_i . \tag{29}$$

An equation similar to (6) can be easily derived, using the new filter, as

$$\overline{\alpha}(\underline{x},t) = \alpha(\underline{x},t) + (\Delta_t^2/4\gamma)\partial^2\alpha(\underline{x},t)/\partial t^2 + \sum_{i=1}^{n} (\Delta_i^2/4\gamma)\partial^2\alpha(\underline{x},t)/\partial x_i^2 . \tag{30}$$

Applying (30), with n=3, to $\overline{\bar{u}_i \bar{c}}$, and dealing with the right hand side of (29) in a conventional Boussinesq fashion we obtain

$$\partial \bar{c}/\partial t + \partial [\bar{u}_i \bar{c} + (\Delta_t^2/4\gamma)\partial^2\bar{u}_i \bar{c}/\partial t^2 + (\Delta_\ell^2/4\gamma)\partial^2\bar{u}_i \bar{c}/\partial x_\ell^2]/\partial x_i$$

$$= \partial [D_i(\partial \bar{c}/\partial x_i)]/\partial x_i . \tag{31}$$

Here $\bar{u}_i$ and $\bar{c}$ are the "smoothed" distributions of $u_i$; and c, depicted by the smooth curves in figure 4.b. The right hand side of (31) models the effects of all the spatial and temporal components of $u_i'$ and c'. Equation (31) theoretically satisfies the requirements listed at the beginning of this section. The equation is, however, too cumbersome and must be solved on a spatially three-dimensional grid. Since the cross-sectional means of $\bar{u}_y$ and $\bar{u}_z$ are negligible for one-dimensional flow, the equation is integrated over the cross-sectional area to obtain a simpler form. Using the Leibnitz rule and the boundary condition for the cross sectional area, this yields

$$\partial \bar{C}A/\partial t + \partial \beta' \bar{U}_i \bar{C}A/\partial x_i + \underbrace{\int_A \{\partial [(\Delta_t^2/4\gamma)\partial^2\bar{u}_i \bar{c}/\partial t^2 + (\Delta_\ell^2/4\gamma)\partial^2\bar{u}_i \bar{c}/\partial x_\ell^2]/\partial x_i\} dydz}_{L_{ic}}$$

$$= \int_A \{\partial [D_i(\partial \bar{c}/\partial x_i)]/\partial x_i\} dydz , \tag{32}$$

in which $\bar{C}$ and $\bar{U}_i$ are the areal means of $\bar{c}$ and $\bar{u}_i$ respectively. For the right hand side of (32), the notion given by Holley and Harleman (1965) is used:

$$\int_A \{\partial [D_i(\partial \bar{c}/\partial x_i)]/\partial x_i\} dydz = \partial [\int_A D_i(\partial \bar{c}/\partial x_i) dydz]/\partial x_i - [D_i(\partial \bar{c}/\partial x_i)]_\sigma$$

$$= \partial [\bar{\bar{D}}_i A(\partial \bar{c}/\partial x_i)]/\partial x_i - [D_i(\partial \bar{c}/\partial x_i)]_\sigma , \tag{33}$$

in which $\bar{D}_i$ is a cross-sectional average of $D_i$. Since $\bar{C}$ is uniform over the cross-section, and since $D_y(\partial \bar{c}/\partial y)$ and $D_z(\partial \bar{c}/\partial z)$ represent lateral diffusion at the boundaries which is zero if no lateral discharges are considered, the right hand side of (33) reduces to its longitudinal direction component. With $\bar{U}_y = \bar{U}_z = 0$, equation (32) becomes

$$\partial \bar{c} A/\partial t + \partial \beta' \bar{U}_x \bar{c} A/\partial x + \int_A (L_{ic}) dydz = \partial [\bar{D}_x A(\partial \bar{c}/\partial x)]/\partial x - [D_x(\partial \bar{c}/\partial x)]_\sigma. \qquad (34)$$

For the special case when the flow is highly turbulent, i.e., $\bar{u}_i$ can be considered uniform over the cross-section, and in reaches where $\bar{c}$ can be safely considered as changing only in the longitudinal direction and time, $\beta'$ is unity and the above equation reduces to

$$\partial \bar{c} A/\partial t + \partial \bar{U}_x \bar{c} A/\partial x + \int_A \{\partial [(\Delta_t^2/4\gamma)\partial^2 \bar{u}_i \bar{c}/\partial t^2 + (\Delta_l^2/4\gamma)\partial^2 \bar{u}_i \bar{c}/\partial x^2]/\partial x_i\} dydz$$

$$= \partial [\bar{D}_x A(\partial \bar{c}/\partial x)]/\partial x - [D_x(\partial \bar{c}/\partial x)]_\sigma. \qquad (35)$$

The last terms in both the left hand side and the right hand side must be expressed in terms of $\bar{U}_x$ and $\bar{C}$, this remains a problem to be solved.

Another suggestion is to take equation (26), in which the term containing the dispersion coefficient $e_x$ models the effects of the y and z fluctuating components, and filter it with respect to the x-t domain using equation (30) with n=1. This yields

$$\partial \bar{c} A/\partial t + \partial \{A[\bar{U}_x \bar{c} + (\Delta_t^2/4\gamma)\partial^2 \bar{U}_x \bar{c}/\partial t^2 + (\Delta_l^2/4\gamma)\partial^2 \bar{U}_x \bar{c}/\partial x^2]\}/\partial x$$

$$= \partial [A(e_x + D_x)\partial \bar{c}/\partial x]/\partial x. \qquad (36)$$

The coefficient $e_x$ still carries the same magnitude as that in (27), but $D_x$ carries much less importance than its counterpart in (27). The filter terms in (36), which are expected to improve the statistical quality of the predictions, do not have a counterpart in (27). Equation (36) contains only one empirical coefficient which is the sum $(e_x + D_x)$.

## REFERENCES

1. Babajimopoulos, C., and Bedford, K.W., "Formulating Lake Models Which Preserve Spectral Statistics," Journal of the Hydraulics Div., ASCE, Vol. 106, No. HY1, Proc. Paper 15137, Jan. 1980, pp. 1-19.

2. Bedford, K.W., Sykes, R.W., and Babajimopoulos, C., "The Turbulent Transport and Biological Structure of Eutrophication Models, Vol.I," Final Report, Project EES 527X, Dept. of Civil Engineering, The Ohio State University, 1978.

3. Bedford, K.W., and Babajimopoulos, C., "Verifying Lake Transport Models with Spectral Statistics," Journal of the Hydraulics Div., ASCE, Vol. 106, No. HY1, Proc. Paper 15138, Jan. 1980, pp. 21-38.

4. Bedford, K.W., and Sykes, R.W., "A Dynamic Water Quality Pollution Model, Vol. I, Structure and Computation," Report for the Ohio EPA, Water Quality Planning and Assessment Div., Columbus, Ohio, 1980.

5. Bedford, K.W., "Spectra Preservation Capabilities of Great Lakes Transport Models," Predictive Abilities of Surface Water Flow and Transport Models, Edited by Hugo Fisher, Academic Press, 1981.

6. Chen, C.L., and Chow, V.T., "Formulation of Mathematical Watershed-Flow Model," Journal of the Engineering Mechanics Division, ASCE, Vol. 97, No. EM3, Proc. Paper 8199, June 1971, pp. 809-828.

7. Cosler, D.J., "Numerical Simulation of Turbulence in a Wind-Driven Shallow Water Lake," M.S. Thesis, The Ohio State University, 1979.

8. Ferziger, J.H., "Higher-Level Simulations of Turbulent Flows," Report TF-16, Dept. of Mechanical Engineering, Stanford University, March 1981.

9. Findikakis, A.N., "Finite Element Simulation of Turbulent Stratified Flows," Ph.D. Thesis, Stanford Univ., November, 1980.

10. Holley, E.R., and Harleman, D.R.F., "Dispersion of Pollutants in Estuary Type Flows," M.I.T. Hydrodynamic Laboratory, Rept. No. 74, T65-02, January, 1965.

11. Keulegan, G.H., "Equation of Motion for the Steady Mean Flow of Water in Open Channels," Journal of Research, U.S. National Bureau of Standards, Vol. 29, Research Paper 1488, July, 1942, pp. 97-111.

12. Kwak, D.,Reynolds, W.C., and Ferziger, J.H., "Three Dimensional Time Dependent Computation of Turbulent Flow," Rept. TF-5, Dept. of Mechanical Engineering, Stanford University, May, 1975.

13. Leonard, A., "Energy Cascade in Large-Eddy Simulation of Turbulent Fluid Flows," Advances in Geophysics, Vol. 18A, 1974, pp. 237-248.

14. Liggett, J.A., "Basic Equations of Unsteady Flow," Chapter 2 in Unsteady Flow in Open Channels, edited by K. Mahmood and V. Yevjevitch, Water Resources Publications, 1975.

15. Love, M.D., "Techniques for Solving Burgers' Equation on a Coarse Mesh with Examples," Private Communications, QMC EP 6023, Dept. of Nuclear Engineering, Queen Mary College, Univ. of London, Oct. 1976.

16. Rodi, W., "Turbulence Models and their Application in Hydraulics: State of the Art Paper," International Association for Hydraulic Research, Secretariat: Rotterdamseweg 185, The Netherlands, 1980.

17. Saint Venant, B., "Théorie du Movement Non-Permanent des Eaux avec Application aux Crues des Rivierès et a L'introduction des Marées dans Leur Lit," Acad. Sci., [Paris] Comptes Rendus, Vol. 73, 1871, pp. 148-154, 237-240.

18. Strelkoff, T., "One-Dimensional Equations of Open Channel Flow," Journal of the Hydraulics Division, ASCE, Vol. 95, No. HY3, Proc. Paper 6557, May 1969, pp. 861-876.

19. Yen, B.C., "Open-Channel Flow Equations Revisisted," Journal of the Engineering Mechanics Division, ASCE, Vol. 99, No. EM5, Proc. Paper 10073, October, 1973, pp. 979-1009.

20. Yen, B.C., "Unsteady Flow Mathematical Modeling Techniques," Chapter 13 in Modeling of Rivers, Edited by H. W. Shen, John Wiley and Sons, 1979.

# HYDRAULICS COMPUTATIONS WITH THE k-ε TURBULENCE MODEL

By Wolfgang Rodi[1]

## ABSTRACT

The paper gives an introduction to the k-ε turbulence model which has been used to solve fluid flow problems in many different areas including, in recent years, hydraulic flow problems. The basic assumptions are described, such as the use of the eddy viscosity/diffusivity concept, the characterization of the local state of turbulence by two parameters and the determination of these from semi-empirical transport equations. A special version of the model for use in depth-average calculations is also introduced briefly. Calculation examples obtained with the standard model are presented for several channel flow situations, and the application of the depth-average model is shown for the problem of turbulent mass exchange between a dead water zone and a mainstream and for a real-live cooling-water discharge into a river. The calculations are always compared with measurements.

## INTRODUCTION

Flows of practical relevance in hydraulics are almost always turbulent, which means that the fluid motion is highly random, unsteady and three-dimensional. Although this motion is governed by the same exact equations which also govern laminar flow, namely the time-dependent Navier-Stokes equations, and although numerical procedures are available to solve these equations in principle, the storage capacity and speed of present-day computers is by far too small to allow a numerical resolution of the small-scale turbulent motion. Hence, the only practical way to determine the flow and temperature or concentration field in water bodies is to solve statistically averaged equations governing mean-flow quantities. These equations are formally obtained by separating the instantaneous values of the various quantities in the original time-dependent equations into mean and fluctuating values and by then averaging the equations. Because of the non-linearity of the original equations, this procedure introduces correlations between various fluctuating quantities. For example, in the Navier-Stokes equations the correlation $\overline{u_i u_j}$ between the fluctuating velocity components $u_i$ and $u_j$ appears which expresses the transport of momentum by the turbulent motion and acts like a stress (turbulent or Reynolds stress). Similarly, in the scalar equation the correlation $\overline{u_i \phi}$ between the fluctuating velocity $u_i$ and the fluctuating scalar $\phi$ appears which represents the heat or mass flux by the turbulent motion. Because of the appearance of these terms, the mean-flow equations are not closed and a turbulence model is necessary to determine the turbulent transport terms before the equations can be solved. A turbulence model is a set of equations for specifying the turbulent transport terms; it is based on hypotheses for the turbulent processes and requires empirical input in form of constants. It does not simulate the details of the turbulent motion but only the effect of turbulence on the mean-flow behavior.

---

[1] Professor, University of Karlsruhe, F.R. Germany, Member ASCE

In many hydraulic flow calculations, the influence of turbulence has been simulated by using a constant turbulent exchange coefficient. Mostly such calculations were for so-called far-field phenomena in large water bodies where the turbulent momentum transport terms, except for the bottom shear, are of little significance and a constant exchange coefficient for the turbulent heat or mass transfer is often sufficiently accurate. In particular, constant exchange coefficients are often used in depth-average calculations; ideally the coefficient should be determined from dye-spreading experiments, but it then accounts also for effects that have nothing to do with turbulence, for example the dye-spreading due to secondary motion. This is discussed in greater detail in Rodi (13). There it is also shown that constant exchange coefficients cannot be used whenever the near field is to be resolved because there the eddy-viscosity and diffusivity can vary strongly across the flow field. Hence, the eddy-viscosity/diffusivity concept is, by itself, not a complete turbulence model as it is usually the task of the turbulence model to also describe the variation of the exchange coefficients over the flow field.

The first real turbulence model was introduced in 1925 by Prandtl (12), and a large number of different models have been proposed since then. Even today, Prandtl's simple mixing-length model is among the most widely used models. It relates the turbulent transport terms directly to the gradients of the mean-flow quantities. As empirical input, the distribution of the mixing length over the flow field needs to be prescribed. Great experience has been gathered in this prescription for shear-layer flows, but the mixing length is difficult to prescribe in more complex flows. The extensive testing of the model has also brought to light its limitations, in particular the lack of universality of the empirical input. One of the main shortcomings of the model is that it is based on the implied assumption that turbulence is in local equilibrium, which means that at each point in the flow turbulent energy is dissipated at the same rate as it is produced so that there can be no influence of turbulence production at other points or at earlier times. Hence, this model cannot account for transport and history effects. It predicts the eddy-viscosity and diffusivity to be zero whenever the velocity gradient is zero, which leads to unrealistic simulations in many cases.

In order to account for history and transport effects, turbulence models have been proposed that employ transport equations for turbulence quantities. The simplest such models solve an equation for the kinetic energy of the turbulent motion, k, which is a measure of the velocity scale of the turbulence. For non-equilibrium shear layers this model is certainly superior to the mixing-length hypothesis, but since it still requires empirical specification of the length scale of the turbulent motion, it is also not suitable for flows more complex than shear layers. Hence, the development has moved on to two-equation models which determine also the length scale from a transport equation. The k-$\varepsilon$ model to be discussed in detail in this paper is of this kind. Two-equation models are the simplest ones that can be used for flows with complex geometries. Higher-order models have been proposed that do not use the eddy-viscosity/diffusivity concept but determine the turbulent transport terms $\overline{u_i u_j}$ and $\overline{u_i \phi}$ from semi-empirical transport equations. These so-called second-order models certainly simulate the physical processes in the most realistic way and can describe turbulent transport against the gradient of the corresponding mean quantity (counter gradient transport) which the eddy-viscosity/diffusivity models cannot, but it should be added that this phenomenon does not occur very often in hydraulic flow situations. Second-order models involve a relatively large number of partial differential equations and are therefore rather complex and expensive of computing time, so that they are still relatively little tested and have not yet reached the state of practical application. They are however important as a starting point for the development of more generally applicable simpler models.

In his review on turbulence models, Rodi (13) concluded that the two-equation k-ε model has been tested most widely for hydraulic flow problems and that it has the widest range of applicability among the relatively simple models. At the present state of development, it appears to be the best compromise between universality and economy as well as ease of use. The present paper gives a brief introduction to this model and shows how well it works for a number of typical hydraulic flow situations. Discussion is restricted here to the standard model which uses an isotropic (scalar) eddy-viscosity and diffusivity and cannot account for buoyancy effects, and to a special version for use in depth-average calculations. Extensions of the model to account for buoyancy effects and turbulence-driven secondary motions can be found in Rodi (13).

## K-ε Model

The task of the k-ε model to be described here is to determine the correlations $\overline{u_i u_j}$ and $\overline{u_i \phi}$ appearing in the mean-flow equations. The model employs the eddy-viscosity/diffusivity concept

$$- \overline{u_i u_j} = \nu_t\left(\frac{\partial U_i}{\partial x_j} + \frac{\partial U_j}{\partial x_i}\right) - \frac{2}{3} k\delta_{ij} \quad , \quad -\overline{u_i \phi} = \Gamma_t \frac{\partial \Phi}{\partial x_i} \text{ with } \Gamma_t = \frac{\nu_t}{\sigma_t} \quad (1)$$

which relates the turbulent stresses to the mean-velocity gradients and the turbulent scalar flux to the gradient of the transported quantity. $\nu_t$ is the eddy-viscosity and the $\Gamma_t$ the eddy-diffusivity; both are not fluid properties but depend on the state of the turbulence. As indicated in Eq. 1, the eddy-diffusivity is assumed proportional to the eddy-viscosity, and the proportionality constant $\sigma_t$ is the turbulent Prandtl-Schmidt number. The eddy-viscosity relation involves not only the velocity gradient but also a term with the turbulent kinetic energy k which is necessary in order to make the turbulent normal stresses (when i=j) sum up to 2k. The basic assumption is now made that the local state of the turbulent motion can be characterized by two parameters, namely the turbulent kinetic energy k and the rate of its dissipation ε. Since, for dimensional reasons, the dissipation rate ε is proportional to $k^{3/2}/L$, where L is the length scale of the energy-containing turbulent motion, the parameter pair k-ε is equivalent to the pair k-L. Once the parameters k and ε have been chosen, the eddy-viscosity $\nu_t$ can then be related to k and ε by dimensional analysis:

$$\nu_t = c_\mu \frac{k^2}{\varepsilon} \quad (2)$$

where $c_\mu$ is a constant. The distribution of the parameters k and ε over the flow field is determined from the following transport equations

$$\frac{\partial k}{\partial t} + U_i \frac{\partial k}{\partial x_i} = \underbrace{\frac{\partial}{\partial x_i}\left(\frac{\nu_t}{\sigma_k} \frac{\partial k}{\partial x_i}\right)}_{} + \underbrace{\nu_t\left(\frac{\partial U_i}{\partial x_j} + \frac{\partial U_j}{\partial x_i}\right)\frac{\partial U_i}{\partial x_j}}_{} - \varepsilon \quad (3)$$

| rate of change | convection | diffusion | P=production | dissipation |

$$\frac{\partial \varepsilon}{\partial t} + U_i \frac{\partial \varepsilon}{\partial x_i} = \frac{\partial}{\partial x_i} \left( \frac{\nu_t}{\sigma_\varepsilon} \frac{\partial \varepsilon}{\partial x_i} \right) + c_{\varepsilon 1} \frac{\varepsilon}{k} P - c_{\varepsilon 2} \frac{\varepsilon^2}{k} \qquad (4)$$

The k-equation 3 is closely related to the exact k-equation which can be derived from the Navier-Stokes equation; the originally appearing production term $-\overline{u_i u_j}\, \partial U_i / \partial x_j$ has been converted with the aid of the eddy-viscosity expression 1, and the diffusion flux of k has been assumed proportional to the gradient of k ($\sigma_k$ being an empirical constant). This equation describes how the rate of change of k is balanced by convective transport by the mean motion, diffusive transport by turbulent motion, production by interaction of turbulent stresses and mean-velocity gradients, and destruction by the viscous dissipation $\varepsilon$. The $\varepsilon$ equation expresses similar physical processes, and it can also be derived in exact form from the Navier-Stokes equation, but such drastic model assumptions have to be introduced that the resulting $\varepsilon$-equation 4 has highly empirical character. All terms on the right-hand side are the outcome of model approximations and contain therefore empirical constants. The model contains altogether six empirical constants whose standard values are given in Table 1. How these constants have been determined from experiments is discussed in a parallel paper (14).

### Table 1: Values of the constants in the k-ε model

| $c_\mu$ | $c_{\varepsilon 1}$ | $c_{\varepsilon 2}$ | $\sigma_k$ | $\sigma_\varepsilon$ | $\sigma_t$ |
|---------|---------------------|---------------------|------------|----------------------|------------|
| 0.09    | 1.44                | 1.92                | 1.0        | 1.3                  | 0.5        |

The standard k-ε model introduced above cannot be applied very near walls where viscous effects become prominent. For this reason, but also because a numerical resolution of the viscous sublayer is computationally expensive, this sublayer is bridged and the boundary conditions are not applied right at the wall but at a point with distance $y_c$ just outside the sublayer. At this point, the velocity components parallel to the wall are assumed to follow the logarithmic law of the wall and the turbulence is assumed to be in local equilibrium, leading to the following relations

$$\frac{U_c}{U_\tau} = \frac{1}{\kappa} \ln(E \frac{U_\tau y_c}{\nu}) \quad , \qquad k_c = \frac{U_\tau^2}{\sqrt{c_\mu}} \quad , \qquad \varepsilon_c = \frac{U_\tau^3}{\kappa y_c} \qquad (5)$$

where c denotes values at point $y_c$ and $U_\tau$ is the friction velocity. In the case of a free surface with significant wind shear, the same conditions apply, that is the surface is treated as a moving wall. In the absence of wind shear, the free surface behaves more like a plane of symmetry where the gradients of all quantities are zero except for the vertical velocity which itself is zero. It seems plausible however that the presence of a free surface should reduce the length scale of turbulence, and this is not accounted for by the symmetry condition for $\varepsilon$. Hence, the following empirical condition is used for $\varepsilon$ at the surface

$$\varepsilon_s = \frac{c_\mu^{3/4} k_s^{3/2}}{\kappa\, ah} \qquad (6)$$

which has the effect of increasing $\varepsilon$ beyond the value that would result from the symmetry condition and tends to reduce the length scale L. The subscript s denotes the surface, a is an empirical constant for which Hossain (6) has determined the value 0.07, and h is the water depth. The boundary condition 6 was conceived and tested for channel flow; it was found to have the right effect but needs further testing under more general situations.

In many surface water problems, the distribution of the flow quantities over the depth is nearly uniform. Such situations are usually simulated with a depth-average calculation method in which depth-average momentum and scalar transport equations are solved. In these equations, depth-average values of the horizontal turbulent transport of momentum and heat or mass appear, which must be determined with the aid of a turbulence model. For this, a depth-average k-$\varepsilon$ model has been developed which basically also uses Eqs. 1 to 4 except that the original quantities are now replaced by their depth-average counterparts and that all vertical gradients are zero. The depth-average k and $\varepsilon$ equations would therefore contain only production terms due to horizontal velocity gradients. The turbulence production due to bed shear is however often also important, and this is accounted for in the depth-average version of the k-$\varepsilon$ model by adding the additional production terms $P_{kv}$ and $P_{\varepsilon v}$ to the k and $\varepsilon$ equations respectively. These production terms are related to the bed friction in the following way (for a detailed discussion see Ref. 14):

$$ P_{kv} = \frac{1}{\sqrt{c_f}} \frac{U_\tau^3}{h} \quad , \quad P_{\varepsilon v} = \frac{c_{\varepsilon 2}\, c_\mu^{1/2}}{c_f^{3/2}\,(e^*\sigma_t)^{1/2}} \frac{U_\tau^4}{h^2} \tag{7} $$

where $c_f$ is the friction coefficient, and $e^* = \Gamma_t/U_\tau h$ is a dimensionless diffusivity whose value is to be provided as empirical input. In wide laboratory flumes $e^* = 0.15$ has been observed, while in natural river flows a value of 0.6 is more appropriate, as discussed further in Ref. 14. It should be added here that so-called diffusion models directly use the dimensionless diffusivity to determine the turbulent heat of mass transport; under conditions where the horizontal velocity gradients are zero, the depth-average version of the k-$\varepsilon$ model becomes identical to a diffusion model.

### Application of the Model

In this section, several applications are presented to typical hydraulic flow situations. The first example concerns developed flow in a wide open channel, and Fig. 1 compares the vertical profiles of velocity U, turbulent kinetic energy k and eddy-viscosity $\nu_t$ with experiments. In these calculations, the surface condition 6 for $\varepsilon$ was applied which leads to an increased $\varepsilon$ near the surface and to a reduction of the $\nu_t$ according to Eq. 2. This reduction in $\nu_t$ brings the calculated eddy-viscosity distribution a long way towards the measured parabolic distribution. A further reduction is necessary however which can be achieved by introducing a surface-damping model simulating the damping of the vertical fluctuations by the surface damping thereby effectively reducing the coefficient $c_\mu$. This model is described in Celik et al (3). Fig. 1 shows that the U and k profiles are in fairly good agreement with the measured ones.

The second example shows Svensson's (17) calculations of wind-induced channel flow. Svensson used the standard k-$\varepsilon$ model and considered the free surface as a moving wall at which the experimental shear stress was prescribed and the conditions 5 were employed. The wind piles up the water which causes a positive longitudinal pressure gradient and a reverse flow near the bottom. All these processes are simulated well by the mathematical model, as can be seen from the comparison of the vertical velocity profile with measurements in Fig. 2.

The next example concerns the three-dimensional calculation of a 180° strongly curved open channel bend. Fig. 3 compares the calculated development of the streamwise velocity profile, the water level at the inner and outer bank and the lateral velocity profile with experiments. It can be seen that the maximum of the streamwise velocity first moves to the inner bank and then to the outer one, that the water level rises at the outer bank and decreases at the inner one, and that the lateral velocity is directed outward near the surface and inward near the bottom. All these complex features are predicted correctly, qualitatively as well as quantitatively. It should

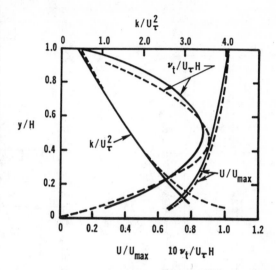

Fig. 1  Vertical velocity, turbulent kinetic energy and eddy viscosity distribution in developed open channel flow, —— predictions (3),--- data (18), ···· data (10).

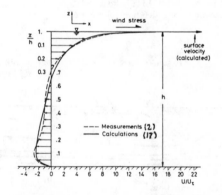

Fig. 2  Velocity profile for wind-induced channel flow.

be mentioned however that this is not so much a merit of the turbulence model because in this case the flow development is governed largely by pressure forces. Therefore, a simpler turbulence model may have given similar results.

The next example involves recirculation and is hence not of the shear-layer type. Such flows would be difficult to simulate with a method that requires an empirical length-scale specification. Fig. 4 shows the application of the standard k-ε model to the flow around a square obstacle placed in a

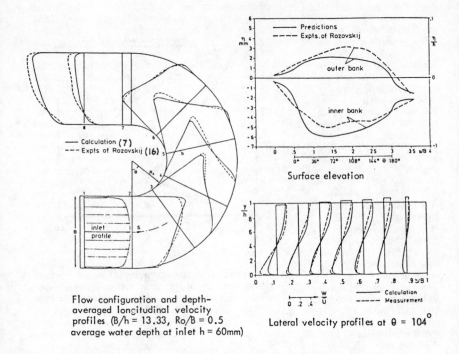

— Calculation (7)
--- Expts of Rozovskij (16)

Predictions
--- Expts. of Rozovskij

outer bank

inner bank

Surface elevation

Flow configuration and depth-
averaged longitudinal velocity
profiles (B/h = 13.33, Ro/B = 0.5
average water depth at inlet h = 60mm)

Lateral velocity profiles at $\theta = 104^{\circ}$

— Calculation
--- Measurement

**Fig. 3  Flow in a 180° open channel bend (7)**

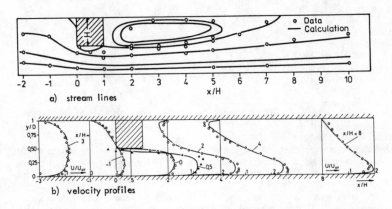

o  Data
—  Calculation

a)  stream lines

b)  velocity profiles

**Fig. 4  Flow around a square obstacle placed in a channel (4)**

closed channel. In this case, the experiments indicate two small recirculation zones in front and on top of the obstacle and a relatively large one behind the block. All three zones are reproduced by the calculation. From the lower part of Fig. 4 it can be seen that the disturbance of the channel profile upstream of the block is well predicted; the size of the recirculation regions on top of and behind the block are also well predicted. The agreement for the velocity profiles in these regions is however not fully satisfactory. Alfrink (1) calculated the flow in dredged trenches with the standard k-$\epsilon$ model. He found that the flow was always predicted qualitatively correct but that the standard constants produced a separation length that was about 30 percent shorter than the measured one. This trend is in agreement with previous calculations of the flow over a backward facing step (see e.g. Ref. 15). When Alfrink changed the empirical coefficient $c_{\epsilon 1}$ from 1.44 to 1.6, the reattachment length was only 10 percent too small. He also obtained satisfactory agreement for the shear stress and the turbulent kinetic energy at various cross sections of the dredged trench. The fact that a higher than the standard value is needed for the coefficient $c_{\epsilon 1}$ points to certain shortcomings of the k-$\epsilon$ model when applied to recirculating flows. Modifications to the standard model have been suggested (see e.g. Ref. 8) that improve the accuracy for recirculating flow calculations.

The last two examples show applications of the depth-average version of the model. First, the "washing out" of a pollutant from a bay or harbor by a mainstream passing by is examined. The flow geometry is sketched in the left part of Fig. 5. As there is no net flow across the dividing streamline, the mass exchange between the dead water zone and the mainstream is due entirely to the turbulent motion. The calculation was started with uniform concentration in the cavity and the concentration development with time was then simulated for various water depths H. The right part of Fig. 5 shows that the dependence of the half-life time $t_{1/2}$ of the average concentration in the cavity on H is predicted well by depth-average model. At small water depths, turbulence is mainly generated by bed friction, while at large depths turbulence is generated by the horizontal velocity gradients in the shear layer springing off the edge of the sudden expansion. Ref. 9 shows that the half-life time in the latter case should be proportional to $B/U_i$ (defined in Fig. 5), and

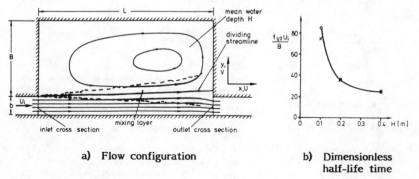

a) **Flow configuration**          b) **Dimensionless half-life time**

Fig. 5   Mass exchange between a mainstream and a dead water zone (9), o—— calculation, x expts.

it can be seen that both calculated and measured $t_{1/2}U_i/B$ approach a constant value at larger H values. The model is therefore capable of accounting properly for the combined effects of turbulence generation due to bed shear and due to horizontal velocity gradients.

The last example shows the application of the depth-average model to a real-life problem, in this case to the near field of the cooling-water discharge from the Karlsruhe power station into the river Rhine. In this near-field situation also, the combined turbulence production due to bed shear and horizontal velocity gradients is very important. The details of the study are given in Pavlovic (11). Fig. 6 compares predicted velocity vectors and isotherms with field measurements. The calculations show a recirculation zone behind the discharge; unfortunately the measurements are not very detailed in this region, but there is one velocity measurement that does indicate a negative velocity in agreement with the calculations. In general, the flow field appears to be correctly predicted. The depth-average isotherms agree also reasonably well with the measurements; the agreement is certainly sufficient for practical purposes. The same model, with the same empirical input, was applied successfully also to cooling-water and sewage discharges into other rivers, and two further examples may be found in the parallel paper (14).

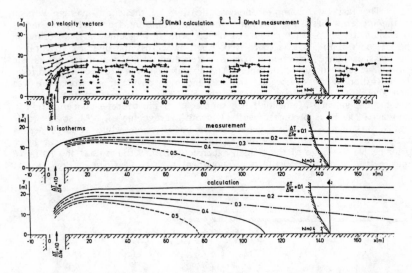

Fig. 6  Velocity vectors and isotherms in the near field of Karlsruhe power station cooling water discharge into the Rhine, calculations Ref. 11, measurements Ref. 5.

## CONCLUDING REMARKS

The paper gave a brief introduction to the k-ε turbulence model which is one of the most widely tested models for hydraulic flow calculations. The model works well for shear-layer situations and is the simplest model that can be used for flows more complex than shear layers. However, for these flows the accuracy of the standard model is generally not as good as for shear layers.

In particular, for recirculating flows the length of the recirculation zone is usually predicted too small. Research is in progres to improve the model for such applications. Extended versions of the model not using an isotropic eddy-viscosity have been developed that can take account of buoyancy effects and turbulence-driven secondary motions in non-circular ducts; these versions extend greatly the applicability of the standard model (see e.g. Rodi (13)). The depth-average model was found to describe well the interaction of turbulence generated by bed shear and by horizontal velocity gradients and is capable of simulating quite different real-live discharge situations with the same empirical constants.

## ACKNOWLEDGEMENTS

This paper was written while the author was a visitor at the University of California at Davis. The support of the University of California is greatfully acknowledged. The author should also like to thank Nancy Nelson for typing the manuscript.

## REFERENCES

1. Alfrink, B.J., "The Computation of Turbulent Recirculating Flow Using Curvilinear Finite Differences. Application of the k-ε Model to the Flow in Dredged Trenches", Report HE/41/81.22, Electricity de France, Chatou, France, 1981.
2. Baines, W.D. and Knapp, D.J., "Wind Driven Water Currents ", ASCE Journal of the Hydraulics Division, Vol. 91, 1965.
3. Celik, I., Hossain, M.S., and Rodi, W., "Simulation of the Influence of a Free Surface on Turbulent Transport", Proc. Symposium on Refined Modelling of Flows, Paris, France, Sept. 1982.
4. Durst, F. and Rastogi, A.K., "Theoretical and Experimental Investigations of Turbulent Flows with Separation", Turbulent Shear Flows 1, Springer-Verlag, Heidelberg, 1979.
5. Grimm-Strele, J., "Messung der Strahlausbreitung im unmittelbaren Nahbereich eines Kühlwasserruckgabebauwerks", Report SFB80/E/200, University of Karlsruhe, 1981.
6. Hossain, M.S., "Mathematische Modellierung von turbulenten Auftriebsströmungen", Ph.D. thesis, University of Karlsruhe, 1980.
7. Leschziner, M.A. and Rodi, W., "Calculation of Strongly Curved Open Channel Flow", Journal of the Hydraulics Division, ASCE, Vol. 105, No. HY10, 1979, pp. 1297-1314.
8. Leschziner, M.A. and Rodi, W., "Calculation of Annular and Twin Parallel Jets Using Various Discretization Schemes and Turbulence-Model Variations", ASME Journal of Fluids Engineering, Vol. 103, 1981, pp. 352-360.
9. McGuirk, J. and Rodi, W., "Calculation of Unsteady Mass Exchange Between a Mainstream and a Dead Water Zone", Proc. 18th IAHR Congress, Cagliari, Italy, 1979.
10. Nakagawa, H., Nezu, I., Ueda, H., "Turbulence in Open Channel Flow Over Smooth and Rough Beds", Proc. Japan Society of Civil Engineers, 241, 1975, pp. 155-168.
11. Pavlovic, R.N., "Numerische Berechnung der Wärme-und Stoffausbreitung in Flüssen mit einem tiefengemittelten Modell", Ph.D. thesis, University of Karlsruhe, 1981.
12. Prandtl, L., "Über die ausgebildete Turbulenz", ZAMM, Vol. 5, 1925, p. 136.

54        APPLYING HYDRAULIC RESEARCH

13. Rodi, W., "Turbulence Models and Their Application in Hydraulics", Book publication of International Association for Hydraulic Research, Delft, The Netherlands, 1980.
14. Rodi, W., "Testing and Calibration of Turbulence Models for Transport and Mixing", Paper at ASCE Hydraulics Division Specialty Conference, Jackson, Mississippi, August 1982.
15. Rodi, W., Celik, I., Demuren, A.O., Scheuerer, G., Shirani, E., Leschziner, M.A., Rastogi, A.K., "Calculations for the 1980-81 AFOSR-HTTM Stanford Conference on Complex Turbulent Flows", Report SFB80/T/199, University of Karlsruhe, July, 1981.
16. Rosovskii, I.L., "Flow of Water in Bends of Open Channels", published by the Academy of Sciences of the Ukrainian SSR, Kiev, 1975, printed in Jerusalem by S. Monson.
17. Svensson, U., "Mathematical Model of the Seasonal Thermocline", Univeristy of Lund, Department of Water Resources Engineering, Sweden, Report No. 1002, 1978.
18. Ueda, H., Möller, R., Komori, S., Mizushina, T., "Eddy Diffusivity Near the Free Surface of Open Channel FLow", Int. J. Heat Mass Transfer, 20, 1977, pp. 1127-1136.

# SOLUTION METHODS FOR TURBULENT FLOW
## VIA FINITE ELEMENTS

David R. Schamber[1], A.M. ASCE

## ABSTRACT

The paper presents the numerical solution of the $k$-$\varepsilon$ turbulence model via the finite element method. Solution strategies using a continuation method and Newton's method with line search are formulated and applied to several one-dimensional turbulent flow simulations. For nonlinear problems, these methods yield a converged solution without the need for a "good" starting approximation to the solution vector.

## INTRODUCTION

In the past decade advances have been made in developing single point closure models which simulate turbulent flows via a set of continuum equations. The $k$-$\varepsilon$ turbulence model is one such model which determines one time scale and one turbulent velocity from a pair of transport equations. These equations, when coupled with the mean-flow mass and momentum conservation equations, produce a set of nonlinear partial differential equations. The finite element discretization of this equation set produces a set of coupled, nonlinear algebraic equations which often fail to converge because, even for simple turbulent flows, the initialization of the field variables falls outside the radius of convergence. Two solution schemes, the continuation method and Newton's method with line search, are presented which produce converged solutions using an initialization vector which would otherwise diverge using the standard Newton method.

The continuation method is one scheme which avoids the initialization problem by embedding the equation set one would like to solve within a system of equations whose solution is known. Appropriate embedding functions pertinent to the finite element solution of the $k$-$\varepsilon$ turbulence model are presented. The line search technique optimizes a relaxation factor in the standard Newton method such that an appropriate objective function is minimized at each iteration cycle. Convergence of the method is thus enhanced. The methods are tested on several one-dimensional, turbulent flow examples. The ultimate goal is to identify solutions techniques which may be appropriate in solving two-dimensional, turbulent flow problems.

---

[1]Assistant Professor, Civil Engineering Department, University of Utah, Salt Lake City, Utah 84112

GOVERNING EQUATIONS

The mass and momentum equations for, steady, turbulent, incompressible flow are (Larock and Schamber, 1981):

$$\frac{\partial U_i}{\partial x_i} = 0 \tag{1}$$

$$U_j \frac{\partial U_i}{\partial x_j} = -\frac{1}{\rho}\frac{\partial P}{\partial x_i} + g_i + \frac{\partial}{\partial x_j}\left[\nu_t\left(\frac{\partial U_i}{x_j} + \frac{\partial U_j}{x_i}\right) - \frac{2}{3}\delta_{ij}k\right] \tag{2}$$

in which $x_i$ = cartesian coordinate component; $U_i$ and P are the mean velocity components and pressure, respectively; $\rho$ = fluid density; $g_i$ = gravitational components; $\nu_t$ = turbulent kinematic viscosity; $\delta_{ij}$ = Kronecker delta; and k = turbulence kinetic energy. The Einstein convention of summing over a repeated subscript is used and the isotropic closure relation for the Reynolds stresses is incorporated into the last term in Eq. 2.

The k-$\varepsilon$ turbulence model is represented by the following pair of transport equations (Larock and Schamber, 1981).

$$U_j \frac{\partial k}{\partial x_j} = P_r - \varepsilon + \frac{\partial}{\partial x_j}\left[\frac{c_\mu}{\sigma_k}\frac{k^2}{\varepsilon}\frac{\partial k}{\partial x_j}\right] \tag{3}$$

$$U_j \frac{\partial \varepsilon}{\partial x_j} = c_{\varepsilon 1}\frac{P_r\,\varepsilon}{k} - c_{\varepsilon 2}\frac{\varepsilon^2}{k} + \frac{\partial}{\partial x_j}\left[\frac{c_\mu}{\sigma_\varepsilon}\frac{k^2}{\varepsilon}\frac{\partial \varepsilon}{\partial x_j}\right] \tag{4}$$

in which $\varepsilon$ = rate of dissipation of turbulence kinetic energy and the production $P_r$ and turbulent viscosity are given respectively by

$$P_r = \left[\nu_t\left(\frac{\partial U_i}{\partial x_j} + \frac{\partial U_j}{\partial x_i}\right) - \frac{2}{3}\delta_{ij}k\right]\frac{\partial U_i}{\partial x_j} \tag{5}$$

$$\nu_t = c_\mu \frac{k^2}{\varepsilon} \tag{6}$$

The constants $c_\mu$ = 0.09, $c_{\varepsilon 1}$ = 1.45, $c_{\varepsilon 2}$ = 1.90, $\sigma_k$ = 1.0 and $\sigma_\varepsilon$ = 1.3 are determined from experimental data and computer optimization (Schamber and Larock, 1980a).

FINITE ELEMENT FORMULATION

The Galerkin finite element forms of Eqs. 1-4 for one-dimensional flow are written for an element domain $\Omega_e$ bounded by the closed contour $\Gamma_e$. Embedding function $H_k$ and $H_\varepsilon$ for the continuation method are included in the expressions. Mathematically, the element residuals are

$$r_U = \int_{\Omega_e} [-N_i \frac{\partial P}{\partial x} - \nu_t \frac{\partial N_i}{\partial z} \frac{\partial U}{\partial z}]d\Omega + \int_{\Gamma_e} N_i \nu_t \frac{\partial U}{\partial z} \ell_z d\Gamma \tag{7}$$

$$r_k = t \int_{\Omega_e} [N_i(P_r - \varepsilon) - \frac{\nu_t}{\sigma_k} \frac{\partial N_i}{\partial z} \frac{\partial k}{\partial z}]d\Omega + t \int_{\Gamma_e} N_i \frac{\nu_t}{\sigma_k} \frac{\partial k}{\partial z} \ell_z d\Gamma + (1-t)H_k \tag{8}$$

$$r_\varepsilon = t \int_{\Omega_e} N_i\{c_{\varepsilon 1} \frac{\varepsilon}{k} P_r - c_{\varepsilon 2} \frac{\varepsilon^2}{k}\}d\Omega$$

$$- t\int_{\Omega_e} \frac{\nu_t}{\sigma_\varepsilon} \frac{\partial N_i}{\partial z} \frac{\partial \varepsilon}{\partial z} d\Omega + t \int_{\Gamma_e} N_i \frac{\nu_t}{\sigma_\varepsilon} \frac{\partial \varepsilon}{\partial z} \ell_z d\Gamma + (1-t)H_\varepsilon \tag{9}$$

in which U = streamwise mean velocity component in the x coordinate direction; z = transverse coordinate; $N_i$ = 8-node biquadratic basis functions; $\ell_z$ = z component of the outward unit normal to the flow boundary $\Gamma$; and t = continuation parameter ($0 \le t \le 1$). Boundary integral terms are evaluated only when $\Gamma_e$ coincides with $\Gamma$. Variables appearing in Eqs. 7-9 have been nondimensionalized with respect to $U_0$ (reference velocity) and $L_0$ (reference length).

The functions $H_k$ and $H_\varepsilon$ are specified to produce a desired initial condition at t = 0. For a constant or linear variation in k and $\varepsilon$ the following forms are appropriate

$$H_k = \int_{\Omega_e} N_i[k_1 \frac{z-z_2}{z_1-z_2} + k_2 \frac{z-z_1}{z_2-z_1}]d\Omega \tag{10}$$

$$H_\varepsilon = \int_{\Omega_e} N_i[\varepsilon_1 \frac{z-z_2}{z_1-z_2} + \varepsilon_2 \frac{z-z_1}{z_2-z_1}]d\Omega \tag{11}$$

Here $k_1$, $k_2$, $\varepsilon_1$ and $\varepsilon_2$ are the specified boundary values at $z_1$ and $z_2$. With these values an initial linear distribution of k and $\varepsilon$ is generated throughout the flow domain. Alternatively, if $k_1 = k_2 = k_0$ and $\varepsilon_1 = \varepsilon_2 = \varepsilon_0$ are specified then a constant field of k and $\varepsilon$ is produced. Other desired initial conditions can easily be created by adjusting the functions in Eqs. 10 and 11.

CONTINUATION METHOD

The assembled forms of Eqs. 7-9 may be compactly written as

$$F(X,t) = 0 \qquad (12)$$

in which F = vector representing the system of nonlinear algebraic equations and X = vector of nodal unknowns for the system. Newton's method solves Eq.12 by solving the following linear equations for $\delta X$

$$A^j \delta X^j = -F(X^j,t) \qquad (13)$$

in which j = iteration counter, $A^j$ = Jacobian of the equation matrix, and $\delta X^j$ = set of incremental corrections to the vector $X^j$. Determination of $X^j$ produces an updated estimate of the solution

$$X^{j+1} = X^j + \alpha \delta X^j \qquad (14)$$

in which $\alpha$ = relaxation factor. For the continuation method $\alpha$ = 1.

The continuation method begins by solving Eq. 13 with $t_o$ = 0. The solution at $t_o$ = 0 is then used as the initialization for solving Eq. 13 at $t_{i+1} = t_i + \delta t_i$ with i = 0. A sequence of solutions (i = 0,1,2,...) at increasing values of t between zero and unity are thus generated. Each previously generated solution is used as the initialization for the next solution. The desired solution is obtained when t = 1. Schamber et al. (1982) detail a predictor-corrector algorithm which selects an appropriate $\delta t$ at each step to avoid divergence of the scheme. Only the results of this scheme are presented here for comparison with the line search technique presented in the following section.

Turbulent flow between smooth parallel walls is examined first. The computational domain consists of six elements which start a small distance from the solid boundary and terminate at the channel center. Law-of-the-wall functions (Schamber and Larock, 1980b) determine the essential boundary conditions for k, $\varepsilon$ and U near the solid boundary. At midchannel symmetry requires the normal derivatives of k, $\varepsilon$ and U to be zero. For fully developed flow the streamwise gradients of these variables are also zero. The continuation solution for this problem is depicted in Figs. 1, 2 and 3. Quantities presented have been nondimensionalized on the half channel width d and shear velocity $U_\tau$ at the wall. The Reynolds number is based on d and the centerline channel velocity. At t = 0 $H_k$ and $H_\varepsilon$ are specified to produce a constant viscosity (Eq. 6) solution i.e. a simple parabolic velocity profile. A total of 9 t-steps are required to produce the desired solution. Only a representative number are plotted in Figs. 1, 2 and 3. From 2-5 iterations are required at each step to achieve convergence and the total number of iterations is 36.

Turbulent asymmetric channel flow has been studied experimentally
by Hanjalic and Launder (1972). One wall in the channel is roughened
with square ribs and the other wall is smooth producing an asymmetric
flow profile. Nine elements are used for this simulation with the
edge nodes displaced from the walls to accommodate the law-of-the-wall
boundary conditions. Essential boundary conditions at the smooth and
rough surfaces are determined from appropriate wall functions (Schamber
and Larock 1980b). Streamwise derivatives for k, $\epsilon$ and U are zero for
fully developed flow. For t=0 $H_k$ and $H_\epsilon$ are specified to produce
linear variations in k and $\epsilon$ which satisfy the boundary conditions.
The continuation solution for this example is shown in Figs. 4, 5 and
6. A total of 7 t-steps are required to give a converged solution.
Only a representative number of profiles are presented in the figures.
At each t-step 2-6 iterations are required to achieve convergence. The
total number of iterations is 29.

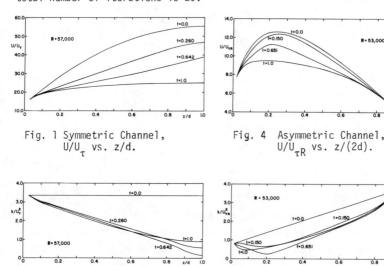

Fig. 1 Symmetric Channel,
$U/U_\tau$ vs. z/d.

Fig. 4 Asymmetric Channel,
$U/U_{\tau R}$ vs. z/(2d).

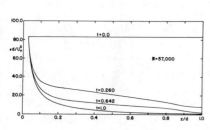 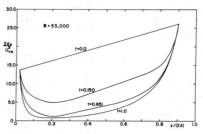

Fig. 2 Symmetric Channel,
$k/U_\tau^2$ vs. z/d.

Fig. 5 Asymmetric Channel,
$k/U_{\tau R}^2$ vs. z/(2d).

Fig. 3 Symmetric Channel,
$\epsilon d/U_\tau^3$ vs. z/d.

Fig. 6 Asymmetric Channel,
$2d\epsilon/U_{\tau R}^3$ vs. z/(2d).

NEWTON'S METHOD WITH LINE SEARCH

Newton's method with line search requires an intial estimate $X^1$ which can be conveniently obtained from Eqs. 13 and 14 with t = 0. At iteration j the search direction $\delta X^j$ is found from Eq. 13. The scalar $\alpha$ in Eq. 14 is selected such that an appropriate objective function $f(X^j + \alpha \delta X^j)$ is minimized with respect to $\alpha$. After minimizing, $X^{j+1}$ is determined according to Eq. 14. Fletcher (1980, p. 99) suggests the following objective function

$$f = F^T(X)F(X) \tag{15}$$

which is simply the sum of the squares of the equation residuals. An appropriate search interval for $\alpha$ is [0,1]. For the upper limit $\alpha = 1$ the quadratic convergence rate of Newton's method is realized. A general search algorithm would also allow negative values of $\alpha$. However, DeVantier and Larock (1982) report that a negative search direction produces a decreasing objective function only when the method is strongly divergent. Thus, negative searches are not implemented in the present algorithm.

Fletcher (1980, pp. 25-29) presents several line search strategies for minimizing f. The more complicated schemes require first derivative information about f while other schemes simply utilize the function itself. These methods generally employ some form of sectioning and possibly interpolation. Evaluating the first derivative of f, $df/d\alpha$, requires an updating of the Jacobian A at each $\alpha$ where $df/d\alpha$ is required. For large problems the amount of computation time would be prohibitive. In order to minimize computing effort a search scheme which utilizes only values of f is adopted.

Initially, a bracket of three $\alpha$-values, $\alpha_1 < \alpha_2 < \alpha_3$, is selected such that $f_2 \leq \min (f_1, f_3)$. This bracket is then contracted by intro-ducing $\alpha_4$, e.g. $\alpha_1 < \alpha_2 < \alpha_4 < \alpha_3$, which creates the new intervals $(\alpha_1, \alpha_2, \alpha_4)$ and $(\alpha_2, \alpha_4, \alpha_3)$. Of these two brackets the one which maintains the condi-tion $f_2 \leq \min (f_1, f_4)$ or $f_4 \leq \min (f_2, f_3)$ is selected as the new interval $(\alpha_1 < \alpha_2 < \alpha_3)$ and the process is repeated. A "golden section search" is adopted in which the ratio $\alpha_3 - \alpha_2 : \alpha_2 - \alpha_1$ is fixed by $\tau:1$ (or $1:\tau$) where $\tau = (1+\sqrt{5})/2$. The new intervals formed by $\alpha_4$ are required to have intervals in the same ratio. The process is terminated at some stage when $\alpha_3 - \alpha_1 \leq b$ and $\alpha_2$ is selected as the approximate minimizing value for f. Accuracy of the search is controlled by the parameter b. The method terminates if the condition $f_2 \leq \min (f_1, f_3)$ fails at any step. However, if the initial bracket fails the test, and $f_3 < f_1$, then the optimizing value $\alpha = \alpha_3$ is selected and updating proceeds via Eq. 14. Ideally, the starting interval $\alpha_1 = 0$ and $\alpha_3 = 1$ is selected. However, in order to ensure realizability of the solution a suitable method for selecting the starting $\alpha_3$ less than unity is now presented.

If the initial estimate to the solution field of k, $\epsilon$ and U is far from the actual solution, negative values of k and $\varepsilon$, which have no physical meaning, may be generated during the iterative algorithm. Experience has shown that negative values quickly produce a diverging result. In order to ensure that these variables are indeed positive the initial starting $\alpha_3$ is selected as follows. At the j iteration a search is made over the $n_k$ unknowns for k and $n_\varepsilon$ unknowns for $\varepsilon$ such that

$$\alpha_k = \min[(k_{min}-k_i^j)/\delta k_i^j] \qquad i = 1,2,\ldots,n_k \qquad (16)$$

$$\alpha_\varepsilon = \min[(\varepsilon_{min}-\varepsilon_i^j)/\delta\varepsilon_i^j] \qquad i = 1,2,\ldots,n \qquad (17)$$

Here $k_{min}$ and $\varepsilon_{min}$ are specified minimum values for the k and $\varepsilon$ fields respectively, while $k^j$ and $\varepsilon^j$ are the appropriate variables from $X^j$, and $\delta k^j$ and $\delta\varepsilon^j$ are the appropriate corrections from $\delta X^j$. Equations 16 and 17 are only applied for those i for which $\delta k_i^j < 0$ and $\delta\varepsilon_i^j < 0$. The starting $\alpha_3$ is then selected from $\alpha_3 = \min(\alpha_k,\alpha_\varepsilon) \leq 1$. If $\alpha_k$ and/or $\alpha_\varepsilon$ fall below some specified minimum, $\alpha_m$, progress in minimizing f will occur at a very slow rate. Therefore corrections via Eq. 14 to k or $\varepsilon$ or both may not be appropriate. A realizability parameter R is then set according to the following conditions: set R = 0 if $\alpha_k \geq \alpha_m$ and $\alpha_\varepsilon \geq \alpha_m$; set R = 1 if $\alpha_k < \alpha_m$ and $\alpha_\varepsilon \geq \alpha_m$; set R = 2 if $\alpha_k \geq \alpha_m$ and $\alpha_\varepsilon < \alpha_m$; and set R = -1 if $\alpha_k < \alpha_m$ and $\alpha_\varepsilon < \alpha_m$. If R = 0, $\alpha_3 = \min(\alpha_k,\alpha_\varepsilon) \leq 1$ and corrections are applied (Eq. 14) to all field variables k, $\varepsilon$ and U. If R = 1, $\alpha_3 = \alpha_\varepsilon$ and corrections are applied only to the $\varepsilon$ and U fields. If R = 2, $\alpha_3 = \alpha_k$ and corrections are applied only to the k and U fields. Finally, if R = -1, $\alpha_3 = 1$ and corrections are applied only to the U field. With R determined and the initial trial value for $\alpha_3$ set, several additional tests are required before the golden section search can be effectively employed.

During the first several j iteration cycles a complete section search seems unreasonable since the initial solution is far from the true solution. Thus the section search is initially suppressed and $\alpha$ is selected based on the criteria $f(\alpha) < f(0)$. If n = total number of equations being solved then $f(0)/n$ represents an average squared measure of error for each equation. For $f(0)/n < \bar{f}$, in which $\bar{f}$ = specified measure of error, the golden section search is implemented. For $f(0)/n \geq \bar{f}$ the section search is suppressed and $\alpha$ is reduced to $\alpha_2 = \alpha_3/\lambda$ in which $\lambda$ = specified reduction parameter ($\lambda \geq 1$). If $f(\alpha_2) < f(0)$ then $X^{j+1}$ is updated via Eq. 14 with $\alpha = \alpha_2$. The update may only affect a portion of the solution field depending on the current value of R. If $f(\alpha_2) \geq f(0)$ and $R \neq -1$ the run is terminated. If $f(\alpha_2) \geq f(0)$ and R = -1 (meaning only U is to be updated) $\alpha_2$ is systematically reduced via $\alpha_2 = \alpha_2/2^m$ for m = 1,2,...10 and at each m the condition $f_2(\alpha_2) < f(0)$ is checked. Upon satisfying this condition, corrections proceed according to Eq. 14 with $\alpha = \alpha_2^m$. If the condition fails for m = 10 the run is terminated. This particular section of the algorithm represents a final effort to align the velocity field with the k and $\varepsilon$ fields whose values are near the limit of $k_{min}$ and $\varepsilon_{min}$. The factor $2^m$ and upper limit of m = 10 have been arbitrarily specified. The parameter $\lambda$ is usually set greater than unity since $\alpha_3$ is often equal to $\alpha_k$ or $\alpha_\varepsilon$ and one would like to gradually approach (over several j iteration cycles) $k_{min}$ and/or $\varepsilon_{min}$. Finally, if f(0) becomes sensibly small the algorithm should set $\alpha$ = 1 to achieve the quadratic convergence rate of Newton's method. This is accomplished by comparing f(0) to a specified minimum value f*. For f(0) $\leq$ f*, $\alpha$ = 1 otherwise the golden section search is implemented.

Several examples are now presented which illustrate the application of the line search method to symmetric and asymmetric channel flow. The results of two runs with differing initial conditions for symmetric channel flow are given in Table I. Initial conditions for the first run

| Table I, Symmetric Channel | | | | | | | |
|---|---|---|---|---|---|---|---|
| Run No. 1 | | | | Run No. 2 | | | |
| Iter. No. | R | $\alpha$ | f | Iter. No. | R | $\alpha$ | f |
| 1→5 | 0 | 0.500 | $10^4$→78. | 1 | 0 | 0.313 | 1800. |
| 6 | 0 | 0.493 | 23. | 2 | 0 | 0.296 | 920. |
| 7 | 0 | 0.219 | 14. | 3 | 0 | 0.284 | 480. |
| 8 | 0 | 0.121 | 11. | 4 | 0 | 0.279 | 260. |
| 9 | 0 | 0.067 | 9.8 | 5 | 0 | 0.293 | 130. |
| 10 | 0 | 0.035 | 9.1 | 6 | 0 | 0.376 | 55. |
| 11 | 1 | 0.224 | 5.5 | 7 | 0 | 0.500 | 16. |
| 12 | 0 | 0.036 | 5.1 | 8 | 0 | 0.500 | 4.5 |
| 13 | 1 | 0.242 | 2.9 | 9 | 0 | 0.500 | 1.3 |
| 14 | 0 | 0.297 | 1.5 | 10 | 0 | 0.500 | 0.4 |
| 15 | 0 | 0.500 | 0.4 | 11→15 | 0 | 1.000* | $0.004$→$10^{-12}$ |
| 16 | 0 | 1.000* | 0.02 | Parameters for run no. 1 & 2: | | | |
| 17 | 0 | 0.584* | 0.004 | $\alpha_m$ = 0.05, $\bar{f}$ = 0.01, b = 0.1, | | | |
| 18 | 0 | 0.400* | 0.001 | f* = $10^{-6}$, $k_{min}$ = $\varepsilon_{min}$ = 0.01, $\lambda$ = 2. | | | |
| 19 | 0 | 0.472* | 0.0004 | | | | |
| 20 | 0 | 0.708* | 0.00004 | | | | |
| 21→23 | 0 | 1.000 | $10^{-7}$→$10^{-13}$ | | | | |
| *Indicates golden section search for f > f*. | | | | | | | |

are depicted in Figs. 1, 2 and 3 for t' = 0. For the second run $k/U_\tau^2$ = 3.33 as shown in Fig. 2 while $\varepsilon d/U_\tau^3$ decreases linearly from 83.3 at z/d = 0.03 to 3.33 at z/d = 1.0. The initial U profile was obtained by solving the momentum equation with the specified k and $\varepsilon$ distributions. For all iterations reported in Table I t = 1.0 in Eq. 13. The total number of iterations, including those used to develop the initial conditions, is 26 and 18 for run no. 1 and run no. 2, respectively.

Several additional test examples identical to run no. 1 but with differing values of $k_{min}$, $\varepsilon_{min}$, and $\lambda$ failed to converge. One run with $k_{min}$=$\varepsilon_{min}$=0.01 and $\lambda$=1 failed to make any progress after 7 iterations.

Another example with $k_{min} = \varepsilon_{min} = 0.2$ and $\lambda = 1.5$ terminated after 10 iterations.

Application of the standard Newton method (i.e. $\alpha=1$ for all iterations and no realizability checks) to run no. 1 & 2 produced a divergent result.

The effect of the parameter R is evident in Table I for run no. 1. Little progress is being made in reducing f in iterations 9, 10 and 12 until R is set to unity for iterations 11 and 13. In both runs it appears that $\bar{f}$ could be increased to allow the golden section search to be implemented at an earlier iteration.

The results of the asymmetric channel flow problem are presented in Table II.

| Table II, Asymmetric Channel | | | | | | | |
|---|---|---|---|---|---|---|---|
| Run No. 3 | | | | Run No. 4 | | | |
| Iter. No. | R | $\alpha$ | f | Iter. No. | R | $\alpha$ | f |
| 1→5 | 0 | 0.500 | 64.→0.4 | 1 | 0 | 0.500 | 85. |
| 6 | 0 | 0.376* | 0.2 | 2 | 0 | 0.484 | 24. |
| 7 | 0 | 1.000* | 0.02 | 3 | 0 | 0.453 | 7.5 |
| 8 | 0 | 0.183* | 0.01 | 4 | 0 | 0.393 | 2.9 |
| 9 | 0 | 0.435* | 0.005 | 5 | 0 | 0.329 | 1.3 |
| 10 | 0 | 0.708* | 0.0006 | 6 | 0 | 0.304 | 0.7 |
| 11→13 | 0 | 1.000* | $10^{-5}$→$10^{-14}$ | 7 | 0 | 0.427* | 0.3 |
| Parameters for run no. 3 & 4: | | | | 8 | 0 | 0.381* | 0.1 |
| $\alpha_m = 0.05$, $\bar{f} = 0.01$, b = 0.1, | | | | 9 | 1 | 0.326* | 0.09 |
| | | | | 10 | 0 | 0.951* | 0.01 |
| $f^* = 10^{-6}$, $\lambda = 2$. | | | | 11 | 0 | 0.562* | 0.002 |
| | | | | 12 | 0 | 0.798* | 0.0001 |
| | | | | 13→15 | 0 | 1.000* | $10^{-6}$→$10^{-14}$ |
| *Indicates golden section search for f > f*. | | | | | | | |

Initial conditions for run no. 3 are depicted in Figs. 4, 5, and 6 with t = 0. For the fourth run the initial conditions for k and $\varepsilon$ begin at the left boundary [z/(2d) = 0.02] shown in Figs. 5 and 6 respectively, and increase parabolically over the first element to the values $k/U^2_{\tau R}$ = 3.662 and $2d\varepsilon/U^3_{\tau R}$ = 26.15 at z/(2d) = 0.05, respectively. The parabolic variation of k and $\varepsilon$ over this first element is such that the slopes of both variables are zero at z/(2d) = 0.05. For z/(2d) $\geq$ 0.05 k and $\varepsilon$ remain constant to the right boundary. This type of initialization may prove useful in a two-dimensional flow problem. The initial U profile for the fourth run was obtained by solving the momentum equation with the specified k and $\varepsilon$ fields. For all iterations reported in Table II t = 1.0 in Eq. 13. The total number of iterations, including those used to develop the initial conditions, is 15 and 18 for run no. 3 and run no. 4, respectively. In run no. 3 $k_{min} = \varepsilon_{min} = 0.01$. For run no. 4 $k_{min} = \varepsilon_{min} = 0.2$. The third run failed to make any progress after 6 iterations if $\lambda = 1$. The fourth run terminated after 6 iterations with $k_{min} = \varepsilon_{min} = 0.01$. Both examples in Table II diverged using the standard Newton method.

PROGNOSIS

The finite element formulation of the $k$-$\varepsilon$ turbulence model has been presented for solution by the continuation method and Newton's method with line search. Both methods work well for all cases reported. The line search algorithm requires from one-third to one-half less total number of iterations compared to the continuation method. This reduction could result in a significant cost savings for any two-dimensional simulation. However, the Newton method with line search appears to be sensitive to the input parameters $k_{min}$, $\varepsilon_{min}$ and $\lambda$. More work is needed here to define a useful range for these parameters. In two-dimensions, an embedding function in the spirit of Eq. 10 and 11 needs to be developed. Possibly the combination of both methods i.e. continuation with line search would result in an efficient yet robust algorithm.

ACKNOWLEDGEMENT

Support of this work by the National Science Foundation under grant CME-8006743 is sincerely appreciated.

REFERENCES

DeVantier, B.A. and Larock, B.E., "Computation of Coupled Turbulent Flow and Sediment Transport," Proc. 4th International Symposium on Finite Element Methods in Flow Problems, Tokyo, Japan, July, 1982.

Fletcher, R., Practical Methods of Optimization - Unconstrained Optimization, Vol. 1, Wiley, 1980.

Hanjalic, K. and Launder, B.E., "Fully Developed Asymmetric Flow in a Plane Channel," J. Fluid Mech., 5, 1972, pp. 301-335.

Larock, B.E. and Schamber, D.R., "Finite Element Computation of Turbulent Flows," Adv. Water Resources, 4, Dec., 1981, pp. 191-197.

Schamber, D.R. and Larock, B.E., "Constant and Variable Eddy Viscosity Flow Simulations in Settling Tanks," Computer and Physical Modeling in Hydraulic Engineering, ASCE, G. Ashton, ed., Aug., 1980a, pp. 345-355.

Schamber, D.R. and Larock, B.E., "Computational Aspects of Modeling Turbulent Flows by Finite Elements, in Computer Methods in Fluids, K. Morgan et al., eds., Pentech Press, 1980b, pp. 339-361.

Schamber, D.R., Larock, B.E. and DeVantier, B.A., "Continuation Methods for the Finite Element Solution of Turbulent Flow," Proc. 4th International Conf. on Finite Elements in Water Resources, Hannover, Germany, June, 1982.

DESIGN OF RETROFIT HYDRO
PLANT FOR WATER QUALITY

by

D. Steven Graham, A.M. ASCE[1]

and

David C. Willer, F. ASCE[2]

## Introduction

Following the 1973 oil embargo a number of Federal
programs were initiated to encourage the development of
small hydro, particularly as retrofit projects at existing
dams. From an environmental viewpoint, such projects are
considered in general to be relatively benign sources of
energy, but certain types of environmental problems
characteristically occur ([1]). Most of these involve water
quality parameters affected by the annual stratification
cycle of reservoirs. The engineering measures used to
meet rather stringent water quality regulations for the
Lake Siskiyou power project are discussed in this paper.
The purpose is to illustrate that retrofit hydro projects
can be designed to meet environmental standards as well as
the nation's need for renewable energy sources.

## Description of Project

The Lake Siskiyou power project is located on Box
Canyon Dam which impounds the Sacramento River to form
Lake Siskiyou near Mt. Shasta, California. See Figure
1. The mean annual flow of the Sacramento River at this
location is 250 cfs. The proposed 5.0 MW project has a
rated capacity of 226 cfs and a rated head of 150 feet.
The Sacramento river passes through the precipitous Box
Canyon immediately below the dam. In order to fit into
this constricted area, the plant will be placed under an
extension of the spillway. For more information on the
project, see reference (2).

---

1. Engineer, Tudor Engineering Company.
2. Vice-President, Tudor Engineering Company, 149 New
   Montgomery Street, San Francisco, California 94105

65

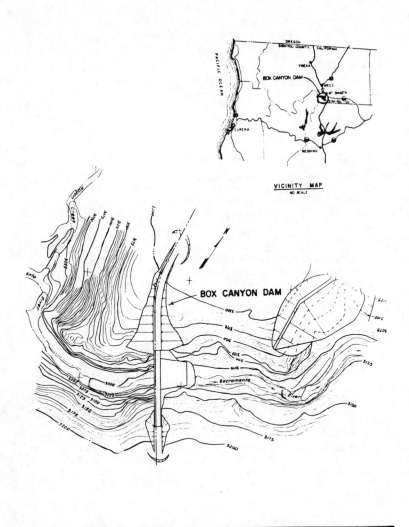

FIGURE 1:   Locaton Map

## Environmental and Other Operating Constraints

Small hydro projects in the West tend to have more rigid external constraints upon development than do those in the east. The Siskiyou project is a typical case. The reservoir was originally funded under California's Davis-Grunsky Act primarily for recreation purposes, with ten percent of active storage dedicated for flood control. Operation of the 26,000 acre-foot reservoir is thus entirely determined by the Davis-Grunsky contract which specifies pool levels and minimum flows for fish (40 cfs). The hydro project therefore has inferior rights to the water and must be designed to be run-of-river despite the physical potential for other modes of operation.

In addition to these constraints on water quantity, severe water quality requirements were also imposed upon the project by the California Department of Fish and Game. The Sacramento River below Box Canyon Dam is considered to be one of the best trout fisheries in the State and a general condition imposed was that the hydro retrofit project should not adversely affect the trout fishery: but rather, if possible, should enhance it. Water quality criteria to achieve these goals were specified in terms of temperature, turbidity, and dissolved oxygen. Ideally, all releases from the hydro project should be at $12.8^\circ C$ (or as close as possible) and have a DO concentration greater than 7.0 mg/l. Turbidity is to be minimized. Because of restrictions on water quantities, these goals have to be met without changing past release patterns.

In general, temperature and DO goals are very difficult to meet simultaneously in discharges from stratified reservoirs. The equation of state of water is such that warmer water is less dense at temperatures above $4^\circ C$, so that stable stratification occurs in summer. Buoyancy forces inhibit vertical eddy diffusivity in the thermocline so that the hypolimnion (lower layer) becomes increasingly depleted in dissolved oxygen until fall turnover. The reduced vertical diffusivity also inhibits warming of the hypolimnion thereby making possible the coldwater releases which permit trout fisheries to become established downstream of these reservoirs. Trout prefer cool water and high levels of DO (typically $13^\circ$ and at least 7 ppm DO are optimal· the saturation concentration of $O_2$ at $13^\circ C$ is 10.6 ppm). In summer the upper levels of a reservoir are oxygen-rich but too warm, while the lower levels are cool but usually deficient in DO.

## Multiport Intake

In order to meet the temperature goals (and the DO goal as well at least some of the time) a multiport intake

will be installed in Lake Siskiyou.  To design the intake
it was necessary to know the stratification
characteristics of the reservoir.  These were measured on
a monthly basis from June through September, 1981 at three
points along the axis of the reservoir.  The sampling
points are depicted in Figure 2, and the August 31 and
September 16, 1981 data are presented in Figures 3 and
4.  It can be seen that the reservoir is approximately
one-dimensional (little spatial variation at a given
depth) and moderately stratified with respect to
temperature.  On August 31 the hypolimnetic DO varied
between 5-6 ppm.  Hypolimnetic temperatures varied between
7-12°C.  Hence, mixing of the upper and lower layers by
selective withdrawal of both can be used to meet the
12.8°C temperature goal and raise DO above hypolimnion
values.  For instance, the values of temperature and DO at
the levels of the present outlets (5m and 43m) on that
date were

| Outlet Depth -m- | T -°C- | DO -ppm- |
|---|---|---|
| 5 | 20 | 7.6 |
| 43 | 8 | 5.3 |

Mixture of 1.5 parts water from 43m to one part from
5m would meet the 12.8°C effluent goal.  This effluent
would also have a DO of 6.68 pm, which is quite close to
the standard.

Note that Lake Siskiyou has a pronounced negative
heterograde oxygen curve (i.e., metalimnetic DO minimum)
[see reference (3)] which would preclude withdrawing
selectively at the 17m level where the 12.8°C water is
located, unless absolutely necessary, since further treatment of the effluent to meet the DO goal would be
required.

Proper design and operation of the multiport intake
requires knowledge of the width of the layer selectively
withdrawn at each port.  This is estimated as 3-4 m for a
withdrawal discharge of 250 cfs in midsummer following the
method of Imberger in Fischer, et al. (4).  The calculations are presented in Appendix 1.  Preliminary port
locations are at depths of 0, 2, 5, 10 and 43m; all but
the 5m and 10m port locations were dictated by California
Department of Fish and Game.

Downstream Water Quality

A survey of water quality downstream of Box Canyon Dam
was made coincidentally with the reservoir survey on

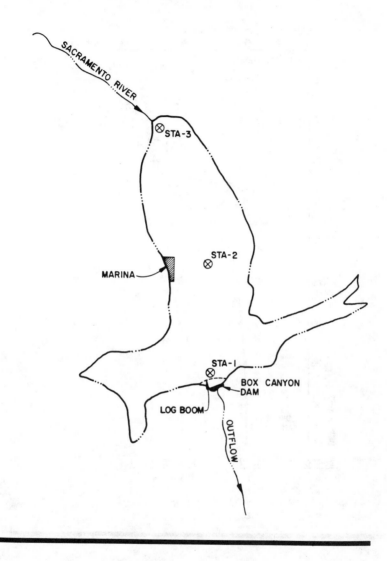

FIGURE 2: Sketch Showing Water Quality Sampling
Locations Lake Siskiyou, 1981

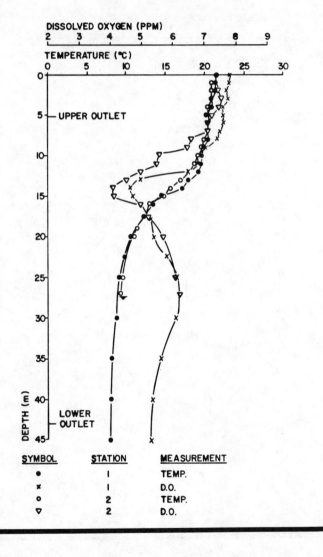

FIGURE 3:   Temperature and Dissolved Oxygen Profiles
            of Lake Siskiyou – August 31, 1981

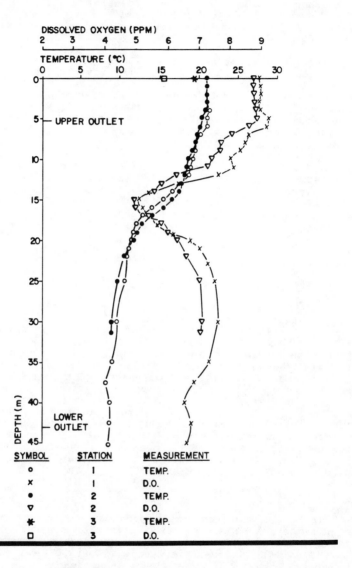

FIGURE 4:   Temperature and Dissolved Oxygen Profiles
            of Lake Siskiyou – September 16, 1981

September 16, 1981. On this date water was released from the lower (43m) Howell-Bunger valve along with a minor spill. After mixture the flow consisted of 62 cfs at 11°C at 1000h.

The DO in the reservoir at 43m depth on this date was 6.25 ppm, but a sample taken in the river below the valve was saturated with oxygen (11.1 ppm). Spraying the water through the air had caused the discharge to become aerated. (This is consistent with data we have taken at other dams.) It is unlikely the water will be so well-aerated after passing through a hydro plant, so mitigation measures might be necessary.

Hydraulic data were measured in the field and a reaeration coefficient was computed based on the formula of Churchill, et al. (9). Calculations indicated a zone of 1650 feet downstream would have DO levels below 7 ppm under the most adverse circumstances. Further data will be taken to confirm that 6.25 ppm is about the minimum DO that can be expected to occur in the hypolimnion at a 43m depth. If so, mitigation will be provided by draft-tube aeration.

## Draft Tube Aeration

On the basis of the single set of field data it appears that the hypolimnetic DO at 43m may not fall below 6 ppm. A relatively inexpensive technology for moderately raising DO levels involves injecting air into the draft tube with a blower below the runner. This is a modification of the original methodology described by Raney and Arnold (6), Raney (7), and Sheppard, et al. (8), and Fox and Harshbarger (9). For a typical September flow of 57 cfs with a 2 mg/l DO deficit, 614 pounds of oxygen per day are required to be added to the effluent. Using an (approximate) 5:1 ratio of air to oxygen and assuming 25 percent transfer efficiency, 6 tons of air must be added per day. This corresponds to a flow of about 113 acfm at NTP. This is approximately 3.3 percent of the water discharge (57 cfs) by volume.

## Summary

The Lake Siskiyou hydro project was constrained by requirements that no changes in flows or reservoir operations occur, effluent temperatures remain near or at 12.8°C as long as possible, and effluent DO levels always be above 7 ppm. To meet these requirements a multiport intake and, if necessary, a draft tube aeration system are being installed.

References

1. Graham, D.S., "Environmental regulations and constraints upon hydro development in California, U.S.A.", Canadian Water Resources Journal, Vol. 6, No. 3, 1981, pp. 295-302.

2. Siskiyou County Flood Control and Water Conservation District, Yreka, California, "Application for License to Authorize Construction of Lake Siskiyou Power Plant, Lake Siskiyou, California", November 1981. FERC Project No. 2796. Prepared by Tudor Engineering Company, San Francisco, California.

3. Gordon, J.A. and Skelton, B.A., "Reservoir Metalimnion Oxygen Demands", Journal of the Environmental Engineering Division, ASCE, Proc. Paper 13437, Vol. 103, No. EE6, December 1977, pp. 1001-1010.

4. Fischer, H.B., et al., "Mixing in Coastal and Inland Waters", Academic Press, New York, New York, 1979.

5. Churchill, M.A., Elmore, H. L. and Buckingham, R.A., "The prediction of stream aeration rates", Journal of the Sanitary Engineering Division, ASCE, Vol. 86, No. SA4, pp. 41-53.

6. Raney, D.C. and Arnold, T.G., "Dissolved Oxygen Improvement by Hydroelectric Turbine Aspiration", Journal of the Power Division, ASCE, Proc. Paper 9707, Vol. 99, No. PO1, May 1973, pp. 139-153.

7. Raney, D.C., "Turbine Aspiration for Oxygen Supplementation", Journal of the Environmental Engineering Division, ASCE, Proc. Paper No. 12890, Vol. 103, No. EE2, April 1977, pp. 341-352.

8. Sheppard, A.R., Miller, P.E., and Buck, C.L. "Prediction of Oxygen Uptake in Draft Tube Aeration Systems", Proceedings of the ASCE Environmental Engineering Specialty Conference, held at Atlanta, Georgia, July 8-10, 1981, 8 pp.

9. Fox, Twana A., and Harshbarger, E. D., "Vacuum Breaber Reaeration Tests, Turbine Discharge Oxygenation Program, Norris Dam", TVA Report No. WR28-1-2-100, Norris, Tennessee, May 1980, 22 pp.

## APPENDIX 1

### ESTIMATION OF THICKNESS OF WITHDRAWAL LAYER
### IN HYPOLIMNION

The following constants are used for Lake Siskiyou:

characteristic length                 $L = 1.8$ km

characteristic width of line sink     $W = 600$ m

characteristic discharge              $Q = 7.1$ m$^3$/s

characteristic temperature gradient   $3.5°C/17$ m

vertical momentum diffusivity         $\varepsilon_v = 5 \times 10^{-6}$ m$^2$s$^{-1}$

thermal expansivity                   $\alpha = 10^{-4}°C^{-1}$

The unit withdrawal discharge, q, is

$$q = Q/W = 0.012 \text{ m}^2\text{s}^{-1} \tag{1}$$

Also

$$\Delta\rho = \alpha\Delta T$$

$$= 3.5 \times 10^{-1} \text{ kg/m}^3 \text{ for } \Delta T = 3.5°C \tag{2}$$

The buoyancy frequency, N, then is

$$N = \sqrt{-\frac{1}{\rho}\frac{\partial\rho}{\partial z}g} \tag{3}$$

$$= 1.42 \times 10^{-2} \quad s^{-1} \tag{4}$$

The Grashof number, $G_r$, is

$$G_r = \frac{N^2 L^4}{\varepsilon_v^2} = 8.48 * 10^{19} \tag{5}$$

The internal Froude number, F, is

$$F = q/(NL^2) = 2.61 \times 10^{-7} \tag{6}$$

The outflow dynamics are governed by a parameter, R, defined as

$$R = F\, G_r^{\,1/3} = 1.15 \tag{7}$$

The expressions for the withdrawal layer half-thickness, $\delta$, to be used is:

$$\delta = 2.0\, L\, F^{\,1/2} \quad \text{for } R \geq 1 \text{ , hence} \tag{8}$$

$$\delta = 1.84 \tag{9}$$

So the withdrawal layer thickness, $2\delta$, is about 3-4m. This means that a single port withdrawing 250 cfs will draw from a layer about 4m thick in midsummer when the hypolimnion is approximately linearly stratified.

FIGURE 1

Location Map and General Geography

# TWO STAGE DROP SPILLWAYS FOR TILLATOBA CREEK, MS

John E. Hite, Jr.[1] and Robert C. Daniel[2]

ABSTRACT: The middle and south forks of Tillatoba Creek have experienced a relatively rapid stage of channel degradation accompanied by severe streambank erosion. A two stage, reinforced concrete grade control structure was designed and constructed by the U. S. Department of Agriculture (USDA), Soil Conservation Service (SCS) on the Middle Fork in 1978.

The low stage structure is designed to convey a one year frequency storm. This structure will have a high degree of submergence during flood flows, thus requiring a long stilling basin. The high stage structure is designed to have a discharge which, when combined with the discharge from the low stage structure, will ensure that the downstream channel is filled before flow through a vegetated earth spillway occurs. The high stage structure is not highly submerged and only a short stilling basin is required.

The South Fork structures, planned for construction in the future, will be very similar to those on the Middle Fork. However, these structures are larger than those normally used by SCS.

Therefore, a model study of the South Fork site was conducted at the U. S. Army Engineer Waterways Experiment Station to verify the design criteria, evaluate the hydraulic performance of the structures, and make modifications to the design, if needed, to improve performance. A 1:25 scale model that reproduced both the high and low stage structures, a 500 foot (152.4 meters) length of the upstream approach and an 1100 foot (335.3 meters) length of the topography downstream from the structures was used for the study. Several modifications were made to the original design to improve hydraulic performance as well as to reduce construction costs.

## MIDDLE FORK STRUCTURE

The SCS constructed a separated two stage Type C straight drop spillway with associated earth dike and vegetated earth spillway on the Middle Fork of Tillatoba Creek in 1978 to control channel degradation. The hydraulic design criteria for the Type C straight drop spillway was developed from model studies by Donnelly and Blaisdell (1965). The hydraulic design was done in accordance with the USDA, SCS National Engineering Handbook, Section II, Drop Spillways. The layout of the Type C spillway is shown in Figure I.

---

[1] Research Hydraulic Engineer, Hydraulic Structures Division, U. S. Army Engineer Waterways Experiment Station, Vicksburg, MS
[2] Civil Engineer, USDA, SCS, State Design Unit, University, MS

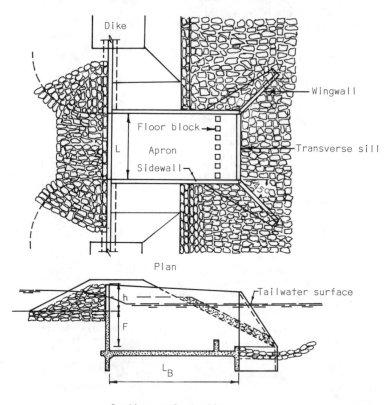

Plan

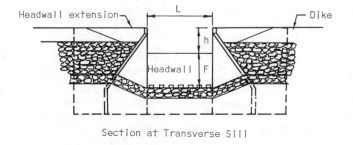

FIG. I.  Layout of Type C Drop Spillway

Hydraulic Design - Due to the large existing channel at the selected site, a single structure was not desirable. An island type structure would allow out-of-bank flow to re-enter the downstream channel below the structure where the water surface in the channel would be several feet below the top bank of the channel and erosion would occur as re-entry of the out-of-bank flow occurred.

Another alternative was to construct a total control type structure where the dikes for the spillway would extend across the flood plain preventing any flow past the structure except what went through the spillway and any associated vegetated earth spillways. However, this alternative would have required a high fill height for the dike and the acquiring of extensive upstream flood easements.

The two stage drop spillway concept overcame the above disadvantages of the single drop spillway. One spillway, referred to as the low stage spillway, was placed in the existing channel. The weir crest was placed at an elevation to tie with a stable grade upstream from the structure. The transverse sill elevation was placed slightly below the assumed ultimate degraded stable channel grade. The weir length was sized to pass the one year frequency storm with the upstream water surface at the top bank elevation. The basin length was computed for the condition of $.7d_c$ submergence of the weir crest. This is a trial and error procedure which is done by computing upstream and downstream tailwater elevations for a given discharge and then determining the degree of submergence of the weir crest. This gives the maximum required basin length for all discharges.

Another spillway, referred to as the high stage spillway, was constructed in the flood plain area slightly separated from the low stage spillway. The weir crest was placed at the flood plain elevation. The transverse sill elevation was placed slightly below the tailwater elevation of the assumed ultimate degraded downstream channel when the upstream water surface was at the weir crest of the high stage spillway. The weir lengths for this spillway and the emergency spillway were determined by a series of flood routings. Several high stage spillway weir lengths were flood routed for the 50 year frequency storm to establish the crest elevation of the emergency spillway. The low stage, high stage and emergency spillway were then flood routed for a moderate hazard freeboard storm to determine the elevation of the top of the dike. Resulting water surface flood elevations downstream and upstream from the structure were evaluated. Major evaluation factors upstream from the structure were land use and damage to the flooded area, and roads in the flooded area for the 50 year storm. Downstream from the structure it was required that the downstream channel be filled before flow occurred through the emergency spillway. Also, the construction cost of the high stage spillway and emergency spillway combination, with associated fill heights, was evaluated. The basin length for the high stage spillway was computed for the 100 year storm discharge since the weir crest submergence was less than $0.7d_c$.

Major structure dimensions were:

|                          | Weir depth h | Weir length L | Vertical drop F | Apron length $L_B$ |
|--------------------------|--------------|---------------|-----------------|--------------------|
| Low Stage Drop Spillway  | 22.5 ft. (6.86 m) | 34.0 ft. (10.36 m) | 13.0 ft. (3.96 m) | 120.0 ft. (36.58 m) |
| High Stage Drop Spillway | 11.5 ft. (3.51 m) | 114.0 ft. (34.75 m) | 8.5 ft. (2.59 m) | 38.0 ft. (11.58 m) |

Emergency Spillway Width:  250 ft. (76.2 m)

Construction - The drop spillways with associated earth dike and emergency spillway was constructed in 1978 at a cost of $1,492,789.

Performance - Performance of the structure has been excellent.  A storm in 1980 created a head of 14 feet (4.27 m) on the low stage structure and 3 feet (0.91 m) on the high stage structure with no apparent damage.  The computed storm discharge was 7,000 cfs (198.2 cms).  Also in 1980, rock riprap slope protection was placed on the inlet channel slopes to protect these areas from erosion created by dispersed soil and channel curvature.  The structure withstood a storm in March 1981 of almost equal magnitude to the 1980 storm.

## SOUTH FORK STRUCTURE

Site conditions at a proposed structure on the South Fork of Tillatoba Creek are very similar to those at the Middle Fork Structure.  The design procedure for this structure was similar to the Middle Fork structure.  Since the original model studies were intended for use on farm-size watersheds and the design criteria is being used on larger structures, the SCS entered into an agreement with the U. S. Army Engineer  Waterways Experiment Station (WES) in 1980 to perform a model study of the South Fork to verify the design criteria, evaluate the hydraulic performance, and make modifications to the design, if needed, to improve the performance.  Specifically, the model study was to determine discharge capacity, stilling basin performance, riprap requirements, and approach and exit channel configurations.

The South Fork project is designed as separated high and low stage structures with an emergency spillway.  Details of the project repro- duced in the model are shown in Figure 2.  The low stage weir crest elevation of 222.0 was set to obtain a stable grade (.0007 ft/ft). The end sill elevation of 212.0 was set at the anticipated degraded elevation in the downstream channel.  The weir length of 29 feet (8.8 m) was sized to carry the bank full inlet channel capacity for an aged condition.  The high stage weir crest elevation of 232.0 was set at the flood plain elevation.  The high stage end sill elevation was set at the tailwater elevation created by the discharge through the low stage structure when the headwater is at the flood plain elevation of 232.0.  The weir length was selected so that the downstream channel

would be filled when the water surface upstream from the structures
would be at the emergency spillway crest elevation of 240.0.

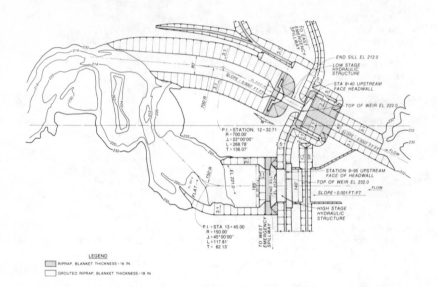

FIG. 2.  Plan View of the Original Design South Fork Project

## MODEL STUDY

The 1:25 scale model shown in Fig. 3 reproduced both the high and low
stage control structures, approximately a 500 ft (152.4 m) length of
the upstream approach and an 1100 ft (335.3 m) length of the topography downstream from the structures.  Also, to study the effect of a
curved trajectory type of drop on the flow conditions in the stilling
basin, a 1:29 scale section model was constructed in a glass-sided
flume.

Several tests were conducted to observe and measure the hydraulic performance of the control structures with the original design and with
various modifications to the structures.  With most of the design
modifications tested, current velocities and patterns, water surface
elevations, and sand scour profiles downstream from the structures were
measured.  Although the sand scour depths measured in the model are not
necessarily representative of the exact depths in the prototype, they
provide a good means for comparing the relative merits of various
design modifications.  The initial tests were conducted to observe and
document flow conditions in the approach area and exit channels of the
low and high stage structures.

FIG. 3.  1:25 Scale Model of South Fork Tillatoba Creek
Drop Structures

Low Stage Structure.  In the approach area to the low stage structure,
flow around the right (looking downstream) approach dike formed eddies
which moved into the entrance channel.  The approach conditions were
improved by shortening the right approach dike.  Flow conditions were
observed as the dike was shortened in 25 ft (7.6 m) increments from the
original length of 125 ft (38.1 m).  A 25 ft (7.6 m) long dike was
determined to be the appropriate length that provided a uniform distribution of flow entering the structure.

The square abutments of the entrance to the low stage drop structure
as originally designed caused separation of flow from the walls at the
weir, and resulted in concentration of flow in the center of the
stilling basin.  Energy dissipation in the stilling basin was adversely
effected by this contraction of flow at the abutments.  Tests were
conducted with wingwalls of several sizes and configurations attached
to the abutments in an effort to reduce this flow contraction.  Wingwalls consisting of four 8 ft (2.4 m) chord sections were found to
uniformly distribute flow and improve energy dissipation in the
stilling basin.  The chord sections (shown in Fig. 4) were used rather
than a curve because of ease of construction in the prototype structure.

The low stage stilling basin was designed for flow conditions that
submerge the weir.  When the tailwater increases, creating a greater
submergence of the weir, flow over the vertical drop plunges at a
greater distance downstream in the basin until it eventually rides the
surface of the tailwater.  Thus, the length of the stilling basin,
100 ft (30.5 m), was longer than a conventional hydraulic jump type

basin.  A 1:29 scale section model, constructed in a glass-sided flume,
was used to determine the tailwater elevation where flow plunged into
the stilling basin and where flow rode the surface of the tailwater.
Results of tests conducted to determine the tailwater elevation where
flow began to ride the surface with various discharges are shown in
Fig. 5.  A curved trajectory was placed downstream from the weir to
cause flow to plunge into the stilling basin with higher tailwater
elevations thus obtaining better energy dissipation.  The shape of this
parabolic trajectory ($x^2 = 40Y$) was based on previous studies conducted
with low head navigation spillways.  This was designated the Type 3
design crest as shown in Fig. 6.  Although the Type 3 design crest
improved flow conditions over the original design, tests were conducted with a longer parabolic curve on the crest, Type 4, in an effort
to further improve energy dissipation.  The shape of this curve
($x^2 = 72Y$) was based on the theoretical trajectory of a free jet with
a design head of 18 ft (5.5 m).  The design head of 18 ft (5.5 m) was
based on the 240 pool elevation with design discharge and the weir
crest elevation 222.  Tailwater elevations at which flow began to ride
the water surface with various discharges are shown in Fig. 5.  There
was a slight improvement of the conditions with this design when compared with the Type 3 design crest, and considerable improvement when
compared with the original design.

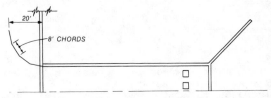

FIG. 4.   Type 7 Approach Wingwalls

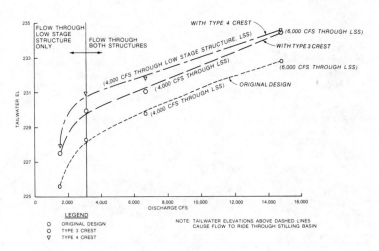

FIG. 5.   Tailwater El vs Discharge

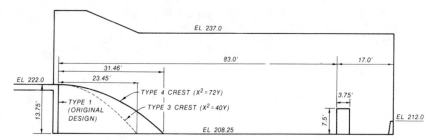

FIG. 6.  Types 1, 3 and 4 Crest Shapes

The riprap plan as originally designed for the low stage structure consisted of riprap with an average diameter ($d_{50}$) of 12 in. (304.8 mm)
and a blanket thickness of 18 in. (457.2 mm). The riprap remained
stable throughout a series of tests consisting of a 5-hr duration with
a discharge of 3100 cfs (87.8 cms) through the low stage structure and
a 5-hr duration with a total discharge through both structures of
14,800 cfs (419.1 cms). Based on these tests it was concluded that the
original plan would provide adequate protection upstream and downstream
from the structure.

Flow conditions were observed in the exit channel with the design flow
of 14,800 cfs (419.1 cms) and a normal tailwater, el 234.3. Eddies
formed in the exit channel in the vicinity of the wingwalls. Sand bars
developed on each side of the channel. It appeared that the strength
of the eddies increased by the wingwalls at the downstream end of the
structure, and the wide exit channel (80 ft) (24.4 m) relative to the
width of the stilling basin (29 ft) (8.8 m). The bottom width of the
exit channel was reduced to 39 ft (11.9 m), 5 ft (1.5 m) wider than the
stilling basin on either side, and the exit wingwalls were removed.
The intensity of the eddies was reduced and since much less excavation
would be needed for the narrow channel, this modification should be
beneficial for the prototype.

High Stage Structure. Small eddies formed off the left approach dike
in the approach area to the low stage structure and moved into the
entrance channel to the high stage structure. A large, low velocity
eddy formed in the area to the right of the structure between the main
levee and the spur dike. They did not adversely affect flow entering
the structure and no revisions were needed in the approach area to the
high stage structure.

There was some concentration of flow at the abutments immediately upstream from the weir which caused concentration of flow into the center
of the stilling basin and exit area. Several modifications to the
abutment were tested in an attempt to reduce the flow contraction.
These modifications included various lengths of rock dikes in the
approach and various types of wingwalls attached to the abutments.
Wingwalls that consisted of five 6 ft (1.8 m) chords with a layout

similar to that shown in Fig. 4 were found to be the minimum length
that effectively reduced contraction of flow.

The stilling basin in the high stage structure was designed for an
8 ft (2.4 m) head on the weir. Therefore, the basin was only 41 ft
(12.5 m) long. A hydraulic jump was maintained in the basin with all
expected discharge and tailwater conditions. Although the stilling
basin as originally designed functioned satisfactorily, a curved
trajectory was added to the weir in an effort to further improve
energy dissipation. The short length of the basin restricted the use
of the equation $x^2$ = 40Y as had previously been tested for the low
stage structure. Thus, a curve based on the parabolic equation
$x^2$ = 20Y was arbitrarily chosen. This was designated the Type 2
design crest. Some improvement of flow conditions was achieved with
this modification. Although the Type 2 design crest improved flow
conditions and reduced scour depths, it would add considerably to
construction cost of the prototype structure because of the long weir.
Also, the occurrence of large flows over the high stage structure is
somewhat infrequent. For these reasons, the curved drop was considered
to be economically unjustified.

The original riprap plan consisted of an average diameter ($d_{50}$) stone
of 12 in. (304.8 mm) and blanket of 18 in. (457.2 mm) thick. Riprap
upstream from the structure was grouted because of the high velocities
immediately upstream from the weir. Also, the riprap on the slope into
the preformed scour hole immediately downstream from the stilling basin
was grouted. The riprap remained stable throughout several tests where
flood hydrographs with a peak discharge of 14,800 cfs (419.1 cms) were
simulated. Although the downstream protection plan was stable, changes
to the exit channel resulted in changes to the original riprap plan.
The preformed scour hole was eliminated since it appeared to concentrate flow toward the center of the exit channel. None of the riprap in
the downstream area was grouted. The riprap remained stable throughout a test conducted with the flood hydrograph.

Flow conditions were observed in the exit channel of the high stage
structure with the design flow of 14,800 cfs (419.1 cms) and the
normal tailwater, el 234.3. Eddies in the exit channel were more intense on the left side (looking downstream) but were not as strong as
those observed in the low stage structure exit channel. In an effort
to reduce the eddies and save on excavation costs, the bottom width of
the exit channel was reduced from 185 ft (56.4 m) to 141 ft (43.0 m)
and the preformed scour hole was eliminated. The downstream wingwalls
were not removed from the structure since they were needed as retaining walls for the levee.

## Summary of Model Test Results

Modifications to the low stage structure developed through the model
studies included 1) shortening the right approach dike, 2) adding
wingwalls at the entrance to the structure, 3) adding a trajectory
curve downstream from the weir crest, 4) removing the downstream wingwalls, and 5) reducing the bottom width of the exit channel. Modifications to the high stage structure included 1) adding wingwalls at

the entrance to the structure, 2) eliminating the scour hole downstream from the structure, and 3) reducing the bottom width of the
exit channel. The two structures were more efficient in passing flow,
especially with higher discharges, after the discussed modifications
were made. The structures, with the modifications recommended in this
study, should perform satisfactorily for all discharge and tailwater
combinations anticipated at the structure. Some scour can be expected
in the areas downstream from the riprap protection after several hours
of operation with large flows. However, this scour should be minimum,
and should not endanger the integrity of the structures.

## ACKNOWLEDGEMENTS

The results reported herein are based on research conducted at the
U. S. Army Engineer Waterways Experiment Station and sponsored by
the Office, Chief of Engineers, and the U. S. Department of Agriculture,
Soil Conservation Service. Permission for the authors to present this
paper does not signify that the contents necessarily reflect the views
and policy of either the U. S. Army Corps of Engineers, the U. S.
Department of Agriculture, or the Soil Conservation Service, nor does
mention of trade names or commercial products constitute endorsement
or recommendations for use.

Appendix I.-References

1. Donnelly, C. A. and Blaisdell, F. W., "Straight Drop Spillway
   Stilling Basin," Journal of the Hydraulic Division, ASCE, Vol. 91,
   No. HY3, Proc. Paper 4328, May, 1965, pp. 101-131.
2. National Engineering Handbook, Section 11, Soil Conservation Service,
   Drop Spillways, USDA-SCS.

Appendix II.-Notation

The following symbols are used in this paper:

$d_c$ = critical depth of flow
L  = length of weir
h  = depth of weir
F  = drop through spillway from crest of weir to top
     of transverse sill
$L_B$ = length of apron
$d_{50}$= rock riprap size, of which 50 percent is finer
     by weight
x  = horizontal distance from the weir to any point
     on the crest shape
Y  = vertical distance from the weir crest to any
     point on the crest shape

# A SITE SPECIFIC STUDY OF A LABYRINTH SPILLWAY

Kathleen L. Houston [1] and Carol S. DeAngelis [2]

## ABSTRACT

The existing structures at Ute Dam do not supply the desired reservoir storage capacity for the future water needs of the area. A labyrinth spillway was selected for the modification of Ute Dam because the labyrinth with a raised crest elevation provides increased reservoir storage and the longer crest length provides greater discharge capability for a fixed spillway width.

Design curves were developed from which a cost effective 14-cycle labyrinth spillway was designed. Hydraulic aspects of the spillway were confirmed by tests with a 1:80 overall scale model and several sectional models. A comparison is made to previously built labyrinth spillways.

## Introduction

Ute Dam is located on the Canadian River 2 mi (3.2 km) west of Logan, New Mexico. The dam is a 121-ft (36.9-m) high earthfill structure with crest elevation 3801 ft (1158.5 m). The existing spillway is an 840-ft (256-m) wide ungated overflow structure with crest elevation 3760 ft (1146 m).

Several alternatives for increasing the reservoir storage capacity at Ute Dam were investigated. Gating the existing crest structure was evaluated but the additional load imposed by the increased water surface would cause instability. Therefore, gated alternatives had to include the cost of a new crest structure. The least cost was estimated at $34,000,000. (All estimated costs based on November 1980 unit prices.) Several ungated alternatives were studied. A labyrinth spillway, combined with raising the dam, was the most economical at $10,000,000.

The labyrinth spillway is a series of triangular or trapezoidal shapes in plan view which increase the effective length of the spillway crest within a fixed spillway width. For a given operating head, higher discharges can be passed over the labyrinth spillway crest than a straight crest. The labyrinth spillway was selected as an economical alternative to increase the storage capacity of the reservoir at Ute Dam. The proposed labyrinth structure is shown on Fig. 1.

---

1/ Hydraulic Engineer, Bureau of Reclamation, Denver, Colorado.
2/ Civil Engineer, Bureau of Reclamation, Denver, Colorado.

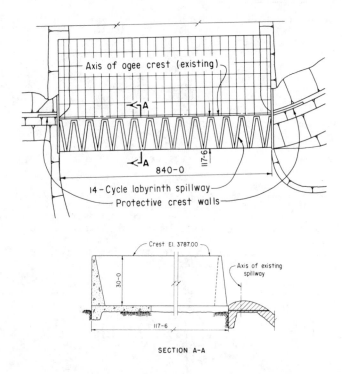

FIG. 1. - Proposed labyrinth spillway structure.

## Design Criteria

The IDF (inflow design flood) has a peak of 590 000 ft$^3$/s (16 707 m$^3$/s) and a 4-day volume of 2 155 000 acre-feet (2.66 x 10$^9$ m$^3$). This flood was also used in the design of the existing structure.

The desired additional reservoir storage of 27 ft (8.2 m) required a crest elevation of 3787 ft (1154.3 m). The maximum reservoir elevation was limited to about 3806 ft (1160 m) to minimize flood damage to private property along the reservoir rim. These criteria are met with a 30-ft (8.2-m) high spillway with a design head, $H_0$, of 19 ft (5.8 m).

Parameters and Initial Design

The hydraulic design of the labyrinth spillway was originally based on design procedures and curves by Hay and Taylor (5). Parameters which affect the performance of the spillway are discussed below and shown on Fig. 2.

The length magnification, $l/w$, and the angle $\alpha$ completely define the shape of a plan form with n being the number of cycles forming the spillway. For a given length magnification, the angle $\alpha$ varies from zero for a rectangular plan form to a maximum for a triangular plan form. The vertical geometry of the spillway is defined by the vertical aspect ratio, $w/P$. The performance of the spillway, $Q_L/Q_{NS}$ is a function of $H/P$, $l/w$, $w/P$, and $\alpha$. The initial design for the Ute spillway complied with the limits defined by Hay and Taylor [(5) pg. 2342] for each parameter.

Other labyrinth spillways are in use at Quincy Dam, Aurora, Colorado (1); Mercer Dam, Dallas, Oregon (2); Woronora and Avon Dams, Sydney, Australia (4); among others. Hydraulic model studies were done on the Navet Pumped-Storage Project, West Indies (8); Mercer Dam, Oregon (3); and Bartletts Ferry Project, Georgia (7). The ranges of some important design parameters for these labyrinth structures are compared with the Ute laybrinth below (symbols as defined on Fig. 2):

| Parameters | Range | Ute labyrinth |
|---|---|---|
| $l/w$ | 1.9-5.0 | 4.0 |
| $H_0/P$ | 0.40-0.72 | 0.63 |
| $H_0$ | 3-7 ft (0.91-2.13 m) | 19 ft (5.79 m) |
| $w/P$ | 1.2-6.0 | 2.0 |
| n | 4-11 | 14 |

For the Ute labyrinth, the trapezoidal plan form was chosen for high efficiency and ease of construction. The $H_0/P$ requirement for the Ute labyrinth spillway was beyond the range of the published design curves for performance of labyrinths. The Hay and Taylor curves were extrapolated for the initial design of a 10-cycle labyrinth with an effective crest length of 2322 ft (708 m). The labyrinth would be placed just upstream of the existing ogee crest.

Because the initial design was based on extrapolated design curves, a hydraulic model study was requested. The model would also be used to determine the approach conditions, pressure loadings on the spillway, the effect of the existing downstream ogee crest on the performance of the labyrinth, and an optimum crest shape.

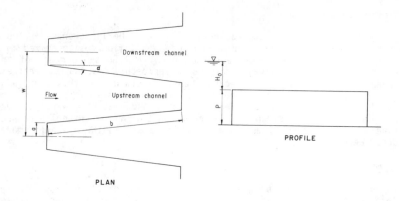

a = half length of labyrinth apex;
b = length of labyrinth side wall;
$C_a$ = discharge coefficient for actual crest section;
$C_s$ = discharge coefficient for the sharp crest;
$H$ = total upstream head over crest, less than $H_0$;
$H_0$ = design head;
$l$ = developed length of one labyrinth cycle = 4a+2b;
$l/w$ = length magnification;
n = number of spillway cycles in plan;
P = spillway height;
$Q_L$ = discharge over labyrinth spillway;
$Q_{Ns}$ = discharge over corresponding linear sharp crested spillway

$$= C_s WH^{3/2};$$

$Q_L/Q_{Ns}$ = flow magnification (measure of spillway performance);
W = width of linear spillway;
w = width of one labyrinth spillway cycle;
w/P = vertical aspect ratio; and
$\alpha$ = angle of side walls to main flow direction.


FIG. 2. - Labyrinth spillway parameters.

Hydraulic Model Study

The model study was conducted in two phases. The first phase included
verification and extrapolation of the Hay and Taylor design curves in a
2.5-ft (0.76-m) wide flume test facility. A 1:80 scale model of the
entire labyrinth spillway, reservoir approach area, and downstream
channel was studied in the second phase. Tests during this phase also
included measuring velocities along the embankment adjacent to the
spillway, testing splitter pier locations, and measuring water surface
profiles and pressures in the upstream and downstream channels, respectively. These results are described in detail in the hydraulic model
study report (6).

Flume Tests

To ensure continuity between the previous and extended range of the
design curves, an initial attempt was made to verify the existing
curves. Tests began with a sharp-crested triangular-shaped plan form
weir of l/w = 3. The results of this test did not confirm those of Hay
and Taylor which led to further flume testing of triangular-shaped weirs
with length magnifications of 2, 4, and 5. The triangular plan form was
emphasized as this is the most efficient. The final results were
significantly different than the work of Hay and Taylor. The differences were particularly noticeable at high H/P values with the lower
values tending towards the ideal case of discharge increasing in exact
proportion to length.

The discrepancy between the two sets of design curves [Hay and Taylor
(5) and revised (6)] appeared to be the difference in upstream head
definition. The discrepancy discovered is attributable to several
factors which are currently under study. This paper will not discuss
these differences but will emphasize the use of the revised curves that
were developed using total head; the measured head plus the velocity
head, $V^2/2g$. The total head, H, is used throughout the study and
design. $H_0$ is the maximum or design head value which also includes
the velocity head. Design curves, shown on Fig. 3, were developed for
n = 2 (two apexes upstream) and w/P = 2.5 with P = 0.50 ft (0.15 m),
w = 1.25 ft (0.38 m), and $0 < H/P \leq 1$.

Ten-cycle Labyrinth Spillway Model

To determine which curves would correctly predict the maximum discharge
in a reservoir situation, the 1:80 scale model of Ute Dam was used. The
Ute 10-cycle labyrinth spillway, as initially designed from Hay and
Taylor's extrapolated curves, was first tested. This test showed that
the maximum discharge could not be passed over the spillway within the
stipulated design head, $H_0$, of 19 ft (5.8 m). Therefore, the total
developed spillway length required would be greater than that predicted
by the Hay and Taylor curves. The design discharge magnification
$(Q_L/Q_{Ns})_{max}$ and design head to crest height ratio, $H_0/P$, when plotted
on the revised design curves gave a length magnification, l/w, of 4.00
instead of 2.74 as predicted by Hay and Taylor's curves. This result
prompted redesign of the spillway based upon the l/w = 4 curve shown on
Fig. 3.

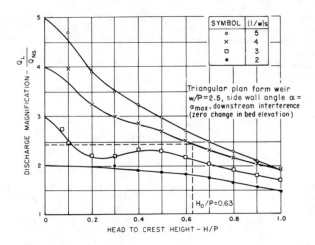

FIG. 3. - Revised design curves.

## Design Procedure

The following design method uses a design head, $H_0$, to size the labyrinth spillway and is based on Hay and Taylor's design procedure with minor modifications:

1. Determine from the site conditions: crest height, P, total channel width, W, and a design head, $H_0$.

2. Estimate the labyrinth discharge $(Q_L)_{max}$ at the design head by a rough flood routing.

3. Determine the sharp-crested linear weir discharge $(Q_{Ns})_{max}$ by the equation on Fig. 2 using $H_0/P$ to find $C_s$ in Fig. 4. Use either w/P curve as they are very similar and the w/P effect of the final design should actually be determined from a model study.

4. Determine $(Q_L/Q_{Ns})_{max}$ and use $H_0/P$ in the design chart (Fig. 3) to find $(1/w)_s$.

5. Select a crest shape and determine its coefficient, $C_a$, for the design head, $H_0$. Fig. 4 shows the curve of experimental data for the Ute crest shape. This curve gives a coefficient of discharge for various values of H/P but this is not to be confused with a discharge coefficient taken in its normal context in the equation: $Q = C_D W H^{3/2}$. The curve on Fig. 4 shows values that have been

derived using this equation with discharge values from the labyrinth rating curve. The equation assumes that the length is constant, but the effective length of the labyrinth spillway is constantly changing with a change in reservoir head. Therefore, the coefficient of discharge shown by this curve reflects this change in length.

6.   Modify $(1/w)_s$ from sharp crest to actual crest section $(1/w)_a$ by $(1/w)_a = \dfrac{(1/w)_s (C_s)}{C_a}$.

It is now necessary to develop a labyrinth rating curve using the following procedure:

7.   Select arbitrary values of H/P between zero and $H_0/P$ and determine $C_a$ and $C_s$ for each value from Fig. 4. ($C_a$ curve on Fig. 4 is specifically for Ute crest shape. Other shapes might require a different curve.)

8.   It is necessary to construct an auxiliary curve on Fig. 3 to determine $(Q_L/Q_{Ns})_{equiv}$ for an equivalent sharp-crested labyrinth weir. For each value of H/P compute:

$(1/w)_{equiv} = \dfrac{C_a}{C_s} (1/w)_a$, then plot $(1/w)_{equiv}$ on Fig. 3 by interpolating between the existing curves.

9.   Determine $(Q_L/Q_{Ns})_{equiv}$ for each H/P value from the curve plotted in Fig. 3.

10.   Determine the labyrinth rating curve by:

$$Q_L = (\tfrac{Q_L}{Q_{Ns}})_{equiv} \times Q_{Ns} = (\tfrac{Q_L}{Q_{Ns}})_{equiv} \times C_s W H^{3/2}$$

11.   Compare the $Q_L$ computed for design head in step 10 with the value of $(Q_L)_{max}$ in step 2. If necessary, adjust the $(Q_L)_{max}$ in step 2 and repeat the procedure. If the $(Q_L)_{max}$ values are very similar, define the geometry of the spillway by selecting the number of cycles, n, and the side wall angle, $\alpha$. With the spillway geometry designed for the total computed length, check that all parameters are within the acceptable design range. If necessary, redefine the spillway geometry.

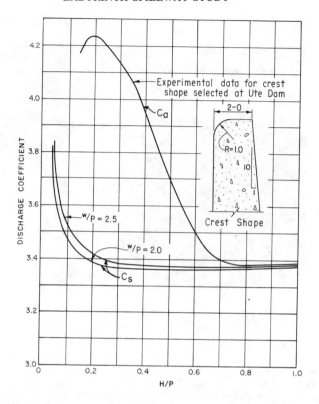

FIG. 4. - Linear weir discharge coefficient curves.

Using this procedure, a 14-cycle spillway was selected for the modification of Ute Dam with an effective crest length of 3360 ft (1024 m)
within the 840-ft (256-m) width. Other parameters and dimensions of the
14-cycle spillway are $H_O/P = 0.63$, $Q_L/Q_{NS} = 2.445$, $l/w = 4$, $w/P = 2$,
b = 114 ft (34.75 m), a = 3 ft (0.91 m), and $\alpha = 12.15°$.

Fourteen-cycle Labyrinth Spillway

The labyrinth spillway described previously was installed in the 1:80
scale model (Fig. 5). The discharge curve developed for the spillway
showed that more discharge could be passed at lower reservoir elevations
than in the 10-cycle spillway.

The high reservoir head caused special hydraulic conditions which
warranted study. The lateral flow along the embankment is channeled by
a protective crest wall into the spillway chute at the downstream apexes
of both end cycles. This lateral flow does not interfere with the flow
over the spillway.

FIG. 5. - Maximum discharge passed by 14-cycle labyrinth spillway model.

Visual observations and water surface profiles showed a significant drop in the water surface immediately upstream of the spillway between cycles. This is attributable to several factors including the area contraction as flow enters the upstream channel, the loss due to the velocity head, and nappe interference. Nappe interference occurs over the upstream apexes of the spillway due to the high reservoir head and small vertical aspect ratio, w/P. As the number of cycles, n, increases for a given overall width, W, w/P is decreased as well as the distance that the labyrinth encroaches into the approach channel. Therefore, by keeping the w/P as small as possible, without lowering the hydraulic efficiency too greatly, the base of the weir will be smaller, which reduces construction costs.

The operating range of low discharges was also of concern. Noise and nappe oscillations have been documented at other sites (9) while operating under low heads. To avoid this problem, a splitter pier will be installed on each spillway side wall 11 ft (3.35 m) upstream of the downstream apexes to supply air and support the nappe. The splitter piers will be 3 ft (0.91 m) tall and become submerged during operation under higher heads.

To assist in the structural design of the labyrinth spillway, pressures were measured on the floor of the downstream channels. The piezometers were located parallel to the side walls and along the centerline. In general, the pressures diminished as the channel width expanded. The pressures were not excessive and ranged from about 6.5 to 17.3 $lb/in^2$ (44.8 to 119.6 kPa) for the full range of discharges.

Conclusions

The labyrinth spillway has many applications. It can be easily adapted to site constraints of new and existing structures. Tradeoffs between hydraulic, structural, and economic considerations are an important part of the labyrinth spillway selection and design. The hydraulic flume tests resulted in the development of a set of revised design curves and a design method. The accuracy of these curves was verified by the further testing of the site specific scale model which required a much higher length magnification than previously predicted. Splitter piers should be used to prevent nappe oscillations under low flow conditions.

REFERENCES

1.  $CH_2M$ Hill, City of Aurora, Colorado, Quincy Terminal Raw Water Storage, 1973.

2.  $CH_2M$ Hill, City of Dallas, Oregon, Additions and Modifications to Mercer Dam, 1971.

3.  $CH_2M$ Hill, City of Dallas, Oregon, Mercer Dam Spillway Model Study, March 1974.

4.  Darvas, Louis A., "Performance and Design of Labyrinth Weirs - Discussion," Journal of the Hydraulics Division, ASCE, August 1971, pp. 1246-1251.

5.  Hay, N., and G. Taylor, "Performance and Design of Labyrinth Weirs," Journal of the Hydraulics Division, ASCE, November 1970, pp. 2337-2357.

6.  Houston, K. L., "Hydraulic Model Study of Ute Dam Labyrinth Spillway," GR-82-7, Bureau of Reclamation, 1982.

7.  Mayer, Paul G., "Bartletts Ferry Project Labyrinth Weir Model Studies," Georgia Institute of Technology, School of Engineering, Atlanta, Georgia, October 1980, February 1981, and July 1981.

8.  Phelps, H. O., "Model Study of Labyrinth Weir - Navet Pumped Storage Project," University of the West Indies, Department of Civil Engineering, St. Augustine, Trinidad, West Indies, March 1974.

9.  Thomas, Henry H., "The Engineering of Large Dams," Part II, p. 480.

# FEASIBILITY STUDY OF A STEPPED SPILLWAY

Marlene F. Young 1/

## ABSTRACT

A model study was conducted to investigate the feasibility of a stepped spillway for Upper Stillwater Dam.

The model findings indicate that the energy contained in the jet is minimal due to the tumbling action induced by the steps. Maximum velocity encountered on the 61-m (200-ft) high spillway is about 11 m/s (36 ft/s). The energy reduction is 75 percent greater than a conventional spillway. The low velocity at the bottom of the dam requires only a 7.6-m (25-ft) long stilling basin to dissipate the remaining energy.

Stepped spillways can be used with dams where the unit discharge is low and stilling basin construction is to be limited or where the use of roller-compacted concrete is considered.

## INTRODUCTION

Upper Stillwater Dam will include two firsts for the Bureau of Reclamation when it is completed in 1986. It will be the first roller-compacted concrete dam constructed by the Bureau and also the first stairstepped spillway of its size in the United States.

The Upper Stillwater damsite is located 120 km (75 mi) east of Salt Lake City, Utah, on the Bonneville Unit of the Central Utah Project. The dam will help regulate the flows of Rock Creek and South Fork of Rock Creek for release into the Strawberry Aqueduct through the 13-km (8.1-mi) long Stillwater Tunnel. The reservoir will be kept full during the summer recreation months. Fluctuations in the reservoir water level will occur only during the winter months.

Techniques used on roller-compacted concrete are similar to those used in highway construction and, according to researchers, could reduce the cost and construction time of mass concrete gravity dams substantially over conventional methods. These techniques involve the use of extruded concrete on upstream and downstream faces retaining a rolled concrete hearting. Roller-compacted concrete has a zero slump and, at Upper Stillwater, will be placed in 300-mm (1-ft) lifts. The concrete used

---

1/ Hydraulic Engineer, Bureau of Reclamation, Denver, Colorado.

has a low cement and high fly ash content, and the dam is built up in thin layers (with a large surface area) across the full width of the valley and continuously compacted by a vibrating roller. In this way, the dam can be raised rapidly without the usual problems of heat generation and differential shrinkage. The facing elements are constructed slightly in advance of the hearting, using a laser-controlled slip form shown on Fig. 1. A test section for Upper Stillwater Dam was constructed during the summer of 1981 to test bonding between lifts and the techniques involved in placing roller-compacted concrete. This test section is shown in Fig. 2. The maximum height of the dam will be 84 m (275 ft), and the crest length 792 m (2600 ft). The initial configuration had a 0.6 to 1 downstream slope along the entire height. The ungated spillway will be 183 m (600 ft) long and 61 m (200 ft) high. The spillway must pass a design flood of 425 $m^3$/s (15 000 $ft^3$/s). The first phase of the model study was for the dam to these specifications.

However, for construction purposes, the top width of the dam was increased from 4.6 to 9.2 m (15 to 30 ft). To accomplish this, the slope of the upper 22 m (72 ft) of the dam was increased to 0.32 to 1 (Fig. 3). Acceptable crest designs were developed for both slopes.


## INVESTIGATION

The purpose of the model study was to determine if the design was feasible and to what degree the stepped spillway affected the energy dissipation.

Two models were constructed to conduct the study. Both were sectional models representing only a small portion of the 183-m (600-ft) crest. One model was built to study only the top few meters of the crest to optimize the tumbling action in the flow as early as possible. A discharge curve was also developed using this model. The second was a 1 to 15 scale sectional model of the entire spillway - from crest to stilling basin. This model was used to size the stilling basin.

### Crest Model

The 1:5 model was set up in a 760-mm (2.5-ft) wide flume. The first crest design tested was for the 0.6 to 1 downstream spillway slope. The crest acted basically as a broad-crested weir. The flow passed through critical depth at the upstream end of the crest and accelerated as it passed downstream. As the flow negotiated the downstream curve and encountered the first step, the underside of the jet impinged on the upstream end of the step and deflected outward, away from the spillway face (Fig. 4). After further experimentation, it was determined that the point where the underside of the nappe was directed was very critical. By adding one additional step with an angled configuration, the location of the impingement was changed and the tumbling action as shown on Fig. 5 was developed. The final dimensions for the 0.6 to 1 crest are shown on Fig. 6.

FIG. 1. - Laser-controlled slip form.

FIG. 2. - Upper Stillwater spillway test section.

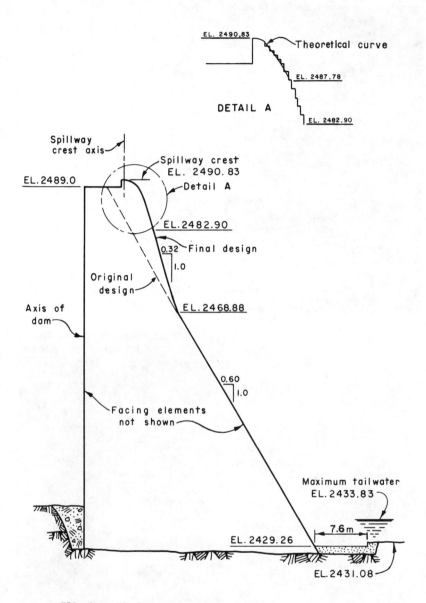

DETAIL A

EL. 2490.83

Theoretical curve

EL. 2487.78

EL. 2482.90

Spillway
crest axis

Spillway crest
EL. 2490.83

Detail A

EL. 2489.0

EL. 2482.90

0.32 Final design

1.0

Original
design

Axis of
dam

EL. 2468.88

0.60

1.0

Facing elements
not shown

Maximum tailwater
EL. 2433.83

7.6 m

EL. 2429.26

EL. 2431.08

FIG. 3. - Final design - Upper Stillwater Dam section.
(Elevations in meters)

FIG. 4. - 0.6 to 1 spillway crest - original design Q = 1.92 m²/s (20.7 ft²/s).

FIG. 5. - 0.6 to 1 spillway crest modified design Q = 1.92 m²/s (20.7 ft²/s).

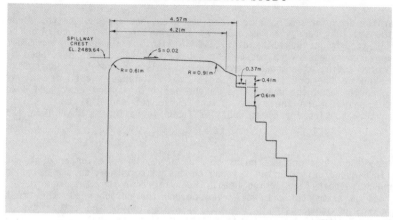

FIG. 6. - Spillway crest dimensions - modified design.

FIG. 7 - Final design - 0.32 to 1 spillway crest - Q = 2.39 m$^2$/s (25.7 ft$^2$/s).

After the decision was made to widen the crest and steepen the top slope
of the spillway face, it was apparent that getting the flow to cling to
a 0.32 to 1 slope would be difficult in light of the earlier studies.
It appeared necessary to reduce the substantial velocities developed by
the crest acting as a broad-crested weir. This was done by dropping the
approach to the crest by 1.82 m (6 ft). A nappe-shaped crest was
designed to match the undernappe of the jet. This shape continued down
to the point where the slope of the curve met the 0.32 to 1 slope.
Using the equation for crest profiles found in Design of Small Dams 1/:

$$y/H_0 = -K \ (x/H_0)^n$$

where y and x designate a point on the curve, Ho = design head, K and
n = inch-pound unit constants found on design curves which are based on
approach velocity and design head. Because, from previous experience,
the design seems to work well even beyond the design head, the crest
shape was designed using 0.9 m (3 ft) as $H_0$, while the actual design
head was 1.07 m (3.5 ft). This was done to ensure that at the middle
range discharges good flow conditions would be developed where previously only marginally acceptable conditions in the crest area had been
encountered.

The nappe-shaped crest is approximated by a series of 0.3- and 0.6-m
(1- and 2-ft) high steps. Near the crest, the downstream tips of the
steps correspond to the theoretical curve and gradually, the downstream
tips impinge into the theoretical jet nappe until the upstream end of
the steps fall on the theoretical curve. The detail shown on Fig. 3
displays this orientation. The curved portion of the spillway ends
6.1 m (20 ft) below the crest. At this point, the slope of the curve is
0.32 to 1 and this slope continues another 15.8 m (52 ft) vertically.
At elevation 2468.9 m (8100 ft), the slope changes abruptly to 0.6 to 1,
which continues the remaining 39.6 m (130 ft) to the stilling basin.

The crest model simulating this new design was constructed in the same
flume as the previous 0.6 to 1 crest, but at a 1 to 10 scale to permit
study of the crest down to the point where the slope of the nappe-shape
becomes 0.32 to 1.

The design worked well; turbulence was developed near the crest and the
jet clung to the extremely steep 0.32 to 1 slope. Fig. 7 shows the
final design operating at the maximum reservoir head. The maximum unit
discharge is 2.39 $m^2$/s (25.7 $ft^2$/s), which is sufficient to pass the
maximum spillway discharge. The developed discharge curve appears on
Fig. 8.

Little splash developed with this design except between heads of 150 and
300 mm (0.5 and 1 ft) when the thin jet springs off the first step and
comes back to the concrete surface at the sixth step. There is no
practical way to alleviate this problem. The tolerances specified in

---

1/ Design of Small Dams, Bureau of Reclamation, pp. 374-75.

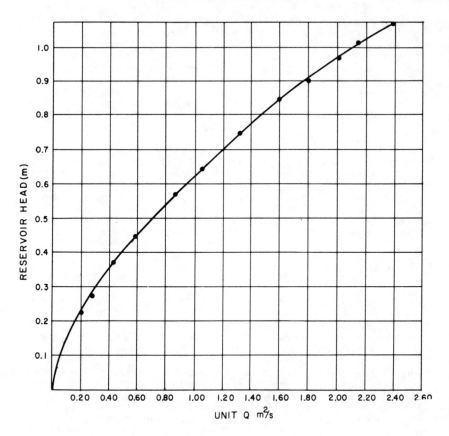

FIG. 8. - Discharge curve - Upper Stillwater spillway final design.

the construction of the prototype spillway crest and top 3 m (10 ft) must be fairly tight because a 25-mm (1-in) prototype variation from the specified dimensions will cause the jet to spring clear of the concrete surface. In this instance, the laser-controlled slip form will be able to work to close tolerances, although it would not prove critical to the design unless a very substantial portion of the spillway step was either below or above the designated curve. Most likely the step will weave slightly under and over the specified dimension along the 183-m (600-ft) wide spillway; therefore, no problem is anticipated.

## Spillway Model

Once the crest shape was optimized, the 1 to 15 scale model from crest to stilling basin was constructed in another, taller flume. The sectional model represented 18.3 m (60 ft) of the 183 m (600 ft) wide crest. The main purpose of this model was to size the stilling basin. Velocities along the length of the spillway were measured and pressures on both the areas exposed to the flow and those where negative pressures might be expected were taken. Also of interest were the flow patterns and pressures where the spillway abruptly changed slope. Necessary wall heights along the sides of the spillway were also determined. At all discharges, the flow down the spillway is well mixed with air and the jet is broken up. At small discharges (0.25 to 0.75 m$^2$/s), the flow strikes each step and loses nearly all of its velocity before accelerating to the next step where the pattern is continued. At larger discharges, the flow appears to speed up near the crest of the spillway until the steps begin to take affect and slow the jet. The velocity appears to remain constant for the remainder of the fall. Two types of velocity measurements were taken to verify these observations.

A high-speed movie was made of the model operating at the maximum discharge with paper squares introduced into the flow. The camera indexed the side of the film at a specific time interval. In this way, a distance-time relationship was established and the velocity of the jet at various points along the spillway could be determined. The velocity measurements made by the high-speed film were spot checked for accuracy by using a pitot tube. Close agreement was found between the two methods.

At the maximum discharge, the flow becomes fully turbulent within 3 m (10 ft) of the crest and begins to slow down. After falling 8 m (26 ft), the velocity is between 8 and 11 m/s (26 and 36 ft/s). The velocity generally stays within this range for the remaining 53 m (175 ft) of fall distance. The energy reduction achieved by the stepped spillway over a conventional smooth spillway of the same height is approximately 75 percent.

The major benefit of the stepped spillway is that flow velocities on the spillway face are kept low due to the tumbling action caused by the steps. This energy reduction over a conventional smooth spillway with a 61-m (200-ft) drop also eliminates the need for a long stilling basin. In this case, with a 183-m (600-ft) wide spillway, this results in an enormous savings in concrete and excavation along with reducing the area over which uplift pressures can develop. Because of the reduced velocities associated with the stepped spillway over those developed by a conventional smooth spillway, the stilling basin for the new design need be only half as long as one for a conventional 61-m (200-ft) high dam. The stilling basin for the Upper Stillwater spillway is 7.6 m (25 ft) in length and quiets the flow beyond the endsill for all discharges.

## CONCLUSIONS

The stepped spillway works well in conjunction with the use of roller-compacted concrete. Because of the energy dissipation produced by the steps, the stilling basin size is minimal. It has specific application to spillways with low unit discharges. Further research will hopefully relate step height versus jet thickness in order to develop more generalized design criteria.

## ACKNOWLEDGMENTS

R. J. Quint and A. T. Richardson, Division of Design, Bureau of Reclamation, are acknowledged for their technical input and helpful suggestions during the course of the model studies.

Hydraulic Applications of a Second-Order Closure
Model of Turbulent Transport

By

Y. Peter Sheng[1]

INTRODUCTION

Eddy-viscosity models have been widely used for the hydraulic
analyses of turbulent transport phenomena in oceans, lakes, and
estuaries. If sufficient data is available to establish the validity
of the required parameters in the subject models, then the predictions
of the models in that particular application give reasonably acceptable
results. However, when sufficient data are not available and the
parameters for a specific application must be extrapolated from much
different situations, the resulting predictions are highly speculative.

For example, sediment transport in coastal waters usually occurs
in highly oscillatory flow with appreciable density stratification.
The flow may also cause bed forms which in turn affect the flow. In
such a situation, large errors could result from the use of standard
eddy-viscosity models since these models do not contain the accurate
physics describing: (1) the time lag between the mean flow gradients
and the turbulent transport; (2) the time-dependent damping of the
turbulent transport due to stable density gradients and the
counter-gradient turbulent transport due to unstable density gradients;
and (3) the partitioning between skin friction drag and profile drag in
the vicinity of an arbitrary roughness element.

This paper highlights a turbulent transport model developed to
make accurate predictions in turbulent flows where data is unavailable
or hard to obtain, using as its strength modeling constants evaluated
in situations far-removed from the flow of application. The basic
turbulent transport model, originally developed by Donaldson and his
associates at A.R.A.P. (1,11,12), involves the retention of the
second-order turbulent correlation equations that affect the mean flow
variables. The added physics contained in the second-order closure
model permit one to directly calculate the phenomena mentioned in the
previous paragraph, without resorting to some ad-hoc eddy viscosity
fixes.

In the following, I will first give a brief description of the
turbulent transport model. I will then discuss three example hydraulic
applications of this model to (1) an oscillatory turbulent boundary
layer, (2) the transport of momentum, heat, and species within a
vegetation canopy, and (3) coastal currents driven by tide, wind, or

---

[1]Consultant, Aeronautical Research Associates of Princeton, Inc.,
P.O. Box 2229, Princeton, N.J. 08540.

density gradient. Emphasis is placed on the first application and detailed comparison with data. In the second example, a canopy model with some preliminary application is presented. In the third example, adaptation of a condensed version of the turbulent transport model to a mesoscale hydrodynamic model of coastal, estuarine, and lake currents will be outlined.

## A TURBULENT TRANSPORT MODEL

The model equations of motion for an incompressible fluid in the presence of both a gravitational and a Coriolis body force, with the mean variables denoted by capitals and the turbulent fluctuations by lower-case, may be written in general tensor notation as follows:

$$\frac{\partial U_i}{\partial t} + U_j \frac{\partial U_i}{\partial x_j} = -\frac{\partial \overline{u_i u_j}}{\partial x_j} - \frac{1}{\rho}\frac{\partial P}{\partial x_i} + g_i \frac{(\Theta - \Theta_o)}{\Theta_o} - 2\epsilon_{ijk}\Omega_j U_k + \frac{\partial}{\partial x_j}\left(\nu \frac{\partial U_i}{\partial x_j}\right) \quad (1)$$

$$\frac{\partial U_i}{\partial x_i} = 0 \quad (2)$$

$$\frac{\partial \Theta}{\partial t} + U_j \frac{\partial \Theta}{\partial x_j} = -\frac{\partial \overline{u_j \theta}}{\partial x_j} + \frac{\partial}{\partial x_i}\left(\kappa \frac{\partial \Theta}{\partial x_i}\right) \quad (3)$$

$$\frac{\partial \overline{u_i u_j}}{\partial t} + U_k \frac{\partial \overline{u_i u_j}}{\partial x_k} = -\overline{u_i u_k}\frac{\partial U_j}{\partial x_k} - \overline{u_j u_k}\frac{\partial U_i}{\partial x_k} + g_i \frac{\overline{u_j \theta}}{\Theta_o} + g_j \frac{\overline{u_i \theta}}{\Theta_o} - 2\epsilon_{ik\ell}\Omega_k \overline{u_\ell u_j}$$

$$- 2\epsilon_{j\ell k}\Omega_\ell \overline{u_k u_i} + 0.3\frac{\partial}{\partial x_k}\left(q\Lambda \frac{\partial \overline{u_i u_j}}{\partial x_k}\right) - \frac{q}{\Lambda}\left(\overline{u_i u_j} - \delta_{ij}\frac{q^2}{3}\right) - \delta_{ij}\frac{q^3}{12\Lambda} \quad (4)$$

$$\frac{\partial \overline{u_i \theta}}{\partial t} + U_j \frac{\partial \overline{u_i \theta}}{\partial x_j} = -\overline{u_i u_j}\frac{\partial \Theta}{\partial x_j} - \overline{u_j \theta}\frac{\partial U_i}{\partial x_j} + g_i \frac{\overline{\theta^2}}{\Theta_o} - 2\epsilon_{ijk}\Omega_j \overline{u_k \theta}$$

$$+ 0.3\frac{\partial}{\partial x_j}\left(q\Lambda \frac{\partial \overline{u_i \theta}}{\partial x_j}\right) - \frac{0.75q}{\Lambda}\overline{u_i \theta} \quad (5)$$

$$\frac{\partial \overline{\theta^2}}{\partial t} + U_j \frac{\partial \overline{\theta^2}}{\partial x_j} = -2\overline{u_j \theta}\frac{\partial \Theta}{\partial x_j} + 0.3\frac{\partial}{\partial x_j}\left(q\Lambda \frac{\partial \overline{\theta^2}}{\partial x_j}\right) - \frac{0.45q\overline{\theta^2}}{\Lambda} \quad (6)$$

$$\frac{\partial \Lambda}{\partial t} + U_j \frac{\partial \Lambda}{\partial x_j} = 0.35\frac{\Lambda}{q^2}\overline{u_i u_j}\frac{\partial U_i}{\partial x_j} + 0.6bq + 0.3\frac{\partial}{\partial x_i}\left(q\Lambda \frac{\partial \Lambda}{\partial x_i}\right)$$

$$- \frac{0.375}{q} \left( \frac{\partial q \Lambda}{\partial x_i} \right)^2 + \frac{0.8 \Lambda}{q^2} g_i \frac{\overline{u_i \theta}}{\Theta_o} \qquad (7)$$

Boussinesque approximation is assumed to be valid such that the only effect of the density stratification is in the gravitational body force term in the momentum equation (1). Equation (3), a diffusion equation for the temperature perturbation, can be written in terms of the density perturbation. For simplicity, diffusion equation for the salinity perturbation or other species concentration is not included here. The overbars in the equation denote ensemble-averaged values.

Many of the right hand side terms in the second-order correlation equations, including the stratification and the rotation terms, are determined precisely and hence did not require any modeling. The last three terms in Equation (4) and the last two terms in Equations (5) and (6) are modeled terms representing the effects of third-order correlation, pressure correlation, and viscous dissipation. Four model coefficients appear in these equations. All the right-hand side terms in Equation (7) had to be modeled. Model constants are determined from analyzing a wide class of flow situations and remain invariant for any new applications. Boundary conditions required for the above equations will be described later in the specific examples.

AN OSCILLATORY TURBULENT BOUNDARY LAYER

Oscillatory turbulent shear flow is encountered in a variety of practical flow situations such as blood flow in arteries, flow past helicopter blades, and oceanic bottom boundary layer under a wave. The important role of wave boundary layer in affecting the suspended sediment concentration in shallow water environments has been quantitatively demonstrated by Sheng (15).

Detailed measurements in oscillatory turbulent boundary layers are scarce. Jonsson and Carlsen (7) measured the detailed flow within an oscillating water tunnel with a fixed bottom, while Keiller and Sleath (9) measured the flow near an oscillating wall. Horikawa and Watanabe (3) measured the bottom boundary layer under a progressive wave in a wave tank. By far, the experiments of Jonsson and Carlsen (I will abbreviate with JC) are still the most comprehensive ones. They considered the more realistic case of a fully turbulent flow over a rough bottom. Jonsson (6) found that flow over a rough bottom becomes fully turbulent when the Reynolds number based on the bottom orbital velocity and the free stream amplitude reaches $10^4$.

In JC's experiment (Figure 1), an 8.39 sec wave with a maximum mean free stream velocity of 2 to 2.22 m/sec and a nearly sinusoidal time variation was imposed on a water depth of 23 cm. Using a micropropeller, they measured the detailed vertical profiles of ensemble-averaged horizontal velocity within the water tunnel at $15^o$ intervals through several wave cycles.

Due to the lack of quantitative understanding of the turbulent transport processes, all existing theoretical analyses of oscillating turbulent boundary layer are based on some ad-hoc eddy viscosity

models.  Multi-layered (8) and time-dependent (6) eddy viscosities had
to be used to achieve reasonable prediction of certain parameters.
Grant and Madsen (2) predicted the velocity profiles of JC's experiment
with reasonable accuracy, but failed to predict the phase relationship
accurately.

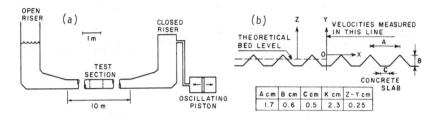

Fig. 1.    Jonsson and Carlsen's oscillatory flow facility:  (a) the
           water tunnel, (b) the bottom roughness elements.

Using the one-dimensional version of the turbulent transport model
described above, we performed a simulation of the oscillatory turbulent
boundary layer measured by JC.  The computational domain extends
vertically from $Z=Z_0=0.077$ cm at the bottom to $Z=17$ cm at the top.  For
simplicity, we assume the mean longitudinal velocity at the top to be
sinusoidal with an amplitude of 2m/sec.  Turbulent correlations at the
top are assumed to be negligible.  A time-periodic horizontal pressure
gradient which balances the time variation of wave orbital velocity at
the top boundary, was imposed at all vertical levels.  At the lower
boundary, all turbulent correlations are assumed to have a zero
gradient, except that the gradient of $\overline{uw}$ balances the horizontal
pressure gradient.  Mean velocities are taken as zero at the bottom.
To avoid the necessity of having to resolve the extremely small
turbulence time scales in the immediate vicinity of the bottom, $\Lambda$ is
assumed to vary linearly with height below a certain height, and is
determined from the dynamic equation (7) from there on.  The nonlinear
inertia terms are neglected, a valid assumption so long as the wave
orbital velocity is much smaller than the phase speed of the wave.  The
model was run for several cycles until the results reached a quasi
steady state, i.e., when results do not change from cycle to cycle.

The mean velocity profiles computed by our model at $\phi=0°$, 45°,
90°, 135°, and 180° are shown in Figure 2.  Excellent agreement between
our results and JC's data was achieved.  At peak amplitude, our model
prediction shows a slightly higher overshoot at the mid-level.  It is
interesting to note that the velocity profiles at 45° and 135° are
quite different.  Adverse pressure gradient is imposed on the flow at
45°, while favorable pressure gradient is imposed on the flow at 135°.
Since the measured free stream velocity is not exactly sinusoidal, the
measured velocity profiles have been normalized for comparison with
model results.

The turbulent shear stress $-\overline{uw}$ computed by our model is also compared with JC's data. As shown in Figure 3, the agreement is very good at $\phi=180^{\circ}$. However, the agreement is not as good at other $\phi$'s. Johnson and Carlsen did not measure the shear stresses directly, but instead computed them indirectly from the momentum equation. Due to the relatively coarse time resolution ($\Delta\phi=15^{\circ}$), errors could be introduced in determining the time derivative of the mean velocities. The computed bottom shear stress does indicate a phase lead of approximately $25^{\circ}$ over the free-stream velocity, which was also measured by JC. Although our model also computes all the other second-order turbulent correlations, no comparison with data could be made since they were not measured.

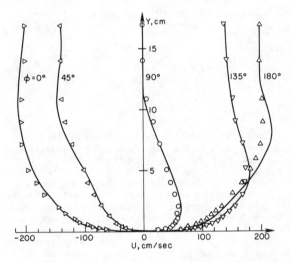

Fig. 2.   Velocities vs. height measured from top of roughness element ( —— model result, ▷◁ ○ ▽△ JC's data).

The phase lags of horizontal velocities at various levels computed with our model compare very well with JC's data (Figure 4). The horizontal velocity near the bottom shows a phase lead of about $25^{\circ}$. Grant and Madsen's eddy viscosity model predicted a much worse phase relationship. As shown in Figure 4, their computed phase lead is actually off by more than 25% near the bottom.

In slowly-varying turbulent boundary layers, there exists a thin layer near the bottom within which the turbulence is at equilibrium with the mean flow gradients and the mean flow variables vary logarithmically with height. This so-called logarithmic layer was also detected in JC's experiment on an oscillatory turbulent boundary layer. However, none of the previously mentioned theoretical analyses were able to predict the thickness of this layer and its variation with time. Figure 5 shows the variation of the log-layer thickness within

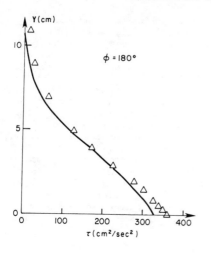

Fig. 3. Turbulent shear stresses vs. height ( —— model result, △ JC's result).

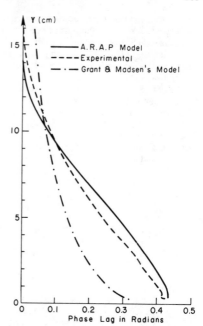

Fig. 4. Vertical profile of phase lag between mean velocities and free stream velocity.

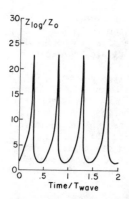

Fig. 5. Computed log-layer thickness as a function of time over two wave cycles.

two wave cycles as computed by our model. The log-layer attains its peak thickness of approximately 25 $Z_o$ shortly after the free-stream velocity reaches the peak amplitude, and almost completely diminishes shortly before the free-stream velocity reaches zero. This is consistent with the velocity profiles shown in Figure 2. For this analysis, the log-layer thickness is defined as the layer within which the turbulent kinetic energy is 95% or more of its wall value. If one defines a log-layer thickness based on the profile of $\overline{uw}$, then the thickness is slightly increased.

Our model also computes the dynamic length scale of turbulence, $\Lambda$, which is a representation of the mean turbulent eddy size. The ratio between this length scale and the total velocity variance, q, represents the time scale associated with the eddy motion. In sediment-laden flows, the relaxation time of sediment particles ($W_s/g$, where $W_s$=settling speed) relative to the time scale of turbulent eddies determines whether the particles follow the turbulent eddy motions or not. For a given particle size, there exists a height below which the sediment particles do not follow the turbulent eddy motions and hence the interaction between the particles and the turbulent eddies has to be considered. In such case, the use of a dynamic turbulence model, as opposed to an eddy-viscosity model, is highly desirable and recommended. Tooby et al. (9) performed an interesting laboratory study on the "vortex trapping mechanism" in affecting the suspended sediment concentration in an oscillatory turbulent boundary layer. To include such a mechanism in a predictive model for sediment concentration, it is essential to consider the interaction between sediment particles and turbulent eddies by means of a turbulent transport model such as ours.

TRANSPORT OF MOMENTUM, HEAT, AND MASS WITHIN A VEGETATION CANOPY

A canopy of vegetation represents a complex lower boundary for hydraulic and atmospheric flows (Figure 6). For flow well above this canopy, it is usually adequate to characterize the boundary in terms of only an aerodynamic roughness, $Z_o$. But when one is interested in the flow within the canopy or immediately above it, a more detailed representation is required.

Flow in vegetated waterways has been modeled empirically (e.g., 4, 10). Although such empirical models may be useful for qualitative flow analysis, a more complete model is required for quantitative estimation of the transfer of momentum, heat, and species within the vegetated environment. Second-order closure models for canopy flow have recently been developed by Wilson and Shaw (20), and Lewellen and Sheng (13). The principal difference between these two models is that the latter consider heat and species transport as well as momentum transport. Lewellen and Sheng also used a more general representation of the drag per unit volume of the vegetation. The model can predict the variation in surface layer heat and species transport as a function of surface Reynolds number, Prandtl number, Schmidt number, and plant area density distribution. Although the basic canopy model was originally designed to aid in the prediction of the dry deposition of gaseous $SO_2$ and particulate sulfate in the atmosphere, it is quite general and hence provides a basic framework for extension to hydraulic applications.

The canopy introduces source and sink terms into the basic conservation equations. The total drag force due to the canopy is composed of a skin friction drag and a profile drag. The skin friction drag forces of the canopy can be estimated by multiplying the shear stress across the laminar sublayer near the leaf surfaces by the total leaf surface area per unit volume. In addition to the skin friction drag, however, a more important pressure drag is generally imposed by the pressure difference between the upwind and downwind surfaces of a leaf or other object in the flow. We take the total drag term due to 'the canopy as

$$D_i = \left[ c_f A_w + c_p A_f (1+U_j^2/q^2)^{1/2} \right] q U_i; \text{ with } c_f = c_1 \left( \frac{\nu}{q\Lambda} \right)^{1/4} \tag{8}$$

where $c_f$ is the skin friction drag coefficient and $c_p$ is the profile drag coefficient. The frontal area per unit volume, $A_f$, and the wetted area per unit volume, $A_w$, appear in Equation (8). These two areas differ at least by a factor of two and in moderate flow conditions when the leaf aligns itself with the flow they can differ by an order of magnitude.

The sink terms in the energy and species equations may be obtained similarly by considering the transfer of heat and species across the sublayer as

$$\dot{Q} = c_f \left( \frac{\kappa}{\nu} \right)^{0.7} A_w q (\Theta - \Theta_s) \tag{9}$$

$$\dot{C} = c_f \left( \frac{D}{\nu} \right)^{0.7} \left[ 1 + c_f \left( \frac{D}{\nu} \right)^{0.7} q R_s \right] A_w q C \equiv c_c A_w q C \tag{10}$$

In summary, a term $-D_i$ is added to the mean momentum equation (1), a term $-\dot{Q}$ is added to the mean energy equation (4), and the mean species equation can be written as

$$\frac{\partial C}{\partial t} + U_j \frac{\partial C}{\partial x_j} = -c_c A_w q C - \frac{\partial(\overline{u_i c} - W_s C)}{\partial x_i} + \frac{\partial}{\partial x_i} \left( D \frac{\partial C}{\partial x_i} \right) \tag{11}$$

Both a source and a sink term need to be added to the Reynolds stress equations:

$$\frac{\partial \overline{u_i u_j}}{\partial t} = 2c_p(U_k^2 + q^2)^{1/2} A_f \overline{u_i u_j} \delta_{ij} - 2c_f A_w q \overline{u_i u_j} \text{ (no sum } i,j) + \dots \tag{12}$$

The first term represents the creation of wake turbulence due to the profile drag, while the second term recognizes that the skin friction can also dissipate the turbulent fluctuation of velocity. The profile drag can also break up the eddies to increase the dissipation, but this is accounted for in the model by introducing an additional constraint for $\Lambda$ which is inversely proportional to the plant area density, i.e., $\Lambda \leqslant \alpha/(c_p A)$. Additional sink terms are added to the temperature and species correlation equations for $\overline{u_i \theta}$, $\overline{u_i c}$, $\overline{\theta z}$, $\overline{c z}$, and $\overline{c \theta}$.

The computational domain extends from z=0 at the bottom to z=2h, twice the canopy height. At the top boundary, the mean variables are specified while the turbulent correlations have zero gradient. At the lower boundary, the mean velocities are zero while the temperature is specified. Gradients of all turbulent correlations are zero, except $\overline{w\theta}$ and $\overline{wc}$ which are given from sublayer relationships similar to Equations (9) and (10).

Detailed measurements of mean flow variables and turbulent correlations within vegetated hydraulic environments are unavailable at the present time. Therefore, for model verification, we used the detailed flow measurements within a corn canopy obtained by Shaw et al. (14).

The vertical profile of plant area density of the canopy is shown in Figure 7. We used the measured distribution of Ah shown in Figure 6 as $A_f h$; $\alpha=0.1$ and $C_p=0.16$. $C_1 A_w/A_f$ was given a value of 1 such that the skin friction drag is about one-third of the profile drag within the canopy, a relation measured experimentally by Thom (18). Based on these parameters, our model predictions agree closely with the measured mean longitudinal velocity Reynolds stress, and standard deviation of longitudinal and vertical velocities (Figure 8). Most of the momentum is absorbed within the upper part of the canopy and little is transported to the ground.

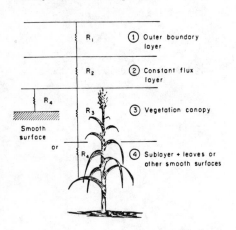

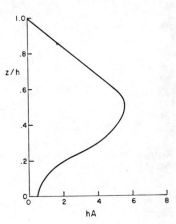

Fig. 6.    Four regions of planetary or oceanic boundary layer in the presence of vegetation canopies. $R_1$, $R_2$, $R_3$, and $R_4$ indicate resistance to species deposition.

Fig. 7.    Profile of plant area density of a corn canopy.

The same basic model was applied to simulate the heat transfer within the corn canopy measured by Shaw et al. in October 1971. The crop changed from 290 cm tall at the beginning of the experiment on

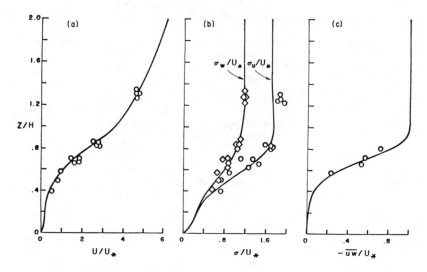

Fig. 8.  Comparison of A.R.A.P. model predictions with data above
a corn canopy.

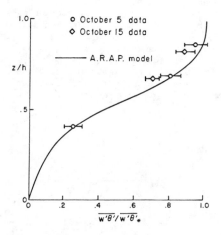

Fig. 9.  Heat transfer within a corn canopy.

October 5 to a noticeably less dense stand of 260 cm height on October
15.  As shown in Figure 9, there is good agreement between the computed
and observed heat transfer rate within the canopy on both dates.

COASTAL, ESTUARINE, AND LAKE CURRENTS

Sheng and Butler (15) developed an efficient three-dimensional, time-dependent numerical model of coastal, estuarine, and lake currents. Special computational features included in the model are: (1) a time-splitting technique which separates the computation of the slowly-varying internal mode (3-D variables) from the computation of the fast-varying external mode (water level and vertically-integrated velocities), (2) an efficient ADI algorithm for the computation of the external mode, (3) a vertically-stretched coordinate that allows the same order of accuracy in the vertical direction at all horizontal locations, and (4) an algebraically-stretched horizontal grid that allows concentration of grid lines in regions of special interest. These features make the model suitable for long-term simulation of the dynamic response of coastal, estuarine, and lake waters to winds, tides, and meteorological forcing.

It remains a challenge for hydraulic engineers and oceanographers to properly resolve the turbulent transport phenomena in such meso-scale circulation models. In Sheng and Butler (15), turbulence parameterization is based on the assumption that the production of turbulence equals the dissipation of turbulence. Quadratic stress laws are assumed at the air-sea interface and the bottom. To improve the predictability of the three-dimensional hydrodynamic model, the turbulent transport model described early in this paper could be utilized. The basic model, as represented by Equations (1) through (7) and appropriate boundary conditions, could be applied to the general three-dimensional time-dependent flow. In such a case, however, the numerical computation of all the dynamic equations represents a formidable task. To keep the problem manageable, I believe a condensed version of the turbulent transport model should be used. Assuming a high Reynolds number local equilibrium, the Reynolds stress and heat flux equations form a set of algebraic relationships between the turbulent correlations and the mean flow derivatives. The turbulent dynamics is carried by the dynamic equation for $q^2 = \overline{u_i u_i}$:

$$\frac{\partial q^2}{\partial t} + U_j \frac{\partial q^2}{\partial x_j} = -2\overline{u_i u_j}\frac{\partial U_i}{\partial x_j} - 2g_i\frac{\overline{u_i \theta}}{\theta_0} + \frac{\partial}{\partial x_i}\left(q\Lambda\frac{\partial q^2}{\partial x_i}\right) \tag{13}$$

and Equation (7) for the turbulent macroscale $\Lambda$. An extra term needs to be added to each of the $\overline{u_i u_i}$ equations to allow Equation (13) to be added without making the system overdetermined. This approximation allows for much better representation of the turbulent boundary layers in the ocean than do the standard eddy-viscosity models, and should be valid so long as the time scale of turbulence, $\Lambda/q$, is less than the time scale of the mean flow.

CONCLUSIONS AND RECOMMENDATIONS

A turbulent transport model suitable for hydraulic applications has been presented. The model gives accurate prediction of the oscillatory turbulent boundary layer measured by Jonsson and Carlsen without having to change any of the model constants. The basic model is being utilized to study the current-wave interaction within the benthic boundary layer (16). A canopy model suitable for atmospheric as well as hydraulic applications is presented, although the effects of additional parameters such as leaf stiffness need to be addressed. By combining the canopy model with a mesoscale ocean circulation model, one can accurately estimate the coastal currents in salt marshes. The basic canopy model may also be carried over for studying the flow and sediment transport in the vicinity of bed forms. Adaption of a condensend version of the complete turbulent transport model into a three-dimensional, time-dependent numerical model of coastal currents is also discussed. The proposed task should lead to significant improvements in the predictability of coastal circulation models.

ACKNOWLEDGEMENTS

This research has been partially funded by the Waterways Experiment Station under contract DACW39-80-C-0087. Work on the canopy model was funded by Electric Power Research Institute under contract RP-1306-1.

APPENDIX I. - REFERENCES

1. Donaldson, C. duP., "Atmospheric Turbulence and the Dispersal of Atmospheric Pollutants," in *Proceedings of Workshop on Meteorology,* American Meteorological Society, ed. by D. A. Haugen, Science Press, ]973, pp. 313-390.

2. Grant, W. D. and Madsen, O. S., "Bottom Friction Under Waves in the Presence of a Weak Current," *NOAA Tech. Rep. ERL-MESA,* 1978, 150 pp.

3. Horikawa, K. and Watanabe. A., "Laboratory Study on Oscillatory Boundary Layer Flow," *Proc. 11th Conf. Coastal Engng.,* Vol. 1, 1969, pp. 467-486.

4. Jackson, G. A., "Effects of a Kelp Bed on Coastal Currents," *EOS, Trans. Amer. Geophys. Union,* Vol. 63, No. 3, 1982, p. 84.

5. Jonsson, I. G., "Measurements in the Turbulent Wave Boundary Layer," *Proc. 10th Congr. Int. Ass. Hydr. Res.,* Vol. 1, 1963, pp. 85-92.

6. Jonsson, I. G., "A New Approach to Oscillatory Rough Turbulent Boundary Layers," *Ocean Engng.,* Vol. 7, 1980, pp. 109-152.

7. Jonsson, I. G. and Carlsen, N. A., "Experimental and Theoretical Investigations in an Oscillatory Rough Turbulent Boundary Layer," *J. Hydr. Res.,* Vol. 14, 1976, pp. 45-60.

8.  Kajiura, K., "A Model of the Bottom Boundary Layer in Water Waves," *Bull. Earthq. Res. Inst.*, Vol. 46, 1968, pp. 75-123.

9.  Keiller, D. C. and Sleath, J. F. A., "Velocity Measurements Close to a Rough Plate Oscillating in Its Own Plane," *J. Fluid Mech.*, Vol. 73, 1976, pp. 673-691.

10. Kouwen, N., Unny, T. E., and Hill, H. M., "Flow Retardance in Vegetated Channels," *J. of the Irrig. and Drain. Div.*, ASCE, Vol. 95, 1969, pp. 6633-6580.

11. Lewellen, W. S., "Use of Invariant Modeling," in *Handbook of Turbulence*, Vol. 1, ed. by W. Frost, T. H. Moulden, Plenum, 1977, pp. 237-280.

12. Lewellen, W. S., "Modeling the Lowest 1 Km of the Atmosphere," *AGARDograph*, No. 267, 1981, 88 pp.

13. Lewellen, W. S. and Sheng, Y. P., "Modeling of Dry Deposition of $SO_2$ and Sulfate Aerosols," *EPRI Report No. EA-1452*, Electric Power Research Institute, Palo Alto, CA, 1980.

14. Shaw, R. H., Silversides, R. H., and Thurtell, G. W., "Some Observations of Turbulence and Turbulent Transport Within and Above Plant Canopies," *Boundary Layer Meteorology*, Vol. 5, 1974, pp. 429-449.

15. Sheng, Y. P., "Modeling Sediment Transport in a Shallow Lake," in *Estuarine and Wetland Processes*, ed. by P. Hamilton, Springer-Verlag, 1980, pp. 299-337.

16. Sheng, Y. P. and Butler, H. L., "A Three-Dimensional Numerical Model of Coastal, Estuarine, and Lake Currents," *Proceedings 1982 Army Numerical Analysis and Computers Conference*, Army Research Office.

17. Sheng, Y. P. and Lewellen, W. S., "Current and Wave Interaction Within the Benthic Boundary Layer," *EOS, Trans. Amer. Geophys. Union*, Vol. 63, No. 3, 1982, pp. 73-74.

18. Thom, A. S., "Momentum, Mass and Heat Exchange of Plant Communities," *Vegetation and the Atmosphere*, Vol. 2, 1975, pp. 57-109.

19. Tooby, P. F., Wick, G. F., and Issacs, J. D., "The Motion of a Small Sphere in a Rotating Velocity Field: A Possible Mechanism for Suspending Particles in Turbulence," *J. Geophys. Res.*, Vol. 82, No. 15, 1977, pp. 2096-2100.

20. Wilson, N. R., and Shaw, R. H., "A Higher Order Closure Model for Canopy Flow," *J. Appl. Meteorol.*, Vol. 16, 1977, pp. 1197-1205.

# APPENDIX II. — NOTATION

| | |
|---|---|
| $A$ | Plant area density |
| $A_f$ | Frontal plant area density |
| $A_w$ | Wetted plant area density |
| $C$ | Mean species concentration |
| $c$ | Species fluctuation |
| $c_c$ | Species diffusion coefficient |
| $c_f$ | Skin friction drag coefficient |
| $c_p$ | Profile drag coefficient |
| $c_1$ | Empirical constant |
| $D$ | Species diffusivity |
| $g$ | Gravitational acceleration |
| $h$ | Height of vegetation canopy |
| $k$ | Nikuradse roughness parameter |
| $P$ | Mean pressure |
| $q$ | Root-mean velocity fluctuation |
| $R_1, R_2, R_3, R_4$ | Resistance to deposition in the four regions shown in Figure 6 |
| $T_w$ | Wave period |
| $t$ | Time |
| $U_i, U_j, U_k$ | Mean velocity components |
| $u_i, u_j, u_k$ | Fluctuating velocity components |
| $U$ | Mean longitudinal velocity |
| $W$ | Mean vertical velocity |
| $W_s$ | Settling speed of particles |
| $x, y$ | Horizontal coordinates |
| $z$ | Vertical coordinate |
| $Z_o$ | Aerodynamic roughness |
| $\alpha$ | Model constant |
| $\delta_{ij}$ | Kronecker delta |
| $\varepsilon_{ijk}$ | Alternating tensor |
| $\Lambda$ | Turbulent macroscale |
| $H$ | Mean temperature |
| $\Theta$ | Temperature fluctuation |
| $K$ | Thermal diffusivity |
| $\delta_u$ | Standard deviation of $U$ |
| $\delta_w$ | Standard deviation of $W$ |
| $\nu$ | Kinematic viscosity |
| $\rho$ | Density |
| $\phi$ | Phase angle |
| $\Omega$ | Earth's rotation |

# TURBULENT DIFFUSION OF MASS IN CIRCULAR PIPE FLOW

by

Anish N. Puri[1], Chin Y. Kuo[2], Raymond S. Chapman[3], A.M.ASCE

## ABSTRACT

An implicit finite difference scheme was used to solve the convective-diffusion equation to predict the steady-state transport of a conservative, neutrally buoyant tracer injected along the centerline into a fully developed turbulent pipe flow. Three different distributions for the radial mass diffusivity have been compared with two independent sets of experimental data. The results indicate that the distribution based on the turbulent kinematic eddy viscosity predicted by k-ℓ model produces the closest agreement between the numerical model predictions and the experimentally observed tracer distribution.

## INTRODUCTION

Many practical engineering problems require an analysis of the mass transport and mixing processes of a tracer when a miscible secondary fluid is continuously injected into a fluid flowing in a circular pipe. In many cases the fluids have approximately the same density, with the injection process so designed that an insignificant amount of momentum is imparted into the flow system. An example of a civil engineering problem involves the prediction of mixing rates and length of "mixing zone" of chemicals injected into pipes carrying contaminated water. Specifically, a suitable length of pipe has to be designed for the chlorination of water in a water treatment plant or for injecting chemicals to neutralize harmful effluents in the outlet pipes of

---

[1]Graduate Research Assistant, Dept. of Civil Engrg. Va. Polytechnic Institute and State University, Blacksburg, Virginia 24061.

[2]Assoc. Prof., Dept. of Civil Engrg., Va. Polytechnic Institute and State Univ., Blacksburg, Va. 24061.

[3]Research Assoc., Dept. of Civil Engrg., Va. Polytechnic Institute and State Univ., Blacksburg, Va. 24061.

industrial plants.  Another example involves the "dilution method" for
the measurement of large flows.  This method is used to determine the
flow rate in a pipe by measuring the difference between tracer concentration at the injection location and a point downstream with dilution
taking place between the two sections.  Both cases require an accurate
determination of the length of the "mixing zone".

Several analytical, numerical and experimental investigations of
turbulent mass transport in pipes have been made over the last decade
(1,2,3,4,5,6).  Most of these investigations adopted the use of either
constant or parabolic radial mass diffusivity distributions (2,4,6).

This paper presents a comparison of a numerically predicted eddy
viscosity profile with uniform and parabolic distributions of radial
mass diffusivity.  A numerical model utilizing these three distributions
was used to predict experimentally measured tracer concentration profiles in turbulent pipe flow.  These investigations indicate that the
numerically predicted distribution more accurately estimates the length
of the "mixing zone" in fully developed turbulent pipe flow.

## MODEL FORMULATION

### Governing Differential Equation

The steady state mass balance equation for a conservative,
neutrally buoyant tracer being injected continuously at the centerline
of a pipe into a fully developed uniform turbulent pipe flow is (5):

$$u\frac{\partial c}{\partial x} = \frac{1}{r} \frac{\partial}{\partial r} (r\, e_r\, \frac{\partial c}{\partial r}) \tag{1}$$

where u = longitudinal velocity, c = concentration of tracer (mass per
unit volume); $e_r$ = turbulent mass diffusivity in the radial (r) direction.  The coordinate system for Eq. 1 is oriented such that x defines
the longitudinal direction along the pipe centerline, with positive x
downstream and with x=0 corresponding to the injection point.  The r
direction is defined with r=0 corresponding to the pipe centerline and
r=R corresponding to the pipe wall, where R = pipe radius.  The boundary
conditions required to solve Eq. 1 for the case of centerline injection
are (2):

$$e_r\, \frac{\partial c}{\partial r} = 0 \quad \text{at} \quad r=0 \quad \text{and} \quad r=R \tag{2}$$

### Velocity Profile

The velocity distribution in a fully developed turbulent pipe flow
was assumed to follow the logarithmic law, which for smooth walled pipes
reduces to (11):

$$\frac{u}{u_*} = 5.75 \ \log_{10} \ (\frac{yu_*}{\nu}) + 5.5 \qquad\qquad (3)$$

in which $u_*$ = shear velocity, $\nu$ = kinematic viscosity of water, y = R-r. Since Eq. 3 is undefined at y=0, the velocity at the pipe wall was prescribed to be zero in accordance with a "no slip" condition. This velocity profile was used in the present investigation in order to compare numerical model predictions with two independent sets of experimental data in which smooth pipes were used.

Kinematic Eddy Viscosity and Radial Mass Diffusivity

The exchange coefficient for momentum, i.e. the kinematic eddy viscosity is defined as:

$$e = \frac{\tau}{\rho \frac{\partial u}{\partial y}} \qquad\qquad (4)$$

in which $\tau$ = shear stress; $\rho$ = mass density of water; and e = kinematic eddy viscosity. If $e_r$ is the radial mass diffusivity, it is possible to define (2):

$$Sc_{turb} = \frac{e}{e_r}$$

in which $Sc_{turb}$ is the turbulent Schmidt number. While expressions for the radial variation of the turbulent Schmidt number have been proposed (5), there is little experimental evidence to support them. Consequently, $Sc_{turb}$ = 1.0 was assumed as an approximation for a neutrally buoyant tracer.

Predictions of the eddy viscosity profile using a numerical model, referred as k-ℓ model, have been made by Taylor, Huges and Morgan (13). Their finite element model utilizes the Navier Stokes equations for turbulent flow and two additional transport equations which depict the spatial variation in both turbulent length scale, ℓ, and kinetic energy, k. The turbulent eddy viscosity is calculated by:

$$e = \rho \ell k^{1/2} \qquad\qquad (6)$$

Tests made using this model show excellent agreement between the predicted eddy viscosity profile and the experimental data of Laufer (7), who presented data for air flow at a Reynolds number Re = $(2U_{CL}R)/\nu$ of 500,000, where $U_{CL}$ = centerline longitudinal velocity. The profile obtained from the predicted nondimensional eddy viscosity distribution of Taylor, Hughes and Morgan is shown in Figure 1.

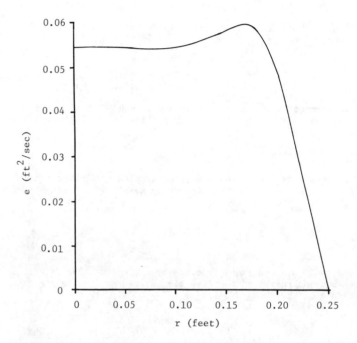

Figure 1.  Input kinematic eddy viscosity profile from Taylor,
Hughes and Morgans (13) nondimensional profile for
model comparison with Lee's (8) experimental results.
and Evan's (2) experimental results.

Several formulas for evaluating a cross-sectional mean $e_r$ have been presented (2,6,12). To provide some basis for comparison, a mean radial mass diffusivity was computed from the predicted profile using the following equation:

$$\bar{e}_r = \int_0^R \frac{2\pi r e_r dr}{\pi r^2} \tag{7}$$

In addition, a parabolic kinematic eddy viscosity distribution was suggested by Evans (2), which is of the form:

$$e = 0.4 \, u_* r \, (1 - \frac{r}{R}) = e_r \tag{8}$$

These three types of distributions were used as alternate descriptions of the radial mass diffusivity profile in the diffusion model, Eq. 1.

Input Concentration Profile

The numerical diffusion model requires an initial concentration profile at the location of the injection point. The distribution was suggested by Evans (2) as:

$$c = \beta \exp \left[ \frac{-(r/R^2)}{\alpha} \right] \tag{9}$$

where $\alpha$ and $\beta$ are constants obtained by fixing the tracer concentration at the outside radius of the injection nozzle at 2 percent of that at the middle of the jet.

COMPUTATIONAL METHODS

Equation 1 with the boundary conditions of Equation 2 and 9 was solved numerically using an implicit finite difference technique described by Murray and Lewis (9). The concentration profile at a particular pipe cross-section was evaluated from the profile at the previous (upstream) cross-section, and the solution was "marched" downstream. The radial grid was kept constant at 100 divisions per radius, and a variable longitudinal grid was employed to reduce computer time.

MODEL VERIFICATION

Numerical simulations were performed in three phases.

(a)  Grid Independence Tests

Grid independence tests were performed by comparing simulations using uniform values of the longitudinal velocity and radial mass diffusivity against a known analytic solution (5). The grid size selected resulted in an excellent agreement between the analytic and numerical solutions. Further grid refinement resulted in no appreciable change in the numerical solution.

(b)  Verification Against Lee's Data

Lee (8) examined the centerline injection of Gentian Violet dye into water flowing in a smooth polyethylene pipe of 3.0725 inch (7.8 cm) diameter D. A centerline velocity $u_{cl}$ of 1.829 ft/sec (55.75 cm/sec) resulted in a Reynolds Number of 50,000. In addition to a measured velocity profile, a measured concentration profile 1 inch (2.54 cm) downstream from the injection point was used as a model input, thus allowing the neglect of the "nozzle effects".

Simulations were performed using the three different types of radial mass diffusivity distributions. Figure 2 shows a comparison between model predictions and experimentally observed values. An exceptionally good agreement with the experimental values is obtained when the predicted profile is used as a model input, while large discrepancies are observed when the uniform or parabolic profiles are utilized. The parabolic profile results in extremely large discrepancies at the pipe centerline, a result which is mainly due to the unrealistic zero centerline value of $e_r$ obtained from Equation 8.

(c)  Verification Against Evans' Data

Evans (2) examined the turbulent diffusion of Nigrosine dye injected along the centerline of a smooth copper pipe of 6 inch (15.24 cm) diameter by a nozzle of 0.086 inch (0.218 cm) diameter. Experimentally measured concentration profiles were presented for Re=50,000 and 100,000 based on a mean longitudinal velocity u. The logarithmic velocity profile given in Eq. 3, and initial concentration distribution of the form given by Eq. 9 were used as model input.

Simulations were performed using the predicted eddy viscosity profile and the uniform $e_r$ defined by Equation 7. Results shown in Figure 3 clearly indicate that the predicted kinematic eddy viscosity profile again results in an excellent agreement with the experimental data, while the uniform profile results in large discrepancies.

SUMMARY

An implicit finite difference scheme was used to solve the convective-diffusion equation to predict the steady-state transport of a conservative, neutrally buoyant tracer injected along the centerline into a fully developed turbulent pipe flow. The analysis is for the case of continuous centerline injection of the tracer into circular pipes.

Three different distributions for the radial mass diffusivity have been compared with two independent sets of experimental data. The results of this study indicate that the distribution based on the

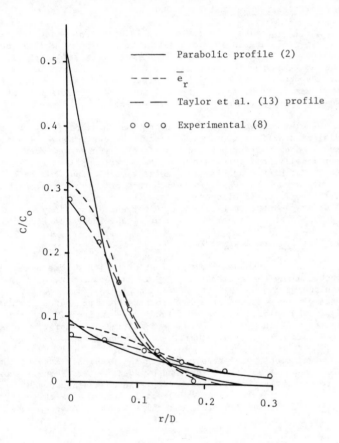

Figure 2.  Comparison of numerical predictions with Lee's (8) experimental results.  $C_o$ = input center-line concentration

turbulent kinematic eddy viscosity predicted by k-ℓ model produces the closest agreement between the numerical predictions and experimentally observed tracer distribution. It was observed that significant discrepancies exist between the numerical predictions and experimental results when the other two theoretical distributions are utilized, with the most severe discrepancies occurring in the "mixing zone" region, i.e. in the upstream region near the injection nozzle. These discrepancies tend to decrease further downstream as the tracer becomes uniformly mixed. Therefore, any attempt to accurately estimate the length of the "mixing zone" using numerical models should utilize an accurate eddy viscosity distribution, predicted by a k-ℓ model such as the one developed by Taylor, Hughes & Morgan or, in a similar way, by a k-ε model as described by Rodi (10) to find the eddy viscosity as a function of the turbulent kinetic energy k and the rate of dissipation ε.

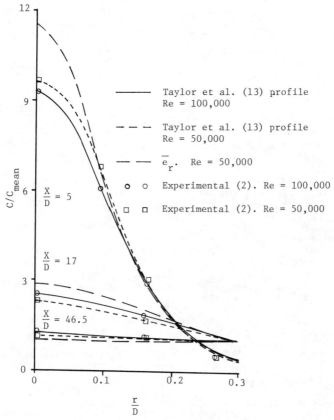

Figure 3. Comparison of numerical predictions with Evans' (2) experimental results.

APPENDIX - REFERENCES

1.  Clayton, C. G., Ball, A. M. and Spackman, R., "Dispersion and
    Mixing During Turbulent Flow of Water in a Circular Pipe," Report
    AERE-R5872, Isotope Research Division, Wantage Research Library,
    Wantage, Berkshire, England, 1968.

2.  Evans, G. V., "A Study of Diffusion in Turbulent Pipe Flow,"
    Journal of Basic Engineering, ASME, Vol. 89D, 1967.

3.  Filmer, R. W. and Yevdjevich, V., "Experimental Results of Dye
    Diffusion in Large Pipelines," Proceedings, 12th Congress of IAHR,
    Vol. 4, Fort Collins, Colorado, 1967.

4.  Flint, D. L., et al., "Point Source Turbulent Diffusion in a Pipe,"
    Journal, AIChe, Vol. 6, 1960.

5.  Ger, A. M. and Holley, E. R., "Comparison of Single-Point In-
    jections in Pipe Flow," Journal of the Hydraulics Division, ASCE,
    102, (HY6), Proc. Paper 12172, June 1976.

6.  Holley, E. R. and Ger, A. M., "Circumferential Diffusion in Pipe
    Mixing," Journal of the Hydraulics Division, ASCE, 104, (HY6),
    Proc. Paper 13663, April 1978.

7.  Laufer, J., "The Structure of Turbulence in Fully Developed Pipe
    Flow," National Advisory Comm. Aeronaut. Tech Reports. No. 1174,
    1954.

8.  Lee, J., "Turbulent Motion and Mixing," Ph.D. Dissertation, the
    Ohio State University, Columbus, Ohio, 1962.

9.  Murray, A. L. and Lewis, C. H., "Three-Dimensional Fully Viscous
    Shock-Layer Flows Over Sphere-Cones at High Altitudes of Attack,"
    Dept. of Aerospace and Ocean Engineering, Virginia Polytechnic
    Institute and State University, VPI-AERO-078, January, 1978.

10. Rodi, W., Turbulent Models and Their Application in Hydraulics,
    IAHR, Netherlands, 1980.

11. Streeter, V. L., Fluid Mechanics, McGraw-Hill Book Co., Inc., New
    York, New York, 1971.

12. Taylor, G. I., "The Dispersion of Matter in Turbulent Flow Through
    a Pipe," Proceedings of the Royal Society, Series A, Vol. 223,
    England, 1954.

13. Taylor, C., Hughes, T. G., and Morgan, K., "A Predictive Model for
    Turbulent Flow Utilizing the Eddy Viscosity Hypothesis and the
    Finite Element Method", Proceedings of the Symposium on Appli-
    cations of Computer Methods in Engineering, Vol. 11, Los Angeles,
    CA, 1977.

# A HYDRAULIC DESIGN METHOD FOR ADDING CAPACITY AND ENERGY OUTPUT WITH FLASHBOARDS

Stefan Manea[1] Ph.D., P.E.
M. ASCE

## ABSTRACT

In numerous circumstances flashboards can provide a simple and economical type of movable crest device which can give rise to additional head and storage above the spillway crest during the low-water season. The existing practice in selection of a particular flashboards arrangement is rather arbitrary; often a sole flashboards section is designed, which would waste a relatively great amount of head and water. A hydraulic design method is developed to provide the most economical flashboards set-up.

## INTRODUCTION

All power dams, small or large, are provided with spillways to discharge the excess flow not turbined or stored in the reservoir. In general, the spillway consists of an unobstructed overflow, the water rising and spilling over the crest as the flow increases. Since the elevation to which headwater may be allowed to rise is limited by "water rights," the crest of the spillway must be limited in elevation to provide sufficient margin for the rise of water surface during floods. Consequently, head and water, which otherwise could be used for power generation or other uses, are sacrificed during the low-water season. Obviously, the impact of the reduced head on the plant output during periods of drought is greater in low-head plants, where a few feet of additional head increase output by a relatively great amount.

Flashboards often can provide a simple and economical type of movable crest device which can give rise to additional head and storage above the spillway crest during the low-water season. History shows that in the USA, flashboards were the most common of all headwater control devices for small dams. Creager, Justin, and Hinds (1944) point out that flashboards up to 10 feet in height have been used for small dams, although a maximum height of about 4 or 5 feet was usually installed.

Recently, Hokenson (1981) points out that many of the spillways in New England and the Midwest were outfitted with flashboards. He also asserts that normally a head increase of up to 5 feet can be provided by using flashboards.

---

[1] Senior Hydraulic Engineer, Gilbert/Commonwealth, Reading, PA

DEVELOPMENT OF AN ECONOMICAL FLASHBOARDS SET-UP

The existing practice in selecting a particular flashboards set-up is rather arbitrary; often a sole flashboards section is designed, which would waste a relatively great amount of head and water.

Rating of the overflow spillway

A rating curve system of the respective spillway is computed for heads which will range between the permanent spillway crest elevation and a few feet above the water surface elevation (WSEL)$_{DS}$ at which the second section of the flashboards is set to drop out of position.

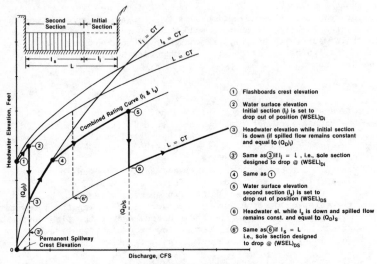

Fig. 1 — Spillway Rating Curve System

To accurately determine the rating curves shown on Figure 1, it is necessary to define properly (for the given spillway as designed or built) the coefficient of discharge, $C_w$, provided in the well-known weir-discharge formula.

$$Q = C_w L H^{3/2} \tag{1}$$

where $Q$ is the discharge in cubic feet per second, $C_w$ is the discharge coefficient, $L$ is the effective length of the crest in feet, and $H$ is the head on the spillway (vertical distance in feet from the crest of the spillway to the reservoir level, i.e. the spillway head measured at a distance upstream from the crest of at least $3H$ where the surface curvature is negligible).

The coefficient of discharge $C_w$ for any spillway shape varies with the head on weir, H.

For spillways designed for WES (Waterways Experiment Station) and other shapes, the $C_w$ factor of Eq. (1) for various heads (within the aforementioned range) can be determined by following the procedure given by Ven Te Chow, 1959, (p. 368, 369).

If the given spillway is of a broad-crested weir type, the $C_w$ value could be taken as equal to 3.09 for negligible boundary shear. With boundary shear along the walls and bottom the $C_w$ value mentioned above should be corrected as described by H. Rouse, 1950 (p. 527, 528).

Linsley and Franzini, 1972 (p. 281, 282) present four shapes of broad-crested weirs experimented by R. E. Horton (1907). If the given spillway conforms to or closely resembles one of these shapes, the $C_w$ values for various heads on weir, H, is advisable to be determined by using the above mentioned literature.

When the head on the broad-crested weir is greater than 1.5 times the length of the weir, experiments have shown that the nappe of the free overfall becomes detached and the weir is in effect a sharp-crested weir.

For sharp-crested weirs, the $C_w$ factor of Eq. (1) is recommended to be determined by using the weir formula of Rehbock (1929) written in the form

$$C_w = 3.27 + 0.40 \frac{H}{h} \tag{2}$$

where h is the height of weir.

An examination of the spillway rating curve system of Figure 1 shows that the headwater level at which the initial section is set to drop out of position $(WSEL)_{Di}$, determines a constant discharge $Q_{Di}$ in the spillway rating curve system regardless of the length of the initial section $1_i$.

The dropping flow* of the second section $Q_{DS}$, for a given dropping water surface elevation $(WSEL)_{DS}$, depends, however, on both the length of the initial section $1_i$, and the length of the second section $1_s$.

Functional relationship between parameters $1_i$, $Q_{DS}$, $(WSEL)_{DS}$

In order to establish a full functional relationship between the above mentioned parameters, the results obtained from a minimum of three graphs of the type of Figure 1 are plotted on a graph of the type presented in Figure 2. Since the parameters $1_i$ and $1_s$ are linearly

---

*The "dropping flow" is the flow corresponding to the set headwater level at which the flashboards drop.

dependent and complimentary within the interval L, for a given $(WSEL)_{DS}$ value, the dropping flow of the second section $Q_{DS}$, can be expressed as a function of either $l_i$ or $l_s$, only.

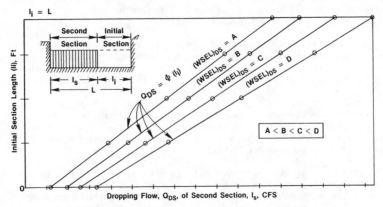

Fig 2 — Relationship Between the Length $L_i$, $Q_{DS}$, And $WSEL_{DS}$

An examination of the graph presented in Figure 2 shows that the parameters $l_i$, $Q_{DS}$, and $(WSEL)_{DS}$ have a linear dependency; the $Q_{DS}$ is directly proportional to the initial section length $l_i$, and the dropping water surface elevation $(WSEL)_{DS}$. On the other hand, as established before, $Q_{Di}$ is a function of the dropping water surface elevation $(WSEL)_{Di}$, only, remaining constant for any initial section length $l_i$. What these mean is that the best flashboards set-up should combine the smallest possible initial section length $l_i$ (since its dropping flow $Q_{Di}$, is constant for any $l_i$) with the highest possible dropping flow of the second section $Q_{DS}$, to reduce the dropping frequency and the time this flashboards section is down.

It can be observed that the $Q_{Di}$ parameter, because it depends solely on $(WSEL)_{Di}$, varies relatively slow with $(WSEL)_{Di}$ (see Figure 1) while the $Q_{DS}$ parameter, because it depends on both $l_i$ and $l_s$ and on $(WSEL)_{DS}$ as well, varies relatively fast with $(WSEL)_{DS}$ (see Figure 1 and Figure 2).

Therefore, to satisfy the previously outlined desideratum (reduced dropping frequency and time period the flashboards are down), the $(WSEL)_{Di}$ should be set as low as possible to provide for the highest possible $(WSEL)_{DS}$ setting. It is advisable to maintain the dropping water surface elevation of the initial section $(WSEL)_{Di}$, within one foot above the flashboards crest elevation.

The value of $(WSEL)_{DS}$ parameter should be taken to the elevation to which headwater may possibly be allowed to rise by "water rights" during normal operating conditions. There might be locations where the land or flowage of water rights is not absolutely limited or very expensive. In such cases the normal operating level determined by $(WSEL)_{DS}$ might be increased even by obtaining additional water and property rights if the head for power, and/or storage are sufficiently valuable to permit the expenditure.

Cost of energy losses because of reduced head while flashboards are down

The amount of the lost energy in the time period while flashboards are down can be determined by using the formula

$$kWh = \frac{Q_t \times He \times e_p \times h}{11.8} \qquad (3)$$

in which:

kWh   = designation of energy expressed in kilowatt-hours;
$Q_t$    = turbined flow rate, cfs;
He    = effective (net) head on turbine, ft.;
$e_p$    = efficiency of the power generating installation = $e_t \times e_g$;
$e_t$    = turbine efficiency, %;
$e_g$    = generator efficiency, %;
$h$     = power plant operating time period, hours;
11.8  = constant = $737/\gamma$ ;
737   = the number of foot-pounds per second in one kilowatt;
$\gamma$ = the specific weight of water under atmospheric pressure at $60^\circ F$ (or $15^\circ C$) = 62.4 lb/ft$^3$.

a)   Turbined flow rate $Q_t$.

The available flow that can be turbined when the flashboards are about to drop or are down is very likely to satisfy the full load turbine(s) capacity, regardless of the type of operation (run-of-river, peaking, etc.) being employed at that time. Consequently, the turbine(s) flow rate at full gate operation may be the most suitable figure to be considered in this type of analysis.

The values of efficiency parameters, $e_t$ and $e_g$, on the same basis, should be those corresponding to full load turbine and generator capacity, respectively.

b)   Effective (net) head on turbine He.

Low-head hydroelectric plants usually are provided with reaction turbines, that is Francis and, especially, propeller turbines as opposed to the impulse turbines which commonly are employed for heads greater than 1000 feet. The effective (net) head He, acting on a reaction-turbine installation is defined in Figure 3.

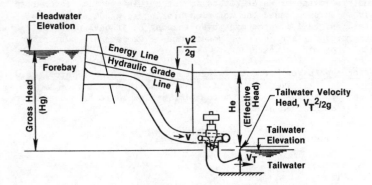

**Fig 3 — Definition Sketch For Effective Head On Turbine Of Reaction Type**

(For impulse turbines, the effective head is roughly determined by subtracting from the elevation of the pressure head plus velocity head at a point immediately upstream from the nozzle, the elevation of the lowest point of the pitch circle of the runner buckets.)

Since for low head the velocity at the draft tube mouth $V_T$, usually does not exceed 6 fps, the tailwater velocity head, $V_T^2/2g$, could be disregarded (in this type of analysis) to expedite the work.

Thus for reaction-turbine developments

$$He = H_g - h_{LP} \qquad (4)$$

where:

$h_{LP}$ = hydraulic losses in penstock system from the entrance to the turbine casing;

$H_g$ = gross head and is given by

$$H_g = HWEL - TWEL \qquad (5)$$

where TWEL is the tailwater elevation.

In order to determine the energy losses because of reduced head while the flashboards are down, the average annual energy based on the headwater elevation during the time the flashboards are up $(HWEL)_{UP}$, should be numerically compared against the one based on the headwater elevation when the flashboards are down $(HWEL)_{DOWN}$.

An examination of the spillway rating curve system of Figure 1 shows that the parameter $(HWEL)_{DOWN}$ varies with the length of the section which is down, "initial" or "second", and the spilled flow. Assuming that the value of the spilled flow would not increase further beyond dropping flow of the respective section, the $(HWEL)_{DOWN}$ is a function of $l_i$ or $l_s$ only and could be determined from the graph of the type presented in Figure 1 by mark ③ while the one for the respective $l_s$ is represented by mark ⑥ .

A full functional relationship between parameters $(HWEL)_{DOWN}$ and $l_i$ and $l_s$ is graphically given in Figure 4 which is developed by plotting $(HWEL)_{DOWN}$ values (given by mark ③ and ⑥ ) obtained from the same graphs of the Figure 1 type used in developing Figure 2.

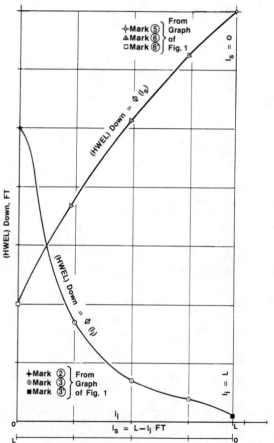

Fig. 4 — Relationship Between the Length $l_i$ and (HWEL) Down; and $L_s$ and (HWEL) Down

Figure 4, $l_i$ or $l_s$ versus (HWEL)$_{DOWN}$ shows that:

- the head water elevation (HWEL) as function of $l_i$, varies between EL.A and EL.B while $l_i$ varies between $l_i = 0$ and $l_i = L$ respectively;

- the head water elevation (HWEL) as function of $l_s$ varies EL.C and EL.D while $l_s$ varies $l_s = 0$ and $l_s = L$, respectively.

Hence, the graphs of Figure 4 are instrumental for evaluation of the headwater elevation during the time the flashboards of any length are down. The headwater elevation during the time the flashboards of the initial section are up is represented by EL.A = (WSEL)$_{Di}$ while the one during the time the flashboards of the second section are up is represented by EL.C = (WSEL)$_{DS}$.

The value of tailwater elevation of Equation (5) can be determined by using the tailwater rating curve (that is supposed to be previously developed with regard to other project requirements) and the total downstream flow $Q_T$ corresponding to a different arbitrary assigned flashboards section length.

The total downstream flow $Q_T$, corresponding to any initial section length $l_i$, is given by

$$Q_T = Q_{Di} + Q_t \qquad (6)$$

in which $Q_T$ is the total downstream flow, $Q_{Di}$ is the spilled flow as defined in Figure 1, and $Q_t$ is the turbined flow rate at full gate operation. It can be observed that for an assigned (WSEL)$_{Di}$, $Q_T$ remains constant for any $l_i$ values, and consequently TWEL would be constant regardless of the length of the initial section $l_i$.

The total downstream flow $Q_T$, corresponding to a different arbitrarily assigned second section length $l_s$, is given by

$$Q_T = Q_{DS} + Q_t \qquad (7)$$

where, $Q_{DS}$ is the spilled flow as defined in Figure 1. The numerical value of $Q_{DS}$ corresponding to a certain length $l_s$ and an assigned (WSEL)$_{DS}$ can be determined by using the graphs of Figure 2.

It can be observed that on the basis of the assumption made (that spilled flow would not increase further beyond dropping flow, i.e. dropping flow is equal to spilled flow) $Q_{Di}$ and $Q_{DS}$ of Equations (6) and (7) do not depend on the flashboards position. Consequently, the value of the TWEL parameter would be the same for either flashboards section position, up or down.

Based on the previously outlined parameters and criteria the lost energy because of reduced head while flashboards are down can directly be determined by substituting Equation (3) for

$$(kWh)_{LOSS} = \frac{Q_t \times \Delta He \times e_p \times h}{11.8} \qquad (8)$$

where He is the lost effective head and is determined as

$$\Delta He = (He)_{UP} - (He)_{DOWN} \qquad (9)$$

in which

$$(He)_{UP} = (HWEL)_{UP} - TWEL - h_{LP}$$

$$(He)_{DOWN} = (HWEL)_{DOWN} - TWEL - h_{LP} \qquad (10)$$

where $h_{LP}$ is constant for the given penstock system as it is determined on the $\frac{h_{LP}}{Q_t}$ basis.

c) Power plant operating time period h.

The number of generating hours should be based in principle on the adopted ratio of generation/nongeneration hours for the respective development. However, for the subject analyses it may be assumed that during the days the flashboards are down the plant would generate continuously.

The parameter h of Equations (3) and (8) which represents the time period while a certain flashboards section length, $l_i$ or $l_s$ is down is a composite time, its component parts being as follows:

Dropping Flow Duration. The average duration, in days, of a certain dropping flow can best be determined by using the flow duration curve (that is supposed to be previously developed with regard to other project requirements) and the total flow $Q_T$, corresponding to the respective - related flashboards section length, as defined by Equation (6) and Equation (7);

Waiting Time Period. The dropping flow duration, determined as previously outlined, represents the cumulated time, in days, during which the respective dropping flow is available within an average water year. The dropping flow is scattered throughout the year and the number of times a certain flashboards section length, $l_i$ or $l_s$, may drop, and the waiting time until the headwater elevation would decrease to the spillway crest proximity, would depend on the flow variation pattern of the respective stream. Both, the dropping number and the waiting time period can be determined by using an average water year tabulation (as presented in the U.S. Geological Survey Water Supply Papers or other sources) of recorded daily data* at or transposed to the site under consideration. A daily hydrograph for the average water year can be used but it often involves more work. The hydrograph is presented in Figure 5 mainly as an explanatory diagram since it better

---

*If recorded data of a shorter observation period is available, this should be considered instead, especially for flashy streams.

visualizes the waiting time period, dropping flow duration components, and other involved parameters, as well.

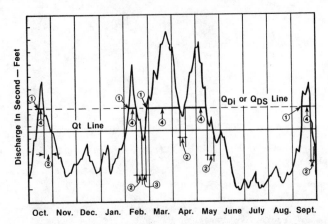

Fig. 5 — Daily Hydrograph for the Average Water Year showing: ① flashboards dropping time; ② waiting time periods; ③ working time period; ④ dropping flow duration component.

The amount of energy lost because of reduced head while flashboards are down can now be determined together with its annual cost.

In order to establish a functional relationship between the annual cost of lost energy and the flashboards section $l_i$ and $l_s$ the computations should be performed for at least three different arbitrarily selected $l_i$ and $l_s$ values. The computations need to be performed separately, and presented in tabular form for each flashboards section.

## Cost of maintenance

The cost of maintenance includes the cost of material for components replacement, and labor to raise the flashboards in position after flow has passed over the spillway crest.

The annual cost of material depends on the type and size of flashboards, section length and the number of times the respective section length drops.

For a given type and size of flashboards, a unitary cost of material in dollars per foot of flashboards length and drop, can be estimated by using regular standards and procedures of cost engineering.

The number of times a certain flashboards section length, $l_i$ or $l_s$, may drop can be determined as previously described under "Dropping Flow Duration" and "Waiting Time Period."

Then the annual cost of materials for a certain section length, $l_i$ or $l_s$ is given by

$$C_m = C_{um} \times l_{i(s)} \times N_{drop} \tag{11}$$

where $C_m$ is the annual cost of materials in dollars, $C_{um}$ is the unitary cost of materials in dollars per foot of flashboards length, $l_{i(s)}$ is one of the pre-selected $l_i$ or $l_s$ in feet, and $N_{drop}$ is the number of times the respective section length may drop during an average water year.

The cost of labor mainly depends on the working time period required to raise the flashboards in position. The working time period depends on the type and size of flashboards, section length, and technique and equipment employed. Here again, like before, for a given type and size of flashboards and employed technique, a unitary cost of labor in dollars per foot of flashboards length can be estimated by using regular standards and procedures of cost engineering.

The annual cost of labor to raise in position a certain section length, $l_i$ or $l_s$, is given by

$$C_l = (C_{ul} + C_{ue}) \times l_{i(s)} \times N_{drop} \tag{12}$$

where $C_l$ is the annual cost of labor in dollars, $C_{ul}$ and $C_{ue}$ are unitary cost of labor and equipment, respecitvely, in dollars per foot of flashboards length.

Both the cost of material and cost of labor should be determined for the same set of $l_i$ and $l_s$ values selected for the computations of the annual cost of lost energy.

For each selected $l_i$ and $l_s$ value, the annual cost of lost energy, annual cost of material and annual cost of labor to raise the flashboards in position are then combined and presented in tabular form, separately, for each flashboards section, $l_i$ and $l_s$. Thus, a functional relationship (between the total annual cost and parameters $l_i$ and $l_s$) of the type $C_i = \Phi(l_i)$ and $C_s \Phi(l_s)$ is now established in tabular form.

In order to establish the most economical flashboards set-up, the tabulated values of $C_i$ and $C_s$, (annual cost related to initial section and second section respectively) obtained as previously outlined, are then plotted on a graph of the type presented in Figure 6.

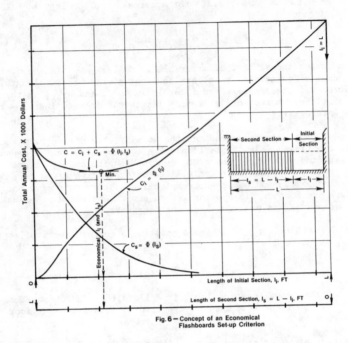

Fig. 6 — Concept of an Economical
Flashboards Set-up Criterion

The summary cost function $C = C_i + C_s = \Phi(l_i; l_s)$ yields a minimum which represents the most economical flashboards set-up which in most cases would bring substantial savings compared to a single flashboards section arrangement.

CONCLUSIONS

In numerous circumstances (especially when the necessity for confining the fluctions of headwater within very narrow limits does not exist) because of their relatively low cost, simplicity of construction and operation, flashboards may prove to be an appropriate movable crest device, especially for small dams. However, the existing practice in design of flashboards does not provide for an efficient flashboards arrangement in terms of head and/or water conservation. The hydraulic design method developed in this paper is intended to alleviate this deficiency.

## References

1. Ven Te Chow, 1959. Open-Channel Hydraulics, McGraw-Hill Book Co., New York.

2. W. P. Creager, Joel D. Justin, and Julian Hinds, 1944. Engineering for Dams, John Wiley & Sons, Inc., New York.

3. Reynold A. Hokenson, 1981. Headwater Enhancement and Encroachment Considerations for the Redevelopment of Existing Hydroelectric Sites. Waterpower '81, An International Conference on Hydropower, Proceedings, Vol. 1.

4. R. K. Linsley and J. B. Franzini, 1972. Water-Resources Engineering, McGraw-Hill Book Co., New York.

5. Hunter Rouse, 1950. Engineering Hydraulics, John Wiley & Sons, Inc., New York.

# DESIGN AID MODELS FOR SMALL HYDRO

By

Albert G. Mercer,[1] M.ASCE
Brent A. Berry [2]
David R. Burkholder [3]
Alan F. Babb,[4] M.ASCE

## ABSTRACT

Innovative approaches to small hydro design are often
dismissed because of the difficulty of predicting flow
behavior. Hydraulic models provide an ideal means of
testing these design innovations. New cost saving
techniques, using plastic foams, can reduce the cost of
modeling to fit small hydro engineering budgets. Several
cases are described showing the application of hydraulic
modeling to actual small hydro projects.

## INTRODUCTION

The hydraulic design of hydroelectric plants customarily
involves unique site and hydraulic conditions that do not
permit duplication of design configurations. The
uniqueness of hydro plants necessitates special design
considerations, particularly with respect to appurtenant
structures such as inlet and outlet works. Hydraulic
models are used routinely in large hydraulic
installations, but small hydro installations are
frequently designed without the use of a hydraulic model.

Small hydro requires much of the same care in design as
large projects and thus engineering costs can be a
significant portion of total project costs. In an attempt
to minimize engineering costs, the design of small hydro
projects often tends to be more conservative. In
addition, less effort is spent on the optimization of flow
characteristics, when compared with larger projects.

---

1  Principal, Northwest Hydraulic Consultants (NHC),
   Vancouver, B. C.
2  Intermediate Engineer, NHC, Vancouver, B. C.
3  Junior Engineer, NHC, Vancouver, B. C.
4  Senior Consultant, NHC, Vancouver, B. C.

Possibilities exist for standardization in small hydro but
the fact remains that most installations have singular
design features whose hydraulic performance have not been
tested. Because of this, design of small scale hydro
projects can also benefit from a hydraulic model study.
This step in the design process is frequently omitted,
however, because of budget and time considerations.
Methods are needed to reduce the time and cost for
conducting model studies for small hydro application.

This can be done in two ways. The first is to reduce the
amount of testing and data collection to only the very
essential elements. The second is to introduce fast and
inexpensive model construction techniques that are
particularly adapted to studying the hydraulic performance
of small scale hydro installations. This paper discusses
these methods and presents four case histories showing
their application to small hydro projects.

MODEL REQUIREMENTS FOR SMALL HYDRO

Hydraulic models provide two important design functions.
The first is a visual conceptualization of flow behavior
and the second is a method to obtain actual numerical
data. Simple observation of flow patterns at specific
flow conditions can often provide insite into the
suitability of a particular design. Collection of data is
a much more time consuming and expensive operation. It is
important therefore, that the need for numerical data be
carefully assessed before it is collected and even before
the model is designed. Many times the hydrology for small
hydro is such that the expense of exceptionally accurate
data collection or model calibration is not justifiable.

If neither accurate data collection nor calibration is
requirement, modifications can be made to the model
design. Usually this means that the model can be made
much smaller. This presupposes that smaller models are
less costly to build and test, which is not always the
case.

Decreasing the model size increases the scaling effects
due to viscosity and surface tension, so that the
importance of these effects must be carefully and
realistically evaluated. But in most situations, even
though the effects of viscosity or surface tension are
significant, the basic flow behavior is maintained and the
resulting observations are valid.

In the case of small hydro, the choice is usually between
a small model or no model and with this choice, some scale
effects are tolerable. However, if small models are
acceptable from a technical standpoint, methods must be
available for producing these models at much reduced costs.

PLASTIC FOAM TECHNIQUES

Plastic foams have made it possible to significantly
reduce costs in modeling both the structures and the
topography.  Structures can be constructed from high
density polyurethane foam.  This is available in precast
slabs which can be shaped very quickly with wood working
tools and takes on an excellent finish.  Fig. 1 shows a
1:150 scale model of a portion of a spillway structure
carved from foam.  It was tested in a laboratory flume to
check the overall configuration.  Another method of foam
construction is to actually cast the foam against a mold.
This was used very successfully in modeling a draft tube
where it was much easier to shape a core and cast the
model than to shape the model itself.  A hard, smooth
surface results.

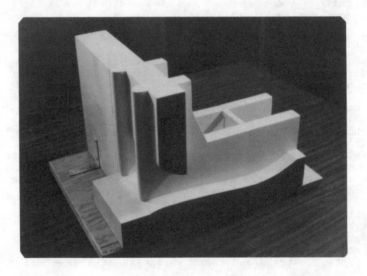

Figure 1. 1:150 Scale Model of a Spillway
Structure Made of Plastic Foam

The more significant application of plastic foam has been
in modeling topography.  The use of carved plastic foam in
architectural models is quite standard, but the carving
process is very wasteful of material and laborious when
applied to models of the size required for hydraulic
testing.  A much better technique is to cast the foam
against a sand mold.  The watertight foam takes on the
exact shape of the sand and a thin coating of the sand is
bonded into the foam to provide a hard granular textured
surface.

Individual panels are cast in a casting box 4m long,
2m wide and 0.6 m deep. The box is partly filled with
sand which is lightly oiled to provide cohesion for
molding. The box has a point gauge carriage with
displacement transducers that detect the position of the
point gauge on a horizontal X-Y plane. On the carriage is
a conventional X-Y plotter with a stylus that is connected
to follow the position of the point gauge. A contour map
of the panel to be molded is placed on the plotter and the
sand is molded by hand, guided by the point gauge. When
the sand contours, traced by the point gauge, match the
mapping, the molding is complete. A reinforcing framework
of wood is placed over the sand surface and the foam is
sprayed on to a thickness of about 0.1 m taking care to
bond in the framework. When the foam panel is lifted off
and turned face up, the panel is complete.

Small models may require only one panel but models
consisting of more than 50 panels, joined together by
foam, have been built. The head box and tailbox, used to
contain the flow at the model inlet and outlet
respectively, can also be cast in foam. Base plates for
the structures can be located in the molded sand so that
they are bonded into the topography at the correct
location and elevation for receiving the modeled
structures. In some cases the modeled structures are
actually set into the sand so that they become bonded
directly into the foam topography.

The principal advantage of this system is that there are
very few intermediate steps between receiving the
topographic maps and molding the sand. All of the data
available on the contours are immediately transferrable to
the model. Other advantages are the ease of making
revisions and the ease of dismantling the model for
storage. The model panels are very light and can be moved
about by hand.

CASE 1 - CHARLOT RIVER POWER DEVELOPMENT

The Charlot River Power Development is located on the
Charlot River approximately 3 km upstream of Lake
Athabasca. The project consists of a 30 m high earth fill
dam with a 10 mw powerplant. The proposed design includes
a rock cut spillway to be constructed on the left (east)
side of the dam. The rock cut is terminated in such a
manner that the discharging water will cascade 20 m down a
talus slope to the river opposite the tailrace and near a
bridge crossing.

A 1:100 scale model was constructed to investigate the
flow in the spillway, particularly any interference with

the powerhouse tailrace or the access bridge. Fig. 2
shows a plan layout of the model which required one
topographic panel.

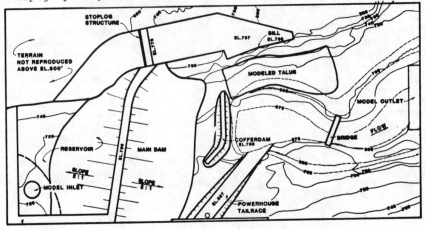

Figure 2.   Plan of 1:100 Scale Charlot River Model

The model was very successful in demonstrating to the
design engineers ways to improve their arrangement.   The
model tests showed that the spillway end sill should be
shortened to avoid impingement on the bridge.   Riprap
protection for the bridge abutments was also sized from
model results.   Over a year after the tests were
completed, the contractor uncovered weak rock and the
model was taken from storage and a spillway alignment
change was developed.

CASE II - KINGSLEY DAM POWERHOUSE RETROFIT

Kingsley Dam was originally built to provide irrigation
storage on the North Platte River near Ogallala,
Nebraska.   Subsequently, it was decided to utilize a
low-level outlet for a retrofit installation of a 50 mw
hydropower plant.   The powerhouse was designed to
discharge into a large concrete stilling basin provided
for the outlet works.   Conventional practice requires a 1
on 6 slope on the channel from the draft tube up to the
river level.

Considerable savings could be achieved if this slope could
be steepened and a model was commissioned to measure
energy losses for steeper slopes.

Fig. 3 shows the arrangement of the 1:50 scale model which
was built at the discharge end of a laboratory flume.   The

draft tube was cast of foam using a shaped core mold and
the basin, representative of existing concrete, was a
single panel cast against a mold made from plywood veneer.

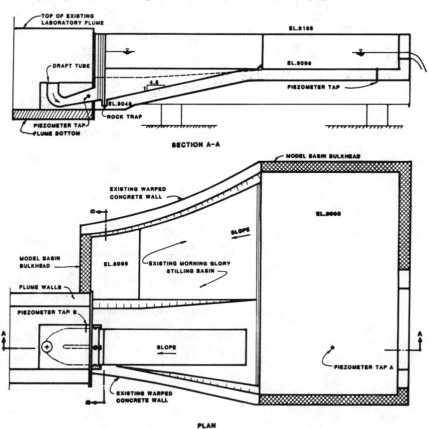

Figure 3.  Plan of 1:50 Scale Kinglsey Powerhouse
           Model Retrofit

The designer ultimately adopted a bottom slope of 1 on
1.75 which caused a head loss of .15 m compared to .05 m
for the 1 on 4.5 slope shown in Fig. 3.

CASE III  WHITEHORSE RAPIDS RETROFIT INTAKE

The existing Whitehorse Rapids Power Development in the
Yukon has a conventional powerhouse, an earthfill dam and

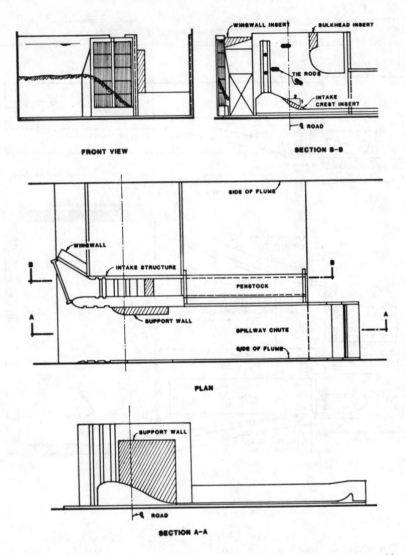

Figure 4.   1:50 Scale Model of White Horse Rapids Retrofit

a spillway with two large bays and one narrow one for fine
flow regulation.   It is planned to use the narrow bay as
an intake for an additional 20 mw retrofit unit.   A number
of unconventional design features were necessary due to

the size limitations of the narrow bay.  A hydraulic model was included as part of the design process to assess the effectiveness of these design features.

Fig. 4 shows the layout of the 1:50 scale model which fitted into a laboratory flume 0.8 m wide and 0.4 m deep. The massive concrete portions were modelled with high density foam and the thinner walls were modelled with clear acrylic plastic to enhance flow visualization.

The purpose of the model was to observe the flow behavior and to recommend geometry changes that would improve the flow.  The model showed a tendency for a surface vortex to form at the intake entrance.  This vortex was eliminated by the bulkhead insert shown in Fig. 4.  Also, an additional support wall was required for the pier because it was inadequate to take the hydraulic loading from the intake.  This support wall interferred with the spillway flow and several configurations were tested to arrive at the best alternative.  Several other minor changes were tested and recommended, eg., to prevent minor separations, etc.  No pressure measurements or flow calibrations were required from this model.

CASE IV   EBC POWERPLANT UTILIZING IRRIGATION FLOW

A study was conducted for a small hydro project in Washington State, utilizing an existing drop in an irrigation canal.  The project consisted of a new lateral trapezoidal canal connecting with an existing canal, an intake structure, penstock, and power plant.

The objectives of the study were to identify, and if necessary, improve on adverse conditions in the flow field

upstream from the penstock entrance.  In particular, any condition that promoted the formation of vortices above the entrance would be considered detrimental to performance and thus unacceptable.  In addition, any surging, excessive turbulence or other condition suggesting inefficiency and vibration potential were also considered to be undesirable.

The model, shown in Fig. 5, was constructed to a scale of 1:18.4 and reproduced a portion of the existing canal, a new lateral canal furnishing water to the powerplant, the intake structure, and a length of penstock.  The canals and a trapezoidal-to-rectangular transition were formed from polyurethane foam, whereas the intake structure and penstock were constructed from transparent acrylic plastic.

Figure 5.   1:18.4 Scale Model of EBC Powerplant Intake

The model was constructed in two weeks and the design was
optimized with structural modifications within one day of
testing during which the design engineers participated
heavily.  The modifications consisted of the addition of
upstream training walls and a sloped roof in the intake
structure to reduce rotation in the approach flow and to
eliminate vortices in the intake structure.

CONCLUSIONS

New materials and modeling construction techniques are
available that make hydraulic model studies compatible
with the engineering budget limitations of small hydro.
These model studies emphasize flow conceptualization and
minimize the collection of hard data in order to remain
cost effective.

# THE CHEREPNOV WATER LIFTER

By Henry Liu, [1] M. ASCE, Michael Fessehaye,[2] and Richard Geekie[3]

## ABSTRACT

A study aimed at understanding and developing the Cherepnov water lifter is near completion at the University of Missouri-Columbia. It is found that the behavior of the lifter can be accurately predicted by solving a set of differential-algebraic equations on computer. The lifter has many advantages such as it is automatic, simple, economical, pollution-free, noiseless and easy to maintain. It has potential applications in hydropower and other hydraulic practices. The paper presents both theoretical and practical aspects of the lifter.

## INTRODUCTION

The Cherepnov lifter is a novel device little-known outside Russia. The device extracts the potential energy of water to lift water. By discharging a portion of the water through the lifter to a lower elevation, the rest of the water goes up. Therefore, water is lifted to a higher elevation without the need of pumps and external energy supply. The lifter may be considered as a wheelless waterwheel or a ramless hydraulic ram. It is simple, economical and easy to construct and maintain. It also conserves energy and is pollution free. It has potential applications in many areas of hydraulics.

A potential application of the Cherepnov lifter is head-augmentation in low-head hydropower. Generally, low-head hydropower is less economical than high-head hydropower because the former requires a larger discharge and hence a larger turbine and penstock than the latter for the same amount of power generated. By using a Cherepnov lifter to augment the head of hydropower, it is possible to reduce the cost of the energy generated, thereby making low-head hydropower more economical. The economics of using a Cherepnov lifter in other hydraulic practices is even more promising.

Under a U.S. Department of Energy contract, the University of Missouri-Columbia is currently studying and developing the Cherepnov lifter. Several modifications of the lifter have been developed. A computer program to aid in the design of the lifters has also been developed. This paper discusses the basic concepts of the Cherepnov lifter and the status of the development work. Practical considerations will also be given. More details of the lifter can be found in (1-5).[4]

---

1.  Professor of Civil Engineering, University of Missouri-Columbia, Columbia, Mo.
2. & 3. Research Assistants, Department of Civil Engineering, University of Missouri-Columbia, Columbia, MO.
4.  Numerals in parentheses represent corresponding items in REFERENCES.

OPERATIONAL PRINCIPLE

As shown in Fig. 1, the Cherepnov lifter is a system of three interconnected water tanks. The intake tank 1 is an open tank, whereas the delivery tank 2 and the drain tank 3 are hermetic. Water enters tank 1 and leaves tanks 2 and 3 in a cyclic (periodic) manner. The water lowered into tank 3 from tank 1 compresses the air in tank 3. The compressed air causes the water in tank 2 to be pushed (lifted) to a higher elevation. Russian literature (1) describes the operation of the lifter in five stages as follows:

Stage 1:  Water fills tank 2. The water level in tank 1 goes down somewhat.
Stage 2:  Tank 2 is filled. The water in tank 1 rises to the top of pipe 5.
Stage 3:  Water flows over the top of pipe 5 and fills tank 3. No water is delivered to the user because the pressure of the air in tank 3 and pipe 6 is less than the back pressure in pipe 7. Valves 9 and 10 are both closed.
Stage 4:  Water continues to fill tank 3. The pressure of the air in the tank has exceeded the back pressure in pipe 7. Valve 10 opens, and the water is delivered through pipe 7.
Stage 5:  Water has reached the top of tank 3, and the float or switch 13 is activated. Vent 12 and drain valve 11 are opened, and the water is drained from tank 3.

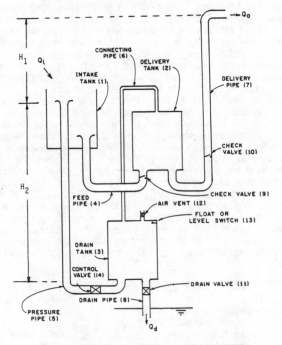

Fig. 1  THE CHEREPNOV WATER LIFTER

Although the foregoing five stages roughly describe the operation of the Cherepnov lifter, they are an oversimplification. As analyzed by Liu (3), and by Liu and Fessehaye (5), a lifter starting with empty tanks requires nine initial stages to bring the system into repeating or equilibrium cycles. Each repeating or equilibrium cycle contains five stages. Therefore, during start-up, the lifter goes through a total of fourteen stages before it starts to repeat itself. The detailed description of each stage together with the corresponding equations is given in (5). Eighty-seven differential and algebraic equations are needed to describe the initial fourteen stages of the operation of the lifter, and thirty-five equations are needed to describe the equilibrium stages (stages 10 through 14).

The study at UMC revealed that many modifications of the design and the operation of the lifter are desired for various purposes. So far, fourteen different modes have been identified and studied, each requiring a different set of equations and computer program (5). Three of these modes will be discussed herein briefly as follows:

Mod 2:          In Mod 2, the drain pipe 8 is not submerged, and the control valve 14 and the air vent 12 do not exist. The system vents naturally through the drain pipe 8 at the end of each drainage stage.

Mod 8:          As in Mod 2, drain pipe 8 is not submerged, and the system vents naturally through the drain pipe. Tank 1 is eliminated. Pipes 4 and 5 are connected directly to a reservoir or lake. Valve 14 is used to control the flow. This valve is electrically or mechanically controlled by the water level in tank 3. Pipe 5 is always full of water. Mod 8 is more suitable than Mod 2 for large lifters that receive water from a large reservoir or lake.

Mod 14:          Same as Mod 8, except the drain pipe 8 is submerged, and the delivery pipe 7 is connected to a large hermetic storage tank partially filled with air. The storage tank is needed for converting the intermittent outflow of the Cherepnov lifter to a continuous and approximately constant outflow which is essential for hydropower applications. This mode of operation is illustrated in Fig. 2.

## EXPERIMENTS

A Cherepnov lifter was built and it is undergong a series of tests in the Hydraulics Laboratory of UMC. The purpose of the tests is to verify the computer models of several modes of lifter operations. While the intake tank is made of plexiglass, the delivery tank and the drain tank are 55-gallon steel drums. The interconnecting piping is transparent flexible vinyl tubings. Tanks 1 and 2 each sits on a hydraulic platform lift, allowing them to be individually raised or lowered. The outflow of the lifter (i.e., the delivery pipe) is suspended on the ceiling of the laboratory with a rope and pulley. The inflow to the system, a steady discharge, is measured by a venturi. The delivery flow, an unsteady discharge, is measured by an electromagnetic flowmeter. The delivery flow can also be calculated from the rate of change of the depth of water in tank 2 from the continuity equation. A pressure transducer is mounted in pipe 6 to measure the air pressure variation in the lifter. There is also a pressure transducer on the bottom of each of the three tanks. The pressure heads measured by these three transducers minus the pressure head of the air in the lifter yield the depths of water in the three tanks. All the pressure transducer readings and the electromagnetic flowmeter readings are recorded on a Model 1508A Honeywell Visicorder.

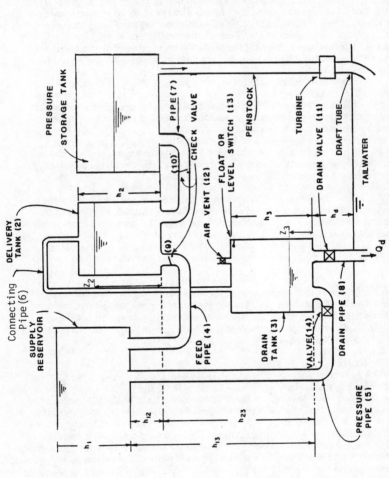

Fig. 2. CHEREPNOV LIFTER USED IN HYDROPOWER WITH HERMETIC STORAGE TANK (MOD 14)

yield the depths of water in the three tanks. All the pressure transducer readings and the electromagnetic flowmeter readings are recorded on a Model 1508A Honeywell Visicorder.

At the time of this writing, only one test run has been conducted. This was done for Mod 2 at an inflow discharge of 0.077 cfs, a head drop of $H_2$ =11.2 ft., and a head rise of $H_1$=8.44 ft. The result of this run is shown in Fig. 3, and the corresponding computer simulation is shown in Fig. 4. Note that only minor differences exist between the test data and the predicted computer result. The computer result is based on headloss coefficients evaluated at a constant discharge. Many more data will be collected in the future.

## ANALYTICAL STUDY

The behavior of the lifter under various conditions can best be investigated using computer simulation. A computer program for each of the fourteen modes of operation mentioned before has been constructed. Several modes and many conditions have been studies; more will be done in the future.

An example of the analytical study is the investigation of the flow through lifters of various sizes for fixed head drop and lift. For instance, Table 1 gives the results for head drop, $H_2$, of 31 ft., and a head rise, $H_1$, of 21 ft. The connecting pipe diameter was varied from 1 to 5 ft. The tank diameter was varied with pipe diameter so as to yield a cycle time within the range of 20-30 seconds. The analysis was performed for both isothermal and adiabatic conditions. Note that the adiabatic condition gives better efficiencies. The efficiency used in Table 1 is defined as

$$E = \frac{V_2 \ (H_1 + H_2)}{(V_2 + V_3) \ H_2} \tag{1}$$

in which $V_2$ and $V_3$ are the volumes of water in each cycle passed through tanks 2 and 3, respectively; and $H_1$ and $H_2$ are respectively the rise and the drop of water going through the lifter as indicated in Fig. 1.

The efficiency E is the ratio of the power available for hydropower generation with and without the use of the lifter. Note that the efficiencies of the lifters given in Table 1 are between 60 and 70%. It is believed that when optimized, over 70% of efficiency will be possible. While the lifters with 1-ft-diameter pipes can deliver water at an average rate of 4 cfs only, the one with 5-ft-diameter pipes can deliver approximately 100 cfs. The delivery discharge is proportional to the square of the pipe diameter--a result that can be expected from the hydraulics of turbulent flow through pipes.

## CONCLUSION

The study shows that the behavior of the Cherepnov lifter can be predicted by solving a set of differential-algebraic equations on a computer. It also shows that there can be many different modes of operation and different designs of the lifter--each best suited for a special purpose.

When used for head augmentation in low-head hydropower, the intake tank and the two standpipes in the tank can be eliminated. Water can be piped into the delivery tank and the drain tank directly from a lake. This reduces cost and increases the efficiency of the lifter. The delivery pipe should be connected to either an elevated open storage tank (when topography allows), or a hermetic storage tank to produce an approximately steady discharge through the turbine.

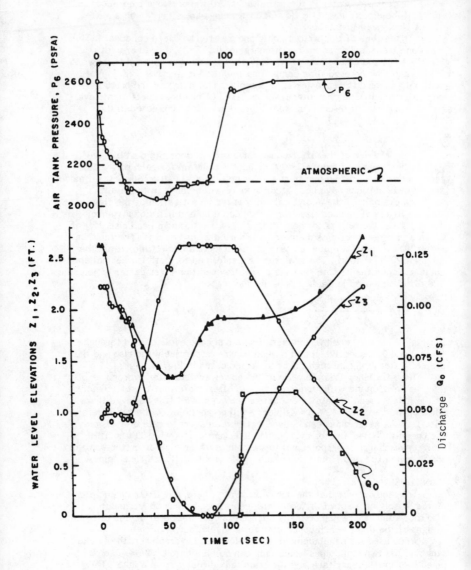

Fig. 3. EXPERIMENTAL RESULT OF LIFTER BEHAVIOR
($Q_i$=0.077 cfs; $H_1$=8.44 ft; $H_2$=11.2 ft)

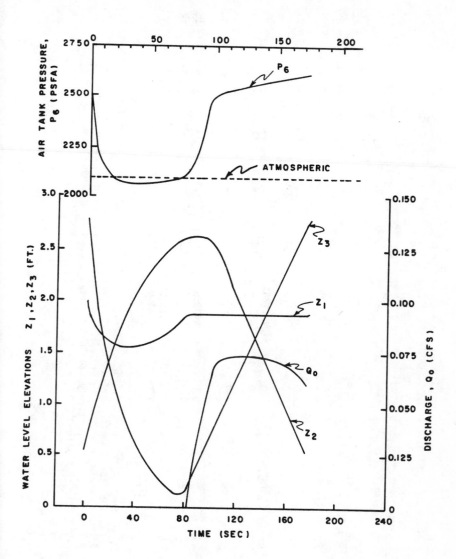

Fig. 4. COMPUTER PREDICTION OF LIFTER BEHAVIOR
($Q_i$=0.077 cfs; $H_1$=8.44 ft; $H_2$=11.2 ft)

TABLE 1. LIFTER PERFORMANCE UNDER MOD 8 ($H_1=21$ ft, $H_2=31$ ft, H=4 ft)

| Run No. | Operating Parameters | | | Performance Parameters | | | | | |
|---|---|---|---|---|---|---|---|---|---|
| | Pipe Diameter (ft.) | Tank Diameter (ft) | | V (ft³) | $V_2$ (ft³) | $V_3$ (ft³) | E (%) | Cycle Time (sec) | $Q_o$ (cfs) |
| | | Tank 2 | Tank 3 | | | | | | |
| 1A | 1.0 | 5.0 | 7.0 | 214.3 | 78.6 | 135.7 | 62 | 22.0 | 3.6 |
| 1B | 1.0 | 5.0 | 7.0 | 222.0 | 87.0 | 135.9 | 66 | 20.0 | 4.4 |
| 2A | 1.5 | 7.0 | 10.0 | 468.5 | 154.1 | 314.3 | 62 | 17.0 | 9.0 |
| 2B | 1.5 | 7.0 | 10.0 | 513.3 | 199.0 | 314.0 | 66 | 16.0 | 12.4 |
| 3A | 2.0 | 12.0 | 18.0 | 2211.0 | 809.0 | 1402.0 | 62 | 21.3 | 38.0 |
| 3B | 2.0 | 12.0 | 18.0 | 2313.0 | 908.0 | 1401.0 | 66 | 18.0 | 50.0 |
| 4A | 3.0 | 16.0 | 24.0 | 4138.9 | 1521.5 | 2617.3 | 63 | 22.0 | 69.3 |
| 4B | 3.0 | 16.0 | 24.0 | 4316.4 | 1699.0 | 2617.3 | 68 | 20.0 | 84.5 |
| 5A | 4.0 | 22.0 | 30.0 | 4405.5 | 1580.1 | 2825.5 | 62 | 20.0 | 79.0 |
| 5B | 4.0 | 22.0 | 30.0 | 4665.4 | 1839.3 | 2826.1 | 67 | 19.0 | 96.8 |
| 6A | 5.0 | 30.0 | 37.0 | 6756.0 | 2456.1 | 4300.8 | 62 | 21.0 | 106.9 |
| 6B | 5.0 | 30.0 | 37.0 | 6981.0 | 2680.0 | 4301.0 | 65 | 20.0 | 134.0 |

Note: 1. Runs 1A, 2A, etc. are for isothermal and 1B, 2B, etc. are for adiabatic conditions.
2. The pipe diameters given in Column 2 are the same for all pipes except the air pipe 6.
3. $H_1$ and $H_2$ are shown in Fig. 1; H (height of tanks) = 4.0 ft; V, $V_2$ and $V_3$ are respectively the total inflow, the flow through tank 2 and the flow through tank 3 during one cycle of operation.

Although this study has been focused on the application of the Cherepnov lifter to head augmentation in low-head hydropower, the same lifter studied can be used in other hydraulic practices where water or wastewater needs to be pumped. In fact, the economics of the lifter for pumping water seems to be even more promising than for hydropower. This is especially true if the use is in a remote area not served by utility. For instance, a small lifter constructed from three standard 55-gallon drums interconnected by 2-inch PVC pipes will cost only a few hundred dollars. Such a lifter consumes no electricity, and it can deliver more than 15 gpm or 21,600 gpd of water--enough to supply the need of several families. The economics of such a lifter, even in places served by utility, is apparent. They only serious limitation of the Cherepnov lifter is topography. At least several feet of drop is needed to make the lifter practical. This limits the use of the lifter either to dam sites or places where there are sudden drops in a creek. Such sites are more common in mountainous or hilly areas.

REFERENCES

1.    Arnovich, G.V. and Shtaerman, E. Ya, "On the Theory of the Cherepnov Water Lifter," Meklanika Zhidhosti i Gaza, Vol. 1, No. 1, Jan/Feb. 1966, pp. 176-178.

2.    Liu, H. and Fessehaye, M., "A Water Lifter That Needs No Energy or Fuel Supply," Proceedings of the 6th Annual UMR-DNR Conference on Energy, Rolla, Missouri, Oct. 1979, pp. 136-138.

3.    Liu, H., "Head Augmentation for Low-Head Hydropower--A Feasibility Study," Proceedings of Waterpower '81--International Conference on Hydropower, Washington, D.C., June 1981, 21 pages.

4.    Liu, H. et al, "Development of Cherepnov Lifter for Water Supply and Small Hydroelectric Power Generation," Proceedings of the 8th Annual UMR-DNR Conference on Energy, Rolla, Missouri, November 1981, 4 pages.

5.    Liu, H. and Fessehaye, M., Theoretical Analysis of the Cherepenov Water Lifter, U.S. Department of Energy Report No. DC-FC07-80ID-12206-1, December 1981, 55 pages.

# ANALYSIS OF RIVER ICE COVER ROUGHNESS

by Hung Tao Shen[1], M.ASCE, and Roger W. Ruggles[2], A.M.ASCE

ABSTRACT

A method for computing the undersurface roughness coefficient of a river ice cover is developed. The method requires that field data be available describing the river discharge, water surface elevation, and the areal extent of the ice cover. This method is capable of treating the bank effect and dealing with partially ice-covered conditions. The reduction of flow area due to the existence of ice cover can also be considered. Using this method, ice cover roughness coefficients for winters between 1974 and 1981 are calculated for a reach of the St. Lawrence River. The calculated results show that the Manning's coefficient of the ice cover can vary substantially from winter to winter. Each winter the roughness coefficient changes from its highest value at the beginning of the winter to its lowest value before the deterioration of the ice cover. The variation of the roughness coefficient during the winter is closely related to the size of the open water area and the ambient air temperature.

INTRODUCTION

Due to the severe winter climate, rivers in the higher latitudes are ice-covered each year for a period of several months. The existence of the ice cover has an important influence on the relation between discharge and water level in the river. The most important effect of the ice cover on channel flow is the additional resistance induced by the ice cover. The correct quantitative understanding of the undersurface roughness coefficient of ice cover is an essential element in the backwater analysis and the hydraulic transient modeling of ice-covered rivers. With the increasing needs in the development of hydropower generation and the need for inland navigation, the resistance characteristics of a river ice cover has recently drawn the interest of an increasing number of researchers (8). Methods for computing ice roughness coefficient from measured velocity profiles are available (3, 5,8,9). This type of analysis, which can be used for evaluating local roughness coefficients, is not adequate for the purpose of computing backwater profiles or hydraulic transient modeling where the gross effects of the ice cover on the Manning's n of a river reach is required. In a recent paper, Witherspoon (10) developed a backwater

---

[1]Associate Professor of Civil and Environmental Engineering, Clarkson College of Technology, Potsdam, N.Y.

[2]Hydraulic Engineer, Acres American, Inc., Buffalo, N.Y., formerly; Research Assistant, Clarkson College of Technology, Potsdam, N.Y.

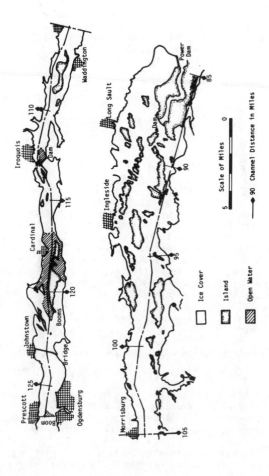

Figure 1.  Ice Cover on St. Lawrence River, February 3, 1979

model to analyze the resistance coefficient of ice cover in the St. Lawrence River. This method can provide accurate estimates of the roughness coefficient of ice cover in fully covered river reaches with relatively small ice thicknesses and negligible bank effects. In this paper, an analytical method is developed for determining Manning's roughness coefficients of the underside of a river ice cover. The method takes into consideration the effect of ice cover thickness, the bank effect, and the effect of partial ice cover over the width. Based on available data on flow and ice cover conditions, this method is applied to a reach of the St. Lawrence River between Cardinal and Iroquois Dam (Figure 1) for winters between 1974-1981.

ANALYTICAL TREATMENT

The unsteady one-dimensional equations of continuity and momentum in an ice-covered channel are (2,4):

$$\frac{\partial A}{\partial t} + \frac{\partial Q}{\partial x} = 0 \tag{1}$$

and

$$\rho \frac{\partial Q}{\partial t} + \rho \left[ \frac{2Q}{A} \frac{\partial Q}{\partial x} - \frac{Q^2}{A^2} \frac{\partial A}{\partial x} \right] = - \rho g A \frac{\partial}{\partial x} \left( z_b + d_w + \frac{P_{i/w}}{\gamma} \right) - \rho g A S_f \tag{2}$$

in which, $\rho$ = density of water; $Q$ = discharge; $A$ = flow area; $g$ = gravitational acceleration; $z_b$ = bed elevation; $d_w$ = depth of flow; $\gamma$ = specific weight of water; $S_f$ = frictional slope; and $P_{i/w}$ = pressure at the ice-water interface. The interfacial pressure, $P_{i/w}$, can be determined from the equation of motion of the ice cover. However, for normal winter flow conditions with a floating ice cover, the interfacial pressure can be approximated by the weight of the ice cover (1). Eq. 2 can be rewritten as:

$$\rho \frac{\partial Q}{\partial t} + \rho \left[ \frac{2Q}{A} \frac{\partial Q}{\partial x} - \frac{Q^2}{A^2} \frac{\partial A}{\partial x} \right] = - \rho g A \frac{\partial}{\partial x} \left( z_b + d_w + \bar{h}_i \right) - \left( p_i \tau_i + p_b \tau_b \right) \tag{3}$$

in which, the equivalent thickness of the ice cover, $\bar{h}_i$, is defined as:

$$\gamma \bar{h}_i = \gamma_i d_s + (1 - p) \gamma_i d_f + p \gamma_w d_f \tag{4}$$

in which, $\gamma_i$ = specific weight of ice; $d_s$ = thickness of the solid ice cover; $d_f$ = thickness of the frazil ice accumulation; and $p$ = porosity of the frazil ice accumulation. The boundary shear term in Eq. 3 can be expressed in terms of the flow and roughness parameters.

For the general case of partially covered cross-sections as shown in Figure 2, the shear resistance term in Eq. 3 can be written as:

$$p_i \tau_i + p_b \tau_b = p_i \tau_i + (p_b \tau_b)_\beta + (p_b \tau_b)_\alpha \tag{5}$$

in which, $\alpha$ and $\beta$ = subscripts which represent portions of flow areas under free surface and ice cover, respectively (Fig. 2).

In terms of Manning's roughness coefficients, Eq. 5 can be written as (6):

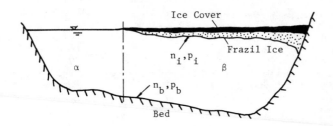

Figure 2.   Flow Cross-Section of Partially Covered Channel

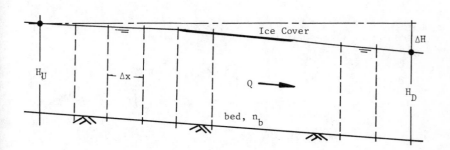

Figure 3.   Longitudinal Profile of a River Reach

$$P_i\tau_i + P_b\tau_b = \gamma n_b^2 [(\frac{Q^2 p^{4/3}}{A^{7/3}})_\beta + (1 + F')^{4/3}(\frac{Q^2 p_b^{4/3}}{A^{7/3}})_\alpha] \tag{6}$$

in which, the parameter F' is defined as

$$F' = (\frac{n_i}{n_b})^{3/2}(\frac{p_i}{p_b})_\alpha \tag{7}$$

where $n_i$ = Manning's coefficient of the ice cover and $n_b$ = Manning's coefficient of the channel bed. The partial discharges $Q_\alpha$ and $Q_\beta$ can be obtained from Eqs. 8 and 9 (6).

$$\frac{Q_\alpha}{Q} = \frac{1}{2}[\frac{A_\alpha R_\alpha^{2/3}}{AR} + (1 - \frac{A_\beta R_\beta^{2/3}}{AR^{2/3}})] \tag{8}$$

$$\frac{Q_\beta}{Q} = 1 - \frac{Q_\alpha}{Q} \tag{9}$$

Since both the free surface case and the fully covered case can be considered as special cases of the partially covered condition, Eq. 6 can be used as a general formula. Terms with the subscript $\beta$ reduce to zero for the fully covered condition and terms with the subscript $\alpha$ reduce to zero for the free surface condition.

For steady state flow, Eqs. 1 and 3 can be reduced to

$$\frac{\partial H}{\partial x} = \frac{1}{gA}[\frac{Q^2}{A^2} \frac{\partial A}{\partial x} - \frac{1}{\rho}(P_i\tau_i + P_b\tau_b)] \tag{10}$$

in which, $H = z_b + d_w + \bar{h}_i$ is the water surface level. By substituting Eq. 6 into Eq. 10, one can obtain:

$$\frac{\partial H}{\partial x} = \frac{Q^2}{gA^3} \frac{\partial A}{\partial x} - \frac{n_b^2 Q^2}{2.208 A^2}[(\frac{p_b}{A})_\alpha^{4/3}(\frac{Q_\alpha}{Q})^2(\frac{A}{A_\alpha}) + (\frac{p_b}{A})_\beta^{4/3}(\frac{Q_\beta}{Q})^2(\frac{A}{A_\beta})(1+F')^{4/3}] \tag{11}$$

The first term on the right-hand side is usually negligible for inland rivers. Eq. 11 can be rewritten in a different form as

$$\Delta H = r(Q)Q^2 \tag{12}$$

in which,

$$r(Q) = - \frac{n_b^2 \Delta x}{2.208 A^2}[(\frac{p_b}{A})_\alpha^{4/3}(\frac{Q_\alpha}{Q})^2(\frac{A}{A_\alpha}) + (\frac{p_b}{A})_\beta^{4/3}(\frac{Q_\beta}{Q})^2(\frac{A}{A_\beta})(1+F')^{4/3}] \tag{13}$$

Eqs. 12 and 13 can be used to determine the undersurface roughness coefficient of ice cover between two water level gaging stations based on flow and ice cover conditions. However, since A and p values vary along the channel with the water surface profile, which is a function of the unknown $n_i$, Eq. 12 is a nonlinear equation which cannot be solved explicitly. An iterative procedure similar to the one used in pipe flow analysis is developed in the following section.

## Determination of Roughness Coefficient

To solve Eq. 12 for $n_i$ using an iterative procedure, a reasonable initial guess is needed. For river reaches which consist of subreaches which are either fully open or fully ice covered as shown in Figure 3, the following difference equation can be obtained from Eq. 12.

$$H_U - H_D = \sum^o \frac{n_b^2 \Delta x \, Q^2}{2.208 \, A^2 R^{4/3}} + \sum^c \frac{n_b^2 \Delta x \, Q^2}{2.208 \, A^2} (\frac{P_b}{A})^{4/3} (1 + F)^{4/3} \tag{14}$$

in which, $H_U$, $H_D$ = water level at the upstream and the downstream ends of the reach, o and c refer to subreaches with free surface and ice covers, respectively. Assuming that the channel cross-sectional area in the reach can be approximated by the average of areas at the upstream and downstream ends, $\bar{A}$, Eq. 14 can be solved explicitly for $n_i$. This $n_i$ value can be used as the first approximation in the iterative procedure:

$$n_i = n_b \{ (\frac{\bar{P}_b}{\bar{P}_i}) [\{\frac{2.208 \, \bar{A}^2 \bar{R}_o^{4/3} (H_U - H_D)}{Q^2 \, n_b^2 \sum^c \Delta x} - \frac{\sum^o \Delta x}{\sum^c \Delta x}\}^{3/4} - 1]\}^{2/3} \tag{15}$$

in which $\bar{R}_o = \bar{A}/\bar{P}_b$, and $\bar{P}_i$ = average width of the channel at the ice-water interface. A similar yet more cumbersome expression can also be obtained to account for partially covered conditions. Since it was found that the iterative scheme will converge to the correct solution even with a wrong initial guess, Eq. 15 will also be used for partially covered conditions.

With the approximate $n_i$ value calculated from Eq. 15, the drop in water surface level in each subreach can be obtained from Eq. 12.

$$-\Delta H = \frac{n_b^2 \, Q^2 \, \Delta x}{2.208 \, A^2} g(n_i) \tag{16}$$

in which,

$$g(n_i) = (\frac{P_b}{A})_\alpha^{4/3} (\frac{Q_\alpha}{Q})^2 (\frac{A}{A_\alpha}) + (\frac{P_b}{A})_\beta^{4/3} (\frac{Q_\beta}{Q})^2 (\frac{A}{A_\beta}) (1 + F')^{4/3} \tag{17}$$

Using Eq. 16 and the measured water level, $H_U$, at the upstream station, the water level at the downstream station can be calculated. If the calculated downstream level $H_D^c$ does not equal the measured value, $H_D$, then modification on the calculated $n_i$ value is required. Let $\Delta n_i$ be the correction required in order to eliminate the difference $\Delta H^c = H_D^c - H_D$, then

$$\Delta H^c = - \sum \frac{\Delta x \, n_b^2 \, Q^2 \, \Delta x}{2.209 \, A^2} \Delta n_i [\frac{dg(n_i)}{dn_i}] \tag{18}$$

or

$$\Delta n_i = \frac{n_i \Delta H^c}{2 \; \Sigma \left[ G \; F' \; (1+F')^{1/3} \right]} \frac{}{\Delta x}$$

(19)

in which,

$$G = \frac{n_b^2 \; Q^2}{2.208 \; A^2} \frac{\Delta x}{\Delta x} \left[ \left(\frac{P_b}{A}\right)^{4/3} \left(\frac{Q_\beta}{Q}\right)^2 \left(\frac{A}{A_\beta}\right) (1 + F')^{4/3} \right]$$

(20)

The repetitive use of Eqs. 19 and 20, until $H_D^c$ converges to $H_D$, will give the correct value of the roughness coefficient $n_i$.

## APPLICATION TO THE ST. LAWRENCE RIVER

The method developed in the preceeding section is applied to a reach in the St. Lawrence River between water level gaging stations at Cardinal, Ontario, and the Iroquois Control Dam (Figure 1). This reach is located downstream of Galop Island. Due to the large flow velocity and the installation of the Galop ice booms, a major portion of the Galop Island reach remains open each winter. With the existence of open-water areas, a large amount of frazil ice is accumulated underneath the ice cover to form hanging ice dams. Since detailed hanging dam profiles are not available, the reduction in channel cross-section area due to hanging dams are not included in the computation of $n_i$. The effect of this reduction in flow area of flow resistance is effectively accounted for as a part of the form loss contributed to the roughness coefficient, $n_i$. The reduction in flow area due to the solid ice sheet is included in the computation. Ice thickness data obtained by the St. Lawrence Seaway Authority, Canada, are used. The areal extent of ice cover is obtained from aerial photographs provided by the St. Lawrence Seaway Development Corporation (7).

Daily roughness coefficients $n_i$ for winters between 1974-1981 are calculated based on mean daily discharge measured at the Iroquois Dam and level data obtained from the Power Authority of the State of New York. Detailed flow, ice cover conditions, and water temperature data are given in Ref. 7. Cross-sectional geometries are obtained from Ref. 4 and summarized in Table 1. It is recognized that $n_b$ for alluvial rivers may vary with flow conditions. Since analysis of flow data in the study reach for periods before the formation of the ice cover and after the disappearance of the ice cover indicated the

Table 1.   Cross-Sectional Geometry at the
Reference Water Level

| Parameter | Cardinal | Iroquois H.W. |
|---|---|---|
| Reference Water Level (IGLD 1955) | 240.80 ft. | 239.90 ft. |
| Width | 2620 ft. | 2620 ft. |
| Wetted Perimeter, $P_b$ | 2690.916 ft. | 2683.1298 ft. |
| Cross-sectional Area | 92900 sq. ft. | 82700 sq. ft. |

variation of $n_b$ is relatively small, a constant $n_b$ value, equal to 0.026 obtained from Red. 4, is used.

A simple computer program is developed to perform the numerical computation. Calculated values of $n_i$ for winters between 1974 and 1981 are presented in Figure 4. This figure shows that $n_i$ values vary between 0.055 to 0.015. The roughness coefficient can vary over a wide range from one winter to another. During each winter the roughness coefficient generally changes from its highest value at the beginning of the ice covered period to its lowest value before the deterioration of the ice cover. In a recent unpublished study, it is found that the time-dependent variation of the ice cover roughness coefficient can be considered as a function which is a linear combination of a monotonically decreasing function during the ice covered period and a fluctuation component which is a function of the size of the open-water area upstream, the air temperature, and the flow velocity.

SUMMARY AND CONCLUSION

In this study an analytical method is developed for analyzing the roughness coefficient of the underside of river ice covers. This method, which takes into consideration the reduction of flow cross-sectional area due to the presence of ice cover and the existence of the roughness coefficient then existing methods. The method is applied to a reach of the St. Lawrence River between Cardinal and Iroquois Control Dam for six winters. The calculated results show that the Manning's roughness coefficient of the ice cover in this reach decreases from its highest value during the formation period to a lowest value before the deterioration period. The variation of $n_i$ during the winter is effected by the variation of air temperature which governs the variation of frazil ice production in the open-water area upstream of the ice cover in this reach.

ACKNOWLEDGEMENTS

This study is supported by the Great Lakes Environmental Research Laboratory, NOAA, Ann Arbor, Michigan, under contract No. NA80RAC0014.

REFERENCES

1. Hirayama, K., "Characteristics of Ice Covered Streams in Conncetion With Water Discharge Measurements," IAHR Ice Symposium, Vol. 2, Lulea, Swedan, Aug. 1978.

2. Landry, S.J., "Frazil Ice Transport and Hanging Dam Formation," M.S. Thesis, Clarkson College of Technology, Potsdam, NY, 1979.

3. Larsen, P.A., "Hydraulic Roughness of Ice Covers," Journal of the Hydraulics Division, ASCE, Vol. No. HY1, Proc. Paper 9498, Jan. 1973.

4. Potok, A.J., "Upper St. Lawrence River Hydraulic Transient Model," NOAA Technical Memorandum, ERL GLERL-24, Ann Arbor, Mich., 1978.

5. Pratte, B.D., "Review of Flow Resistance of Consolidated Smooth and Rough Ice Covers," Canadian Hydrology Symposium: 79, Proceedings, National Research Council of Canada, May 1979.

6. Shen, H.T., and Ackermann, N.L., "Wintertime Flow Distribution in River Channels," Journal of Hydraulics Division, ASCE, Vol. 106, No. HY5, May 1980.

7. Shen, H.T., Ruggles, R.W., "Ice Production in the St. Lawrence River Between Ogdensburg and Massena," Report No. DTSL55-80-C-CO330-A, U.S. Department of Transportation, Washington, D.C., Aug. 1980.

8. Tsang, G., and Beltaos, S., Proceedings of Workshop on Hydraulic Resistance of River Ice, Environment Canada, Burlington, Canada, 1980.

9. Uzuner, M.S., "The Composite Roughness of Ice Covered Streams," Journal of Hydraulic Research, Vol. 13, No. 1, Jan. 1975.

10. Witherspoon, D.F., "Hydraulic Resistance of Ice Cover in the International Rapids Section of the St. Lawrence River," Proceedings of Workshop on Hydraulic Resistance of River Ice, Environment Canada, Burlington, Canada, 1980.

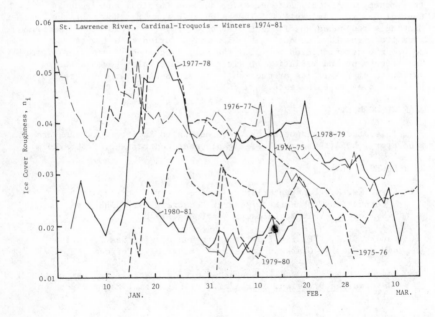

Figure 4. Variations of Manning's Roughness Coefficient of Ice Cover

BARGE TOWS AND SUSPENDED SEDIMENT IN LARGE RIVERS

by Nani G. Bhowmik[1] and J. Rodger Adams[2], Members, ASCE

ABSTRACT:  Tow traffic in inland waterways resuspends sediment and
moves the sediment in a lateral direction.  An extensive investigation
of the effects of tow traffic on the movement and suspension of
sediments in the Illinois and Mississippi Rivers was conducted.  Depth-
integrated suspended sediment samples were collected for a total of 41
tow passage events from both the rivers.  Data were collected when the
ambient suspended sediment concentrations were low (about 100 ppm) and
also when high (about 300 to 500 ppm).  Analyses of the data indicated
that the suspended sediment concentrations are increased by as much as
20 percent for up to 90 minutes after a tow passes the sampling
transect.  The concentration increase is more pronounced in the channel
border areas than in the tow wake.  The increased concentration is more
significant when the ambient suspended sediment concentration is low.
It was also observed that successive tow passages at time intervals of
less than 90 minutes result in extended periods of increased sediment
concentration.

INTRODUCTION

     Depth-integrated suspended sediment samples were collected
following barge tow passage on the Illinois River near Hardin, Illinois
and on the Mississippi River near Mozier, Illinois, to measure the
effect of commercial navigation on suspended sediment.  This is the
first use of U.S. standard sediment samplers for a navigation effects
study.
     Previous studies by Karaki and VanHoften (6) and Link and
Williamson (7) obtained qualitative data from infrared aerial
photographs.  Johnson (5) obtained suspended sediment and turbidity
data using a pumped sampler on the Illinois and Upper Mississippi
Rivers.  He concluded that tow traffic increases the suspended sediment
concentration and that the resuspended sediments were transported
laterally.  Analysis of the lateral gradation of bed material in the
Illinois River (2) and the Kaskaskia River (4) shows that fine sand,
silt, and clay are not common in the navigation channel, but do occupy
the river bed in channel border areas.  Resuspension and lateral
movement by barge tows contribute to this bed material sorting.

     The investigation summarized here has been reported in detail by
Bhowmik, et al. (1).  Field procedures are described, typical data are

---

[1]Principal Scientist, Illinois State Water Survey, Champaign, IL
[2]Associate Hydrologist, Illinois State Water Survey, Champaign, IL

presented, and a statistical analysis of tow passage events is
presented here.


FIELD DATA COLLECTION

   **Surveying.** The survey team was responsible for: (1) determining the track and the average velocity of each tow, (2) developing a
means of repositioning the sampling boats precisely and quickly, and
(3) developing appropriate communications systems.

   A site survey defined the shorelines adjacent to the sampling area
and the locations of the three suspended sediment sampling stations.
Tow tracking began as soon as the entire length of the vessel was
visible from both survey stations (see Fig. 1). Each theodolite was
zeroed on the opposite station, and horizontal angles were measured
simultaneously from each station to a previously agreed-upon point on
the tow. An angle to the bow from each station was measured and, 30
seconds later, an angle to the stern from each station was measured. A
running account of the tracking operation and the angular data was
recorded on microcassette recorders and later transcribed.

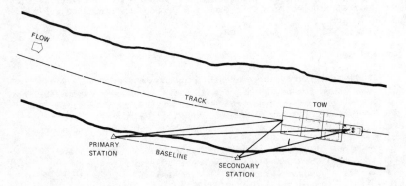

Figure 1.   Tow Tracking Procedure


   **Sampling Boat Positioning.**--The scheme shown in Fig. 2 was used to
allow the sampling boats to reoccupy the sampling sites several times
each day without tedious direction from the survey stations. Two posts
were erected on shore about one hundred feet apart and on the sample
transect line, which was near the primary survey station. Boat C was
located in the main channel, boat A was in shallower water (the channel
border area), and boat B was located midway between boats A and C. At
some convenient distance upstream, another index post was erected in a
clear area on shore. Between this post and the water, three color-
coded foresite posts were set in positions which aligned with each of
the sampling boats. This part of the system had a delta shape. Each
boat could be maneuvered to its sampling position at the intersection
of the transect line and its color-coded line.

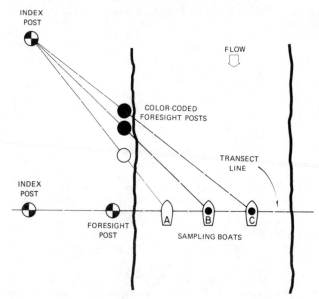

Figure 2.   Delta Post Positioning System

**Suspended Sediment Sampling.** The DH-59 sampler was used where water depths were less than 18 feet.  The P-61 and P-72 samplers were used in boats B and C where depths were over 18 feet.  The samplers and the standard operating procedures are described by Guy and Norman (3).

Sampling during Tow Passage Events.--When the bow of a barge tow crossed the study transect, boat A gave a signal to initiate an event. Boat A was usually in position on the transect when the tow passed. Thus, boat A began sampling immediately, while boats B and C could not begin sampling until the tow had passed and they could safey maneuver into position.  During the first thirty minutes after a tow had passed, depth-integrated suspended sediment samples were taken every two minutes.  For the next thirty minutes, samples were taken at five-minute intervals.  During the last thirty minutes of the event, a sample was collected very ten minutes.  If another tow arrived, the sampling routine was restarted at two-minute intervals and continued until 90 minutes after the second tow had passed.

RESULTS OF SUSPENDED SEDIMENT SAMPLING

**Illinois River.** Data were collected in July and September 1980 at Hadley's Landing, 13.2 miles (21.2 km) above the confluence of the Illinois and Mississippi Rivers.  The river is 800 feet (245 meters) wide with a maximum depth of 18 feet (5.5 meters) at normal

pool elevation. A depth of 15 feet (4.6 meters) or more for safe
navigation occupies about 600 feet (185 meters) of the total width.
The main channel bed material is sand with a median diameter of 0.019
inches (0.45 mm). The bed material at one-half depth is sand, with a
median diameter of 0.035 inches (0.90 mm), on the outside of the bend,
and silt, with a median diameter of 0.0015 inches (0.038 mm), on the
inside of the bend (2). Fifteen different towboats were observed in
sixteen events. Towboat horsepower ranged from 1530 (1140 kw) to 5850
(4360 kw) and averaged 2919 (2180 kw). Speeds averaged 4.3 mph (6.9
km/h) for upbound tows and 5.0 mph (8.0 km/h) for downbound tows. The
average number of barges per tow was 10.6.

Typical plots of suspended sediment concentration in ppm versus
time are shown in Fig. 3. The suspended sediment concentration is much
more variable and generally above the background values for 30 to 90
minutes following tow passage. Successive tow passages extend the time
when concentrations are higher and more variable. Tow tracks were
determined for 14 of 16 events at Hadley's Landing. Distance from
shore to the tow centerline varied between 300 and 560 feet (90 and 170
meters), with an average of 457 feet (139 meters). Just upstream, tows
came considerably closer to the shore on the outside of the sharp bend
at the north end of Twelve Mile Island.

Two geometric parameters are very important for navigation in
constricted waterways: the blocking factor and the depth-to-draft
ratio. The blocking factor is the ratio of the channel cross-sectional
area to the submerged cross-sectional area of the vessel. The blocking
factor varied between 12 and 56. Depth-to-draft ratios were between 2
and 9.

**Mississippi River.** Data were collected in April 1981 at Rip
Rap Landing, river mile 265.1 (426.6 km), and in May 1981 at Mozier
Landing, river mile 260.2 (418.8 km). The river channel is similar at
the two sites. Both are on bends. The channel width is about 1600
feet (490 meters) and the maximum depth is about 35 feet (10 meters).
The channel deepens rapidly on the left bank (outside of the bend),
remains over 15 feet (4.6 meters) deep for about 750 feet (230 meters),
and has a border area about 800 feet (245 meters) wide with depths less
than 10 feet (3 meters). The bed material is medium sand with a median
diameter of 0.017 inches (0.42 mm).

Nineteen events involving 23 tow passages were observed on the
Mississippi River. Towboat horsepower ranged from 1275 (950 kw) to
6450 (4810 kw) and averaged 3773 (2815 kw). Speeds averaged 5.5 mph
(8.9 km/h) for upbound tows and 8.8 mph (14.2 km/h) for downbound tows.
The average number of barges per tow was 12.0.

Fig. 4(a) shows six events in one day at Rip Rap Landing.
Fig. 4(b) shows four widely-spaced events at Mozier Landing. This was
a time of rapidly dropping stage and discharge, and there is a definite
difference in background concentrations: an average of 341 ppm at
station A, 397 ppm at station B, and 510 ppm at station C.

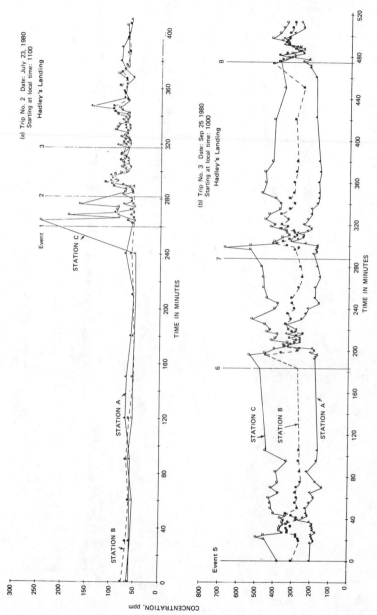

Figure 3.    Typical Suspended Sediment Data on the Illinois River

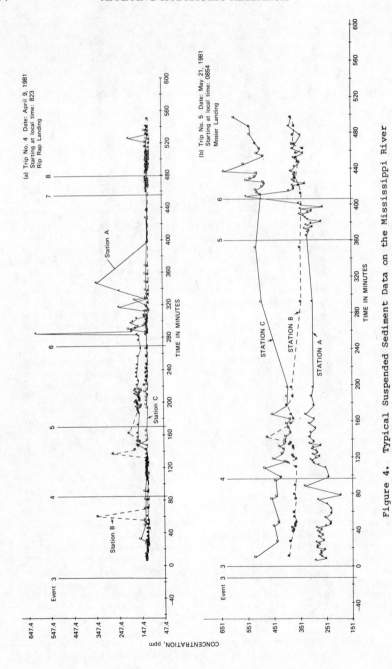

Figure 4. Typical Suspended Sediment Data on the Mississippi River

At the Rip Rap Landing transect, the average distance from shore to the tow track was 350 feet (101 m) and varied from 75 feet (23 m) to 700 feet (213 m). At the second survey station, the tows were turning away from the shore and the distance to the sailing line averaged 460 feet and varied from 300 feet (91 m) to 650 feet (198 m). The blocking factor varied between 25 and 114 and was 21 for a double tow passage event. Depth-draft ratios were between 4 and 18.

At Mozier Landing, the sailing line distance at the transect varied from 350 feet (107 m) to 960 feet (293 m) and averaged 730 feet (223 m). At the secondary survey station the sailing line distance varied from 400 to 725 feet (122 to 221 m) and averaged 590 feet (180 m). The blocking factor varied between 34 and 150 and the depth-draft ratio varied between 3 and 18.

**Statistical Analysis.** The mean and standard deviation were calculated for the background suspended sediment concentration at each sampling location. The same statistics were also calculated for each tow passage event in four different ways: (1) all samples from the event, (2) 90-minute time periods, (3) 60-minute time periods, and (4) until the concentration returned within one standard deviation of the ambient. The last three schemes combine closely spaced events.

For the five events in July 1980 at Hadley's Landing, the daily average background suspended sediment concentration was about 65 ppm. The incremental increase in average concentration during an event was: station A--maximum 34 ppm, minimum 6 ppm, average 19 ppm; station B--maximum 25 ppm, minimum -4 ppm, average 15 ppm; and station C--maximum 68 ppm, minimum 14 ppm, average (excluding 68 ppm value) 19 ppm. The ratio of the mean concentration during an event to the background concentration varied from 0.93 to 1.49. The average ratio was 1.28 at station A, 1.24 at station B, and 1.28 at station C. For the conditions during this trip a ratio of event to background concentration of about 1.25 and an increase in concentration of about 16 ppm were representative of the effect of tow traffic on suspended sediment.

During September 1980, the background sediment concentration was much higher than in July and the trip averages were 192 ppm for station A, 267 ppm for station B, and 348 ppm for station C. The incremental increase in average concentration during events was: station A--maximum 92 ppm, minimum 6 ppm, average 38 ppm; station B--maximum 64 ppm, minimum 8 ppm, average 32 ppm; and station C--maximum 95 ppm, minimum 38 ppm, average 28 ppm. The ratio of event to background concentration varied from 0.90 to 1.48. This ratio averaged 1.18 at station A, 1.11 at station B, and 1.07 at station C. The background concentration was highest at station C in the center of the navigation channel, and lowest at station A in the channel border. The effect of tow passage on suspended sediment concentration was greatest at station A and least at station C. The comparisons of events and background means by the t test are given in Table 1 for the Illinois River suspended sediment data. These results indicate the variability of suspended sediment data.

At Rip Rap Landing, the average background concentration was about 118 ppm. The incremental increase in average concentrations during a

TABLE 1.--Results of t Test of Suspended Sediment Events
on the Illinois River

| Event | Station A | | Station B | | Station C | |
|---|---|---|---|---|---|---|
| | t | P* | t | P | t | P |
| (1) | (2) | (3) | (4) | (5) | (6) | (7) |
| (a) Trip 2, July 23-24, 1980 | | | | | | |
| 4 | 1.67 | .20 | 2.91 | .01 | 1.11 | - |
| 5 | 1.28 | .20 | 3.62 | .01 | 2.35 | .01 |
| (b) Trip 3, September 23-26, 1980 | | | | | | |
| 1 | 2.16 | .05 | 3.44 | .01 | 2.75 | .05 |
| 2 | 1.28 | - | 0.89 | - | 4.04 | .01 |
| 3 | 1.73 | .05 | 0.83 | - | 0.45 | - |
| 4 | 4.30 | .01 | 4.88 | .01 | 0.92 | - |
| 5 | 0.33 | - | 0.46 | - | 0.16 | - |
| 6 | 1.52 | .20 | 0.63 | - | 0.99 | - |
| 7 | 0.72 | - | 1.40 | .20 | 0.47 | - |
| 8 | 3.23 | .01 | 1.60 | .20 | 1.47 | .20 |
| 9 | 0.46 | - | 2.22 | .05 | 1.12 | - |

* Probability level at which the event mean is
significantly different from the background mean.

tow passage event was: station A--maximum 67 ppm, minimum 4 ppm,
average 26 ppm; station B--maximum 40 ppm, minimum -3 ppm, average
12 ppm; and station C--maximum 12 ppm, minimum -22 ppm, average -1 ppm.
The ratio of event to background mean concentration varied from 0.82 to
1.58. This ratio averaged 1.22 at station A, 1.11 at station B, and
1.00 at station C.

Data collection at Mozier Landing took place during a period of
high water and sediment flows. The incremental increase in mean
suspended sediment concentration during events was: station A--maximum
73 ppm, minimum -39 ppm, average 18 ppm; station B--maximum 41 ppm,
minimum -29 ppm, average 11 pm; and station C--maximum 59 ppm, minimum
-69 ppm, average 3 ppm. The ratio of event to background mean concentrations ranged between 0.87 and 1.24. This ratio averaged 1.04 at
station A, 1.03 at station B, and 1.00 at station C. At station A the
average ratio of means corresponded to an increase of 14 ppm if the
background concentration was 340 ppm. On the first day, the average
ratio corresponded to an increase of 14 ppm at station B for an event.
On the second and third days, this increase was 11 ppm at station B.
The average ratio of 1.00 at station C did not increase the mean
concentration during an event.

The results of the statistical comparison of event and background
means for the Mississippi River suspended sediment data are given in
Table 2. The results are similar to those on the Illinois River.

TABLE 2.--Results of t Test of Suspended Sediment Events
on the Mississippi River

| Event | Station A | | Station B | | Station C | |
|---|---|---|---|---|---|---|
| | t | P* | t | P | t | P |
| (1) | (2) | (3) | (4) | (5) | (6) | (7) |
| (a) Trip 4, Rip Rap Landing, April 8-10, 1981 | | | | | | |
| 3 | 2.18 | .05 | 0.62 | - | 2.36 | .05 |
| 4 | 2.29 | .05 | 2.27 | .05 | 0.67 | - |
| 5 | 16.49 | .01 | 3.31 | .01 | 0.83 | - |
| 6 | 1.40 | - | 0.86 | - | 0.66 | - |
| 9 | 1.95 | .10 | 1.92 | .10 | 1.96 | .10 |
| 10 | 2.25 | .05 | 2.04 | .10 | 2.59 | .05 |
| (b) Trip 5, Mozier Landing, May 20-22, 1981 | | | | | | |
| 3 | 3.52 | .01 | 2.85 | .05 | 0.15 | - |
| 4 | 0.08 | - | 2.46 | .05 | 2.77 | .05 |
| 6 | 2.75 | .10 | 0.36 | - | 2.38 | .05 |
| 7 | 2.47 | .05 | 1.24 | - | 0.76 | - |
| 8 | 2.02 | .10 | 0.95 | - | 4.64 | .01 |
| 9 | 1.50 | .20 | 2.46 | .05 | 6.50 | .01 |

* Probability level at which the event mean is
significantly different from the background mean.

CONCLUSIONS

New data on the effects of commercial tow traffic on suspended
sediment were obtained on both the Illinois and Mississippi Rivers
during periods with ambient suspended sediment concentrations at low
(about 100 ppm) and high (300 to 500 ppm) levels.

The effect of tow passage on suspended sediment concentrations is
highly variable in time and space. An increase in suspended sediment
was observed in most events, though statistical analyses were
inconclusive. The data support the general validity of the hypothesis
that sediment is moved in a lateral direction. Specific points are:

(1) Tow passage increases suspended sediment concentrations.

(2) The increase in concentration is greater in channel border
areas than in the navigation channel.

(3)  The increase is more significant when the ambient suspended sediment concentration is low.

(4)  The concentration is increased for 60 to 90 minutes after a tow passes.

(5)  The effects of tow passage are greater on the Illinois River than on the Mississippi River. This is consistent with the differences in channel dimensions.

REFERENCES

1.  Bhowmik, N., Adams, R., Bonini, A., Guo, C-Y., Kisser, D., and Sexton, M., "Resuspension and Lateral Movement of Sediment by Tow Traffic on the Upper Mississippi and Illinois Rivers," Contract Report 269, Illinois State Water Survey, Sept., 1981.

2.  Bhowmik, N., Schnepper, D., Hill, T., Hullinger, D., and Evans, R., "Physical Characteristics of Bottom Sediments in the Alton Pool, Illinois Waterway," Contract Report 263, Illinois State Water Survey, July 10, 1981.

3.  Guy, Harold P., and Norman, Vernon W., "Field Methods for Measurement of Fluvial Sediment," Techniques of Water-Resources Investigations of the United States Geological Survey, Bood 3, Chapter C2, United States Government Printing Office, Washington, D.C., 1970.

4.  Herricks, E.E., and Gantzer, C.J., "Effects of Barge Passage on the Water Quality of the Kaskaskia River," Environmental Engineering Series No. 60, Civil Engineering Studies, University of Illinois, Urbana, Illinois, December 1980.

5.  Johnson, J.H., "Effects of Tow Traffic on the Resuspension of Sediments and on Dissolved Oxygen Concentrations in the Illinois and Upper Mississippi Rivers under normal Pool Conditions," Technical Report Y-76-1, U.S. Army Engineer Waterways Experiment Station, Environmental Effects Laboratory, Vicksburg, Mississippi, 1976, 181 pp.

6.  Karaki, S., and VanHoften, J., "Resuspension of Bed Material and Wave Effects on the Illinois and Upper Mississippi Rivers Caused by Boat Traffic," CER 74-75SKJV9, Colorado State University, Ft. Collins, Colorado, Nov. 1974.

7.  Link, L.E., Jr., and Williamson, A.N., Jr., "Use of Automated Remote Sensing Techniques to Define the Movement of Tow-Generated Suspended Material Plumes on the Illinois and Upper Mississippi Rivers," Army Engineer Waterways Experiment Station, Vicksburg, Mississippi, 1976, 71 pp.

WAVES GENERATED BY RIVER TRAFFIC

by Nani G. Bhowmik[1] and Misganaw Demissie[2], Members, ASCE

ABSTRACT: An excessive amount of bank erosion along a number of waterways in Illinois and surrounding states exists at the present time.
Erosion of stream banks attracts public attention because it causes
property losses and degrades water quality along stream courses. Among
the main causes of bank erosion along navigable rivers are waves generated by river traffic.

To investigate and collect data on waves generated by river traffic
on the Illinois and Mississippi Rivers, six field trips were taken to
four test sites. The maximum wave heights measured in the field ranged
from a low of 0.1 foot to a high of 1.08 foot.

The measured maximum wave heights were compared to those expected
on the basis of existing predictive equations and the correlations were
found to be low. Multivariate regression analyses for correlation
between the measured values and the important hydraulic and geometric
parameters which were felt to influence the generation of waves resulted
in an equation which predicts maximum wave heights fairly well.

INTRODUCTION

An excessive amount of bank erosion along a number of waterways in
Illinois and surrounding states exists at the present time. Along some
reaches of the Illinois River, it is estimated that 75 percent of the
banks are being eroded away by waves generated by river traffic and wind
(3). Similar types of bank erosion problems also exist along the
Mississippi and Ohio Rivers. Erosion of stream banks reduces property
value, results in permanent loss of real estate, increases turbidity of
streams, and accelerates the silting of reservoirs or backwater lakes
along the stream course. Among the main causes of bank erosion along
navigable rivers are waves generated by river traffic and wind. In
order to prevent the erosion of stream banks by waves, an understanding
of the characteristics of the waves generated by river traffic and wind
is needed to evaluate their impacts on the shoreline.

To investigate and collect data on waves associated with river
traffic on the Illinois and Mississippi Rivers, six field trips were
taken to four study sites. Wave data were collected for a total of 59

[1]Principal Scientist, Illinois State Water Survey, Champaign, IL 61820
[2]Associate Professional Scientist, Illinois State Water Survey,
Champaign, IL 61820

events.  The characteristics of the tows, such as length, width, draft, and speed were also determined during each event.

This paper discusses the instrumentation used during field data collection, the data collection procedures, and the wave data obtained.


BACKGROUND

As a vessel moves on or near the free surface of a water body, it generates a disturbance in the flow field.  The flow around the hull of the vessel is accelerated due to changes both in magnitude and direction.  The flow in front of the bow is decelerated until it reaches the stagnation point (where the velocity is zero) at the bow because of the blockage of the flow area by the vessel.  These accelerations and decelerations result in corresponding changes in pressure and thus water level elevation.  In areas where the flow is accelerated, the pressure and thus the water level elevation drops, and vice versa.  Waves are generated at the bow, stern, and any points where there are abrupt changes in the vessel's hull geometry to cause disturbance in the flow field.  As the vessel moves forward with respect to the water, the energy transferred to the water from the vessel generating the disturbance is carried away laterally by a system of waves.  In general the system of waves will consist of two sets of diverging waves and one set of transverse waves.  The diverging waves move forward and out from the vessel, while the transverse waves move in the direction of the vessel.  The transverse waves meet the diverging waves on both sides of the vessel along two sets of lines called the cusp lines which form a 19°21' angle with the sailing line for a point disturbance moving at a constant velocity in an initially still, deep, and frictionless fluid (9).  Sorensen (9) has shown that the general wave pattern generated by a model hull in deep water agrees well with the wave pattern described above except for a small change in the cusp angle.

Since waves are generated both at the bow and stern of a vessel, they interact with each other at some distance from the vessel.  If the waves generated at the bow and stern are in phase, i.e., if the crest and trough of one set coincide with the other, they tend to reinforce each other, resulting in higher waves.  If the waves are out of phase, they tend to cancel each other, resulting in relatively smaller waves.

In deep water the wave heights generally increase with increasing velocity, except at certain velocities where the bow and stern waves tend to cancel each other.  The wave heights then decay with distance from the vessel as the total energy per wave is distributed over a larger area (2,6,7,8,9).

When the channel is narrow in the lateral dimension so as to affect the flow pattern around a vessel, the waves generated will be higher than those generated in unrestricted waters by the same vessel moving at the same speed.  This is because of a significant reduction in the flow area and the associated higher accelerations of flow around the vessel.  If in addition to being narrow the channel is shallow, the combined effect will result in more complex flow conditions and much higher wave heights (9).

From laboratory and field observations, some investigators have
developed empirical equations for predicting river traffic-generated
wave heights based on channel and vessel parameters.  Balanin and Bykov
(1) used the vessel velocity and a modified blockage factor as the
primary variables to develop the following equation for estimating the
wave height in the vicinity of a ship.

$$H = 2.5\left(\frac{V^2}{2g}\right)\left[1 - \left(1 - 1\bigg/\!\left(4.2 + \frac{A_c}{A_m}\right)^{0.5}\right)\left(\frac{A_c/A_m - 1}{A_c/A_m}\right)^2\right] \tag{1}$$

where H = wave height in feet; V = vessel velocity in ft/sec; g = gravitational acceleration in ft/sec$^2$; $A_c$ = the cross-sectional area of
the channel; $A_m$ = bxD = the submerged cross-sectional area of the
vessel in square feet; b = the width of the vessel in feet; and D = the
draft of the vessel in feet

Another equation for estimating maximum wave height is given by
Hochstein in Ref. 10 as follows:

$$H_{max} = 0.0448 \ V^2 \left(\frac{D}{L}\right)^{0.5}\left(1 - \frac{A_m}{A_c}\right)^{-2.5} \tag{2}$$

where $H_{max}$ = maximum wave height in feet and L = length of the
vessel in feet.  All other variables are as defined before.

The main difference between the two equations other than their form
is the inclusion of the vessel's length in Eq. 2.

INSTRUMENTATION AND DATA COLLECTION

Two different systems were utilized to measure wave height.  The
first system was a staff gage and a movie camera, and the second system
was an electronic wave gage connected to a mini-computer.

In the first system, a 5-ft (1.5 m) staff gage was installed in the
river.  A movie camera located on shore was then used to take pictures
of the water surface elevations at the staff gage during an event.  The
wave heights were read from the movie films with the aid of either a
movie editor or projector.

The second system included a new electronic wave gage, with sensors
exposed 0.05 ft (1.56 cm) apart, a mini-computer, and an interface
between the wave gage and the computer.  The wave gage is connected to
the interface by a cable.  The interface in turn is connected to the
computer.  A master program which controls the sampling frequency and
duration is read into the computer, which is activated during an event.
The wave data are finally stored on cassette tapes for further processing and analysis.

The cross section and velocity data at the test sites were taken
according to the procedure described by Buchanan and Somers (5) for

stream gaging.  Velocities were measured at 0.2 and 0.8 of the depth at each vertical.

During each event all the pertinent information, such as the vessel type, size, draft, distance from shore, and direction of movement, was recorded on data sheets.  The distance of vessels from the shore was measured by theodolite or range finders.  The speed of vessels was determined by determining the time it takes the vessels to cross two cross sections which are spaced a known distance from each other.  A detailed description of the instruments and the data collection procedure is given in Ref. 4.

DISCUSSION OF DATA

Summary of Field Trips

There were a total of six field trips to collect wave data.  Four of the field trips were to the Illinois River, and two were to the Mississippi River.  The locations of the test sites on the Illinois and Mississippi Rivers are shown in Fig. 1.

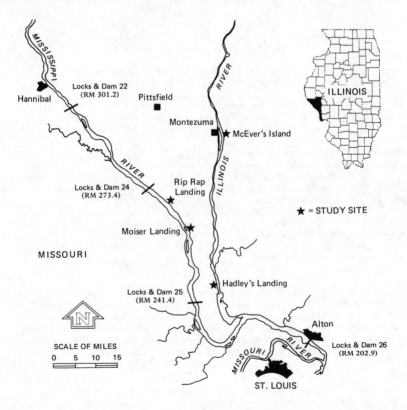

Fig. 1.  Location of study sites on the Mississippi and Illinois Rivers

Trips 1, 2 and 6 were to Hadley's Landing on the Illinois River, river mile 13.2. Trips 3 and 4 were to the Mississippi River at Rip Rap Landing (river mile 265.1) and Mosier Island (river mile 260.2), respectively. Trip 5 was to McEver's Island on the Illinois River

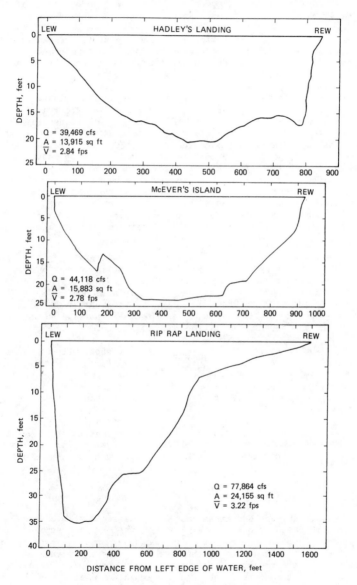

Fig. 2.  Cross-sectional profiles of the study sites

(river mile 50). The cross-sectional profiles for Hadley's Landing,
McEver's Island, and Rip Rap Landing are shown in Fig. 2. The profile
for Mosier Island is very similar to that for Rip Rap Landing.


Wave Data

Most of the wave data collected during the field trips were generated by tows, with little generated by recreational and other river
traffic. This was because of the low frequency of recreational vessels
during the field trips.

The wave patterns generated by tows in restricted channels are much
more complex than those generated by streamlined vessels traveling in
open and deep waters. Even though the diverging and transverse waves
are generated both at the bow and stern, there are also surge waves
behind the tows generated because of the displacement of a large portion
of the water in the river by the loaded barges. In some instances the
surge waves totally predominate over the other types of waves. There is
also a narrow band of disturbed water surface behind the towboat resulting from the discharging of the propeller near the water surface. The
water surface fluctuation caused by the propeller jet seems to be higher
than the waves which reach the shore when observed behind the tow. This
water surface fluctuation is, however, dissipated in the middle of the
channel before it reaches the shore.

An example of a tow-generated wave is shown in Fig. 3. The wave
data were collected at the Hadley's Landing test site on the Illinois
River during a passage of a downstream-bound tow with 15 loaded barges,
traveling at a speed of 8.54 ft/sec. In this wave pattern it is
possible to identify the bow, stern, and the towboat stern waves as
shown in the figure. During this event the maximum wave height, which
is 0.39 feet, was generated by the bow of the tow.

The kind of waves generated by a single towboat without barges is
indicated in Fig. 4. The wave pattern is significantly different from
that generated by tows pushing barges in that it consists of only a
couple of sharp, well-defined waves and dies out quickly. However, the
maximum wave height is 0.89 ft, which is higher than most of the waves
generated by tows.

The maximum wave heights for all 59 events during the six field
trips were determined from plots similar to those in Fig. 3. The
maximum wave heights ranged from a low of 0.1 ft to a high of 1.08 ft.


Comparison of Data with Theory

The maximum wave heights measured in the field were compared with
wave heights calculated with Eqs. 1 and 2, which were discussed in the
background section. The equations were made dimensionless by dividing
both sides of the equations by the draft, D. The agreements between the
equations and the measured data were not very good. The correlation
coefficients between the measured and calculated maximum wave heights
are .69 and .80 for equations 2 and 1, respectively.

In an effort to develop an empirical equation which could better predict the measured wave heights, a multivariate regression analysis was carried out between the measured maximum wave heights and the important parameters which were felt to influence the generation of waves. After many combinations of variables were tried, the following equation, which predicts the non-dimensional maximum wave height based only on the draft Froude number, $F_D$, was found to give the best result:

$$\frac{H_{max}}{D} = 0.133 \, F_D \qquad\qquad (3)$$

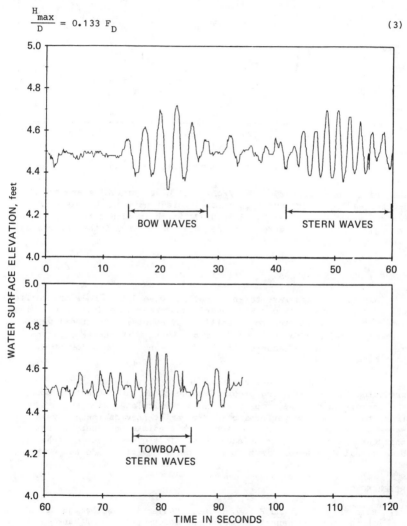

Fig. 3.   Wave pattern generated by a tow

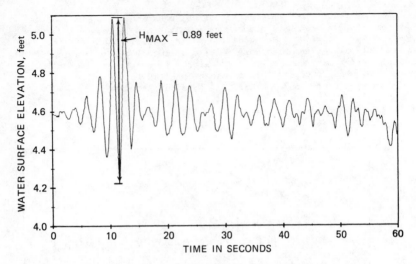

Fig. 4.  Wave pattern generated by a towboat

The draft Froude number, $F_D = V/\sqrt{gD}$.  All other variables are as
defined earlier.  The correlation coefficient between the measured
maximum wave heights and Eq. 3 is 0.87, which is much better than those
for Eq. 1 and 2.  The spread of the data points around the regression
line (Eq. 3) is shown in Fig. 5.


SUMMARY AND CONCLUSIONS

    Six field trips were taken to collect wave data.  Four of the field
trips were to the Illinois River, and two were to the Mississippi River.
Wave data were collected for a total of 59 events.  Additional wave data
were collected during the passage of a towboat without barges.  The
maximum wave heights measured in the field ranged from a low of 0.1 ft
to a high of 1.08 ft.

    Comparison of the measured maximum wave heights with those
predicted through existing predictive equations was not satisfactory.
The correlations between the measured and calculated wave heights were
found to be low.  By performing a multivariate regression analysis, it
was possible to obtain an equation which predicts wave heights better
than the previously existing equations.  In the new equation, the non-
dimensional wave height is a function of the draft Froude number only.


REFERENCES

1.  Balanin, V. V., and Bykov, L. S.,  "Selection of leading dimensions
    of navigation channel sections and modern methods of bank
    protection," proceedings of the 21st International Navigation
    Congress, PIANC, Stockholm, Sweden, 1965.

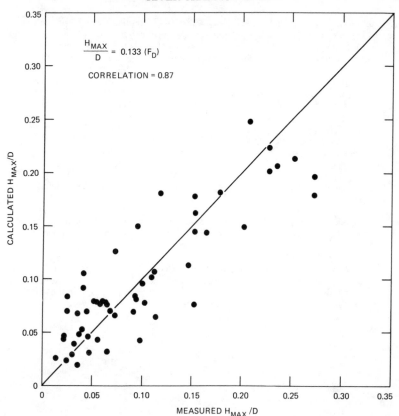

Fig. 5.   Comparison of the regression equation with the measured maximum wave heights

2.   Bhowmik, N. G., "Development of criteria for shore protection against wind-generated waves for lakes and ponds in Illinois," University of Illinois Water Resources Center Research Report 107, Urbana, Illinois, 1976.

3.   Bhowmik, N. G., and Schicht, R. J., "Bank erosion of the Illinois River," Illinois State Water Survey Report of Investigation 92, Urbana, Illinois, 1980.

4.   Bhowmik, N. G., Demissie, M., and Guo, C. Y., "Waves generated by river traffic and wind on the Illinois and Mississippi Rivers," University of Illinois Water Resources Research Center Report No. 167, Urbana, Illinois, 1982.

5.  Buchanan, T. J., and Somers, W. P., "Discharge measurements at
    gaging stations," Book 3, Chapter A7, Techniques of Water-Resources
    Investigations of the United States Geological Survey, United
    States Government Printing Office, Washington, D.C., 1969.

6.  Das, M. M., "Relative effect of waves generated by large ships and
    small boats in restricted waterways," Report No. HEL-12-9, Univer-
    sity of California, Berkeley, California, 1969.

7.  Das, M. M., and Johnson, J. W., "Waves generated by large ships and
    small boats," proceedings of the 12th Conference on Coastal
    Engineering, Chapter 138, Washington, D.C., 1970.

8.  Johnson, J. W., "Ship waves in shoaling waters," proceedings of the
    Eleventh Conference on Coastal Engineering, Chapter 96, London,
    England, 1968.

9.  Sorensen, R. M., "Ship generated waves," Advances in Hydroscience,
    editor V. T. Chow, Vol. 9, Academic Press, New York and London,
    1973.

10. U.S. Army Corps of Engineers, Huntington District, "Gallipolis
    Locks and Dam replacement, Ohio River, phase I, advanced engineer-
    ing and design study," General design memorandum, appendix J,
    vol. 1, Environmental and Social Impact Analysis, 1980.

# INTERPLAY OF AMBIENT TURBULENCE AND JET MIXING

Steven J. Wright, A.M. ASCE[1]

ABSTRACT

   The problem of defining the effect of ambient turbulence on jet mixing is considered. In conventional theories, this is accomplished by assuming that there is additional entrainment due to ambient turbulence and that this can be handled by adding a term accounting for this effect to ordinary jet entrainment relations. Two different possible formulations which relate directly to ambient diffusion analyses are presented. Numerical simulations are presented which show the effect of these terms to be important even if the jet discharge velocities are an order of magnitude larger than ambient velocities. Experimental results are presented for cases with relatively low jet discharge velocities which indicate that mixing is not nearly as rapid as implied by these models. It is shown that a jet analysis that does not include the ambient turbulence formulations is better for predicting the jet spreading rates.

INTRODUCTION

   The assessment of environmental impact from wastewater effluents discharged into rivers has typically been done on the basis of one-dimensional longitudinal dispersion analyses. Fischer (3) has shown that considerable distances (on the order of tens of kilometers for large rivers) are required before the one-dimensional approach with constant dispersion coefficient is valid. Even two-dimensional diffusion models consider the effluent to be well-mixed in the vertical. These assumptions cannot be expected to be valid in the near field where influences such as temperature and velocity differences between the discharged and receiving fluids may have a significant impact upon the mixing behavior. These factors, along with turbulent diffusion processes, control the near field spreading that will need to be estimated in order to define an initial condition for simplified far field diffusion or dispersion models.

   For the case of submerged discharges, buoyant jet theory has been developed to account for the effect of density or velocity differences on the mixing phenomena. A considerable effort has been expended on the problem of buoyant jets in flowing receiving fluids (e.g., see the review by Jirka, et al (4)). However, most of these theories do not account for the effect of the ambient turbulence or the presence of system boundaries; the former problem is considered herein. It is

---

   [1]Asst. Prof. Civil Engineering, The University of Michigan, Ann Arbor, MI.

assumed that there exists a region where neither buoyant jet analyses (without accounting for ambient turbulence) nor ambient diffusion/dispersion theories are completely valid. What is required is an adequate transition between the two types of analysis over a region that may extend for a few meters up to a few hundred meters in a typical application. The investigation described herein was performed for the purpose of making predictions in this transition region. Conventional theories accounting for the effect of ambient turbulence are reviewed and results of a hydraulic model study are presented to determine the validity of these formulations. It is noted that this analysis is not advanced as the only possible physical occurrence in the near field zone, but for relatively low discharge rates such as may be typical of wastewater discharges, the general formulation should be appropriate.

The general problem for which this analysis was developed was that of a wastewater effluent discharged from a multiport diffuser. In the specific application, discharge is from nozzles on risers at equal spacing along the buried diffuser aimed only slightly above the horizontal in the downstream direction. Hydraulic constraints on head loss through the diffuser dictated conditions where the design discharge velocities were only somewhat larger than the receiving fluid velocity. Conventional analyses for ambient diffusion and jet mixing revealed that ambient diffusion calculations would predict much more rapid mixing than the jet effects. However, existing information indicated that significant temperature differences could be presented between the discharge effluent and the river. Proper analysis of the problem appeared to require both a consideration of buoyant jet and ambient diffusion effects; the analysis presented is an attempt to resolve the problem.

MIXING MODELS

The specific problem consists of discharge from a series of round nozzles at arbitrary angle to the horizontal into a flowing ambient fluid. The effluent discharge is continuous with concentration $C_o$ of an arbitrary contaminant; the other discharge variables are the discharge per nozzle $Q$, nozzle diameter $D$, temperature difference between ambient and discharged fluids which can be translated to a density difference $\Delta\rho$, and spacing between nozzles $S$. The river is described by a constant velocity $U_A$ and depth $H$; turbulent flow is assumed throughout the flow field. Simplified mixing models are discussed to describe this problem.

Ambient Diffusion Models

An ambient diffusion calculation considers the mass loading at the source as the parameter of major importance in describing the initial condition. Ogata (5) presents a solution for the maximum concentration for a steady release from a source if longitudinal diffusion is neglected

$$C_m/C_o^* = 1 - \exp\left(-D^2 U_A/16\, D_r X\right) \tag{1}$$

Here $C_m$ is the maximum concentration at any position $X$, $D_r$ is a radial diffusion coefficient, and $C_o^*$ is a source concentration assuming no difference in velocity between the source and the ambient fluids. Continuity consideration relate $C_o^*$ and the actual source concentration by

$$C_o^* U_A \frac{\pi}{4} D^2 = C_o Q \tag{2}$$

For values of the exponent $D^2 U_a/D_r X$ small (large distances downstream), this solution can be approximated by the point source solution

$$C_m \approx \frac{C_o Q}{4\pi D_r X} \tag{3}$$

Assuming that the point source solution is valid, an additional consideration is that the vertical and lateral diffusion coefficients are not equal (Fischer (3)). The point source solutions for constant vertical and lateral diffusion coefficients $D_z$ and $D_y$ respectively is given as

$$C = \frac{C_o Q}{4\pi x \sqrt{D_y D_z}} \exp\left(- \frac{u_A y^2}{4D_y X} - \frac{u_A z^2}{4D_z X}\right) \tag{4}$$

where z is a vertical coordinate and y a lateral one defined relative to the plume axis. The effect of vertical boundaries and additional discharges at a spacing S are accounted for with image sources. The diffusion coefficients are estimated from the relations

$$D_z = a_1 u_* h \tag{5}$$

$$D_y = a_2 u_* h \tag{6}$$

where $u_*$ is the shear velocity for the ambient flow, $a_1$ is a constant on the order of 0.1 and $a_2$ according to Fischer may be expected to lie within the range of 0.2 to 0.8 for typical rivers.

Buoyant Jet Mixing

Any mixing effects due to discharge angle, velocity differences, and density effects are most easily estimated with a buoyant jet analysis by an integral approach. A standard approach to this problem was employed; complete details are not presented, but it is essentially the same approach outlined by Jirka, et al (4). Gaussian profiles for concentration, etc. of the form

$$C(X,r) = C_m(X) \exp\left[-(r/\lambda bu)^2\right] \tag{7}$$

are assumed for the problem where $b_u$ is a characteristic width of the velocity profile and $\lambda$ is a multiplier to account for differences in velocity and concentration profiles. The system of ordinary differential equations must be solved numerically for most applications and standard methods are available for these purposes. Zone of flow establishment corrections are required and turbulent closure is achieved through the integrated continuity equation by specification of the entrainment relation. The specific one utilized was

$$E = 2\pi b_u \left[ 0.5 U_A \sin\theta + (.057 + .488 \frac{g'b_u}{u_s^2}) u_s \right] \tag{8}$$

where $g'$ is the local maximum value of the effective gravity force, $u_s$ is the magnitude of the jet velocity above the ambient in the axial direction and $\theta$ is the local angle with the horizontal.

The entrainment constants are estimated from previous experiments in stagnant receiving fluids. This expression does not consider the effect of ambient turbulence on any dilution of discharge concentrations. The addition of terms to account for this influence has been suggested in several instances. There are two general approaches to the formulation of this additional entrainment term $E_a$. One is to follow the analysis of Slawson and Csanady (10) assuming that the diffusion coefficient is constant in the ambient fluid. Since the jet must eventually approach the state of being passively advected with the ambient velocity, the previous relations may be used to derive a formulation. The relationship between local volume flux $q$ within the jet and concentrations with the assumed gaussian profiles is obtained for conserved tracer flux as

$$\frac{C_o}{C_m} = \frac{\lambda^2}{1+\lambda^2} q/Q \tag{9}$$

Since from Eq. 2

$$\frac{C_o}{C_m} = 4\pi \frac{\sqrt{D_y D_z}}{Q} x \tag{10}$$

mathematical manipulation yields the result

$$\frac{dq}{dx} = Ea = \frac{1+\lambda^2}{\lambda^2} 4\pi \left( D_y D_z \right)^{1/2} \tag{11}$$

Combining the results in Eqs. 4 and 6 yields the final expression

$$Ea = 2\pi \alpha_4 H u_* \tag{12}$$

where $\alpha_4$ is a collection of all of the relevant constants. This equation simply yields the result that if the jet is moving exactly with the ambient velocity, the jet mixing model will predict the same rate of concentration decrease as the ambient diffusion model described earlier. Thus, there is a direct transition between buoyant jet mixing and ambient diffusion implied in the model, except that the effect of boundaries is not included in the jet formulation.

A second formulation for the additional entrainment term is obtained by the same reasoning except that Taylor's (11) classical paper on diffusion is employed to allow for a nonconstant diffusion coefficient. Analyses presented by Ooms (6), Petersen and Cermak (7) and others utilize this approach and this is the most accepted formulation.

Application of Taylor's analysis will result in an expression of the form

$$Ea = 2 \frac{\sqrt{2}}{\lambda} u' b_u \tag{13}$$

where $u'$ is the rms value of the turbulent intensity and is difficult to predict theoretically. The constant factor in this equation comes from the definition of the variance in the gaussian distribution given by Eq. 7. The shear velocity $u_*$ is often substituted for $u'$ in the formulation to yield

$$Ea = 2\pi\alpha_5 b_u u_* \tag{14}$$

These two velocity scales are of the same magnitude in open channel flow (8) and the approach appears reasonable. By inspection of the two results, it can be seen that Eq. 14 will predict lower dilution rates than Eq. 12 if the jet radius is much smaller than the depth of flow but that as the two lengths approach each other, Eq. 14 will predict the larger entrainment due to the normally larger entrainment coefficients in the expression.

Application of Jet Model

Although it is fairly intuitive what the qualitative differences between the different entrainment relations will be, it is instructive to examine the magnitudes of the effects predicted. Benedict and Preston (1) has presented a similar sort of analysis for surface buoyant jets and obtain many of the same conclusions. For the purpose of comparison, a series of different numerical solutions were made. These are referred to as type A (no ambient diffusion term in the entrainment relation), type B (Eq. 14 added to Eq. 8 to give the total entrainment rate), and type C (Eq. 12 used in place of Eq. 14). The major quantity of interest in the simulation is the velocity difference between the discharged and receiving fluids; therefore simulations with discharge in the same direction as the ambient fluid with no density differences were made. Other simulation variables were $D = 0.5$, $u_* = 0.067$, $H = 10$, $\alpha_4 = 1.0$, and $\alpha_5 = 1.0$. These latter values were selected as being only of the general order of magnitude required in the above development. Any consistent set of units are valid here; the numbers were selected to be representative of the experimental study described below. Finally, discharge velocities of 1.25, 10, and 100 were made; the results of these are given in Fig. 1. It is obvious from these results that unless the discharge velocity is well in excess of the ambient velocity (e.g. 100 times), the addition of the ambient diffusion terms to the buoyant jet formulation will have a significant impact upon the computed results in the first few tens or hundreds of meters where jet mixing effects are important. Given typical ranges of parameter variation, it is clear that this question of ambient turbulence influence is important for a large fraction of applications.

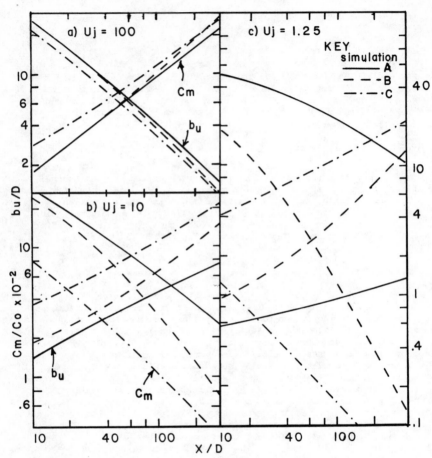

Figure 1.   Comparison of Three Different Entrainment Formulations
for (a) $U_j = 100$; (b) $U_j = 10$; (c) $U_j = 1.25$.

EXPERIMENTAL INVESTIGATION

The experimental study was performed because it was hypothesized
that the effect of ambient turbulence was not nearly as great as predicted by either of the approaches described.   Previous experiments
on surface buoyant jets by Weil and Fischer (12) indicate that this may
be the case.   Shirazi and Davis (9) followed a similar development for
including the effect of ambient turbulence in the surface jet formulation and it was found that the coefficient $a_2$ in Eq. 6 would have to
be approximately 0.02 to account for the observed influence on the mixing characteristics.   Since this is far less than the range of values
given for the case of pure ambient diffusion, the implication is that

the effects of ambient and jet mixing may not be additive as in the
formulations presented.

The specific problem studied was for a wastewater effluent discharge into the Detroit River near its discharge into Lake Erie. The
river is about 3 m deep and has a velocity of approximately 0.3 m/sec
at the proposed outfall diffuser location. As previously mentioned,
it was desired to minimize head losses in the diffuser to allow gravity
flow, so a preliminary design was for nozzles of 25.4 cm diameter aimed
at 20 degrees above the horizontal in the downstream direction. The
range of discharge velocities for the design discharge down to initial
flows associated with the early phases of the system operation were
from 2.3 to 0.2 m/sec. Note that the smallest velocities are actually
less than the river velocity and the largest is only about seven times
the river velocity, clearly in the range where ambient turbulence
effects predicted by the numerical solutions will be noticeable.
During the winter months, data from existing sewage treatment plants
indicates that the effluent may be as much as $15^{\circ}C$ warmer than the
river temperature. A series of tests were performed with a 1/25 scale
model of a single port discharge with the maximum density difference
for several discharge velocities to observe the mixing behavior as a
function of distance from the source. The density differences were
simulated by adding sodium chloride to the discharge solution and
performing the problem upside down; this is a common approach in laboratory investigations. A series of photographs were taken of the dyed
jets from which trajectories and widths can be estimated; dilutions
will be measured at a later date. The results scaled from the photographs and visual observations provide the present results. The 0.3 m
wide by 15 m long tilting flume has a Manning n of approximately 0.009
determined from previous investigations and flow conditions scaled from
the prototype values specified above were used. The coefficients $a_1$
and $a_2$ in Eqs. 5 and 6 were assumed to have the values 0.1 and 0.2,
respectively. Jet Reynolds Numbers ranged from 580 to 4700 and the
ambient flow Reynolds Number based upon the depth was about 9000. There
may have been some viscous effects present,especially at the lowest jet
discharges, but the flows were observed to be turbulent.

RESULTS

The most interesting results came from the visual observations
and cannot easily be quantified. It was observed that dye passively
injected into the ambient flow mixed much more rapidly over the depth
than did the jet discharges. This was due to the much larger scale
of the ambient turbulence and mixing over the depth did not imply that
there were not significant concentration gradients remaining as very
"patchy" dye fields resulted. The finer jet structure yielded more
efficient local mixing, but somehow prevented the large scale ambient
turbulence from interacting with it. The jet discharges mixed over the
depth only after impingement upon the bottom (surface in the prototype)
and the jet structure was rapidly destroyed; the ambient turbulence
then rapidly completed the mixing process. Comparisons of the numerical
simulations for the three situations discussed earlier also clearly
indicated the lack of ambient influence. The visual width of the
buoyant jet was estimated to be 1.7 $b_u$ from previous results from

Fan (2) and is compared against the predictions in Fig. 2 for two of
the tests.  Comparisions are somewhat obscured by the lack of a good
zone of flow establishment correction in the numerical model (extension of the source for 6.2 diameters at the initial angle) but the

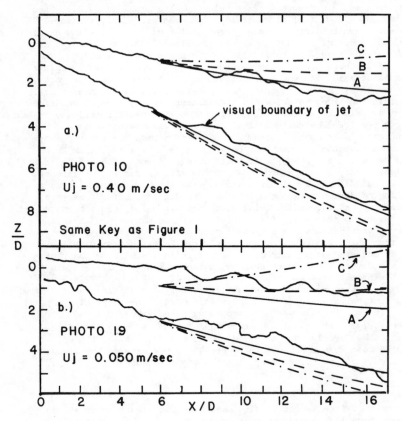

Figure 2.  Comparison of Jet Spread with Different Numerical
           Predictions.

conclusions are fairly obvious.  The type A simulations, namely the
ones without any ambient turbulence corrections do the best job of
reproducing the observed jet widths.  The results in Fig. 2
for a jet velocity less than the ambient value and represents an
extreme test of the concepts discussed.  For the sake of comparison,
Fig. 3 presents the result of an ambient diffusion experiment (dye
crystals suspended on a plate dissolving into the flow).  The point
source solution in Eq. 4 with the same definition of visual width and
estimated diffusion coefficients does a fairly good job of reproducing
the growth rate of the diffusing cloud.  On the other hand, a jet
calculation with no ambient turbulence term and velocity ratio

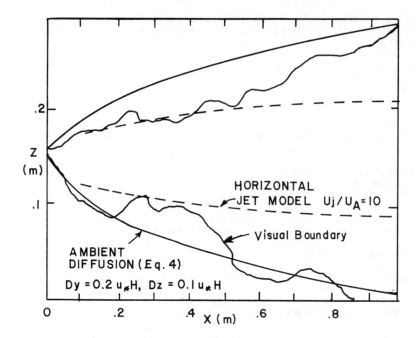

Figure 3.  Spreading by Ambient Diffusion Compared to
Different Models.

$U_j/U_A$ = 10 considerably underpredicts the growth rate.  These results
imply more rapid spreading (and smaller time average concentrations)
when there is no jet structure present.

SUMMARY AND CONCLUSIONS

Further testing is being conducted to define the problem further
but the preliminary results indicate some fairly important concepts.
The first is that the processes of ambient diffusion and turbulent jet
entrainment are not additive, an argument previously advanced without
verification by Jirka, et al (4).  The problem appears to be one of
the interaction of two different types of turbulence, each with their
own characteristic scales and these cannot be assumed to be simply
added together.  The only other experimental evidence that the author
is aware of for this type of problem is that given by Petersen and
Cermak (7) and they also conclude that their numerical model including
ambient turbulence entrainment is unsatisfactory when the jet velocities
approach ambient levels.  Together with the Weil and Fischer (12)
results, all available data implies that the effect of the ambient
turbulence is far less than expected by present theoretical concepts.
In the absence of an alternative formulation, it is recommended to
leave out the ambient turbulence term altogether from the analysis.

When the discharge impinges upon a free surface or reaches a stable position in a stratified fluid, the jet structure is lost and ambient turbulence may then take over. This is an incomplete argument since eventually ambient turbulence must overwhelm jet mixing in the absence of these natural transitions, but it is at a much further distance than predicted by present theories.

REFERENCES

1. Benedict, B.A. and Preston, J.W., "Sensitivity of the Prych-Davis-Shirazi Model," Proceedings, Hydraulics Div. Spec. Conf. 1980, Chicago, IL, pp. 322-332.

2. Fan, L.N. "Turbulent Buoyant Jets into Stratified or Flowing Ambient Fluids," W.M.Keck Lab. Rept. KH-R-15 Calif. Inst. of Tech. Pasadena, CA. 1967.

3. Fischer, H.B., "Longitudinal Dispersion and Turbulent Mixing in Open-Channel Flow," Annual Review of Fluid Mech. Vol.5, 1973 p.59-78.

4. Jirka, G.H., Abraham, G. and Harleman, D.R.F., "An Assessment of Techniques for Hydrothermal Prediction," R.M. Parsons Lab. Rept. 203, Mass. Inst. of Tech. Cambridge, MA., 1975.

5. Ogata, A., "Two-Dimensional Steady State Dispersion in a Saturated Porous Medium," J.Research U.S. Geol. Surv. Vol. 4, 1976 pp.277-284.

6. Ooms, G., "A New Method for the Calculation of the Plume Path of Gases Emitted by a Stack," Atmospheric Environment, Vol. 6, 1972.

7. Petersen, R.L. and Cermak, J.E., "Plume Rise for Varying Ambient Turbulent, Thermal Stratification and Stack Exit Conditions - A Numerical and Laboratory Evaluation," Fifth Int.Conf. on Wind. Eng. 1979, Fort Collins, CO.

8. Sedimentation Engineering ASCE Manual No. 54, ASCE 1975.

9. Shirazi, M.A. and Davis, L.R., "Workbook of Thermal Plume Pre-diction, Vol. 2, Surface Discharge" U.S. EPA Report, EPA-R2-72-0056, 1974.

10. Slawson,P.R. and Csanady, G.T., "The Effect of Atmospheric Conditions on Plume Rise," J.Fluid Mech. Vol. 47, 1971, pp 33-49.

11. Taylor, G.I., "Diffusion by Continuous Movements," Proc. London Math. Soc. Vol. 20, 1921, pp. 196-211.

12. Weil, J. and Fischer, H.B., "Effect of Stream Turbulence on Heated Water Plumes," J. Hydraulics Div. ASCE, Vol. 100, 1974, pp. 951-970.

# MIXING IN PUMPED STORAGE RESERVOIRS

by

Philip J. W. Roberts[1], A. M. ASCE and P. Reid Matthews[2]

## ABSTRACT

Preliminary results of experiments on horizontal round buoyant jets of small or neutral buoyancy into a linearly stratified ambient fluid are presented. The gross characteristics of the flow are discussed, and results given from which they can be predicted for arbitrarily specified conditions.

It is shown that flows of this type occur in pumped-storage reservoirs, and an example of application of the results to Carters Lake in Georgia is presented. The entrainment caused by the inflow jet and the gross jet characteristics are predicted.

## Introduction

Because of mixing induced by discharges in pumped-storage projects, these types of reservoirs have temperature stratifications which may be significantly different from natural lakes. This can occur either in the lower reservoir during power generation, or the upper during pump-back. As either of these reservoirs may be used for other purposes, such as water supply, and as the water released may have different temperatures, and hence downstream ecological effects, it is a matter of considerable importance to be able to predict the mixing in pumped-storage reservoirs caused by the inflows.

The design of the inflow channels often causes the inflow to behave

---

[1] Asst. Prof., School of Civ. Engrg., Ga. Inst. of Tech., Atlanta, Ga. 30332

[2] Graduate Research Asst., School of Civ. Engrg., Ga. Inst. of Tech., Atlanta, Ga. 30332

as a jet.  A number of flow possibilities exist, depending on boundary
proximity, inlet channel design, temperature of the inflow relative to
the receiving water, and the stratification of the receiving water.
Roberts (3) identified 16 possible jet flow types, and found that no one
mathematical model could be used to predict mixing for all situations.
A case which arises frequently, however, corresponds to the discharge of
a horizontal high momentum jet of low or neutral buoyancy into a stratified ambient fluid.  Studies exist of horizontal buoyant jets into
stratified ambient fluid, and mathematical models are available, for
example, Fan and Brooks (1), to predict their behavior.  The assumptions
used in these models break down, however, when the source buoyancy
becomes low, and fail completely for a jet of neutral buoyancy.  No experimental studies of this situation exist.  For these reasons, a fundamental experimental study of horizontal round buoyant jets of small of
neutral buoyancy discharging into a linearly stratified ambient fluid
was undertaken.  The purpose of this paper is to present preliminary
results of this study and to demonstrate their application to an
operating pumped-storage reservoir.

Analysis

A definition diagram of the problem under consideration is shown in
Figure 1.  A horizontal round jet of diameter d discharges a fluid of

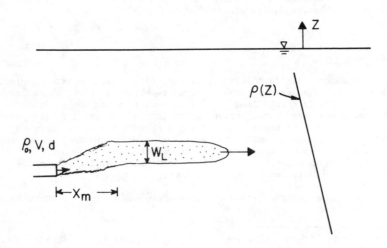

Fig. 1  Definition Diagram

density $\rho_o$ into an ambient fluid having a density $\rho_a$ at the level of the
jet and a linear density profile.  Because of the ambient stratification, the jet rises to some height, and the turbulence is damped out,
collapsing in a horizontal length $x_m$.  After collapse, the jet flows
horizontally in a layer of thickness $W_L$.  Because the jet is turbulent,
ambient fluid is entrained into the jet and a circulation pattern set up
in the reservoir.  The entrained flow does not come from the whole reservoir depth, but from a thin layer of width $W_e$, in a manner similar to
that of selective withdrawal flows.

For small density differences, the jet can be characterized by its
kinematic fluxes of volume, Q, momentum, M, and buoyancy, B, where:

$$Q = \frac{\pi}{4} d^2 v \tag{1}$$

$$M = \frac{\pi}{4} d^2 v^2 \tag{2}$$

$$\text{and} \quad B = \frac{\pi}{4} d^2 v \left( \frac{\rho_a - \rho_o}{\rho_a} \right) g \tag{3}$$

where g is the acceleration due to gravity.  The ambient stratification
can be characterized by $\varepsilon$, where

$$\varepsilon = - \frac{g}{\rho_a} \frac{d\rho}{dz} \tag{4}$$

The square root of $\varepsilon$ is the commonly used buoyancy frequency.

Using Q, M, B, and $\varepsilon$, it is possible to form several length scales,
the most useful for present purposes being:

$$\ell_Q = \frac{Q}{M^{\frac{1}{2}}} = \sqrt{\frac{\pi}{4}} \, d \tag{5}$$

$$\ell_M = \frac{M^{3/4}}{B^{\frac{1}{2}}} \tag{6}$$

$$\text{and} \quad \ell_\varepsilon = \left( \frac{M}{\varepsilon} \right)^{\frac{1}{4}} . \tag{7}$$

This is similar to the appraoch used by Wright (4), Fischer, et al.
(2), and others.  Any dependant variable can be expressed in terms of
ratios of these length scales, for example:

$$x_m, \ W_1, \ W_e = f(Q, M, B, \varepsilon) \tag{8}$$

becomes, after a dimensional analysis:

$$\frac{x_m}{\ell_\epsilon} \; , \; \frac{W_1}{\ell_\epsilon} \; , \; \frac{W_e}{\ell_\epsilon} = f\left(\frac{\ell_Q}{\ell_\epsilon} \; , \; \frac{\ell_M}{\ell_\epsilon}\right)$$

(9)

Because the jet collapses at some distance from the source, the total volume flux, $\mu$, in the jet asymptotes to a constant value. Thus:

$$\mu = f(Q, M, B, \epsilon)$$

(10)

or

$$\frac{\mu\epsilon^{\frac{1}{4}}}{M^{3/4}} = f\left(\frac{\ell_Q}{\ell_\epsilon} \; , \; \frac{\ell_M}{\ell_\epsilon}\right)$$

(11)

## Experimental Procedure

To investigate the functional form of the relationships expressed by Eqs. 9 and 11 a systematic series of experiments was performed. The experiments were performed in a plexiglass walled laboratory tank 8 feet long, 4 feet wide and 4 feet deep. Three jets were used: 1/4 inch, 1 inch, and 4 inch diameter. The tank was stratified with salt, and the resulting stratifications used were nominally 12 inches for the 1/4 inch and 1 inch nozzles, and 22 inches for the 4 inch nozzle. Two nominal stratification strengths corresponding to $\epsilon$ of 0.12 and 1.2 sec.$^{-2}$ were used, and the resulting stratifications were always very closely linear. Consideration of typical pumped reservoir conditions suggested a useful experimental range would be:

$$0.1 < \frac{\ell_Q}{\ell_\epsilon} < 2$$

For highly buoyant jets, ie. $\ell_M/\ell_\epsilon \ll 1$, the results of Fan and Brooks (1) can be used. It was therefore decided to run tests at $\ell_M/\ell_\epsilon$ of 2 and $\infty$, the latter being neutrally buoyant.

An overhead camera was used to photograph the experiments, and a mirror at the side of the tank gave a simultaneous side view of the jet. The collapse lengths $x_m$ and $W_1$ were measured from the photographs, and $W_e$ from the fluctuations in salt concentration in the jet measured by the conductivity probe. The induced circulation in the tank, and hence the entrained flow was measured from the timed movement of potassium permanganate dye streaks.

## Results

A typical photograph of a neutrally buoyant jet is shown in Fig. 2. Near the nozzle, the jet behaves like simple momentum jet in a homogeneous fluid. Further from the source, however, the turbulence is damped, and eventually the jet collapses. Following collapse, the flow intrudes horizontally as a density current.

The results for volume flux entrainment, Eq. 11 are shown in Fig. 3 for $\ell_M/\ell_\epsilon = 2$ and $\infty$. It was found that in this range, the buoyancy

Fig. 2. Side view photograph of typical experimental jet.

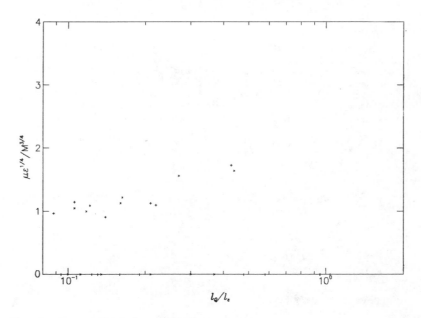

Fig. 3. Measured total volume flux in jet.

flux had no measurable effect on the magnitude of the entrained flow, but it did affect its location.

For non buoyant jets, the dependent variables of Eqs. 9 and 11 reduce to $\ell_Q/\ell_\varepsilon$. If $\ell_Q/\ell_\varepsilon \ll 1$, however, the effect of $\ell_Q$ is confined to the nozzle region, and the results become independent of this parameter. For larger values of $\ell_Q/\ell_\varepsilon$, the source volume flux becomes more important. This trend is apparent in Fig. 3. The results for $W_1$, $W_e$ were found to be not strongly dependent on $\ell_Q/\ell_\varepsilon$, and average values of the results were:

$$\frac{W_L}{\ell_\varepsilon} = 0.92 \tag{12}$$

and
$$\frac{W_e}{\ell_\varepsilon} = 1.4 \tag{13}$$

Values of $x_m$ were difficult to estimate, but a reasonable value is:

$$\frac{x_m}{\ell_\varepsilon} = 3.0 \tag{14}$$

## Applications

Carters Lake in North Georgia is a pumped storage project with a shallow lower regulation pool which is usually well mixed, and an upper deep reservoir·which is usually stratified. When one unit is operating during pumpback, flows of about 3000 c.f.s. are discharged through a rectangular nozzle 20.5 ft. deep by 14 ft. wide. The nozzle invert is about 100 ft. below the reservoir surface. The source buoyancy flux varies through the year from slightly negative to slightly positive.

On 2 August 1979, the pumpback flowrate was 3365 c.f.s., and $\varepsilon$ was $2.34 \times 10^{-3}$ sec.$^{-2}$. Substitution into Eqs. 1, 2, 5, and 7 yields:

$$Q = 3365 \text{ ft}^3/\text{s}$$
$$M = 3.95 \times 10^4 \text{ ft}^4/\text{s}^2$$
$$\ell_Q = 16.9 \text{ ft.}$$
$$\ell_\varepsilon = 64.1 \text{ ft.}$$
$$\ell_Q/\ell_\varepsilon = 0.26$$

From Fig. 3, we find
$$\frac{\mu\varepsilon^{\frac{1}{4}}}{M^{3/4}} = 1.6,$$

or $\mu$ = 20,000 ft$^3$/s.

The dilution in the jet, S = $\mu/Q$ = 6.

The collapse length of the jet, $x_m$ predicted by Eq. 14 is 192 ft; the thickness of the collapsed layer, $W_1$, from Eq. 12 is 59 ft., and the entrained flow is drawn from a layer of thickness $W_e$, given by Eq. 13, of 90 ft.

These experimental results are now being incorporated into a numerical reservoir model to predict the yearly stratification cycle of a pumped-storage reservoir. Results of these applications will be published later.

## ACKNOWLEDGEMENT

The work reported here was funded by the U. S. Army Corps of Engineers, Waterways Experiment Station. Thanks are expressed to Mark Dortch and Dennis Smith of WES for their support and excellent liaison work.

## REFERENCES

1. Fan, L. N., and Brooks, N. H., (1969), "Numerical Solutions of Turbulent Buoyant Jet Problems", W. M. Keck Laboratory of Hydraulics and Water Resources, California Institute of Technology, Report No. KH-R-18.

2. Fischer, H. B., et al. (1969), "Mixing in Inland and Coastal Waters", Academic Press.

3. Roberts, P. J. W., (1981), "Jet Entrainment in Pumped-Storage Reservoirs", Technical Report E-81-3, prepared by Georgia Institute of Technology Atlanta, Ga., for the U. S. Army Engineer Waterways Experiment Station, CE, Vicksburg, Miss.

4. Wright, S. J., (1977), "Mean Behavior of Buoyant Jets in A Crossflow", J. Hydraulics Div., ASCE, 103 (HY5): 499-513.

# MIXING IN SURFACE JETS UNDER DISTORTED MODEL CONDITIONS

by

Benjamin R. Roberts[1] and Robert L. Street, M. ASCE [2]

ABSTRACT

The behavior of distorted jets in physical models is a key aspect of model performance and often strongly influences predictive capability. In this study, formal nondimensionalization of the spatially-averaged Navier-Stokes equations and corresponding transport equations for the small-scale turbulence is used to obtain key dimensionless groups. The magnitude of each group and its dependence on model distortion is analyzed. Laboratory experiments in a variable distortion modeling facility that test and amplify the dimensional analysis results are reported. Quantitative relations between distortion and jet parameters, such as vertical penetration, half-width, surface temperature, and turbulence intensity, are obtained.

## 1. INTRODUCTION

When distorted physical models are used to simulate the turbulent mixing and advection in buoyant or isothermal jets, incomplete similitude leads to model behavior which is inconsistent with prototype behavior. Specifically, if Froude scaling is used, failure to maintain the jet Reynolds number as a constant influences turbulence intensities and advection rates, while changes in jet geometry (jet aspect ratio) caused by distortion influence the geometry of turbulent structures. The rates of jet mixing in the vertical and transverse directions are typically not consistent between model and prototype (see Neale and Hecker, 1972 and Ryan and Harleman, 1973). Consequently, the dilution, spreading rates, and geometry of the distorted model jet are not accurate representations of prototype conditions.

The precise relationships between near-field jet behavior and distortion have not previously been quantified experimentally. Moretti and McLaughlin (1977) describe the apparent effects of distortion on a mid-density model jet entering a stratified pond. The experimental works of Shirazi (1973) and others were evaluated by Haggstrom (1978) in order to predict near-field jet behavior in distorted models. He concluded that errors in predicting near-field

---

[1]  Consulting Assistant Professor, Dept. of Civil Engineering, Stanford University and Supervising Engineer, Anderson-Nichols and Co., Inc., Palo Alto, CA  94303.

[2]  Professor of Fluid Mechanics and Applied Mathematics, Stanford University, Stanford, CA  94305

jet behavior may be large, but that available experimental data do not adequately address the effects of distortion.

An ongoing research program at Stanford is focused on distortion impacts and the dynamics of turbulence in buoyancy-influenced flows. The present paper deals with (1) analytic results which describe the impact of jet distortion on turbulence production and dissipation and (2) experimental results which quantify the relationships between distortion and jet parameters.

## 2.   SIMILITUDE CRITERIA

Commonly accepted similitude criteria for physical model design require that (1) the Froude number (Fr) be the same in model and prototype, (2) the densimetric Froude number (Fd) be the same in model and prototype if buoyant forces are important, and (3) the Reynolds number (Re) be maintained large enough to ensure fully turbulent conditions in the model. Full dynamic similarity requires, however, that the Reynolds number also be the same in model and prototype and that the same scale factor be applied to all model dimensions. The use of a distorted model precludes the satisfaction of these last requirements and results in non-similar model behavior.

Numerous model/prototype comparisons (e.g., Ackers and Joffrey, 1972 and Neale, 1971) have demonstrated that under conditions where advection and buoyancy dominate the transport process (i.e., far-field), good agreement between model and prototype can be obtained despite incomplete similarity. Considerable evidence of incorrect model behavior under conditions where turbulence and mixing dominate transport (i.e., near-field) have also been presented (see Moretti and McLaughlin, 1977, Haggstrom, 1978, and Fischer and Holly, 1971).

Kuhlman and Chu (1981) show that, for a buoyant surface jet in a crossflow, smaller jet Reynolds numbers result in considerably less transverse and vertical mixing in the jet near-field, but have minimal effect on far-field conditions. Goldschmidt and Bradshaw (1981) found that turbulence levels at a jet entrance have a significant effect on jet spreading rates in the near-field.

## 3.   ANALYTIC STUDY RESULTS

Fractional analysis of the governing equations for three-dimensional turbulent flow was undertaken. They represent instantaneous values of momentum, heat balance, continuity, pressure, and state. By spatial averaging, the instantaneous quantities are separated into averaged and fluctuating (turbulent) components. Two interrelated sets of equations arise: the averaged equations representing large scale motions and a corresponding set of turbulence equations representing small scale motions. The turbulence equations are a variant of the Reynolds stress transport closure adopted by Launder, et al. (1974) and are obtained by manipulating the averaged momentum and heat balance equations. Each equation is made dimensionless by assigning a characteristic scale to each variable and defining a dimensionless variable as the ratio of the actual variable to the characteristic scale. The dimensionless equations, after simplifying assumptions, are shown in Figure 1. The equations are written in tensor notation with: i,j,k - tensor subscripts; c, $C_i$ - turbulent and averaged velocity scales; $\ell$, $L_i$ - turbulent and averaged length scales; D - model depth; g - gravitational acceleration; H', H - turbulent and averaged temperature scales; p - pressure; u, $U_i$ - turbulent and averaged velocity; x,y,z - cartesian coordinates; $Rs_i$ - ratio of the ith length scale to the horizontal length scale; $\rho$ - density; $\theta$ - temperature; $\tau$ - time scale; and the overbar indicates spatial averaging.

Many of the terms in the dimensionless equations contain dimensionless groups which include Reynolds Number = Re = $C_1 D/\nu$ ; Euler Number = $\rho_o C_1{}^2/\psi$; Prandtl Number = Pr = $\nu/K$; Froude Number = Fr = $C_1/\sqrt{gD}$; Densimetric Froude Number = Fd = $C_1/\sqrt{gD\Delta\rho/\rho_o}$; Turbulent Reynolds Number = Re' = c $\ell/\nu$; Turbulent Euler Number = $\rho_o c^2/\psi'$; Turbulent Froude Number = Fr' = $c/\sqrt{g\ell}$; and Turbulent Densimetric Froude Number = Fd' = $c/\sqrt{g\ell \alpha H'/\rho_o}$. The traditional groups represent the force ratios used in physical model design, while the turbulent groups are analogs of these for the fluctuating components. Here $\psi'$ and $\psi$ = turbulent and averaged pressures, $\nu$ = kinematic viscosity, K = thermal diffusivity and $\alpha$ = bulk expansion factor. Any change in characteristic scales, as occurs when a model is distorted in the vertical dimension, influences the size of the dimensionless groups in the equations.

Selection of characteristic scales for the averaged quantities is based on model dimensions, inflow velocity, inflow temperature, and stratification. Characteristic scales for the fluctuating quantities are obtained through order of magnitude arguments and experimental data (see Tennekes and Lumley, 1972) and are related to the averaged scales through dimensionless ratios. We choose to include the larger, transportive turbulent structures in the averaged quantities. Between these large scales and the very small scale, dissipative structures lies a broad range of structure sizes forming (often) an "inertial subrange" and permitting (always) a cascade of energy from the transportive scales to the dissipative scales. The cut-off between averaged and turbulent scales is the largest scale which is effectively isotropic, only weakly transportive, and within the inertial subrange. Based on criteria for isotropy (Tennekes and Lumley, 1972) dimensionless scale ratios are obtained as $\ell/L_1 = 10^{-2}$ and $c/C_1 = 10^{-1}$.

Substitution of characteristic scales into the dimensionless equations produces order of magnitude estimates for each term in an undistorted model and in a distorted model with vertical scales ten times larger. The traditional similitude criteria (equal Pr, Fr, and Fd) have been maintained between the two models. The results, as summarized in Figure 1, provide insights into the effect of vertical scale distortion on turbulence. First, in the averaged momentum and heat balance equations, diffusion is negligible for the distorted case but dominates the undistorted case because Re is small. Second, no influence of distortion on convection, Reynolds stress or other terms in these equations is indicated. Third, the shear and buoyant generation terms dominate the Reynolds stress transport and turbulent heat flux equations for the undistorted model, but are not dominant in the distorted model. Fourth, viscous dissipation is small in both cases, but is influenced by distortion.

**Averaged Momentum**

$$\frac{\partial u_i}{\partial T} + u_j\frac{\partial u_i}{\partial x_j} = -\frac{1}{Eu\,Rs_i^2}\frac{\partial p}{\partial x_i} + \frac{1}{Fr^2 Rs_i^2}\frac{\partial}{\partial x_i}\int_z^{\zeta} p\,dx_3 + \frac{1}{Re\,Rs_j^2}\frac{\partial^2 u_i}{\partial x_j^2} - \frac{\delta_{i3}}{Fd^2 Rs_i^2}\Delta\Theta - \frac{c^2 L_1}{C_{in}^2\ell\,Rs_i}\frac{\overline{\partial u_i u_j}}{\partial x_j}$$

| | Time derivative | Convection | Pressure | Free surface | Diffusion | Buoyancy | Reynold stress | Condition |
|---|---|---|---|---|---|---|---|---|
| | 1 | 1 | 1 | 1 | $10^2$ | 0 | 1 | Undistorted |
| | 1 | 1 | 1 | 1 | $10^{-2}$ | 0 | 1 | 10:1 Distorted |

**Averaged Heat Balance**

$$\frac{\partial\Theta}{\partial T} + u_j\frac{\partial\Theta}{\partial x_j} = \frac{1}{Pr\,Re\,Rs_j^2}\frac{\partial^2\Theta}{\partial x_j^2} - \frac{c\,H'\,L_1}{C_{in}\,H\,\ell}\frac{\overline{\partial u_j\theta}}{\partial x_j}$$

| | Time derivative | Convection | Diffusion | Turbulent heat flux | Condition |
|---|---|---|---|---|---|
| | 1 | 1 | $10^{-1}$ | 1 | Undistorted |
| | 1 | 1 | $10^{-3}$ | 1 | 10:1 Distorted |

**Reynolds Stress Transport**

$$\frac{\overline{\partial u_i u_j}}{\partial t} + u_k\frac{\overline{\partial u_i u_j}}{\partial x_k} = -\frac{\ell C_{in} Rs_q}{L_1 C\,Rs_k}\overline{u_j u_k}\frac{\partial U_i}{\partial x_k} - \frac{\ell C_{in} Rs_i}{L_1 C\,Rs_k}\overline{u_i u_k}\frac{\partial U_j}{\partial x_k} + \frac{1}{Re}\frac{\overline{\partial u_i}}{\partial x_k}\frac{\partial u_j}{\partial x_k} + \frac{1}{Eu'}\frac{p}{\rho}\frac{\partial u_i}{\partial x_j} + \frac{\ell C_{in}^2 H'}{L_1 C^2 H\,Rs_3\,Fd^2}u_j\theta - \frac{\overline{\partial u_i u_j u_k}}{\partial x_k} - \frac{\delta_{ik}}{Eu'}\frac{\overline{\partial u_j p}}{\partial x_k} - \frac{\delta_{jk}}{Eu'}\frac{\overline{\partial u_i p}}{\partial x_k}$$

| | Time derivative | Convection | Shear generation | Viscous dissipation | Pressure scrambling | Buoyant generation | Diffusive transport | Condition |
|---|---|---|---|---|---|---|---|---|
| | 1 | 1 | 10 | $10^{-1}$ | 1 | 10 | 1 | Undistorted |
| | 1 | 1 | 1 | $10^2$ | 1 | 1 | 1 | 10:1 Distorted |

**Turbulent Heat Flux**

$$\frac{\overline{\partial u_i\theta}}{\partial t} + u_k\frac{\overline{\partial u_i\theta}}{\partial x_k} = -\frac{\ell H'}{L_1 H'}\frac{Rs_q}{Rs_k}\overline{u_i u_k}\frac{\partial\Theta}{\partial x_k} - \frac{\ell C_{in} Rs_q}{L_1 C\,Rs_k}\overline{u_k\theta}\frac{\partial U_i}{\partial x_k} + \frac{1}{Pr\,Re'}\frac{\partial\theta}{\partial x_k}\frac{\partial u_i}{\partial x_k} + \frac{1}{Eu'}\frac{p}{\rho}\frac{\partial\theta}{\partial x_k} + \frac{\ell c^2 H'}{L_1 c^2 H\,Rs_3\,Fd^2}\theta^2 - \frac{\overline{\partial u_i\theta u_k}}{\partial x_k} - \frac{\delta_{ik}}{Eu'}\frac{\overline{\partial\theta p}}{\partial x_k}$$

| | Time derivative convection | Mean field generation | Dissipation | Pressure scrambling | Buoyant generation | Diffusive transport | Condition |
|---|---|---|---|---|---|---|---|
| | 1 | 10 | $10^{-1}$ | 1 | 10 | 1 | Undistorted |
| | 1 | 1 | $10^{-2}$ | 1 | 1 | 1 | 10:1 Distorted |

Figure 1: Dimensionless equations and order of magnitude estimates for the undistorted and distorted models

Distortion significantly reduces the generation (or suppression) of turbulence by shear and buoyant forces. The dominance of generation terms in the undistorted model, but not in the distorted model implies that changes in turbulent interactions (e.g., energy transport downscale or eddy sizes and shapes) may also occur. Reduced generation will result in lower turbulence intensities (relative to the averaged flow) in the distorted model and reduced influence of turbulence on the averaged flow. Hence, dimensionless mixing will be less in a distorted model jet than in a prototype or undistorted model.

## 4.  DESCRIPTION OF EXPERIMENT

The Variable Distortion Facility (VDF) was designed to represent a "typical" lake or cooling pond of roughly kidney-shaped plan form, a trapezodial cross-section, and with a rectangular surface jet inflow and submerged withdrawal. A detailed description of the apparatus, instrumentation and experimental data is given in Roberts (1981). The VDF dimensions are 10 m maximum length by 4.6 m maximum width with adjustible depths from 0.15 m to 1.5 m. This ten-fold change in depth allows a shallow "undistorted" model to be compared with a 10:1 distorted model. All horizontal model dimensions are invariant, while all vertical dimensions scale with the depth. Three distortions were used: 10:1, 2:1, and 1:1. Jet entrance geometry (aspect ratio) changed from square in the undistorted model to a vertical slot in the 10:1 distorted model. Discharge temperatures and velocities were set so that Fr and Fd were maintained, while Re was kept in the turbulent range in the near-field region. Three different temperature regimes were studied: isothermal; buoyant jet into an initially isothermal model and mid-density jet into a linearly stratified model. The following additional parameter values obtain for the heated inflow case at three distortions.

| Parameter | | Model | |
|---|---|---|---|
| Heated Inflow | Undistorted | Distorted | |
| 1:1 | 2:1 | 10:1 | |
| Discharge Velocity $U_j$ (m/s) | 0.071 | 0.100 | 0.224 |
| Discharge Temperature Rise $(H_j-H_L)$( C) | 8 | 8 | 8 |
| Discharge Geometry $(W_j \times D_j)$(mm x mm) | 25 x 25 | 25 x 50 | 25 x 250 |
| Jet Reynolds Number $(Re_j)$ | $1.8 \times 10^3$ | $5 \times 10^3$ | $5.6 \times 10^4$ |
| Jet Froude Number $(Fr_j)$ | 0.142 | 0.142 | 0.142 |
| Densimetric Froude  Number $(Fd_j)$ | 3.65 | 3.65 | 3.65 |
| Characteristic Time $\tau = (=L_L/U_j)$(s) | 141 | 100 | 44.6 |

Here the subscript L refers to overall scales and j refers to jet scales. Key dimensionless numbers are defined as: $Re_j = U_j D_j/\nu$; $Fr_j = U_j/\sqrt{gD_j}$; $Fd_j = U_j/\sqrt{g(\Delta\rho/\rho)D_j}$. The isothermal condition uses the same jet Froude number as the heated jet case. The stratified condition incorporates a total stratification temperature difference of 8°C, an initial jet temperature 4°C above minimum lake temperature, and a densimetric Froude number of 5.06.

Data collection during model runs included temperature and two-dimensional velocity (longitudinal-vertical plane) profiles in the jet near-field and temperature profiles at 26 locations in the far-field. Only temperature and velocity profiles within the near-field jet region are discussed here. Temperature profiles were taken at the jet centerline and in the jet mixing region 75 cm from the discharge using thermistor probes. Crossed hot film probes obtained velocity profiles at the same locations. Rotamine dye studies were used to visualize the trajectory, spreading rates, and general turbulence structure of the jets and to locate the probes within the jet.

5.  EXPERIMENTAL RESULTS

The three distortions have different velocity, depth, and time scales and were not made under identical thermal conditions (due to variations in ambient lab temperatures); dimensionless quantities are used to report the results. These quantities are: Temperature -- $\theta^* = (\theta - \theta_L)/(\theta_i - \theta_L)$; Depth -- $D^* = D/D_L$; Mean Velocity -- $U^* = U/U_{max}$; Turbulent Intensity -- $u^* = u/U_{max}$; and Time -- $T^* = T/\tau$, where $\theta$, D, U, u, and T are the measured values at a given point and $U_{max}$ is the maximum mean velocity in the profile.

Although profiles of velocity and temperature were taken for the nine experiments, only the profiles obtained during the heated inflow runs are presented. They are representative of distortion effects for the isothermal and stratified profiles as well. The dimensionless temperature and velocity profiles (Figures 2 and 3) show the depth of penetration of the jet and extent of vertical mixing at $T^* = 7$. The discharge channel (and initial jet depth) extends down to $D^* = 0.17$. A substantial decrease in dimensionless vertical mixing accompanies distortion. However, the downward penetration of temperature is considerably larger than that for velocity, indicating different rates of diffusion for heat and momentum. A consistent relationship between distortion and jet penetration is clear. The turbulent intensity profiles (Figure 4) show a similar decrease in penetration. The integral of these profiles gives a rough measure of the strength of turbulent fluctuations relative to the mean flow in the jet at each distortion and shows a clear, consistent decrease in the strength of these dimensionless fluctuations with increasing distortion.

Spreading decreases with increasing distortion. As shown in Figures 5 and 6, this distortion impact is consistant for each thermal regime, but varies with the strength of the buoyant forces. Maximum dimensionless vertical penetration ($d^*_{max}$) is the depth (normalized by maximum model depth) at which the mean longitudinal velocity becomes essentially zero. Half-width is one-half of the total jet width normalized by discharge width. Half-depth appears to be proportional to $R_D^{-0.4}$ ($R_D$ is the distortion ratio), while half-width is proportional to $R_D^0$, $R_D^{-0.2}$, and $R_D^{-0.3}$ for the isothermal, heated jet, and stratified cases, respectively. With the exception of the 2:1 isothermal model values, data scatter is small. Similarly (Fig. 7), dimensionless turbulence intensities for the heated case are proportional to $R_D^{-0.24}$ while dimensionless centerline surface temperatures are proportional to $R_D^{+0.21}$. There appears to be a direct relationship between the strength of turbulent motions and the amount of mixing (dilution) that occurs in the jet.

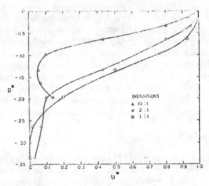

Figure 2  Dimensionless Mean Longitudinal Velocity Profiles at Jet
Centerline and $T^* = 7$ for Heated Discharge Experiments

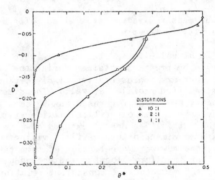

Figure 3  Dimensionless Temperature Profiles at Jet Centerline and
$T^* = 7$ for Heated Discharge Experiments

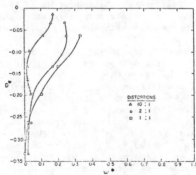

Figure 4  Dimensionless Vertical Turbulent Intensity Profiles at
Jet Centerline and $T^* = 7$ for Heated Discharge Experiments

6.  DISCUSSION

Dimensionless jet mixing rates are strongly impacted by model
distortion, despite maintenance  of Froude number and densimetric
Froude number similarity.  The evidence supports the concept of a
power law relationship between model distortion and key jet
characteristics.  There appears to be a direct correspondance between
dilution and the strength of turbulent fluctuations in the jets, as
exemplified by the inverse power laws for jet surface temperature and
jet turbulence intensities.

Fractional analysis results suggest linear reductions in
dimensionless turbulence generation with distortion.  However, since
dimensionless turbulent energy dissipation and turbulent interactions
also decrease in response, observed turbulence intensities are less
sensitive to distortion than an abstract examination of the
dimensionless equations suggests.  This is borne out by the
experiments.  Indeed, Haggstrom (1978) concludes that measured jet
parameters such as centerline temperature are overpredicted by a
factor of $R_D^{0.5}$, while jet vertical spreading rates are underpredicted
by the same factor.  In distorted model experiments Crickmore (1972)
observed that transverse mixing in a heated discharge was exaggerated
while vertical mixing was underpredicted.  Other distorted model
experiments indicate varying effects of distortion on surface
temperatures and jet spreading (see, for example, Hindley, et al
1971)  It is, however, difficult to compare different experimental
findings, because variations in outlet geometry, stratification, jet
Reynolds number and other parameters cause changes in jet behavior
independent of distortion.

Distortion of a model jet, assuming Froude number similarity,
causes several changes in jet characteristics which can be related to
mixing.  Although jet Reynolds number increases with distortion, the
increased turbulence generation due to shear does not maintain
dimensionless turbulence intensities high enough to compensate for
distortion in the vertical dimension.  Under buoyant jet conditions,
increases in turbulence generation by buoyant forces are not
sufficient to compensate for increases in advection (due to higher
inertial forces).  The net results are decreases in dimensionless
transverse and vertical spreading of the jet and higher jet
temperatures and mean velocities within the jet.

7.  CONCLUSIONS

Quantitative relationships between distortion and jet mixing were
obtained and can be linked to fundamental changes in the physics of
turbulence production.  However, the effects of changes in jet
Reynolds number and distortion of jet entrance geometry have not been
independently quantified.  Both changes accompany model distortion and
each change, by itself, modifies the dynamics of jet turbulence.  It
will be necessary to create experiments which represent them
separately.  Experiments at other distortions will also be needed to
further verify the power law relationships obtained.

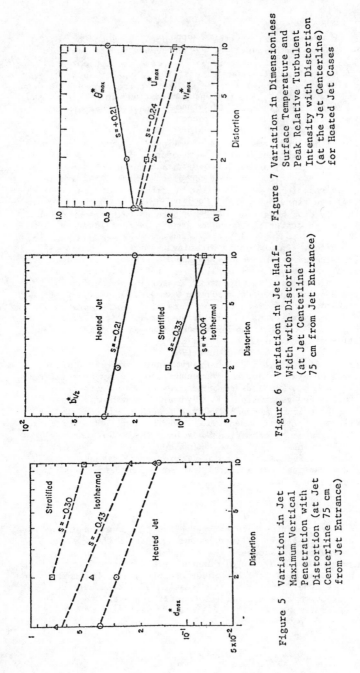

Figure 5  Variation in Jet Maximum Vertical Penetration with Distortion (at Jet Centerline 75 cm from Jet Entrance)

Figure 6  Variation in Jet Half-Width with Distortion (at Jet Centerline 75 cm from Jet Entrance)

Figure 7  Variation in Dimensionless Surface Temperature and Peak Relative Turbulent Intensity with Distortion (at the Jet Centerline) for Heated Jet Cases

ACKNOWLEDGMENT

This material is based upon work supported by the National
Science Foundation under Grant Nos. ENG-7713880 and CME-7921324.

REFERENCES

Ackers, P. and Joffrey, L. J. "The Applicability of Hydraulic Models
to Pollution Studies", Paper 16, Mathematical and Hydraulic
Modeling of Estuarine Problems, Proceedings of Symposium held at
the Water Pollution Research Laboratory, London, 1973.
Crickmore, M. J. "Tracer Tests of Eddy Diffusion in Field and Model",
J. Hydraulics Div. Proc. ASCE, Vol. 98, 1972, pp. 1737-52.
Fischer, H. B. and Holly, E. R. "Analysis of the Use of Distorted
Models for Dispersion Studies", Water Resources Research, Vol. 7,
pp. 46-51.
Haggstrom, S. Surface Discharge of Cooling Water: Effects of
Distortion in Model Investigations, Report Series A.3, Chalmers
University of Technology, Goteborg, Sweden. 1978.
Goldschmidt, V. W. and Bradshaw, P. "Effects of Nozzle Exit
Turbulence on the Spreading (or Widening ) Rate of Plane Free
Jets", Joint Bio-E., Fluids E., Applied Mech. Conf., ASME Boulder,
CO, Preprint 81-FE-22, June 1981.
Hindley, P. D., Miner, R. M. and Cayot, R. F. "Thermal Discharge: A
Model-Prototype Comparison", J. Power Div., ASCE, P04, 1971, pp.
783-98.
Kuhlman, J. M. and Chu, Li-C. "Reynolds Number and Ambient Turbulence
Effects on the Nearfield Mean Characteristics of a Laboratory Model
Thermal Plume", Joint Bio-E., Fluids E., Applied Mech. Conf., ASME
Boulder, CO, Preprint 81-FE-7, June 1981.
Launder, B. E. "On the Effects of a Gravitational Field on the
Turbulent Transport of Heat and Momentum", J. Fluid Mech., Vol. 67,
1974, pp. 569-81.
Moretti, P. M. and McLaughlin, D. K. "Hydraulic Modeling of Mixing in
Stratified Lakes", J. Hydraulics Div., ASCE, Vol HY4, 1977, pp.
367-79.
Neale, L. C. "Chesapeake Bay Model Study for Calvert Cliffs", J. Power
Div., ASCE, Vol. P04, 1971, pp. 827-839.
Neale, L. C. and Hecker, G. E. "Model Versus Field Data on Thermal
Plumes from Power Stations", Int. Symp. on Stratified Flows, IAHR,
Novosibirsk, 1972.
Roberts, B. R. "Impact of Vertical Scale Distortion on Turbulent
Mixing in Hydraulic Models", Ph.D. Dissertation, Stanford
University, June 1981.
Ryan, P. J. and Harleman, D. R. F. An Analytic and Experimental Study
of Transient Cooling Pond Behavior, Rep. No. 161, Parsons Lab.,
M.I.T., January, 1973.
Shirazi, M. "Some Results for Experimental Data on Surface Jet
Discharge of Heated Water", First World Conf. on Water Resources,
Chicago, 1973.
Tennekes, H and Lumley, J. L. A First Course in Turbulence, The MIT
Press, Cambridge, 1972.

# MASS TRANSFER ACROSS A DEAD ZONE BOUNDARY

by

Linda S. Weiss,[1], A.M.ASCE, and Chin Y. Kuo,[2], A.M.ASCE

## ABSTRACT

Mass transfer coefficients across the boundary of the dead zone caused by a sudden open channel expansion were determined by means of a laboratory experiment.  Based on statistical analysis, the coefficients significantly correlate with the Froude number and the Reynolds number for the case of salt tracer build-up and release from the dead zone. The coefficients were found to have the same range as the values reported by Valentine and Wood for bottom cavities and Westrich for side pockets.  The coefficients were quite independent of the location within the dead zone for the flow condition and channel configuration investigated.  Both the first order and the second order kinetics assumption were made to find the mass transfer coefficients.

## INTRODUCTION

Recent research has been performed to study the pollutant transport process in rivers with dead zones.  The "dead zone" or "stall region", based on Prandtl's boundary layer concept, has been described as a region of backflow, when compared to the main flow direction (5).  Some examples of these separated regions are sudden increases or decreases in cross-sectional areas in a channel, groins, bridges, river bends, etc.

By determining the mass transfer rate across dead zone boundary, it is possible to incorporate the dead zone effects into the study of pollutant transport in waterways with various flow conditions and channel geometric configurations.  For example, Valentine and Wood (6,7,8) have studied the dispersion process in terms of stationary eddy

---

[1]Graduate Student, Dept. of Civil Eng., Univ. of Illinois, Champaign-Urbana, Il.; formerly Graduate Student, Dept. of Civil Eng., Virginia Polytechnic Institute and State University, Blacksburg, Va.

[2]Assoc. Prof., Dept. of Civil Eng., Virginia Polytechnic Institute and State Univ., Blacksburg, Va.  24061.

structures or bottom cavities, adjacent to the channel bed. Westrich (9) has presented a model which simulates an adjacent recirculating region, or side pocket, of rectangular shape. Applied to the non-uniformities in channel bottoms and riverbanks of natural streams, these analyses have provided a means to study the rate at which the pollutants are trapped or released in bottom cavities and side pockets. In terms of mathematical modeling of pollutant transport, the transfer process can be included in the source or sink term of the one-dimensional advective diffusion equation.

This paper represents an analysis of pollutant build-up and release in the dead zones of an open channel expansion by means of a laboratory experiment. The investigation focuses on the mass transfer of salt tracer between the main flow of the open channel and the dead zone caused by a sudden channel expansion. The main purpose of the study is the determination of the transfer coefficient K for varying flow conditions and at different locations within the dead zone if the co-efficients are found to be non-uniform within a zone. Assumptions of first-order or second-order kinetics will be examined based on the equation of conservation of mass of dispersant similar to the one developed by Valentine and Wood (7).

## THEORETICAL CONSIDERATIONS

Valentine and Wood (6,7) have developed a differential equation to describe the interchange of tracer mass between the dead zone and the main flow. This equation, applied to the parameters used in the sudden channel expansion, has the following form:

$$\frac{dC}{dt} = \frac{KV_o}{E} (C_o - C) \tag{1}$$

in which $C_o$ = constant salt concentration in the main flow; $C$ = salt concentration in the dead zone at any time $t$; $E$ = expansion width; $V_o$ = mean upstream channel velocity; and $K$ = mass transfer coefficient. In this equation, first order kinetics assumption is made that the concentration varies linearly with respect to time. Dividing both sides of Eq. (1) by $C_o$, integrating, and then solving for the non-dimensional K,

$$K = \frac{E}{V_o(t_2 - t_1)} \ln \frac{1-(C/C_o)_1}{1-(C/C_o)_2} \tag{2}$$

Thus, by plotting $\ln(1-C/C_o)$ versus $t$, a dimensionless K may be evaluated by multiplying the average slope (obtained from the plotted curves) by the value $E/V_o$. Terms with subscripts 1 and 2 refer to time $t_1$ and $t_2$.

If a second-order kinetics is postulated with K assumed as a constant, the basic second-order differential equation, where K is non-dimensional, is

$$\frac{dC}{dt} = \frac{KV_o}{C_o E} (C_o - C)^2 \tag{3}$$

with the parameters defined as before. Dividing both sides of Eq. (3) by $C_o$, integrating, and solving for K,

$$K = \frac{E}{V_o} \frac{1/(1-C/C_o)_2 - 1/(1-C/C_o)_1}{t_2 - t_1} \tag{4}$$

Thus, by plotting $1/(1-C/C_o)$ versus t, one can obtain K values by multiplying the average slope from the plotted curves by $E/V_o$.

## EXPERIMENTAL INVESTIGATION

Dimensions used in the construction of the flume were based on those reported by Abbott and Kline (1,2) in their reports which investigated reattachment lengths for dead zones resulting from sudden expansion. The flume was constructed of 3/8 in (0.95 cm) Plexiglas to permit visualization. City water was allowed to flow through a 2 in. (5.08 cm) pipe into the head tank and was disposed of after use. The flume was composed of three main parts (see Fig. 1): (i) a head tank, (ii) a 12 in. (30.48 cm) wide and 8 ft (2.44 m) long inlet section with straight parallel walls, and (iii) an 8 ft (2.44 m) long test section adjustable in expansion width from 0 to 6 inch (0 to 15.24 cm). All joints were smooth and sealed with DMC solvent adhesive and reinforced with fasteners. The flume was leveled to an approximately horizontal position by means of a transit.

A Leeds and Northrup 7075-1 Conductivity Monitor, which provided continuous liquid monitoring with output capabilities, was used with a Gould 200 Recorder to measure the salt concentration. A Leeds and Northrup 4905 conductivity probe was used, since it is a temperature-compensated flow-through cell. The probe was placed at different locations within the dead zone. A probe was also placed in the main stream to measure $C_o$ for comparison purposes. Flow disturbance due to probe placement was insignificant in terms of shape and reattachment length of the dead zone.

Flow rates were determined by taking volumetric measurements per unit time. Rating curves of downstream water depth $y_{down}$ versus flow rate Q for various downstream channel widths were developed for use during the experiments. Flow rates varied from approximately 0.02 cfs (566.34 $cm^3$/sec) to 0.15 cfs (4247.53 $cm^3$/sec). For the flow rates used, a Reynolds number ($N_R = Ry_{up}/\nu$) range of approximately 1000 to 4600 was obtained. (R, $y_{up}$ and $\nu$ are hydraulic radius, upstream water depth and kinematic viscosity, respectively). This Reynolds number range includes values in both the upper transitional and turbulent regions. The transitional range for open-channel flow can be assumed to be $N_R$ = 500 to 2000 (3). Froude numbers $N_F = V_o/\sqrt{gy_{up}}$ ranged from

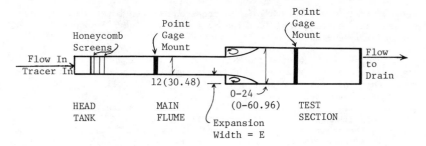

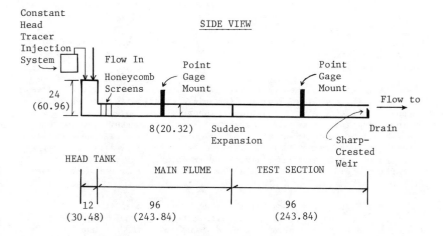

Figure 1. Schematic diagram for experimental set-up, all
dimensions in inches (centimeters)

approximately 0.013 to 0.077 (g is acceleration due to gravity). These
values, all in the subcritical range, are low. Higher Froude number
ranges will be studied in the future.

In order to determine the length of the dead zone for placement of
the probe, a visualization technique was employed in which a line of
methylene blue dye was placed into the flow on the wall upstream from
the sudden expansion. Step heights of 2 in. (5.08 cm), 3 in. (7.62 cm),
and 5 in. (12.70 cm) were used. These step heights provided for both
the symmetrical (2E/b<0.5) and unsymmetrical cases (2E/b>0.5) as
described by Abbott and Kline, although major emphasis was placed on the
symmetrical cases (b is the upstream channel width). Salt concentration
increase or decrease as a function of time was measured at different
locations within the dead zone, with one location set for one experiment
run.

## RESULTS

From Eqs. (2) and (4), K values were determined and plotted as a
function of Froude number and Reynolds number, for both tracer build-up
and release from the dead zone. Examples of this type of plot are
depicted in Figs. 2 and 3.

Statistical Analyses were performed using the "Statistical Analysis
Systems" (SAS) computer program. In order to determine the general
trend of the calculated K values, a linear regression analysis (4) was
employed to find the straight line that best fit the set of data points.
The Pearson product-moment correlation coefficient, r, and an overall
f-test was performed to determine how closely the data 'fit' a straight
line, and if there was a significant correlation between the mass
transfer coefficient K and either the Froude number or Reynolds number
for the data.

For a Froude number range less than 0.1, the mass transfer coefficient K was found to increase with increasing Froude number for
tracer build-up in the dead zone, and K was found to decrease with
increasing Froude number for tracer release (from a larger negative
number to a smaller negative number, where the negative sign indicates
releasing, the direction of mass transfer being out of the dead zone).
Similarly, for a Reynolds number range of approximately 1000 to 4600,
the mass transfer coefficient K was found to increase with increasing
Reynolds number for tracer flushing from the dead zone. Statistical
analysis determined that there was a significant correlation between K
and both Froude number and Reynolds number for the cases of tracer
build-up and release from the dead zone.

Analyses were made assuming both first-order and second-order
kinetics. The non-dimensional mass transfer coefficients were found to
have larger numerical values for the second-order assumption, although
the basic trends were similar to those of a first-order kinetics assumption. Statistical analysis determined, however, that the values
obtained for K using the first-order kinetics assumption were more
significantly correlated with Froude number and Reynolds number for both

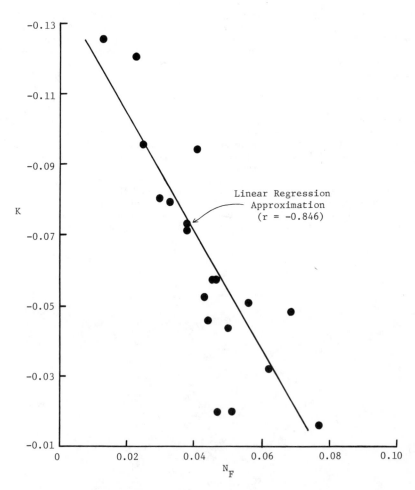

Figure 2.  Typical effect of Froude number on mass transfer
coefficient.  First-order kinetics, tracer release
case, $2E/b = 0.33$.

tracer build-up and release than were the values obtained using the second-order kinetics assumption.

Non-dimensional K values ranged from approximately 0.01 to 0.11 for the first-order kinetics assumption for tracer build-up (-0.01 to -0.11 for tracer release) as compared with Valentine and Wood's K=0.02 for bottom cavities and Westrich's K=0.019 for side pockets.  For the second-order kinetics assumption, non-dimensional K values ranged from approximately 0.02 to 0.50 for tracer build-up (-0.02 to -0.50 for tracer release).

An individual f-test was performed to examine if location within the dead zone played a significant role in the determination of K.  It was found that the value of K was independent of the location within the dead zone for both tracer build-up and release, for the range of Froude numbers and Reynolds numbers employed in the experiment.  Extensions to this study would include a higher Reynolds number and Froude number range for the experimental run.

As a concluding remark, one can increase the understanding of the exchange process across a dead zone boundary through the determination of mass transfer coefficient K.  The knowledge obtained in this study, and other similar studies will improve the water quality modeling capabilities for a river by incorporating the dead zone effect.

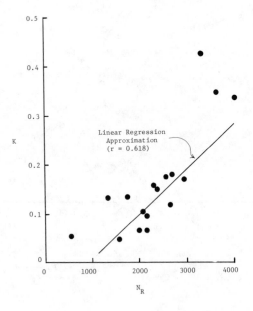

Figure 3.  Typical effect of Reynolds number on mass transfer
            coefficient.  Second-order kinetics, tracer build-up
            case, 2E/b = 0.33.

REFERENCES

1. Abbott, D. E., and Kline, S. J., "Experimental Investigation of Subsonic Turbulent Flow Over Single and Double Backward Facing Steps," Journal of Basic Engineering, Trans. ASME, Vol. 84, Sept., 1962, pp. 317-325.

2. Abbott, D. E. and Kline, S. J., "Theoretical and Experimental Investigation of Flow Over Single and Double Backward Facing Steps," Report MD-5, Stanford University, Stanford, California, June, 1961.

3. Chow, V. T., Open-Channel Hydraulics, McGraw-Hill Book Company, New York, N.Y., 1959.

4. Mendenhall, W., Introduction to Probability and Statistics, Duxbury Press, Massachusetts, 1975.

5. Prandtl, L., and Tietjens, O., Fundamentals of Hydro and Aero Mechanics, McGraw-Hill Book Company, New York, N.Y., 1974.

6. Valentine, E. M., and Wood, I. R., "Experiments in Longitudinal Dispersion with Dead Zones," Journal of the Hydraulics Division, ASCE, Vol. 105, No. HY8, Aug., 1979, pp. 999-1016.

7. Valentine, E. M., and Wood, I. R., "Longitudinal Dispersion with Dead Zones," Journal of the Hydraulics Division, ASCE, Vol. 103, No. HY9, Sept., 1977, pp. 975-990.

8. Valentine, E. M., and Wood, I. R., Comment on "Experiments in Longitudinal Dispersion with Dead Zones," by E. M. Valentine and I. R. Wood, Journal of the Hydraulics Division, ASCE, Vol. 107, No. HY3, March, 1981, pp. 373-375.

9. Westrich, B., "Simulation of Mass Exchange in Dead Zones for Steady and Unsteady Flow Conditions," Proceedings of International Symposium on Unsteady Flow in Open Channels, Newcastle-upon-Tyne, England, 1976.

# SMALL SCALE HYDROELECTRIC PROJECTS IN ALASKA

by

Donald R. Melnick, (1) M. ASCE,
Glenn S. Tarbox, (2) M. ASCE, and
Howard E. Lee, (3) M. ASCE

## ABSTRACT

Development of hydroelectric projects for small communities in rural Alaska presents challenges to the designer which are
not typically encountered in the hydraulic design of small hydroelectric projects located elsewhere. Due to the remoteness of project
sites from the load centers, lack of a power transmission grid system, and small size of the electrical loads to be served, small scale
hydroelectric projects in rural Alaska require special attention with
respect to selection of the optimum project size to match system
loads. These projects also require detailed analysis of system operating characteristics to assure proper interface. Such considerations extend beyond what might normally be required in the design of
small scale hydroelectric projects and are more often only a critical
factor in the sizing and design of much larger installations. This
paper outlines the details of sizing and design of the hydraulic
structures and equipment for two small scale hydroelectric projects
which are currently being developed in Southeast Alaska for the communities of Sitka and Ketchikan. Both projects, when combined with
existing hydroelectric generating resources, are designed to meet increasing electrical system capacity and energy requirements for these
communities for the next 5 to 10 years. Although the two projects
are quite similar in concept, each site has unique physical constraints which dictated consideration of different designs for many
of the individual project features. The Green Lake Project began
generating power to serve the City and Borough of Sitka in March
1982, and the Swan Lake Project, which is presently under construction, is scheduled to deliver power to the City of Ketchikan beginning in early 1984.

## INTRODUCTION

Sitka and Ketchikan are electrically isolated communities. Each has its own municipal electric utility system which operates a combination of diesel and hydroelectric power plants to supply
the bulk of the electricity used within the community. During the

---

(1)  Partner, R. W. Beck and Associates, Seattle, Washington
(2)  Associate, R. W. Beck and Associates
(3)  Principal  Engineer,  R.  W.  Beck  and  Associates

mid 1970's both communities began to explore the feasibility of adding additional hydroelectric facilities to their systems to enable them to meet long-term capacity and energy growth projections. Both Projects are the largest generating resources in their respective communities.

Figure 1 shows the relative locations of the two Projects in relation to the communities they serve.

The Green Lake Project is located approximately 12 air miles south of Sitka along an access road that was extended 7 miles through rugged terrain, as the first phase of Project construction, during the latter half of 1979. As shown in Figs. 2 and 3, the Project consists of a 210-foot-high double-curvature concrete arch dam, a 1,900-foot long power tunnel, a powerhouse at tidewater containing two 11,300-horsepower vertical shaft Francis turbines connected to synchronous generators, and approximately 7.4 miles of new 69-kV transmission line to connect the Project with the City's existing transmission network. Located at the mouth of Green Lake, the dam provides an active reservoir storage of 90,000 acre-feet above minimum pool and creates a maximum power head of 400 feet. The power conduit includes a vertical intake on the upstream face of the dam and a concrete-lined power tunnel driven through the hillside to the powerhouse on a 14% grade. The powerhouse, located about 30 feet north of the Vodopad River outlet on Silver Bay, is set into a hillside excavation and anchored to the rock slope. The turbines discharge into separate draft tubes which are connected to a single short tailrace that conveys flows from the power plant into the bay. Photographs of the nearly completed Green Lake Project are shown in Figs. 6 and 7 on pages 9 and 10.

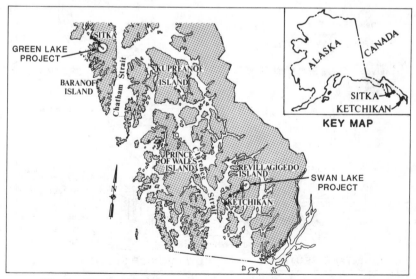

Fig. 1 - Green Lake and Swan Lake Projects - Location Map

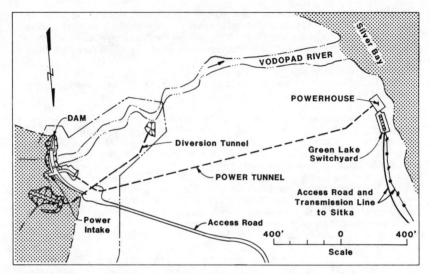

Fig. 2 - Green Lake Project - Site Plan

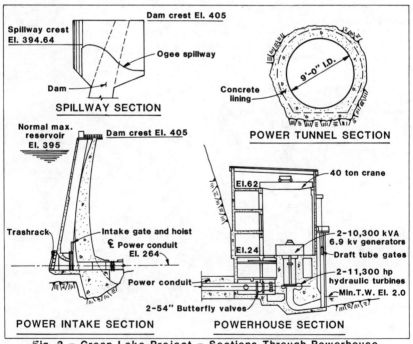

Fig. 3 - Green Lake Project - Sections Through Powerhouse,
Power Tunnel, Intake and Spillway

The Swan Lake Project is located approximately 22 air miles northeast of Ketchikan near the head of Carroll Inlet. The Project includes a 174-foot-high double-curvature elliptical concrete arch dam located on Falls Creek about 0.8 mile downstream from the outlet of the existing Swan Lake, a 2,300-foot-long power tunnel, a tidewater powerhouse containing two 13,300-horsepower vertical shaft Francis turbines connected to synchronous generators, and approximately 30 miles of 115-kV transmission line to deliver Project power to the load center in Ketchikan. The dam provides an active reservoir storage of 86,000 acre-feet above minimum pool and creates a maximum power head of 330 feet. The power conduit includes a vertical intake structure located on the right dam abutment and a concrete-lined power tunnel excavated through rock on an 11% grade leading to the powerhouse. The powerhouse, located adjacent to Falls Creek at its outlet on Carroll Inlet, is designed to convey flows through twin draft tubes into a short tailrace leading to the inlet. Design details of the dam, spillway, powerhouse, and power tunnel are similar to those shown for the Green Lake Project. Figure 4 illustrates several features of the Swan Lake Project. As construction is in its early stages, no photographs of the completed features are available at this time.

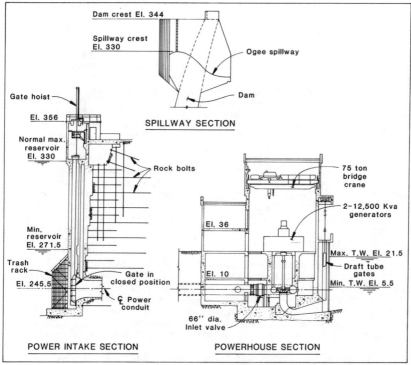

**Fig. 4 - Swan Lake Project - Sections Through Powerhouse, Power Intake and Spillway**

HYDRAULIC DESIGN

   1.  Dam and Power Plant Sizing

         Both Project sites are located in relatively small drain-
age basins within the heavily forested mountainous island terrain
that typifies much of Southeast Alaska.  The Projects are located on
streams for which limited streamflow records existed at the time feasibility studies were undertaken, and therefore required comprehensive studies of regional precipitation and runoff records to develop
estimates of expected basin runoff.  A computer model was utilized to
correlate precipitation and flow records to arrive at meaningful
estimates of long-term basin yields as well as seasonal distributions
of runoff for use in selecting optimum sizes for the dams and power
plants and for estimating the generating potentials of the projects.

         Because of flow and head limitations, most small scale
projects can fill only a small fraction of the load they are to
serve, and are therefore a minor part of the electrical generation
system.  Typically, a small scale hydroelectric project has constraints on its development which limit its capacity and energy potential to a finite range, so that a given project cannot be sized to
meet system load requirements but, rather, can only provide a small
increment of power.  By contrast, Sitka and Ketchikan both have an
abundance of good potential hydroelectric sites to choose from and
were faced with a wide range of hydroelectric resource options to
allow them to displace their existing expensive diesel-fired generating resources and at the same time provide additional power to supply
a major portion of their projected long-term demand and energy requirements.  Initial site selection surveys were performed to screen
out sites which had the least potential for meeting the desired objectives.  The selected sites were then subjected to detailed feasibility investigations to determine the optimum development to match
projected power requirements for the two individual communities.
Economic considerations played a major role in the process of selecting dam heights, reservoir storage volumes and power plant capacities
for each Project.  Optimization studies were performed using a computer model which simulated reservoir operation under varying load
conditions.  Each Project was designed to provide firm power generation to closely match the projected monthly demand pattern for its
respective community.  Each power plant was sized to provide system
dependable capacity which, when combined with other hydroelectric
resources in the system, would exceed projected system demands at the
on-line date of the Project, thereby allowing deferral of generation
from existing diesel-electric generating plants in the system.  As
electrical loads continue to grow in the future, the diesel units
will be called on only during peak load periods, to supplement hydroelectric generation.  Although the system electrical loads and resource requirements are different for Sitka and Ketchikan, the similarities in characteristics permit the use of only one set of demand
and energy curves to depict typical conditions for both communities.

         As shown in Fig. 5, when the Green Lake Project was
brought on-line in March 1982, the City and Borough of Sitka's dependable generating potential was increased from 15.1 MW to 28.6 MW.

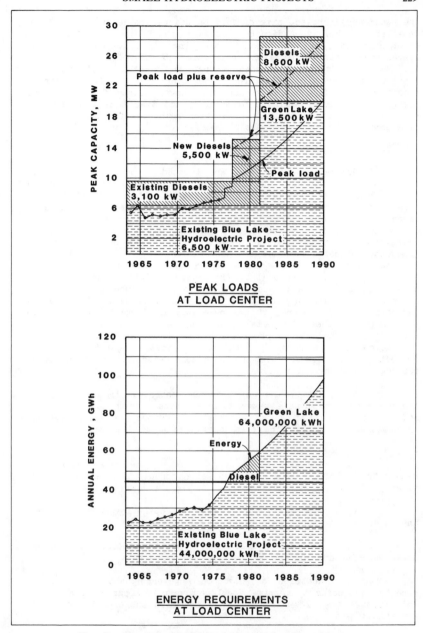

Fig. 5 – System Electrical Loads and Resource
Requirements: Sitka – Green Lake

The recommended Project size of 16.5 MW was determined as part of the
detailed feasibility studies conducted for the Project, and was based
on a comparison of net benefits over the first 10-year period of
operation for installed plant capacities ranging from 15.6 MW to
18.7 MW. The Project displaces generation from the City's more expensive to operate diesel-fired generating units and essentially
eliminates its dependence on oil-fired power plants for electrical
energy production. Both the Green Lake and Swan Lake Projects are
sized to serve as the primary generation resources for their respective communities for many years to come.

2.  System Frequency Regulation

Because both Projects are located in remote areas, each
plant is fully automated so that it can be remotely operated from a
central control station near the load center. The remoteness of the
sites, coupled with the relatively large size of the power plants in
comparison to the electrical generation systems in the two communities the Projects are designed to serve, added a somewhat unique
design constraint to these Projects. Both systems operate isolated
from interconnected transmission networks and, by nature of the independent operation of the generating units in the systems, neither
system enjoys the high quality frequency regulation experienced by
power systems which are interconnected to large power grids. Hydroelectric generating stations have inherently poor regulating characteristics and, therefore, it was necessary in formulating the basis
of design for these Projects to develop specific features for increasing their ability to provide a higher degree of system frequency
regulation than is normally required for small scale hydroelectric
projects. The following elements were considered for improving Project regulating characteristics:

a.  Addition of synchronous continuous bypass valves in
the powerhouse to bypass water to compensate during load rejection
and acceptance, allowing the governor to improve frequency response
by transferring water flow from the turbine to the valve or vice
versa;

b.  Addition of a surge tank to the power conduit to
reduce the effective length of the water conduit and water starting
time;

c.  Decreasing the wicket gate operating time to permit
the turbines to respond more rapidly to load changes;

d.  Increasing the power conduit diameter to decrease
water velocity and thereby reduce pressure transients;

e.  Addition of electronic governors to reduce speed
rise by responding to machine acceleration as well as machine speed.
This enables the wicket gates to be moved in anticipation of speed
deviations; and

f.  Increasing the rotational inertia of the generator
to reduce speed change for a given load change.

In both cases, the recommended design was to incorporate the latter three elements into the Projects.  Additional capital costs for increasing the tunnel diameters were partially offset by a corresponding increase in hydraulic efficiency due to lower operating velocities.  Providing electronic governors and adding rotational inertia to the generators also increased Project capital costs, but did not have the significant cost impact that water wasting bypass valves or expensive surge tanks would have on long-term power revenues.  Also, while a decrease in the wicket gate operating time would permit the turbines to respond more rapidly to load changes, benefits were found to be limited by the fact that for a short period of time after the wicket gate opening is increased, the power output of the turbine actually decreases due to a pressure drop while the water column is accelerating.  The reverse happens when the wicket gate opening is decreased.  Decreasing the wicket gate operating time also results in higher pressures in the water conduit as a result of the rapid changes in water velocity.  Designing the power conduit for a larger pressure change would have added significantly to the cost of the power conduit.

### 3.  Principal Project Statistics

As previously mentioned, the two Projects are similar in design concept and arrangement due to striking similarities in the two Project sites.  For the same reason, many of the hydraulic features of the Projects are similar in design as well.  Table 1 following shows the principal statistical data for the two Projects to demonstrate their similarities.

### Table 1 – Comparison of Project Statistical Data

|  | Green Lake | Swan Lake |
|---|---|---|
| **BASIN HYDROLOGY** | | |
| Drainage Area Above Dam (sq.mi.) ... | 28.2 | 36.5 |
| Average Annual Runoff at Dam Site (AF) ........................ | 227,000 | 335,000 |
| PMF Inflow Volume (AF) ............ | 43,000 | 38,700 |
| **PROJECT POWER DATA** | | |
| Avg. Annual Energy Generated (kWh) . | 64,900,000 | 88,000,000 |
| Annual Firm Energy, Delivered (kWh) | 44,200,000 | 66,700,000 |
| Installed Capacity (MW) ........... | 16.5 | 22.5 |
| Dependable Capacity, Delivered (MW) | 13.7 | 18.0 |
| Normal Maximum Power Pool Elevation | 390 | 330 |
| Active Storage Capacity in Reservoir (AF) ................... | 90,000 | 86,000 |

continued on next page

### Table 1 (cont) – Comparison of Project Statistical Data

| | Green Lake | Swan Lake |
|---|---|---|
| **HYDRAULIC STRUCTURES AND EQUIPMENT** | | |
| Dam ................................ | Concrete arch | Concrete arch |
| Spillway .......................... | 100-foot ungated over-flow weir | 100-foot ungated over-flow weir |
| Plunge Pool ....................... | Natural still-ing pool | Excavated in rock to El 170 |
| PMF Peak Outflow (cfs) ............ | 32,600 | 20,600 |
| Power Intake Type ................. | Vertical single level with wheeled gate; on dam | Vertical single level with wheeled gate; on right abutment |
| Power Tunnel Length/Diam. (feet) ... | 1,900/9.0 | 2,300/11.0 |
| Powerhouse Type ................... | Above-ground; concrete | Above-ground; concrete |
| Turbine Type, Size ................ | Two vertical shaft Francis; rated output 8,430-kW each unit at best gate | Two vertical shaft Francis; rated output 10,140-kW each unit at best gate |
| Turbine Discharge at Full Gate and Rated Head, Each Unit (cfs) ...... | 363 | 550 |
| Turbine Inlet Valves, Type, Size ... | Two 54-inch butterfly | Two 66-inch butterfly |

Fig. 6 – Green Lake Hydroelectric Project – Upstream View of Dam

**Fig. 7 – Green Lake Hydroelectric Project – Interior View of Powerhouse**

CONCLUSIONS

        Although the Green Lake and Swan Lake Projects can be con-
sidered to fall within the classification of small scale hydro, the
factors considered in designing these projects and bringing power on-
line are as complex as they are for most major hydroelectric installations.  Furthermore, integrating these Projects into small, electrically isolated utility systems in Alaska posed design constraints
not normally encountered in the development of small scale hydroelectric projects located elsewhere in the country.  The two Projects,
although quite similar in concept, are tailored to their individual
sites and demonstrate many subtle design differences.  Both Projects
are designed to meet the unique system requirements of their respective communities and will serve the people of Alaska for many years
to come.

Retrofitting Power Plants
into a Distribution System


By C. F. "Ted" Voyles[1], Member ASCE


## ABSTRACT

        In the mid-1970's The Metropolitan Water District
of Southern California (MWDSC) embarked on an ambitious
program of retrofitting small hydroelectric power plants
into the pipelines of its wholesale water transmission and
distribution system. Techniques were developed to identify
economical sites and to determine optimum turbine sizing
and selection. Fourteen sites have been determined to be
technically and economically feasible for turbine/generator
installation, with additional sites under study. For the
most part the sites are adjacent to existing pressure-
control structures and take advantage of excess pressure
(energy) that previously had been dissipated and wasted.
The program is half way completed with the seventh plant
expected to become fully operational in July 1982. When
the $97 million program is completed in late 1984,
460 million kilowatt-hours (kWhr) of pollution-free energy
will be produced annually from the installed capacity of
77 megawatts (MW). The average payoff period to recover
all costs associated with this program is about 15 years.
(Table 1.)


## INTRODUCTION

        The MWDSC operates a large-diameter (3- to 17-
foot ID) transmission and distribution pipeline system
(approximately 700 miles in length) throughout the coastal
plain of six counties in Southern California. The service
area contains a population of 12 million people. Imported
water served by this system is pumped through the 240-mile-
long Colorado River aqueduct and the 440-mile-long
California aqueduct across intervening mountain ranges and
stored in terminal reservoirs which are approximately 1500
feet above sea level. Flow from these terminal reservoirs
and throughout the entire distribution system is by gravity
requiring several multiple-valve pressure-regulating stations in order to reduce the pressure and control the flow
within the service area which ranges from sea level to 1200
feet above sea level. (Chart 1.)
        The facilities selected for small hydro develop-
ment range from the first and smallest, a 1250-hp pump/

[1]Assistant to Chief of Operations
The Metropolitan Water District of Southern California
Los Angeles, California

| POWER PLANT | TYPE OF TURBINE[1] | TYPE OF GENERATOR[2] | CONSTRUCTION CONTRACT AWARD DATE | ESTIMATED ON-LINE DATE | INSTALLED CAPACITY (MW) | DESIGN NET HEAD (FT) | DESIGN FLOW (CFS) | ESTIMATED CAPITAL COST[3] ($ MILLION) | PREDICTED YEARLY GENERATION[4] (MILLION KW-HRS) | NO. OF HOMES PLANT CAN SERVE[5] | BARRELS OF OIL SAVED WITH PLANT IN OPERATION[6] PER DAY | PER YEAR | ACTUAL YEARLY GENERATION[7] (MILLION KW-HRS) |
|---|---|---|---|---|---|---|---|---|---|---|---|---|---|
| **PHASE 1** | | | | | | | | | | | | | |
| GREG AVENUE | P | SN | N/A | DEC. 1979 | 1.0 | 200 | 63 | $ 0.4 | 4.5 | 1,500[8] | 358 | 12,900[8] | 8.4 |
| LAKE MATHEWS | R | IN | NOV. 1978 | AUG. 1980 | 4.9 | 90 | 625 | 7.0 | 18.6 | 4,800[8] | 1168 | 42,200[8] | 27.4 |
| FOOTHILL FEEDER | R | IN | JAN. 1979 | APR. 1981 | 9.1 | 180 | 600 | 8.7 | 61.3 | 10,800 | 258 | 94,300 | |
| SAN DIMAS | I | SN | JUNE 1979 | JUNE 1981 | 9.9 | 400 | 300 | 7.7 | 68.2 | 12,100 | 287 | 104,900 | |
| YORBA LINDA FDR. | I | SN | SEP. 1979 | NOV. 1981 | 5.1 | 200 | 140 | 6.6 | 33.5 | 6,000 | 141 | 51,500 | |
| TOTAL PHASE 1 | | | | | 30.0 | | | 30.4 | 186.1 | 35,200 | 837 | 305,800 | |
| **PHASE 2** | | | | | | | | | | | | | |
| SEPULVEDA CANYON | I | SN | JAN. 1980 | JULY 1982 | 8.6 | 300 | 475 | $ 8.7 | 56.2 | 9,900 | 237 | 86,500 | |
| VENICE | R | SN | MAY 1980 | MAY 1982 | 10.1 | 280 | 475 | 7.1 | 60.0 | 10,600 | 253 | 92,300 | |
| TEMESCAL | R | IN | JAN. 1981 | SEP. 1982 | 2.8 | 135 | 300 | 7.9 | 18.0 | 3,200 | 76 | 27,700 | |
| CORONA | R | IN | JAN. 1981 | NOV. 1982 | 2.8 | 135 | 300 | 7.5 | 18.0 | 3,200 | 76 | 27,700 | |
| TOTAL PHASE 2 | | | | | 24.3 | | | $31.2 | 152.2 | 26,900 | 642 | 234,200 | |
| **PHASE 3** | | | | | | | | | | | | | |
| PERRIS | R | SN | SEP. 1981 | MAR. 1983 | 7.9 | 160 | 650 | $ 7.4 | 40.3 | 7,100 | 170 | 62,000 | |
| RIO HONDO | R | IN | MAR. 1982 | AUG. 1983 | 1.9 | 220 | 110 | 6.5 | 12.3 | 2,200 | 52 | 18,900 | |
| COYOTE CREEK | R | IN | SEP. 1982 | MAR. 1984 | 3.1 | 218 | 175 | 6.8 | 19.6 | 3,500 | 83 | 30,200 | |
| TOTAL PHASE 3 | | | | | 12.9 | | | $20.7 | 72.2 | 12,800 | 305 | 111,100 | |
| **PHASE 4** | | | | | | | | | | | | | |
| RED MOUNTAIN | R | SN | JAN. 1983 | JULY 1984 | 5.9 | 232 | 360 | $ 7.1 | 37.9 | 6,700 | 160 | 58,300 | |
| VALLEY VIEW | I | SN | MAY 1983 | SEP. 1984 | 4.1 | 430 | 130 | 7.5 | 13.6 | 2,400 | 57 | 20,900 | |
| TOTAL PHASE 4 | | | | | 10.0 | | | $14.6 | 51.5 | 9,100 | 217 | 79,200 | |
| GRAND TOTAL | | | | | 77.2 | | | $96.90 | 462.0 | 84,000 | 2,001 | 730,300 | |

1. I = IMPULSE   R = REACTION   P = PUMP   F = FLOWTHROUGH
2. IN = INDUCTION   SN = SYNCHRONOUS
3. "ESTIMATED CAPITAL COST" BASED ON JANUARY 1982 DOLLARS AND INCLUDES DISTRICT UTILITY INTERCONNECTION COSTS.
4. "PREDICTED YEARLY GENERATION" IS BASED ON AVERAGE YEAR IN 10-YEAR PERIOD.
5. "NUMBER OF HOMES PLANT COULD SERVE" IS BASED ON THE AVERAGE RESIDENTIAL ENERGY USE IN THE SOUTHERN CALIFORNIA EDISON SERVICE AREA = 5,650 KW-HR/YR.
6. ONE BARREL OF OIL IS EQUIVALENT TO 650 KW-HRS.
7. ACTUAL YEARLY GENERATION IS GIVEN ONLY FOR THOSE PLANTS WHICH HAVE BEEN ON LINE FOR AT LEAST ONE YEAR.
8. "NUMBER OF HOMES PLANT COULD SERVE" AND "BARRELS OF OIL SAVED..." ARE BASED ON "ACTUAL YEARLY GENERATION" FOR THOSE PLANTS WHICH HAVE BEEN ON LINE FOR AT LEAST ONE YEAR.

TABLE I   2/82

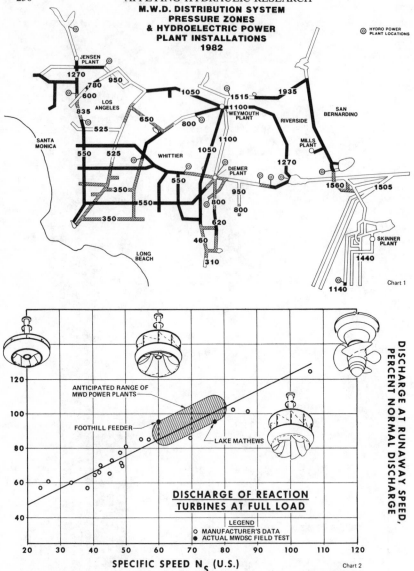

Chart 1

Chart 2

motor which was converted to a 1-MW turbine/generator by modification of the equipment and reversing flow direction to a recently completed 10.1-MW unit which operates between two high-pressure, closed-conduit systems (u.s. 400 psi, d.s. 220 psi).

The pipelines were designed and constructed prior to any thought of utilizing turbine/generators for energy

dissipation. It was necessary to limit turbine selection
to those with hydraulic characteristics which minimize the
potential for sudden flow changes since the resultant surge
and water hammer effects would be very critical.
     To assure the continued high degree of reliabil-
ity of water service, it was necessary to design each
facility to maintain nearly constant water flow and pressure in the service area with no noticeable effects as a
result of electrical system disturbances or turbine/
generator equipment malfunction. Provisions have been made
for locally automatic flow transfer to occur upon electrical system load rejection so that the water distribution
system's operation will be essentially unaffected when a
turbine/generator goes off line. These installations are
all designed for unmanned, unattended operation and are
remotely monitored and operated through the existing supervisory control system used by MWDSC in operating its water
distribution facilities. The designers were requested by
the Operations staff to provide sufficient local automatic
controls so that a new hydro-plant would appear as "just
another valve" to the remote operator.

                HYDRAULIC DESIGN CONSIDERATIONS

          As a general rule the MWDSC feeders and pipelines
were designed following the standard practice of selecting
an economic size based on an expected friction factor
existent at that point in time when the design flow rate
must be conveyed. Ideally, then whenever the design flow
occurs in combination with the design friction factor, all
the available head will be dissipated.
          The design flow was determined from water demand
projections including an allowance for a peak monthly flow
of 130% of the annual average flow. The minimum monthly
flow in a feeder which has reached its design condition is
about 70% of the annual average flow rate or about 50% of
its design flow. Current peaking conditions on the MWDSC
system are about 140% and 65% of annual average flow for
maximum month and minimum month respectively.
          The friction factors used in design were expected
to prevail after 30 to 40 years of operation; however, flow
tests indicate that nearly every pipeline still has maximum
carrying capacity of 10% + above design flow.
          Thus, there is hydraulic energy to be dissipated
during both the early years of operation of MWDSC transmission lines and during all months except the maximum even
when the design condition has been attained.
          The relationship between flow and head at a
typical MWDSC regulating facility is such that the maximum
power available for conversion from hydraulic to electrical
energy occurs when the pipeline is operating at 58% of the
maximum flow. Remember that head vs flow is an inverse,
nonlinear relationship, e.g., as flow increases from zero
to maximum, head available decreases from maximum to zero.
Preliminary studies had indicated that the operating limits
for turbine/ generator equipment would permit operation

238     APPLYING HYDRAULIC RESEARCH

whenever the flow in the pipelines is between about 25% and
75% of the maximum flow.
        The problem of sizing and designing a turbine/
generator installation to capture a portion of the wasted
hydraulic energy under the above operating conditions that
would be cost-effective was tackled by MWDSC Planning and
Engineering staffs.
        A 15-year economic pay back period was used to
evaluate each potential site.  The anticipated flow vs.
head relationship was developed for each month of the
15-year study period.  Operating characteristics of
turbine/generator machinery were then applied, an optimum
size determined, the total number of kilowatthours computed
and an initial cost estimated.  If the project indicated a
break-even point within 15 years after using a conservative
estimate for the value of energy and allowing for operation
and maintenance costs, then final design was initiated.
        Protection of the pipeline and structures from
the effects of water hammer associated with nearly instantaneous flow changes when a generator separates from the
power system and a reaction-type turbine goes to runaway
speed (approximately 180% to 200% of normal speed) was
necessarily given top priority in equipment design and
selection.
        A decision was made early in the preliminary
design phase to limit turbine/generator selections to those
which are intrinsically safe from a water hammer or surge
standpoint and accept lower efficiency and/or higher capital costs.  Thus, wherever possible, an impulse-type
(Pelton wheel) turbine is used because loss of electric
load causes no disturbance to the water system.  The use of
the reaction-type (Francis) turbine is unavoidable wherever
the downstream water level is much above the axis of the
turbine.  This is the most common case in the MWDSC pressure
pipeline system.  The turbine/generator specification set
requirements for equipment to be capable of going from
normal speed to runaway speed with a flow change not greater
than plus or minus 10 percent of design flow.  Additionally, the reaction-type turbine/generator must be capable
of remaining at runaway speed for a minimum of 30 and a
maximum of 95 minutes to allow sufficient time for the
water to be safely transferred away from the turbine back
through the parallel pressure-reducing station utilizing
the existing slow-operating valves.  From a detailed engineering analysis, it was determined that a reaction-type
turbine of a specific speed of about 75 would give a zero
flow change at runaway.  In an effort to err on the conservative side, the turbine specifications were written with a
goal of achieving specific speeds in the range of 60 to 80.
(See Chart 2.)  The adverse consequences of a slight flow
reduction are more amenable to water system operations than
those associated with an increase in flow upon turbine
runaway.
        Automatic flow transfer is provided at each site
to accomplish the flow transfer from the turbine to the
bypass regulating station.  Many sites are equipped with a
programmable controller to accomplish this function.  The

programmable controller is also being utilized for start-up and to assist remote operators in making flow changes.

## CASE HISTORIES

### Greg Avenue Power Recovery Plant

The Greg Avenue Power Plant, with a capacity of 1.0 MW, the first of the MWDSC energy recovery facilities was dedicated and put into initial service on December 27, 1979. It is unique in that no purchase of hydrogeneration equipment was required. This facility was initially constructed for the purpose of pumping water to serve the most westerly portion of the MWDSC service area. A series of pumps and surge tanks were installed for that purpose. As MWDSC distribution system was expanded, the need for this pumping station was eliminated and in fact flow was reversed in this feeder by removing 2 of the 3 strings of pumps and replacing them with sleeve valves for pressure reduction. Two 1250-hp centrifugal pumps in series were left as standby to provide the capability of again reversing the flow through the feeder for emergency service.

Since a centrifugal pump and a reaction turbine are physically similar, the conversion of the Greg Avenue facility to hydrogeneration required only reversing the direction of flow of water through one pump to make it a turbine and open the bypass around the second pump. The synchronous electric motor which operated the pump became a generator. Some electrical and mechanical changes were necessary to permit the pump/motor to operate safely as a turbine/generator. The generator rotor was strengthened structurally to withstand a higher sustained runaway speed, the exciter was replaced to permit bidirectional operation and a complete overhaul of all mechanical equipment was accomplished. Hydro-pneumatic surge tanks, originally provided for protection of the suction and discharge lines at the pump station, were fortunately left in place when the pump station was converted to a pressure control facility. The same surge tanks are now being used when operating in the generating mode to reduce hydraulic surges following any loss of electrical load. Upon loss of the load, the unit is disconnected from the line; the discharge valve is slowly closed and the flow of water simultaneously transferred to the existing throttling valves. The conversion of the Greg Avenue facility to power recovery was made without sacrificing the standby pumping capability. Consequently, Greg Avenue Power Plant was the first to generate and the least expensive in cost per kilowatt. It operates at a near continuous flow rate of 60 cfs, dropping the pressure at the end of a 10-mile-long 4.0-feet ID concrete pipeline from 200 psi to about 100 psi. The revenues from this plant covered all costs including operation and maintenance after 24 months of operation.

### Lake Mathews and Foothill Feeder Power Recovery Plants

These two power recovery facilities are located at opposite ends of the MWDSC water distribution system.

The Lake Mathews Power Plant is located at the base of the
terminal reservoir for the Colorado River Aqueduct.  The
Foothill Feeder Power Plant is located at the base of the
terminal reservoir of the West Branch of the State Water
Project.  Reaction-type turbines were used with induction
generators.  The upstream head varies seasonally such that
at the end of the maximum summer flow period the available
head is at its lowest point.  Both power plants feed long,
large-diameter water transmission lines.

The Upper Feeder conveys water from the Lake
Mathews power plant afterbay to a water treatment plant 35
miles away and varies in diameter from 9-1/2 to 13-foot ID.
The 8-foot ID Lower Feeder also conveys water from the same
afterbay to a second water treatment plant about 24 miles
away.  Both feeders are a combination of concrete and steel
pipe and were designed on falling hydraulic grade lines
with spillway protection at the downstream end to preclude
overpressurization.

The Foothill Feeder conveys water from Castaic
Lake to a treatment plant approximately 15 miles away.  The
pipeline has a 17-foot inside diameter and is of concrete
construction.  There is no afterbay at the Foothill Power
Plant since the pipeline downstream of the turbine must be
pressurized to maintain water deliveries.

The Lake Mathews Power Plant produces 4.9 MW at
625-cfs design flow and dissipating a net head of 98 feet.
The Lake Mathews unit operates over a flow range of 350 cfs
to 750 cfs.  When the total outflow from Lake Mathews
exceeds 750 cfs, the additional flow is diverted around the
turbine.  The turbine must be shut down and all flow bypassed through the existing bypass structure whenever the
total outflow is less than 350 cfs.

Two turbines of equal size were installed in the
Foothill Feeder Power Plant.  Each turbine has a capacity
of 4.5 MW when the 300 cfs design flows through the unit at
net head of 206 feet.  The operating range for this power
plant is from 110 cfs to 660 cfs, with bypassing required
similar to that described for the Lake Mathews facility.

San Dimas, Yorba Linda, and Sepulveda Canyon Power Recovery
Plants

At these three locations, it was possible to use
an impulse-type turbine since the discharge from the turbine could be to atmosphere with a downstream water level
below the turbine's axis.

The San Dimas Power Plant produces 9.9 MW at a
design flow of 300 cfs and a net head drop of 410 feet.
Water is supplied from a reservoir 2.5 miles upstream via
an 8.0-foot diameter concrete pipeline.  This power plant
operates at flows between 100 cfs and 350 cfs.  Provision
was made for adding a second unit since there is excess
energy to be dissipated even when the pipeline is operating
at its maximum capacity.

At the Yorba Linda site, 5.1 MW are produced at
design flow of 340 cfs with a net head of 200 feet.  The
minimum operating flow at this plant is 60 cfs.  It is

located at the end of a 15.0-mile-long, 10-to-8-foot ID
concrete pipeline. This power plant discharges directly
into the inlet channel of a water treatment plant.
The Sepulveda Canyon Power Plant produces 8.6 MW
with a net head of 245 feet at design flow of 470 cfs. The
minimum operating flow is 100 cfs. This power plant is
located 16.5 miles downstream from a treatment plant and is
served by an 12.5-to-8-foot ID concrete pipeline.

## Venice Power Recovery Plant

The 10.1 MW synchronous generator unit is driven
by a reaction-type turbine that operates at a design flow
of 471 cfs and a design net head of 250 feet. This power
plant operates between an upstream pressure zone of approximately 400 psi and a downstream pressure zone of about
220 psi. The potential for and dire consequences of water
hammer at this location required very serious consideration.
The downstream pressure zone is made up of a series of
interconnected pipelines feeding the central portion of the
MWDSC service area. Eight miles upstream is Sepulveda
Canyon Power Recovery Facility which has on the downstream
side two large steel reservoir tanks with a combined capacity of 40 acre-feet. These provide the only source for
balancing flow variations due to fluctuations caused by the
operation of the Venice and Sepulveda Power Plants. The
time required to initiate start-up and to fully load the
power plant is 70 minutes. Thus the programmable controller
installed for automatic flow transfer during power system
outage also is an appreciated operator aid in the day-to-day
flow changes that are necessary.

## SUMMARY

Metropolitan's transmission and distribution
system is operated from five area control centers through
an existing supervisory control system. This relatively
slow reacting and very forgiving wholesale water delivery
system is being transformed considerably by the installation
of quickly responding hydrogeneration equipment. Valve
operating speeds of concern to the operator in the past
were in the order of 30 minutes (or higher) from fully
opened to fully closed. A 50 cfs per hour rate of flow
change has been used as operating criteria to minimize risk
of surges. However, a turbine/generator upon loss of
electric load will shut down within a period of 8 to
35 minutes, thus quick reaction is necessary for the operators, as well as reliable redundant backup systems to
accommodate the locally automatic flow transfer function at
each site. The systems have been tested out and are performing as designed. The reliability appears to be very
high on the automatic flow transfer system, thus permitting
the MWDSC to continue the tradition of very high reliability
in service and at the same time, add the very complex
sophistication of the small hydro power recovery facilities
on its existing transmission and distribution pipelines.

# A PASSIVE FISH SCREEN FOR HYDROELECTRIC TURBINES

George J. Eicher [1]

ABSTRACT

The author designed a fixed screen to bypass fish from the intakes of hydroelectric turbines. It is entirely within a turbine penstock. It utilizes Johnson wedgewire screen with 2 mm bars and 2 mm spaces and is in the penstock at 19° to the flow with bars parallel to the flow. A volume of 400 cfs goes through 250 square feet of screen resulting in approach velocities of 1.6 fps. Flow over the screen is 5 fps. A 30-inch pipe bypasses 50 cfs at 10 fps. The main screen tilts 45°, so that back-flushing cleans it. Test fish put through the system were collected in a trap. No mortalities were found. A small number had lost scales, but this probably was from sources other than the screen. During a year's operation, the designed cleaning system was at times used, although not generally by necessity. Cost was about $100,000, of which $15,000 was for the screen material.

In 1979, the author was commissioned by Portland General Electric Company to design a method of bypassing water around a turbine installation at Willamette Falls on the Willamette River in Oregon. Because of an extreme lack of space in which to install conventional traveling screens, it was decided to attempt a new departure, a fixed self-cleaning screen within the penstock itself that would divert water and fish around the turbine placement and eventually to tailwater. To be bypassed was the last unit of a line of 13 turbines in a row parallel to the forebay flow. The trash racks in front of the other 12 units had been modified to provide a louver array guiding the fish past them and eventually through a wing gate and into the No. 13 unit. It had been planned to remove this turbine and use the 400 cfs normal supply for it as a bypass flow.

Simply opening the wicket gates after removal of the turbine, however, would have resulted in an intolerable

---

[1] President, Eicher Associates, Inc., Portland, OR

condition. Fish could not be expected to survive
passage through openings gated to prevent the 400 cfs from
becoming a much larger quantity in absence of the turbine.
With this in mind, a screen was conceived that would divert
the fish into an existing 30-inch bypass pipe carrying 50
cfs into a trap and holding box for counting and examination.

Ordinary wire mesh screen would have been
unsuitable for this purpose because it clogs too easily,
particularly in the Willamette River which carries much
natural detritus such as small wood particles, sticks and
an abundance of leaves. A slope of the screen of 19° to
the flow was planned but, even with this advantage,
impingement of debris on wire mesh was an obvious certain-
ty. A perforated plate would have maintained itself in a
cleaner condition, but would not have sufficient open area
to pass the 400 cfs in the allowable space. Additionally,
it also tends to clog to a certain extent.

The only useful alternative was the Johnson
wedgewire screen, which has come into use for fish screen-
ing within the last decade. Although most such use has
been of cylindrical configurations, a few have used
inclined plane installations, usually atmospheric. Mr.
Richard Krcma, of the National Marine Fisheries Service,
had experimented with use of flat plane wedgewire screens
as partial diverters of water and fish into gate well
slots on one of the Columbia River dams. These were set
at about 45° to the flow and employed approach velocities
of 4-6 fps. Although these proved moderately successful,
some impingement of fish and debris occurred, probably
because of a lack of sufficient flow sweeping the screen
in relation to the rather high velocity into it. The
wedgewire screen has been in use for several decades,
having been originally developed to screen groundwater
inflows to deep wells where the screen could not be
cleaned. The principle is that debris particles, once
they pass the wedge shoulders, immediately proceed on
through the opening without clogging. Although clogging
can occur in the wedgewire screen, it is much less than in
other types.

It was decided to use this in a pressure situation; that is, all of the screen was completely submerged
and subject to relatively even pressure throughout its
area as induced by the turbine supply flow. Used was a
relatively fine opening with 2 mm wide parallel bars
spaced 2 mm apart. Research has found that this relatively narrow spacing tends to clog less than wider slots.
The greater.part of the screen was set at an angle of 19°
to the flow, which was parallel to the bars. Compromises
with space and peculiarities of the existing penstock and
turbine chamber were required. Ideally, the 19° slope

would have continued to the roof of the chamber so that a constant transportation flow of 5 fps into a funnel-shaped bypass would have resulted. Unfortunately, space, location of the existing bypass and irregularities in the chamber interior precluded this. Every available square inch was required to keep the approach velocity into the screen at 1.5 fps. A humpback configuration over the turbine and down to the bypass resulted in an enlarged area over the screen, which halved the transportation velocity to 2.6 fps in the downslope area.

At this point an obvious conclusion was reached that, if the flow into the turbine chamber were completely screened to fish, why remove the turbine? It was allowed to stay.

Three methods of cleaning the screen were planned. Central of these was pivoting of the larger approach section on an axle some 45° so that backflushing occurred with the turbine in operation. This was to be accomplished by an air-operated piston, but temporarily until general worth of the system was proven, a chain falls drawing chains at either end of the screen through pipes to tilt the array was used. Shutdown of the turbine was required. Had this proved insufficient, it was planned to provide air blasts under the screen to dislodge debris. As a last resort, the wicket gates would be slammed shut to provide a back jump of water to loosen material.

In practice, it has been found that the tilting option has been more than adequate. Tubes from below and above the screen communicate to a gage reading that shows head differential on the screen. Only occasionally has this exceeded 2-5 inches. The screen has been in operation over 16 months and has seldom required cleaning of any kind. Once a month, however, plant operators cycle the screen tilt as a training procedure.

The approach velocity into the screen violates a criterion of 0.5 fps for the smallest salmonid fry. However, this criterion is based on Canadian research of some 20 years ago reflecting the velocity that a 1-inch fry salmon could withstand for one hour in front of a mesh screen at right angles to the flow before either finding a bypass on one side or impinging on the screen. Such criteria are not applicable to this type of pressure screen, because th 19° slope coupled with the 5-6 fps transportation flow over the extremely smooth surface funneling into the bypass means that the fish are physically and involuntarily swept into the bypass in a very few seconds after passing over the leading edge. Swimming ability is meaningless. An indicator of efficiency is the behavior of leaves, even limp specimens of which pass

through the system without impingement.  Since leaves are
much poorer swimmers than the smallest fish, obviously
what passes them should pass any fish.

Many thousands of wild fish have been processed
in the trap after passing through the system, and a large
known number of marked test fish have been released into
the penstock ahead of the screen and examined after
recovery in the trap.  All of these passed through the
installation and into the trap in a matter of seconds.
Although no mortalities were found, a small proportion,
principally of hatchery fish, had lost scales in some of
the tests.  The source of this loss was not determined.  A
number of sources were suspect, such as the rough interior
and many protrusions including the chains mentioned within
the turbine chamber; the weld bead around the bypass exit,
a gate valve in the 30" pipe, and principally the humpback
screen between the bypass and trap used to separate most
of the water from the fish.  Unlike the turbine screen,
which is the smoothest area within the system, this screen
is of rough wire mesh which many of the fish strike at a
velocity of over 10 fps on emergence from the bypass pipe.
After passing over the crest of the hump, the fish slide
over the dry side into a trough at the bottom.  Many have
been observed going down tail first, which virtually
assures a certain amount of descaling.  Research involving
introduction of the fish between the bypass pipe and the
trap has indicated that most of the descaling occurs after
the fish leave the turbine casing and enter the trapping
arrangement.

Cost of this system was roughly $100,000, of
which $15,000 covered the screen itself.  This was a
difficult retrofit problem involving concrete removal and
field assembly of the screen within the chamber.  A
somewhat larger screen planned for another turbine at
Willamette Falls has been estimated at somewhat over
$100,000, but this includes a long bypass pipe, an improv-
ed trap utilizing a wedgewire screen and a lighted viewing
window.  In new construction, costs will obviously be
lower than for retrofits.

This next screen will be a considerable improvement over that described.  In addition to improved trapping, it will employ a single smooth 18° slope funneling
into a 10" bypass pipe flowing about 65 cfs.  Transportation velocities over the oval-shaped screen array within a
smooth round penstock will accellerate from 5.5 fps at the
leading edge to 7 fps as the bypass is entered.  Plans
have also been made for rectangular shapes in square
penstocks.

This system offers some distinct advantages over any other type of hydroelectric or thermal plant intake screen.

1. Although unit costs of the screen material are high, total costs are small because of amount used and other considerations.

2. Because the penstock or intake conduit supports the screen, no elaborate support structures are required upstream of the entrance.

3. Because of the pressure principle of operation with the screen completely submerged at all times, no forebay-matching controls are necessary and all of the screen is completely utilized at all times.

4. Placement entirely within the penstock or intake conduit relieves the necessity for installation space in the environs, which frequently is scarce.

5. Operation and maintenance are negligible because of lack of moving parts and self cleaning.

6. Complete submergence in an area of swift water movement virtually precludes icing, a problem of many screens.

The author has applied for a patent on this type of device which is pending.

TWO-DIMENSIONAL ANALYSIS OF BRIDGE BACKWATER

By Jonathan K. Lee[1], A. M. ASCE, David C. Froehlich[1], A. M. ASCE,
J. J. Gilbert[2], and Gregg J. Wiche[3]

ABSTRACT

A two-dimensional finite-element surface-water flow model was used
to study backwater and flow distribution at a highway crossing of a
complex flood plain. The model was run with and without the roadway
embankments in place, and the impact of the highway was determined by
comparing "before" and "after" results.

Model results show that upstream from the roadway maximum backwater
at the west edge of the flood plain is greater than maximum backwater
at the east edge. Downstream from the roadway, backwater occurs at
the east edge of the flood plain and drawdown at the west edge. The
study shows that the two-dimensional flow model is capable of simulating
the significant features of steady-state flow in a complex river-flood-
plain system.

INTRODUCTION

In April 1979 and April 1980, major flooding on the lower Pearl
River in Louisiana and Mississippi caused extensive damage to homes
located on the flood plain in the Slidell, La., area. The April 2,
1980, flood overtopped Interstate Highway 10 (I-10), interrupting traffic for several hours. Constrictions created by highway embankments,
together with other physical features of the flood plain, caused significant lateral variations in water-surface elevation and flow distribution during the April 1980 flood.

The two-dimensional finite-element surface-water flow modeling
system FESWMS is capable of simulating both lateral and longitudinal
velocities and variations in water-surface elevation, highly variable
flood-plain topography and vegetative cover, and geometric features such
as highway embankments, dikes, and channel bends. Geometric features of
different sizes are easily accommodated within a single finite-element
network. An earlier version of the modeling system FESWMS was used
successfully by Lee (4) and Lee and Bennett (5) to study the impact on
flood stages of a proposed highway crossing of the flood plain of the
Congaree River in South Carolina. In order to determine whether FESWMS

---

[1]Hydrologist, U.S. Geological Survey, Gulf Coast Hydroscience Center,
NSTL Station, Miss. 39529.
[2]Civil Engineer, U.S. Geological Survey, P. O. Box 6642, Baton Rouge,
La. 70806.
[3]Hydrologist, U.S. Geological Survey, P. O. Box 6642, Baton Rouge, La.
70806.

can be used effectively to analyze steady-state flow in large multichannel flood plains, the U.S. Geological Survey, in cooperation with
the Louisiana Department of Transportation and Development, used the
model to determine the effect of the I-10 crossing on water-surface
elevations and flow distribution during the April 1980 flood on the
Pearl River. The April 1980 flood was chosen for study because it is
the maximum flood of record and because it is a 56-year flood, close to
a 50-year design flood.

This paper summarizes the application of FESWMS to the Pearl River
and illustrates the usefulness of the two-dimensional model in analyzing
steady-state flow with both lateral and longitudinal variations. Simulation of the 1980 flood with I-10 in place (model calibration) and
without I-10 in place is described. Backwater and drawdown caused by
the roadway are discussed.

MODEL DESCRIPTION

The core of the modeling system FESWMS is a two-dimensional
finite-element surface-water flow model based on the work of Norton
and King (3,6,7,8). Around this core, the Geological Survey has
developed pre- and postprocessing programs which make the system
accessible to the user.

Two-dimensional surface-water flow in the horizontal plane is
described by three nonlinear partial-differential equations--two for
conservation of momentum and one for conservation of mass (3). The
dependent variables are the depth-averaged velocity components, u and
v, in the x and y directions, respectively, and the depth, h. The
surface-water flow equations account for energy losses through two
mechanisms: bottom friction and turbulent stresses. The momentum
equations use a two-dimensional form of the Chézy equation to model
bottom friction and Boussinesq's eddy-viscosity concept to model turbulent stresses. Boundary conditions consist of the specification of
flow components or water-surface elevations at open boundaries and
zero flow components or zero normal flow at all other boundaries. For
a time-dependent problem, initial conditions must also be specified.

In FESWMS, quadratic basis functions are used to interpolate the
velocity components, and linear basis functions are used to interpolate
depth on triangular, six-node, isoparametric elements (mixed interpolation). Model topography is defined by assigning a ground-surface
elevation to each element vertex and requiring the ground surface to
vary linearly within an element. The finite-element model requires the
specification of a constant Chézy coefficient and a constant symmetric
turbulent-exchange, or eddy-viscosity, tensor over each element.

Flow components are specified at inflow boundary nodes, and water-
surface elevations are specified at outflow boundary nodes. In this
study, zero normal flow was specified at all lateral boundaries.
Isoparametric elements permit the use of smooth, curved lateral
boundaries. Improved accuracy is obtained by using such boundaries,
together with the specification of zero normal flow there, in the
mixed-interpolation formulation of the surface-water flow equations
(1,3,9,10).

Galerkin's method of weighted residuals, a Newton-Raphson iteration
scheme, and a frontal solution algorithm using out-of-core storage (2)
are used to solve for the nodal values of the velocity components and
depth.  The time derivatives are handled by an implicit finite-difference scheme; in the application reported here, however, only the steady-
state forms of the equations were solved.

DESCRIPTION OF THE STUDY AREA

The reach of the Pearl River flood plain studied in this paper is
shown in Fig. 1.  The study reach, approximately 12 miles long, is
bounded on the north by old U.S. Highway 11 and Interstate Highway 59
(I-59) and on the south by U.S. Highway 90.  The eastern and western
boundaries are natural bluffs at the edge of the flood plain, where
ground-surface elevations rise abruptly to 5 to 15 ft NGVD in the
southern part of the study reach and to 15 to 25 ft NGVD in the northern
part.

The major channels in the study reach are the Pearl, East Middle,
Middle, West Middle, and West Pearl Rivers, and Wastehouse Bayou.  The
Pearl flows along the east side of the flood plain, and the West Pearl
along the west side.  In the northern part of the study reach, the West
Pearl River is the largest channel in the flood plain.  Near
Gainesville, Miss., the channel of the Pearl becomes the largest and
remains the largest to the mouths of the river system.

Flood-plain ground-surface elevations range from 1 ft NGVD in the
southern part of the study area to 15 ft NGVD in the northwestern part.
Between the upstream boundary and I-10, ground-surface elevations are
higher near the West Pearl River than on the east side of the flood
plain.  Except near Highway 90, the flood plain is covered by dense
woods, mixed with underbrush in many places.  Near Highway 90, coastal
marsh predominates.

Flow enters the study reach through the old Highway 11 bridge
opening at the Pearl River, through the I-59 opening at the West
Pearl River, and through numerous small openings in the old Highway 11
embankments.  The I-10 crossing, about 4.4 miles long, spans the
flood plain in an east-west direction in the middle of the study
reach.  There are bridge openings at the Pearl, Middle, and West Pearl
Rivers, with lengths of 4,980, 770, and 2,240 ft, respectively.  The
embankment between the Pearl and Middle Rivers is about 0.8 mile long,
and the embankment between the Middle and West Pearl Rivers is about
2.1 miles long.  The elevation of the roadway is between 12 and 13 ft
NGVD.  Flood plain elevations near I-10 range from 1 to 3 ft NGVD.
Flow leaves the study reach through five openings in Highway 90.

SIMULATION OF THE APRIL 1980 FLOOD

Data Collection and Analysis

A large amount of hydrographic and topographic data was collected
and analyzed for use in modeling the April 1980 flood.  Gage-height
records collected by the Geological Survey at Pearl River, La., at the
upper end of the study reach, and by the U.S. Army Corps of Engineers at

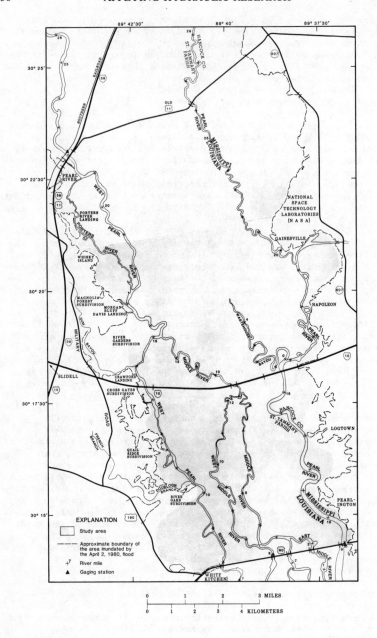

FIG. 1.--Study Area of Lower Pearl River Basin near Slidell

Pearlington, Miss., at the lower end of the study reach, were used to
justify a steady-state analysis. Approximately 200 high-water marks
within and near the study area were located by the Geological Survey as
the flood waters receded. The high-water marks were examined for
validity and grouped for use in establishing model boundary conditions
and calibrating the model. Discharge measurements made by the
Geological Survey and the Corps of Engineers at old Highway 11, I-59,
I-10, and Highway 90 during the 1980 and earlier floods were assembled
for the same purposes.

Approximately 50 miles of longitudinal channel profiles were
obtained for the significant channels in the study reach, and 73 representative and special-purpose cross-section surveys were made to further
define channel geometry. Detailed topographic data were obtained at and
near the bridge openings. Infrared aerial photographs of the study area
were obtained for use in determining vegetation type and density. The
collected data were supplemented by historic hydrologic data and
Geological Survey topographic maps.

Network Design

The network, shown in Fig. 2, was designed to closely represent
the highly nonuniform boundary of the area inundated by the April 1980
flood. The upstream boundary was located at old Highway 11 and I-59,
and the downstream boundary was located at Highway 90, where hydraulic
conditions could be reasonably well specified. Modifications to the
model near the I-10 crossing were assumed to have little effect on the
boundary conditions because both the upstream and downstream boundaries
were at least one flood-plain width distant from the highway crossing.
Smooth, curved-sided elements were used along all lateral boundaries,
at which zero normal flow was specified.

After the boundaries were defined, the study area was divided into
an equivalent network of triangular elements. Subdivision lines
between elements were located where abrupt changes in vegetative cover
or topography occurred. Each element was designed to represent an
area of nearly homogeneous vegetative cover. In areas where velocity,
depth, and water-surface gradients were expected to be large, such as
near bridge openings and in areas between overbanks and channel bottoms,
network detail was increased to facilitate better simulation of the
large gradients by the flow model.

The use of elements with aspect ratios greater than one made it
possible to design the network with fewer elements than would have been
required otherwise. Such elements were used primarily in defining
river channels. The longest element side was alined with the assumed
flow direction, in which velocity and depth changes would typically be
small. Element aspect ratios were kept to a maximum of about 10.

The complex geometry of the flood plain of the Pearl River was
modeled in detail. Most prototype lengths and widths were realistically
represented in the model; however, in order to reduce the number of
elements in the network, several approximations were made. Only large
channels were included in the network. Prototype channel cross sections
were represented in the model by either triangular or trapezoidal cross

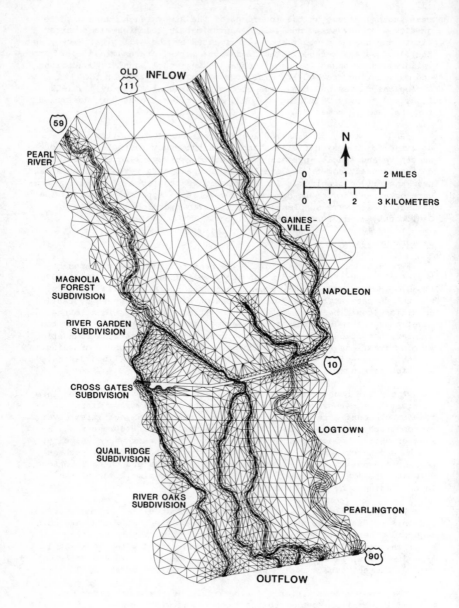

FIG. 2.--Finite-Element Network with I-10 in Place for the
April 2, 1980, Flood

sections with cross-sectional areas equal to the measured areas. Some meandrous channel reaches with relatively small flows were replaced with artificially straightened, but hydraulically equivalent, reaches. The width of simulated stream channels was kept to a minimum of 200 ft.

In its complete state, the network contained a total of 5,224 triangular elements and 10,771 computational node points, requiring the simultaneous solution of 23,697 nonlinear algebraic equations.

## Boundary Conditions

The discharge distribution at the upstream boundary was based on the peak discharge of 174,000 cfs measured at I-10 on April 2, 1980, and on previous discharge measurements at the bridge openings in old Highway 11 and I-59. Inflow was concentrated at the old Highway 11 bridge across the Pearl River and the I-59 bridge across the West Pearl River. Flow into the study reach through numerous small openings in old Highway 11 was represented as continuous inflow between the east edge of the flood plain and the Pearl River and between the Pearl and West Pearl Rivers. Water-surface elevations at the downstream boundary were based on high-water marks near the five bridge openings in Highway 90.

## Model Adjustment

The model adjustment process consisted of two parts: the adjustment of empirical model coefficients (model calibration) and the adjustment of model boundary conditions, network detail, and ground-surface elevations on the basis of additional information obtained during the study.

On the basis of previous finite-element simulations, the values of all components of the eddy-viscosity tensor were initially set at 100 lb-s/sq ft for all elements in the network. Numerical experiments indicated that once the values of these coefficients were set high enough to insure convergence, the solution was much less sensitive to changes in their values than to changes in the values of the Chézy coefficients. Because of a lack of information about their correct values and to avoid convergence problems, the values of all components of the eddy-viscosity tensor were maintained at 100 lb-s/sq ft throughout the study for all elements in the network.

Once the values of the eddy viscosities were fixed, preliminary calibration work focused on determining the values of Chézy coefficients. Nominal values were selected for initial use with the model on the basis of the infrared aerial photographs of the flood plain and field inspection. In making both the initial estimates of the Chézy values and subsequent modifications to them, care was taken to insure that the assigned values were reasonable and mutually consistent. The Chézy value assigned to a channel element in an artificially straightened reach was derived from the value for the corresponding natural or unstraightened reach on the basis of the equation $C_s = C_n(L_s/L_n)^{1/2}$, where C is the value of the Chézy coefficient (feet to the one-half power per second), L is the length of the reach (feet), and the subscripts s and n denote straightened and natural, respectively.

A series of simulations was conducted to determine the relative effect on water-surface elevations of changes in the values of the Chézy coefficients of both overbank and channel elements. Computed water-surface elevations were most sensitive to changes in the value of the Chézy coefficient of the wooded flood plain. Changes in the Chézy values of the channel elements had little or no effect on computed water-surface elevations except for channel reaches carrying a significant percentage of the total flow. Such reaches included the Pearl River between I-10 and Highway 90 and reaches located a few thousand feet upstream and downstream from bridge openings. Computed water-surface elevations were also moderately sensitive to the values of the Chézy coefficients of the overbank areas under the three I-10 bridges.

Preliminary calibration consisted of matching as closely as possible all observed high-water marks as well as measured discharges at the three bridge openings in I-10.

Appropriate adjustments to the values of the Chézy coefficients gave close agreement between computed and observed data in most cases. In several areas, however, a lack of agreement between model results and observations made it necessary to check the location and elevation of a few high-water marks and study previously overlooked local topographic features. On the basis of the results of the early simulations and the additional observations, modifications were made to model boundary conditions, network detail, and model ground-surface elevations. During this adjustment process, it was observed that computed water-surface elevations along the upstream model boundary were quite sensitive to changes in the upstream discharge distribution and that the distribution of discharge among the three I-10 bridge openings was affected significantly by flood-plain ground-surface elevations at and near the three I-10 openings.

After these adjustments were completed, minor adjustments to the values of the Chézy coefficients were needed for final calibration of the model. The final Chézy values were 22 $ft^{1/2}/s$ for the wooded flood plain, 28 to 35 $ft^{1/2}/s$ for the marsh-grass areas, 21 to 40 $ft^{1/2}/s$ for the overbank areas under the three I-10 bridges, and 85 to 115 $ft^{1/2}/s$ for the unstraightened channels. Computed flow depths range from 2 to 23 ft for the wooded flood plain, from 4 to 10 ft for the marsh-grass areas, from 4 to 9 ft for the overbank areas under the I-10 bridges, and from 5 to 47 ft for the unstraightened channels. On the basis of these depths, values of the Manning n corresponding to the final Chézy values range from 0.077 to 0.114 $ft^{1/6}$ for the wooded flood plain, from 0.055 to 0.074 $ft^{1/6}$ for the marsh-grass areas, from 0.046 to 0.098 $ft^{1/6}$ for the overbank areas under the I-10 bridges, and from 0.021 to 0.033 $ft^{1/6}$ for the unstraightened channels.

Computed flow depths average about 21 ft in the channels and about 8 ft on the flood plain. Most cross-sectional average channel velocities are between 1 and 3 fps. Somewhat higher velocities occur at several of the bridge openings. The average velocity on the flood plain is about 0.7 fps.

Comparison of Simulated and Observed Values

    The computed water-surface elevation is in close agreement with the
elevation of the observed high-water mark or marks at most of the 45
locations where high-water marks were available.  The root mean square
difference between the computed and observed values is 0.18 ft.  The
computed water-surface elevations are within ±0.3 ft of the elevations
of the high-water marks at all but four locations, and at these four
locations, the computed elevations are within ±0.5 ft of the observed
values.

    The discharge measurements made at the I-10 bridge openings on
April 2, 1980, were also used in model calibration.  The errors in computed discharge at the bridge openings at the Pearl, Middle, and West
Pearl Rivers, as a percent of the measured discharge at each opening,
are 7, -10, and -7, respectively.

    The model simulates accurately the observed shift of flow from the
west side of the flood plain to the east side between the upstream
boundary and I-10.  At the upstream boundary, 56% of the inflow was
estimated to pass through the bridge opening at the West Pearl River,
but at I-10, 59% of the measured discharge passed through the bridge
opening at the Pearl River.  In the calibrated model, 63% of the computed discharge passes through the Pearl River bridge opening at I-10.

SIMULATION OF THE APRIL 1980 FLOOD WITHOUT THE I-10 EMBANKMENTS IN PLACE

    The finite-element network used to simulate the April 1980 flood
was modified to represent conditions prior to roadway construction, and
the hydraulic impact of the I-10 embankments was determined by comparing
"before" and "after" results.

Network Modifications

    Elements were added in the areas occupied in the original network
by the I-10 embankments.  Elsewhere, the two networks were identical.
Model ground-surface elevations at and near the highway embankments were
changed to the elevation of the surrounding natural flood plain.  The
Chézy coefficients corresponding to the new elements and the elements
formerly located in overbank areas under the I-10 bridges were assigned
the value 22 ft$^{1/2}$/s, the value used in both "before" and "after"
simulations for the wooded flood plain.  Upstream and downstream boundary conditions were the same as those used in the simulation with the
highway embankments in place.

Results of the Simulation

    Without I-10 in place, the flow shift from the west side of the
flood plain to the east side does not occur as far upstream as with I-10
in place.  Flow is reduced 41% at the Pearl River bridge opening, 81% at
the Middle River opening, and 67% at the West Pearl River opening.
The computed discharge across the areas occupied by the I-10 embankments
is 95,200 cfs.  As expected, water-surface elevations upstream from the
I-10 site are lower without the highway embankments in place.

BACKWATER AND DRAWDOWN CAUSED BY THE I-10 EMBANKMENTS

A map of backwater and drawdown was obtained by subtracting nodal water-surface elevations computed without the roadway in place from the corresponding nodal water-surface elevations computed with the roadway in place.  Lines of equal backwater and drawdown are shown in Fig. 3. The 1.2-foot to 2.0-foot lines form a "mound" north of I-10 between the Pearl River and the west edge of the flood plain.  Upstream from the roadway, maximum backwater at the west edge of the flood plain (1.5 ft) is greater than maximum backwater at the east edge (1.1 ft), but backwater decreases more rapidly in the upstream direction along the west edge than along the east edge.

The highway embankments cause higher water-surface elevations downstream from the roadway in the eastern part of the flood plain and lower water-surface elevations downstream in the western part.  Backwater ranging from 0.6 to 0.2 ft extends more than a mile downstream from the Pearl River bridge opening in I-10 at the east edge of the flood plain. A large area of drawdown extends from the downstream side of the highway embankment between the Middle and West Pearl Rivers to the west edge of the flood plain.  Drawdown of 0.2 ft or more occurs along approximately 2 miles of the west edge of the flood plain downstream from I-10.  The lateral variations in backwater and drawdown are due in part to the relatively greater constriction of the flow in the western part of the flood plain and in part to the topography of the flood plain.  These results show that FESWMS is capable of simulating both longitudinal and lateral variations in backwater and drawdown.

SUMMARY AND CONCLUSIONS

The two-dimensional finite-element surface-water flow modeling system FESWMS was used to study the effect of I-10 on water-surface elevations and flow distribution during the April 2, 1980, flood on the Pearl River near Slidell, La.  A finite-element network was designed to represent the topography and vegetative cover of the study reach. Hydrographic data collected for the 1980 flood were used to adjust the flow model to simulate the flood event as closely as possible.  The finite-element network was then modified to represent conditions prior to roadway construction, and the hydraulic impact of I-10 was determined by comparing "before" and "after" results.

Upstream from the roadway, maximum backwater at the west edge of the flood plain (1.5 ft) is greater than maximum backwater at the east edge (1.1 ft).  Backwater ranging from 0.6 to 0.2 ft extends more than a mile downstream from the Pearl River bridge opening in I-10 at the east edge of the flood plain, and drawdown of 0.2 ft or more occurs along approximately 2 miles of the west edge of the flood plain downstream from I-10.

The capability of the modeling system FESWMS to simulate the significant features of steady-state flow in a complex multichannel river-flood-plain system with variable topography and vegetative cover was successfully demonstrated in this study.  These features included lateral variations in discharge distribution and backwater or drawdown.

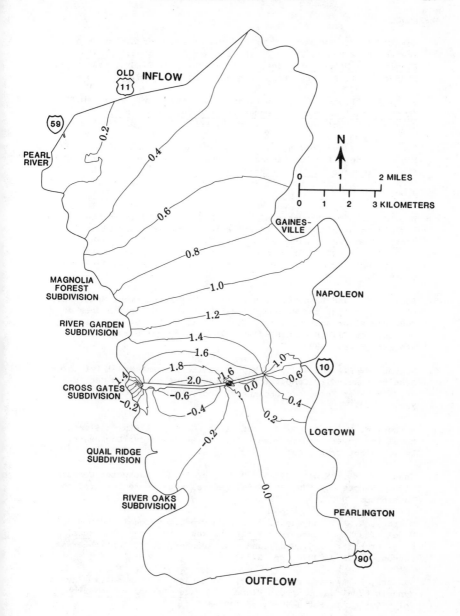

FIG. 3.--Backwater and Drawdown with I-10 in Place for the
April 2, 1980, Flood, Interval 0.2 ft

REFERENCES

1.  Gee, D. M., and MacArthur, R. C., "Development of Generalized Free
    Surface Flow Models Using Finite Element Techniques," Finite
    Elements in Water Resources, C. A. Brebbia, W. G. Gray, and G. F.
    Pinder, eds., Pentech Press, London, England, 1978, pp. 2.61-2.79.

2.  Hood, P., "Frontal Solution Program for Unsymmetric Matrices,"
    International Journal for Numerical Methods in Engineering,
    Chichester, England, Vol. 10, No. 2, Feb., 1976, pp. 379-399.

3.  King, I. P., and Norton, W. R., "Recent Applications of RMA's Finite
    Element Models for Two Dimensional Hydrodynamics and Water Quality,"
    Finite Elements in Water Resources, C. A. Brebbia, W. G. Gray, and
    G. F. Pinder, eds., Pentech Press, London, England, 1978, pp. 2.81-
    2.99.

4.  Lee, J. K., "Two-Dimensional Finite Element Analysis of the Hydrau-
    lic Effect of Highway Bridge Fills in a Complex Flood Plain,"
    Finite Elements in Water Resources, S. Y. Wang, C. A. Brebbia,
    C. V. Alonso, W. G. Gray, and G. F. Pinder, eds., University of
    Mississippi, School of Engineering, University, Miss., 1980, pp.
    6.3-6.23.

5.  Lee, J. K., and Bennett, C. S., III, "A Finite-Element Model Study
    of the Impact of the Proposed I-326 Crossing on Flood Stages of the
    Congaree River near Columbia, South Carolina," U.S. Geological
    Survey Open-File Report 81-1194, U.S. Geological Survey, NSTL
    Station, Miss., 1981.

6.  Norton, W. R., and King, I. P., "A Finite Element Model for Lower
    Granite Reservoir, Computer Application Supplement and User's
    Guide," Water Resources Engineers, Inc., Walnut Creek, Calif.,
    Mar., 1973.

7.  Norton, W. R., King, I. P., and Orlob, G. T., "A Finite Element
    Model for Lower Granite Reservoir," Water Resources Engineers,
    Inc., Walnut Creek, Calif., Mar., 1973.

8.  Tseng, M. T., "Finite Element Model for Bridge Backwater Computa-
    tion," Report No. FHWD-RD-75-53, Federal Highway Administration,
    Washington, D.C., Apr., 1975.

9.  Walters, R. A., and Cheng, R. T., "A Two-Dimensional Hydrodynamic
    Model of a Tidal Estuary," Finite Elements in Water Resources,
    C. A. Brebbia, W. G. Gray, and G. F. Pinder, eds., Pentech Press,
    London, England, 1978, pp. 2.3-2.21.

10. Walters, R. A., and Cheng, R. T., "Accuracy of an Estuarine Hydro-
    dynamic Model Using Smooth Elements," Water Resources Research,
    Vol. 16, No. 1, Feb., 1980, pp. 187-195.

# ANALYSIS OF ALTERNATIVES FOR REDUCING BRIDGE BACKWATER

By Gregg J. Wiche[1], J. J. Gilbert[2], and Jonathan K. Lee[3], A. M. ASCE

ABSTRACT

A two-dimensional finite-element surface-water flow modeling system was used to study the effect of two alternative modifications for improving the hydraulic characteristics of a highway crossing of a complex flood plain. Both alternatives reduce backwater and average velocities on the overbanks and in the channels, and both eliminate roadway overtopping.

The study shows that the modeling system FESWMS is a useful tool for analyzing both structural and nonstructural modifications of highway crossings of complex flood plains. In particular, the model is capable of simulating both longitudinal and lateral changes in discharge distribution and backwater due to the modifications.

INTRODUCTION

In April 1979 and April 1980, major flooding on the lower Pearl River caused more than $12 million in property damage to homes located on the flood plain in the Slidell, La., area. The 1980 flood overtopped Interstate Highway 10 (I-10) between Slidell and Bay St. Louis, Miss., and forced the closing of the highway while the flood crest passed.

The U.S. Geological Survey, in cooperation with the Louisiana Department of Transportation and Development (LDOTD), used the two-dimensional finite-element surface-water flow modeling system FESWMS to study the effect of I-10 on water-surface elevations and flow distribution during the April 1980 flood, as reported by Lee et al. (2). FESWMS was then used to analyze proposed modifications of the I-10 crossing to determine whether they reduced backwater, eliminated roadway overtopping, and reduced bridge-opening velocities for a flood of the magnitude of the April 1980 flood.

---

[1]Hydrologist, U.S. Geological Survey, P. O. Box 6642, Baton Rouge, La. 70806.
[2]Civil Engineer, U.S. Geological Survey, P. O. Box 6642, Baton Rouge, La. 70806.
[3]Hydrologist, U.S. Geological Survey, Gulf Coast Hydroscience Center, NSTL Station, Miss. 39529.

This paper summarizes the application of FESWMS to two alternative modifications of the I-10 crossing. The simulation of backwater and flow distribution at the I-10 crossing for the two alternatives is discussed, and the results are compared with those for the existing crossing. The paper illustrates the usefulness of the model in analyzing modifications of highway crossings of complex multichannel river-flood-plain systems.

DESCRIPTION OF THE STUDY AREA

The reach of the Pearl River studied is located in the lower part of the basin on the Mississippi-Louisiana border. Only the study area's middle part, indicated by shading in Fig. 1, is considered in the analysis presented in this paper. The full study reach is approximately 12 miles long. It is bounded on the north by old U.S. Highway 11 and Interstate Highway 59 (I-59) and on the south by U.S. Highway 90. The eastern and western boundaries are natural bluffs at the edge of the flood plain, where ground-surface elevations rise abruptly to 5 to 15 ft NGVD in the southern part of the study reach and to 15 to 25 ft NGVD in the northern part.

Within the study reach, the axis of the flood plain is south-southeast. Flood-plain ground-surface elevations range from 1 ft NGVD in the southern part of the study area to 15 ft NGVD in the northwestern part. Between the upstream boundary and I-10, ground-surface elevations are higher near the West Pearl River than on the east side of the flood plain. Except near Highway 90, the flood plain is covered by dense woods, mixed with underbrush in many places. Near Highway 90, coastal marsh predominates.

The major channels in the study reach are the Pearl, East Middle, Middle, West Middle, and West Pearl Rivers, and Wastehouse Bayou. The Pearl flows along the east side of the flood plain, and the West Pearl flows along the west side. In the northern part of the study reach, the West Pearl River is the largest channel in the flood plain. Near Gainesville, Miss., the channel of the Pearl becomes the largest and remains the largest to the mouths of the river system.

The I-10 crossing, about 4.4 miles long, spans the flood plain in an east-west direction. There are bridge openings at the Pearl, Middle, and West Pearl Rivers, with lengths of 4,980, 770, and 2,240 ft, respectively. The embankment between the Pearl and Middle Rivers is about 0.8 mile long, and the embankment between the Middle and West Pearl Rivers is about 2.1 miles long.

Flow enters the study reach through the old Highway 11 bridge open-ing at the Pearl River, through the I-59 opening at the West Pearl River, and through numerous small openings in the old Highway 11 embank-ments. Flow leaves the study reach through five openings in Highway 90.

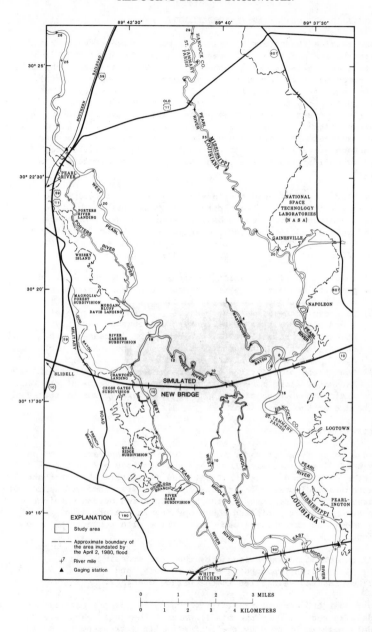

FIG. 1.--Study Area of Lower Pearl River Basin near Slidell

IMPACT OF I-10 DURING THE APRIL 1980 FLOOD

   To determine the effect of I-10 on water-surface elevations and
flow distribution near the I-10 crossing during the April 1980 flood,
a finite-element network was designed to represent the topography and
vegetative cover of the study reach, as reported by Lee et al. (2).
The model requires the specification of flow components at upstream
boundary nodes and water-surface elevations at downstream boundary
nodes.  The upstream and downstream boundaries were located about one
flood-plain width upstream and downstream, respectively, of the I-10
crossing, so that the upstream discharge distribution and the downstream
water-surface elevations could be assumed to be not significantly
affected by modifications made to the model near the middle of the
network.

   The flow model was adjusted to simulate the April 1980 flood as
closely as possible.  Then the finite-element network was modified to
represent conditions prior to roadway construction, and the hydraulic
impact of I-10 was determined by comparing "before" and "after" results.

   The effects of I-10 may be summarized as follows:

   (1) I-10 caused flow to shift from the west side of the flood
       plain to the east side farther upstream than would have
       occurred without the highway in place.
   (2) Upstream of I-10, maximum backwater at the west edge of the
       flood plain (1.5 ft) was greater than maximum backwater
       at the east edge (1.1 ft), but backwater decreased more
       rapidly in the upstream direction along the west edge of the
       flood plain than along the east edge.
   (3) Downstream of I-10, the highway embankments caused significant
       backwater along more than a mile of the east edge of the flood
       plain and an area of drawdown along several miles of the west
       edge of the flood plain.

ALTERNATIVES STUDIED

   The alternatives for modifying the I-10 crossing were chosen on
the basis of discussions with LDOTD and observations made during
calibration of the model for the April 1980 flood (2).  Two of these
alternatives are discussed in this paper.  The first involved improving
the hydraulic characteristics of the three existing bridge openings,
and the second involved placing a new opening in the crossing.  Model
results were used to evaluate each alternative with respect to three
objectives:

   (1) reducing backwater caused by the I-10 crossing,
   (2) eliminating overtopping of the roadway, and
   (3) decreasing velocities in the bridge openings.

   To reduce the cost of preliminary evaluation of the alternatives,
the model was run for only the middle part of the full-reach network.
The middle part is bounded by the east-west lines crossing the full-
reach network approximately 2 miles upstream and 1 mile downstream of

I-10.  Boundary conditions for the alternative simulations were obtained from the results of the calibration simulation.  These boundary values were not allowed to vary in the alternative simulations.  Thus, results of the alternative simulations differ slightly from results that would have been obtained with the full-reach model had it been used.

ALTERNATIVE 1 SIMULATION

During the model adjustment process reported in (2), it was observed that computed water-surface elevations were moderately sensitive to the values of the Chézy coefficients of the overbank elements within the three I-10 bridge openings and to model ground-surface elevations within and near the bridge rights-of-way.  Alternative 1, developed on the basis of these observations, involved modifications to the overbank areas within the bridge rights-of-way. These modifications included the removal of spoil left after construction and natural levees along the channels and the clearing of brush and trees.

To simulate alternative 1 with FESWMS, model ground-surface elevations of overbank areas within the rights-of-way at the three I-10 bridge openings were lowered to the natural elevation of the surrounding flood plain.  The Chézy coefficients of overbank elements in the bridge rights-of-way were assigned a value of 40 ft$^{1/2}$/s, which is recommended by Chow (1) for a vegetative cover of short grass with no brush.  The network for alternative 1 is shown in Fig. 2.

After FESWMS was run for alternative 1, the nodal water-surface elevations from the simulation without I-10 in place were subtracted from the corresponding water-surface elevations computed for alternative 1 to produce the lines of equal backwater and drawdown shown in Fig. 3(a).  Maximum backwater of 1.7 ft occurs on the upstream side of the I-10 embankment between the Middle and West Pearl Rivers in this simulation.  The maximum backwater in the calibration simulation occurred at the same location and was 2.1 ft.  Because I-10 was overtopped by only a few inches in the April 1980 flood, alternative 1 eliminates the possibility of roadway overtopping for a flood of the magnitude of the 1980 flood.  Maximum backwater at the east edge of the flood plain is 0.7 ft and occurs 0.8 mile below Napoleon.  Maximum backwater at the west edge of the flood plain is 1.1 ft near Crawford Landing.  The corresponding values for the calibration simulation were 1.1 ft and 1.5 ft at the east and west edges, respectively.  The water-surface gradients in the Middle and West Pearl River openings are reduced from the calibration values in this simulation but remain larger than the gradient in the Pearl River opening.

The computed discharges at each of the three bridge openings are given in column (3) of Table 1.  Corresponding discharges for the calibration simulation are given in column (2).  Most of the increase in discharge of 6,600 cfs at the Middle River opening is captured from the Pearl River opening.  The increase in discharge on the left overbanks of the Pearl and West Pearl Rivers is a result of the lower ground-surface elevations there within the bridge rights-of-way.

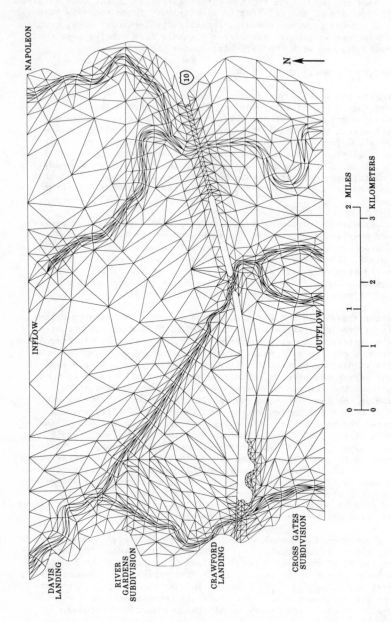

FIG. 2.--Finite-Element Network for Alternative 1

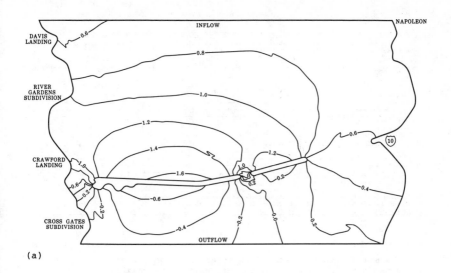

(a)

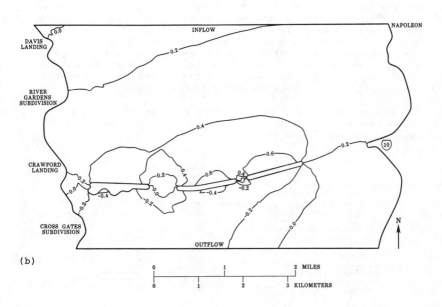

(b)

FIG. 3.--Backwater and Drawdown for Alternative 1 (a) and
Alternative 2 (b), Interval 0.2 ft

TABLE 1.--Computed Discharges at the I-10 Bridge Openings for Existing Conditions and Modifications of the Highway Crossing

| Opening subarea (1) | Discharge with existing conditions (calibration), in cubic feet per second[a] (2) | Discharge for alternative 1, in cubic feet per second[b] (3) | Discharge for alternative 2, in cubic feet per second[b] (4) |
|---|---|---|---|
| | (A) Pearl River | | |
| Left overbank | 23,600 | 24,900 | 17,400 |
| Channel | 50,200 | 40,600 | 39,100 |
| Right overbank | 36,100 | 36,200 | 26,400 |
| Total | 110,000 | 102,000 | 82,900 |
| | (B) Middle River | | |
| Total | 27,000 | 33,600 | 17,500 |
| | (C) West Pearl River | | |
| Left overbank | 10,000 | 11,800 | 5,660 |
| Channel | 16,900 | 14,100 | 10,600 |
| Right overbank | 11,000 | 11,000 | 6,620 |
| Total | 37,900 | 36,900 | 22,900 |
| | (D) New bridge opening | | |
| Total | --- | --- | 41,000 |
| | (E) Total[c] | | |
| | 175,000 | 172,000 | 164,000 |

[a]Discharges are obtained from a full-reach simulation (2).
[b]Discharges are obtained from a middle-reach simulation.
[c]Because there are only about half as many equations for conservation of mass as there are for conservation of momentum in either the x or y direction, conservation of mass is not satisfied exactly by the finite-element model. This accounts for the differences between the total computed discharges at I-10 and the upstream inflow.

The average velocities at the three openings are given in column (3) of Table 2. Corresponding velocities for the calibration simulation are given in column (2). The average velocities decrease in each of the three openings. At the Pearl and West Pearl Rivers, the average velocities decrease in the main channels and on all of the overbanks. Alternative 1 reduces backwater without an increase in velocity in the openings because the increase in the cross-sectional flow areas of the openings more than compensates for the reduction in resistance in the bridge rights-of-way. The lower velocities indicate that alternative 1 reduces the potential for scour at the bridge openings.

Alternative 1 shows that a nonstructural approach can result in reduced backwater and lower bridge-opening velocities, even with increased discharge through an opening.

TABLE 2.--Computed Average Velocities at the I-10 Bridge Openings for Existing Conditions and Modifications of the Highway Crossing

| Opening subarea (1) | Velocity with existing conditions (calibration), in feet per second[a] (2) | Velocity for alternative 1, in feet per second[b] (3) | Velocity for alternative 2, in feet per second[b] (4) |
|---|---|---|---|
| | (A) Pearl River | | |
| Left overbank | 1.5 | 1.2 | 1.2 |
| Channel | 3.6 | 2.9 | 2.9 |
| Right overbank | 1.8 | 1.5 | 1.5 |
| Total | 2.3 | 1.8 | 1.8 |
| | (B) Middle River | | |
| Total | 4.2 | 4.0 | 2.9 |
| | (C) West Pearl River | | |
| Left overbank | 2.3 | 1.6 | 1.5 |
| Channel | 3.5 | 2.8 | 2.3 |
| Right overbank | 1.4 | 1.0 | 0.9 |
| Total | 2.2 | 1.6 | 1.5 |
| | (D) New bridge opening | | |
| Total | --- | --- | 2.0 |

[a]Average velocities are obtained from a full-reach simulation (2).
[b]Average velocities are obtained from a middle-reach simulation.

ALTERNATIVE 2 SIMULATION

Alternative 2 involved placing a new 2,000-foot opening in the I-10 embankment between the Middle and West Pearl Rivers.  Ground-surface elevations were reduced to 0.0 ft NGVD within the right-of-way between the ends of the bridge, and a rectangular area centered about the roadway was cleared.  The cleared area was 1,000 ft wide and 3,000 ft long, with the long side of the clearing parallel to the roadway.  Conditions at the other three bridges were the same as in the calibration run.

To simulate alternative 2 with FESWMS, the finite-element network was modified to include the new bridge opening (Fig. 4).  By utilizing the model's capability of including elements in the flow domain or excluding them from it and automatically assigning a zero-normal-flow boundary condition between included and excluded elements, it was possible to design a single network for simulating either a 4,000-foot, a 3,000-foot, or a 2,000-foot opening in I-10.  Only the 2,000-foot opening was analyzed in this study.  Ground-surface elevations at nodes located in the bridge right-of-way were set at 0.0 ft NGVD.  The Chézy coefficients of the elements in the rectangular cleared area were assigned the value 40 $ft^{1/2}/s$.

Alternative 2 produces a major change in the direction of the velocity field.  The velocity field is alined in a southerly direction throughout the reach modeled here, whereas the velocity field obtained from the calibration simulation was alined in a south-southeasterly direction.  In the calibration simulation, a significant shift of flow from the west side of the flood plain to the east side occurred upstream of I-10.  Alternative 2 significantly reduces this flow shift upstream of I-10.  Convergence toward the I-10 openings and divergence back onto the flood plain occurred in a 3-mile-long reach centered about I-10 in the calibration run, whereas it occurs in a 0.6-mile-long-reach in this simulation.

Lines of equal backwater and drawdown for this simulation are shown in Fig. 3(b).  Maximum backwater of 0.8 ft occurs on the upstream side of I-10 between the Pearl and Middle Rivers.  Thus, alternative 2 prevents overtopping of the roadway.  Backwater along the east edge of the flood plain is 0.3 ft between the upstream boundary and I-10.  Maximum backwater at the west edge of the flood plain is 0.3 ft near Crawford Landing.  Alternative 2 virtually eliminates backwater upstream of Davis Landing at the west edge of the flood plain.  Alternative 2 greatly reduces the mound of backwater that extended more than a mile downstream of the Pearl River bridge opening on the east side of the flood plain in the calibration simulation.  The drawdown that existed on the west side of the flood plain in the calibration simulation is also reduced.  The water-surface gradients at the bridge openings are less in alternative 2 than in alternative 1.  The water-surface gradient at the Middle River is greater than that at the other three openings.

The computed discharges at the three bridge openings are given in column (4) of Table 1.  Alternative 2 reduces the discharge 25%, 35%, and 40%, at the Pearl, Middle, and West Pearl Rivers, respectively, as

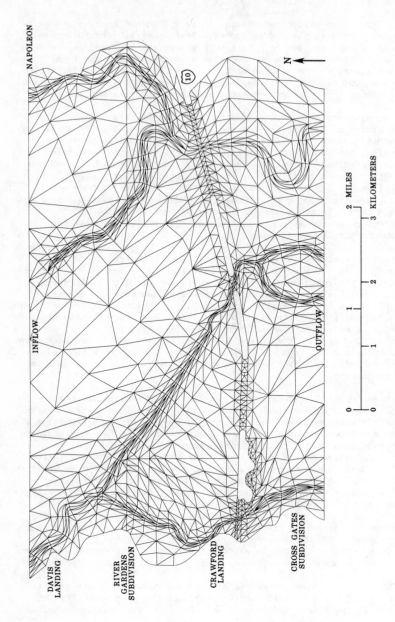

FIG 4.—Finite-Element Network for Alternative 2

# REDUCING BRIDGE BACKWATER 269

a percentage of the calibrated discharge at each opening. Twenty-five percent of the total discharge passes through the new bridge opening.

The average velocities at the three openings are given in column (4) of Table 2. Alternative 2 reduces the average velocities in all of the openings. The largest reduction in average velocity (1.3 fps) occurs at the Middle River opening.

Alternative 2 shows that a structural approach involving a new bridge opening can provide a greater reduction in backwater than a nonstructural approach such as alternative 1. By restoring a more uniform distribution of discharge across a flood plain, backwater can be reduced to a fraction of its former value, and velocities in existing openings can be significantly reduced.

SUMMARY AND CONCLUSIONS

The two-dimensional finite-element surface-water flow modeling system FESWMS was used to study the effect of two alternatives for improving the hydraulic characteristics of the I-10 crossing of the flood plain of the Pearl River near Slidell, La. The analysis utilized the model's capability to simulate changes in flood-plain topography, flood-plain vegetative cover, and highway-embankment geometry.

Compared to the existing highway crossing, both alternative modifications reduce backwater and bridge-opening velocities. Both eliminate roadway overtopping. The evaluation of alternatives such as the two presented here will enable LDOTD to formulate a plan to improve the hydraulic characteristics of the I-10 crossing.

Some generalizations can be made on the basis of the results of this study. Both nonstructural and structural modifications of highway crossings of wide flood plains can have significant beneficial effects on the hydraulic characteristics of such crossings. A nonstructural approach can result in reduced backwater and lower bridge-opening velocities, even with increased discharge through an opening. A structural approach can provide a greater reduction in backwater than a nonstructural approach and significantly reduce velocities in the existing bridge openings.

This study shows that the modeling system FESWMS is a useful tool for analyzing both structural and nonstructural modifications of highway crossings of complex flood plains. In particular, the model is capable of simulating both longitudinal and lateral changes in discharge distribution and backwater due to the modifications.

REFERENCES

1. Chow, V. T., Open-Channel Hydraulics, McGraw-Hill, Inc., New York, N. Y., 1959, p. 113.

2. Lee, J. K., Froehlich, D. C., Gilbert, J. J., and Wiche, G. J., "Two-Dimensional Analysis of Bridge Backwater," Proceedings, Applying Research to Hydraulic Practice, American Society of Civil Engineers, Aug., 1982.

Scouring Effects of Water Jets Impinging on
Non-Uniform Streambeds

Walter C. Mih*, M. ASCE

ABSTRACT

The effects of impingement of high velocity water jets on a non-uniform streambed were studied through laboratory experiments and theoretical analysis. The result of this research is used to develop a method for the removal of fine silt from the salmon spawning gravel to enhance the survival of salmon. Biologists have found the average egg-to-fry survival rate to be less than 10 percent in silted gravel while survival in cleaned coarse gravel is nearly 80 percent.

The non-uniform streambed used in the laboratory tests had very wide gradation with a uniformity coefficient, defined as $d_{60}/d_{10}$, equal to 75. The variable of the water jets were: jet velocities range from 20 to 70 ft. per sec; jet diameters were 0.5, 1.0 and 1.5 inches; jet angles were 45°, 60° and 90° from the horizontal; and jet positions were from 0 to 21 in. above the streambed surface. The water depth was constant at 2 ft. After jetting and scouring reached a steady state, the scoured depth, armored depth, cleaned depth, defined as the depth where fine silt was flushed out below the gravel surface, of the streambed were measured. To facilitate the measurements on armored and cleaned depths, half jet nozzles were used and postioned such that the transparent flume wall was the plane of symmetry. The armored depth and cleaned depth below the gravel surface could be readily observed and measured through the transparent wall.

Through theoretical analysis, measured data were correlated well with the use of a dimensionless number which is the ratio of drag force on a gravel particle to the net gravitation force of the same particle.

INTRODUCTION

Salmon fisheries are an important economic and social resource in the Pacific Northwest. The fascinating life cycle of the anadromous fish has been subjected to intensive studies (3). After two to four years maturing in the ocean, their remarkable homing instinct brings them back from the trackless expanse of the ocean, through numerous tributaries and river miles to a shallow natal stream for their final ritual.

---

*Professor, Albrook Hydraulics Laboratory, Department of Civil and Environmental Engineering, Washington State University, Pullman, WA 99164-3001

Salmon Spawners choose clean gravel whenever possible, generally ranging in size from 1/8 in. (3mm) to 4 in. (100mm). The fertilized eggs are buried in the stream gravel 6 in. (15 cm) to 18 in. (45 cm) below the streambed depending on size and species of the fish. During more than a 100-day period between spawning and fry emergence from the gravel, the stream gravel must have adequate interstitial flow to supply oxygen for the developing eggs. If large amounts of fine silt or mud choke the intragravel spaces, the interstitial water flow and, hence, the oxygen supply will be reduced resulting in a high mortality for young salmon.

In a review of all available methods for gravel cleaning in natural streams, it was concluded that the best method would consist of a mobile machine traveling in the shallow stream and utilizing a row of high velocity water jets to flush out the fine silt from the gravel. The silt would then be removed from the stream by a suction pump. Before such a machine could be designed, it was necessary to measure the flushing effect of impinging water jets on non-uniform streambeds. The study of impinging water jets on streambeds has other applications in the field of hydraulic engineering; erosion of streambeds by water jets issuing from hydraulic outlets, weirs, and spillways as well as erosion by secondary flows around bridge piers and river bends. High velocity jets have been successfully used in waterway dredging for loosening hard-packed soil.

Most of the studies dealing with jet impingement on sand beds have been empirical due to the complex nature of the process. Rouse (7) conducted the first systematic study on scouring of nearly uniform fine sand by vertical submerged water jets.

Rajaratnam and Beltaos did extensive studies on jet impingement on solid boundaries (2) and on loose boundaries (5, 6). Horizontal beds of uniform sand and polystyrene spheres were scoured by vertical jets (5) and by horizontal air and water jets (6). For the asymptotic steady state, the scoured geometries were correlated by a Froude number modified by the density ratio between the solid particle and jetting fluid called the densimetric Froude number. For large impinging height defined as more than a distance of 8.3 nozzle diameters, the characteristic length changes to the nozzle diameter. These findings are in general agreement with the two flow zones of a free jet as described by Albertson et al (1) who determined that the potential core extends 6.2 nozzle diameters downstream which is called the zone of flow establishment. Further downstream the zone is referred to as established flow. Kobus and Westrich (4) studied the scouring geometries of steady and pulsating jets impinging on a bed of uniform sand.

All the studies mentioned considered sands of uniform particle size. This study appears to be the first dealing with impingement of high velocity water jets on a non-uniform bed of mixed fine sand and coarse gravel of wide gradation. It should be mentioned that the "armor action," the paving of coarse gravel at the scoured surface, can only occur in bed material having a wide gradation.

EXPERIMENTS

The jet impinging tests were performed on spawning-size gravel placed in a 4 ft (1.2 m) wide, 5 ft (1.5 m) high, and 30 ft (9 m) long

test flume in the Albrook Hydraulics Laboratory. Fig. 1 is the definition sketch of the experiment. The experimental streambed compares
closely to average salmon spawning beds in the coastal streams of the
state of Washington. White fine sand less than 0.3 mm was mixed with
the river gravel to simulate the fine silt found in the silted spawning
gravel. The results of the sieve analysis for the mixture are given in
Fig. 2.

There was a distinct color difference between the fine white sand
and darker river rocks used. The cleaned area at the streambed surface
can be easily determined by visual inspection. The cleaned area is
where the white sand had been removed by the jetting action.

For each test the fine sand and gravel was thoroughly mixed,
compacted, and leveled. The jet nozzle was positioned at the desired
angle and height above the streambed surface. Water was slowly added to
the test flume to a depth of 24 in. (61 cm). Then the water jet was
started and increased to a desired velocity. The excess water from the
jet was spilled through an end weir to maintain the water depth in the
flume at 24 in. (61 cm) throughout the test. There was practically no
streamwise flow. The scouring action reached a steady state in about 4
minutes of jetting. However, the jetting was allowed to continue for
10 minutes to assure that the steady state was reached. At the end of
10 minutes the jet stopped, the flume was slowly drained, and the scouring geometries were measured.

The important scouring geometrical parameters are cleaned depth,
D, armored depth A and scoured depth S. Cleaned depth is the lowest
point of the cleaned gravel where fine sand was removed to the original
streambed surface. For a round jet placed in the middle of the flume,
excavation to the bottom of the cleaned zone was necessary for the measurement of cleaned depth. However, the time involved was excessive and
accurate determination was not easy. To overcome these difficulties, a
half-jet (semi-circular jet nozzle) was used. The flat side plate of
the half-jet was placed next to the transparent flume side wall which
acted as the plane of symmetry as shown in Fig. 1. The measured cleaned
depth of the half-jet which was made through the transparent side wall
was found to be the same as for the round jet placed in the middle of
the flume. The armored depth is the distance from the lowest point of
gravel larger than 1 in. (2.54 cm) to the original streambed surface.
The demarcation between the armored zone and the smaller cleaned gravel
was made by visual inspection through the transparent side wall.

ANALYTICAL CONSIDERATIONS

The important forces acting on an individual solid particle
moved by the impinging jet flow are the drag force in the direction of
the fluid velocity vector and a net downward force resulting from the
gravity and the buoyancy forces on the solid particle. On the particle
in contact with the stationary streambed, there is also a frictional
force. However, the frictional force depends upon the drag and the
gravitational forces already mentioned. Hence, the ratio of the drag
force to the net gravitational force becomes an important parameter in
predicting the motion of a solid particle produced by the impinging jet
flow. Assuming solid particles are spherical in shape for the purpose
of analysis, the fluid drag force, $F_d$, on a solid sphere of diameter,
$d_s$, is

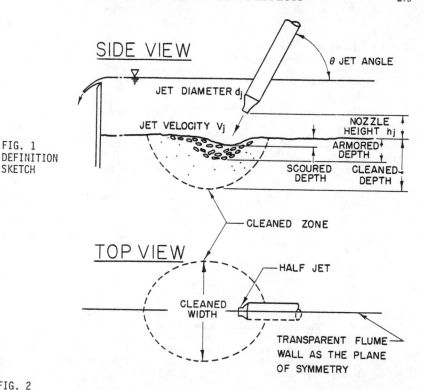

FIG. 1
DEFINITION
SKETCH

FIG. 2
SIEVE
ANALYSIS

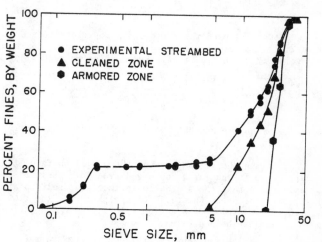

$$F_d = C_d \frac{\pi}{4} d_s^2 \frac{\rho V^2}{2} \qquad (1)$$

where $C_d$ is drag coefficient, $\rho$ is fluid density, and V is relative velocity between the fluid and the solid particles. For the incipient motion of the solid particle along the stationary streambed where the velocity of the solid particle is zero, the V will be the velocity of fluid.

The net gravitational force, $F_g$, of the solid particle is

$$F_g = \frac{\pi}{6} d_s^3 \, g \Delta \rho \qquad (2)$$

where g is gravitation acceleration and $\Delta \rho$ is the density of gravel in excess of density of the fluid. The dimensionless ratio of the drag force to the net gravitational force $F_d/F_g$, is an important parameter in this process.

$$\frac{F_d}{F_g} = \frac{3C_d}{4} \frac{V^2}{gd_s(\frac{\Delta \rho}{\rho})} \qquad (3)$$

The drag coefficient $C_d$ is a function of Reynolds number and the particle shape. For certain particle shapes and range of Reynolds number, $C_d$ is nearly a constant. The square root of the ratio

$$\frac{V^2}{gd_s(\frac{\Delta \rho}{\rho})}$$

in Eq. 3 is called the densimetric Froude number. However, the author feels that the densimetric Froude number may not be the best descriptive name. It is generally used in the study of stratified flows where there is a density difference between two fluids. For this study, the density difference is between a fluid and a solid. The author prefers to call it the Impingement Number I.

$$I = \frac{V}{\sqrt{gd_s(\frac{\Delta \rho}{\rho})}} \qquad (4)$$

Rouse (7) used the ratio $V_j/V_m$ to relate to the scoured depth, where $V_j$ is the velocity of water jet at the nozzle exit and $V_m$ is the fall velocity of the sediment. By equating $F_d$ and $F_g$, $V_m$ is equal to

$$\sqrt{gd_s(\frac{\Delta \rho}{\rho})\frac{4}{3C_d}} \text{, then } \frac{V_j}{V_m} = \frac{V_j}{\sqrt{gd_s(\frac{\Delta \rho}{\rho})\frac{4}{3C_d}}} = I_j \sqrt{\frac{3C_d}{4}} \qquad (5)$$

where $I_j$ is the impingement number at the jet nozzle, $I_j = \dfrac{V_j}{\sqrt{gd_s(\frac{\Delta \rho}{\rho})}}$.

Between the jet nozzle and the gravel surface, the flow field is very close to that of a submerged free jet. Beltaos and Rajaratnam (2) have shown that the stagnation effect of a solid plate extends $1.2d_j$ in front of the plate. In this experiment, the flow was observed entering the gravel bed at a high velocity; the stagnation effect should be small and can be neglected.

The impingement number at the streambed surface $I_i$ is defined as

$$I_i = \frac{V_i}{\sqrt{gd_s(\frac{\Delta\rho}{\rho})}} \tag{6}$$

where $V_i$ is the centerline velocity of the free jet flow at the streambed surface.

The flow field of a submerged free jet was formulated by Albertson (1). For a circular jet, the velocity field has two zones according to the downstream distance. First, the zone of flow establishment (potential core) is in the region where $h_j < 6.2\,d_j$, and the centerline velocity $V_i$ is equal to the velocity at the jet nozzle.

$$V_i = V_j \text{ and } I_i = \frac{V_j}{\sqrt{gd_s(\frac{\Delta\rho}{\rho})}} = I_j \tag{7}$$

Second, the zone of established flow is in the region where $h_j < 6.2\,d_j$, and the centerline velocity decreases to

and

$$V_i = 6.2\frac{d_j}{h_j}V_j$$

$$I_i = 6.2\frac{d_j}{h_j}\frac{V_j}{\sqrt{gd_s(\frac{\Delta\rho}{\rho})}} = 6.2\frac{d_j}{h_j}I_j \tag{8}$$

The "input functions" of the impinging zone are $I_i$ and $d_i$, the impingement number and impinging diameter at the streambed surface. The boundary of a jet flow can be defined as the locus of the point of inflection (this is the maximum velocity gradient) for the longitudinal velocity which corresponds to the region of maximum intensities of turbulence and shear. For the zone of flow establishment ($h_j < 6.2\,d_j$), the jet boundary so defined is parallel to the jet centerline such that the impinging diameter remains the same as the nozzle diameter, hence $d_i = d_j$. Further downstream in the zone of established flow ($h_j < 6.2\,d_j$), based upon the same criterion for the jet boundary, the jet expands downstream at an included angle of 9.2° which is derived from Tangent 9.2° = 1/6.2; hence, impinging diameter $d_i = h_j/6.2$ for the zone of established flow.

Table 1.  Impingement Diameter $d_i$ &
Impingement Number $I_i$ at the Streambed Surface

| Flow Zone | Impingement Mode | Jet Flow Angle | Impingement Diameter $d_i$ | $I_i$ Impingement Number at the Streambed Surface | |
|---|---|---|---|---|---|
| Flow establishment or potential core $h_j < 6.2\,d_j$ | Short | 0° | $d_j$ | $\dfrac{V_j}{\sqrt{gd_s(\frac{\Delta\rho}{\rho})}} = I_j$ | $\dfrac{d_j}{d_i}I_j$ |
| Established flow $h_j < 6.2\,d_j$ | Long | 9.2° | $\dfrac{1}{6.2}h_j$ | $6.2\dfrac{d_j}{h_j}\dfrac{V_j}{\sqrt{gd_s(\frac{\Delta\rho}{\rho})}}$ | |

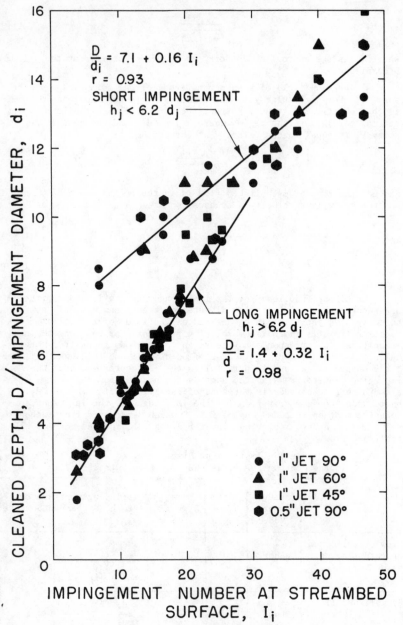

Fig. 3. Cleaned Depth, D.

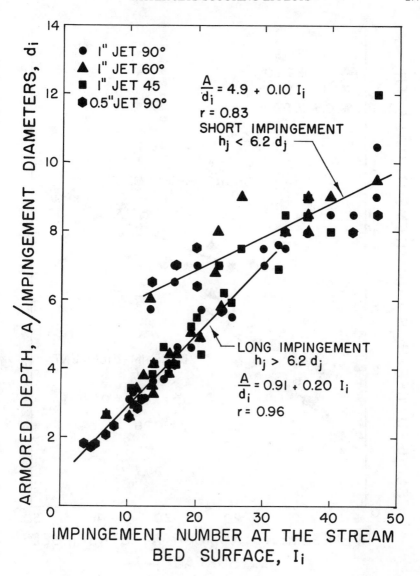

Fig. 4. Armored Depth, A

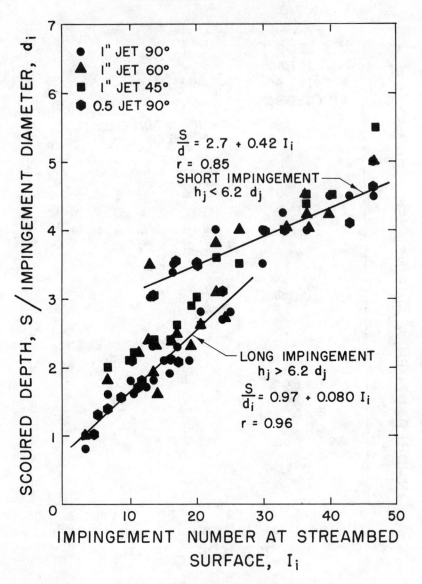

Fig. 5. Scoured Depth, S

The experimental measurements of the effect of flushing (cleaned depth, armored depth and scoured depth) are normalized by $d_i$ and correlated with the impingement number at the gravel surface $I_i$. For the experimental streambed of wide gradation, the representative particle size is chosen at a size 50 percent finer by weight:

$d_s = d_{50} = 0.5$ in.(12.7 mm). For the gravel used, $\frac{\Delta\rho}{\rho} = 1.7, \sqrt{gd_s(\frac{\Delta\rho}{\rho})} =$ 1.5 fps (0.45 m/sec) and $V_m = 2.8$ fps (0.85 m/sec)

## DISCUSSION AND CONCLUSION

Figures 3 through 5 present the experimental results of the normalized cleaned depth, armored depth and scoured depth versus the impinging number $I_i$ at the streambed surface. These figures show that data correlate well for different nozzle diameters. On these figures, all data are correlated according to two distinct zones of impingement, namely, short impingement ($h_j < 6.2\,d_j$) and long ($h_j > 6.2\,d_j$) impingement. The reason for these two distinct impingements is due to the two zones of jet flow downstream of the jet nozzle. For short impingement which is in the zone of the potential core, the jet flow has zero angle of diffusion. For long impingement, which is in the zone of established flow, the flow has a 9.2 degree angle of diffusion.

## ACKNOWLEDGEMENT

Funding support for this study was provided by the Washington State Department of Fisheries, the National Marine Fisheries Service, the State of Washington Water Research Center and the Albrook Hydraulics Laboratory, Washington State University, Pullman, Washington.

## APPENDIX I.--REFERENCES

1.  Albertson, M. L., Dai, Y. B., Jensen, R. A., and Rouse, H., "Diffusion of Submerged Jets," Transactions of the American Society of Civil Engineers,Vol. 155, 1950, pp. 639-664.

2.  Beltaos, S., and Rajaratnam, N., "Impinging of Axisymmetric Developing Jets," Journal of Hydraulic Research, International Association for Hydraulic Research, Vol. 15, No. 4, 1977, pp. 311-326.

3.  Hasler, A. D., Underwater Guideposts--Homing of Salmon, Univ. of Wisconsin Press, Madison, 1966, 155 pp.

4.  Kobus, H., Leister, P., and Westrich, B., "Flow Field and Scouring Effects of Steady and Pulsating Impinging on a Movable Bed," Journal of Hydraulic Research, International Association for Hydraulic Research, Vol. 17, No. 3, 1979, pp. 175-192.

5.  Rajaratnam, N., and Beltaos, S., "Erosion by Impinging Circular Turbulent Jet," Journal of the Hydraulics Division, ASCE, Vol. 103, No. HY10, Proc. Paper 13287, Oct., 1977, pp. 1191-1205.

6.  Rajaratnam, N., and Berry, B., "Erosion by Circular Turbulent Wall Jets," Journal of Hydraulic Research, International Association for Hydraulic Research, Vol. 15, No. 3, 1977, pp. 277-289.

7.  Rouse, H., "Criteria for Similarity in the Transportation of Sediment," Proceedings of Hydraulics Conference, Univ. of Iowa, Engineering Bulletin 20, 1940, pp. 33-49.

Erosion by Unsubmerged Plane Water Jets

by Nallamuthu Rajaratnam[1], M.ASCE

Abstract

This paper presents the results of an experimental study on
the erosion of sand beds by plane water jets with minimum depth of tailwater. The characteristics of the scour hole in the asymptotic state
have been correlated with a densimetric (type of) Froude number. A
comparison has also been made between the erosion caused by free and
submerged plane water jets.

Introduction

Considering the erosion of sand beds by water jets, starting
with the pioneering work of Rouse (2), a number of studies have been
made on the erosion caused by plane and circular turbulent water jets,
in the impinging as well as the wall jet modes (Ref. 1 gives a list of
these investigations). Limiting the present discussion to erosion of
sand beds by impinging plane turbulent water jets, if the jet is fully
submerged as shown in Fig. 1(a), based on a recent study by the
author(1) and the earlier investigations, the following general comments
could be made. The maximum depth of erosion $\varepsilon_m$ occurring under the jet
increases linearly with log t where t is the time from the start of
erosion, for a considerable part of the erosion process and for larger
times, the eroded bed profile and its characteristic lengths reach a so-
called asymptotic (or end) state. It has been found (1) that for large
impingement heights, $\varepsilon_{m\infty}/H$ is mainly a function of the parameter
$E = F_o/\sqrt{H/2b_o}$ where $\varepsilon_{m\infty}$ is the maximum depth of erosion in the asymptotic state, H is the height of impingement (see Fig. 1(a)), $F_o$ is (a
kind of) densimetric Froude number equal to $U_o/\sqrt{gD\Delta\rho/\rho}$ $U_o$ is the jet
velocity (at the nozzle), g is the acceleration due to gravity, D is
the mean particle size, $\Delta\rho$ is the difference between the density of
water $\rho$ and that of the sand and $2b_o$ is the nozzle thickness. If $\varepsilon_{m\infty}'$
is dynamic erosion depth (measured with the jet on whereas $\varepsilon_{m\infty}$ is
measured after the jet is stopped and is referred to as the static
erosion depth), the ratio $\varepsilon_{m\infty}'/\varepsilon_{m\infty}$ could be larger than one and increases
with E. A distinct ridge is formed at the end of the scour hole and its
characteristics in terms of H are again functions of mainly the parama-
ter E. The profile of the eroded bed, in the scour hole part, has been
found to be similar.

---

1. Prof., Dept. of Civil Engrg., Univ. of Alberta, Edmonton, Alberta
Canada.

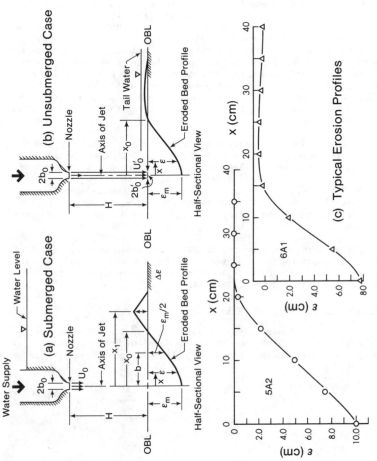

Fig. 1.   Erosion By Unsubmerged Plane Jets – Definition Sketches and Typical Erosion Profiles.

In the present study, the depth of tailwater was reduced to the minimum possible value (as required by the flow to leave the test channel, which was generally about 1 cm) so that the impinging jet becomes an unsubmerged or free-falling jet and the erosion produced by these unsubmerged plane jets was studied and the results are presented in this paper.

## Experiments and Experimental Results

The experiments were performed in rectangular plexiglass flume 0.15 m wide, 0.3 m deep and 1.8 m long, placed inside a larger flume. The jet was produced by a well designed nozzle of thickness $2b_o$ of 2.54 mm which could be positioned at different heights of impingement H above a sand bed of thickness of 0.15 m. The jet velocity $U_o$ was obtained by means of a total head tube of 1 mm diameter. The flow to the nozzle was produced by a pump. Each experiment was run for a period of at least 24 hours before the asymptotic erosion profile was obtained and some experiments were run for even larger times. The asymptotic (static) erosion profile was measured with a point gauge whereas the maximum dynamic scour was obtained visually using the grid-lines drawn on the side of the flume.

On the whole 21 experiments were performed with two sand sizes with the mean particle size D = 1.2 mm (10-20 sieve size) and 2.38 mm (8-12 sieve size). The impingement height H was varied from 2.92 cm to 14.35 cm and the jet velocity at the nozzle $U_o$ was varied from 2 to 3.4 m/sec. The details of these experiments (done in 6 series) are given in Table-1 along with the maximum depth of scour $\varepsilon_{m\infty}$ and the half-width of the scour hole $x_{o\infty}$ as well as the dynamic scour $\varepsilon'_{m\infty}$.

Two typical eroded bed profiles are shown in Fig. 1(c). One significant aspect of these erosion profiles for unsubmerged jets is the absence of the characteristic ridge which forms prominently for submerged jets (see also Fig. 1a & b).

## Analysis of Experimental Results

With reference to Fig. 1(b), if $U'_o$ and $2b'_o$ are respectively the velocity and width of the jet at the level of the (original) uneroded bed, using the Bernoulli and continuity equations, it can be shown that

$$U'_o = \sqrt{U_o^2 + 2gH} \qquad (1)$$

and

$$b'_o = U_o b_o / U'_o \qquad (2)$$

Considering the maximum depth of erosion, occurring under the jet, in the asymptotic state, one could write

Table-1 Erosion by Plane Water Jets With Minimum Tailwater Depth

| Expt | $U_o$(m/s) | H(cm) | $2b_o$(mm) | D(mm) | $U'_o$(m/s) | $2b'_o$(m m) | $F'_o$ | $\varepsilon_{moo}$(cm) | $\varepsilon_{noo}$(cm) | $\varepsilon'_{moo}/\varepsilon_{noo}$ | $x_{ooo}$(cm) | $F_o$ | $E=\dfrac{F_o}{\sqrt{H/2b_o}}$ | $U_o b_o/\nu$ |
|---|---|---|---|---|---|---|---|---|---|---|---|---|---|---|
| 1A1 | 2.36 | 2.92 | 2.54 | 1.2 | 2.48 | 2.42 | 17.84 | 8.23 | 8.64 | 1.05 | 15.2 | 16.98 | 5.01 | 2796 |
| 1A2 | 2.29 | 2.92 | 2.54 | 1.2 | 2.41 | 2.41 | 17.34 | 6.87 | 7.37 | 1.07 | 15.2 | 16.47 | 4.86 | 2888 |
| 1A3 | 2.42 | 2.92 | 2.54 | 1.2 | 2.54 | 2.42 | 18.27 | 7.95 | 8.38 | 1.05 | 15.2 | 17.41 | 5.13 | 3052 |
| 1A4 | 2.42 | 2.92 | 2.54 | 1.2 | 2.54 | 2.42 | 18.27 | 7.77 | 8.38 | 1.08 | 14.2 | 17.41 | 5.13 | 3052 |
| 2A1 | 2.15 | 6.60 | 2.54 | 1.2 | 2.43 | 2.25 | 17.50 | 5.33 | 7.59 | 1.42 | 15.2 | 15.47 | 3.03 | 2712 |
| 2A2 | 2.22 | 6.60 | 2.54 | 1.2 | 2.49 | 2.26 | 17.91 | 7.62 |  |  | 16.3 | 15.97 | 3.13 | 2800 |
| 2A3 | 2.47 | 10.67 | 2.54 | 1.2 | 2.86 | 2.19 | 20.50 | 9.03 | 10.16 | 1.13 | 16.3 | 17.77 | 2.74 | 3113 |
| 3A1 | 2.34 | 10.67 | 2.54 | 1.2 | 2.75 | 2.16 | 19.78 | 7.16 | 8.13 | 1.14 | 16.3 | 16.83 | 2.60 | 2951 |
| 3A2 | 2.43 | 14.35 | 2.54 | 1.2 | 2.95 | 2.09 | 21.22 | 7.77 | 9.14 | 1.18 | 16.3 | 17.48 | 2.33 | 3065 |
| 3A3 | 2.18 | 14.35 | 2.54 | 1.2 | 2.75 | 2.01 | 19.78 | 6.50 | 7.37 | 1.13 | 16.3 | 15.68 | 2.09 | 2749 |
| 3A4 | 2.66 | 4.19 | 2.54 | 1.2 | 2.81 | 2.40 | 20.22 | 8.99 | 10.92 | 1.22 | 21.6 | 19.14 | 4.71 | 3355 |
| 4A1 | 3.43 | 4.19 | 2.54 | 1.2 | 3.55 | 2.45 | 25.54 | 11.28 | 13.46 | 1.19 | 25.4 | 24.68 | 6.08 | 4326 |
| 4A2 | 3.45 | 6.22 | 2.54 | 1.2 | 3.62 | 2.42 | 26.04 | 11.02 | 13.46 | 1.22 | 27.9 | 24.82 | 5.02 | 4351 |
| 4A4 | 2.44 | 7.11 | 2.54 | 1.2 | 2.71 | 2.29 | 19.50 | 7.01 | 9.40 | 1.34 | 15.8 | 17.55 | 3.32 | 3077 |
| 5A1 | 3.33 | 10.92 | 2.54 | 1.2 | 3.64 | 2.32 | 26.19 | 12.34 | 13.46 | 1.09 | 30.5 | 23.96 | 3.65 | 4200 |
| 5A2 | 2.93 | 10.92 | 2.54 | 1.2 | 3.28 | 2.27 | 23.60 | 10.21 | 11.94 | 1.17 | 24.4 | 21.08 | 3.21 | 3695 |
| 5A3 | 2.93 | 9.14 | 2.54 | 1.2 | 3.22 | 2.31 | 23.17 | 10.47 | 11.68 | 1.12 | 22.9 | 21.08 | 3.51 | 3695 |
| 5A4 | 2.30 | 9.27 | 2.54 | 1.2 | 2.67 | 2.19 | 19.21 | 6.71 | 8.64 | 1.29 | 15.2 | 16.55 | 2.74 | 2901 |
| 6A1 | 3.20 | 9.27 | 2.54 | 2.38 | 3.47 | 2.34 | 17.70 | 7.93 | 10.16 | 1.28 | 14.2 | 16.33 | 2.70 | 4036 |
| 6A2 | 3.32 | 7.11 | 2.54 | 2.38 | 3.52 | 2.39 | 17.96 | 9.14 | 10.67 | 1.17 | 19.3 | 16.94 | 3.20 | 4187 |
| 6A3 | 3.39 | 5.33 | 2.54 | 2.38 | 3.54 | 2.43 | 18.06 | 9.30 | 11.18 | 1.20 | 19.6 | 17.30 | 3.78 | 4276 |

$$\varepsilon_{m\infty} = f_1 [U_o', b_o', \rho, g\Delta\rho, D, \nu] \tag{3}$$

where $\nu$ is the kinematic viscosity of water and $f_1$ denotes a function.

Using the Pi theorem, Eq. 3 could be reduced to

$$\frac{\varepsilon_{m\infty}}{b_o'} = f_2 \left[ F_o' = \frac{U_o'}{\sqrt{g\frac{\Delta\rho}{\rho}D}}, \frac{U_o'2b_o'}{\nu}, \frac{2b_o'}{D} \right] \tag{4}$$

If the Reynolds number equal to $U_o' 2b_o'/\nu$ is of the order of a thousand, the effect of viscosity on the jet diffusion and erosion appears to be negligible. If $2b_o'/D$ is of the order of unity or larger, it appears that $2b_o'/D$ will not be important. In the light of these comments, Eq. 4 may be reduced to

$$\frac{\varepsilon_{m\infty}}{b_o'} = f_3 [F_o'] \tag{5}$$

Similarly, the length scale $x_{o\infty}$ representing the lateral extent of the scour hole (see Fig. 1b) can be given by the expression

$$\frac{x_{o\infty}}{b_o'} = f_4 [F_o'] \tag{6}$$

The experimental results for $\varepsilon_{m\infty}/b_o'$ and $x_{o\infty}/b_o'$ are shown in Fig. 2(a) and Fig. 2(b) plotted individually against $F_o'$. It appears that in both these cases, the data could be satisfactorily represented by straight lines passing through the origin and these lines are described by the equations

$$\frac{\varepsilon_{m\infty}}{2b_o'} = 1.82 \, F_o' \tag{7}$$

and

$$\frac{x_{o\infty}}{2b_o'} = 4.2 \, F_o' \tag{8}$$

If $\varepsilon_{m\infty}'$ is the depth of dynamic scour, as measured with the jet on, the variation of $\varepsilon_{m\infty}'/\varepsilon_{m\infty}$ with $F_o'$ is shown in Fig. 2c, which indicates considerable scatter about a mean line representing a value of about 1.2. It should be mentioned that measured values of $\varepsilon_{m\infty}$ should be considered very approximate.

At this stage, it is interesting to compare the erosion caused by the free-falling on unsubmerged jets with the erosion caused by the corresponding submerged jets. Using the results of Ref. 1 as representing the erosion caused by submerged plane jets, if $\varepsilon_{m\infty*}'$, $\varepsilon_{m\infty*}'$ and $x_{o*}$ are the characteristic erosion lengths for the

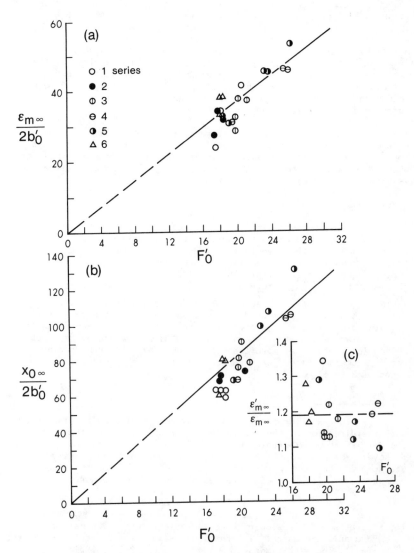

Fig. 2. Length Scales for Erosion – Unsubmerged Plane Jet.

corresponding submerged jet, then the variation of the ratios $\varepsilon_{moo}/\varepsilon_{moo}\ast$; $\varepsilon'_{moo}/\varepsilon'_{moo}\ast$ and $x_{Ooo}/x_{Ooo}\ast$ with the parameter $E = F_O/H/2b_O$ where $F_C=U_O/\sqrt{gD \frac{\Delta\rho}{\rho}}$ is shown in Fig. 3a to c. From Fig. 3(a) it can be said that for $E$ in the range of 2 to 6, $(\varepsilon_{moo}/\varepsilon_{moo}\ast)$ increases first with $E$ reaching a value of about 1.7 at $E\simeq4$ and then decreases to approach unity at $E\simeq6$. A somewhat similar behaviour is noted in Fig. 3(c) for the ratio $(x_{Ooo}/x_{Ooo}\ast)$ whereas in Fig. 3(b) it is seen that the results for the ratio of the dynamic scour scatter around the horizontal line having a value of unity. Hence it appears that for $E$ in the range of 2 to 6, the unsubmerged plane jet produces a deeper and wider scour hole than the corresponding (deeply) submerged jet.

Another interesting difference in the erosion caused by unsubmerged and submerged plane jets is that for the submerged case, at the downstream end of the scour hole, a characteristic ridge will form whereas for the unsubmerged case, such a ridge is generally absent (or is hardly noticable).

## Conclusions

Based on a preliminary experimental study on the erosion of sand beds by unsubmerged plane water jets, the following conclusions can be drawn. In the asymptotic (or end state), the maximum depth of scour $\varepsilon_{moo}$ and the lateral extent of the scour hole $x_{Ooo}$ in terms of the width of the jet at the impinging level $2b'_O$ are functions of mainly the parameter $F'_O = U'_O/\sqrt{gD\frac{\Delta\rho}{\rho}}$ where $U'_O$ is the velocity of the jet at the impinging level, g is the acceleration due to gravity, $\Delta\rho$ is the difference between the mass density of sand and $\rho$ the mass density of water and D is the mean particle size. It has also been found that the depth as well as the lateral extent of scour caused by an unsubmerged jet is somewhat more than that caused by the corresponding deeply submerged plane jet when the parameter $E = F_O/\sqrt{H/2b_O}$ is in the range of 2 to 6.

## Acknowledgements

The experiments for this study were performed by Alan Lee with the direction of the author in the Grad. Hydraulics laboratory of the University of Alberta, Edmonton. The author is thankful to him and S. Lovell for his help in the experimental work and to the Natural Sciences and Engineering Research Council of Canada for the financial assistance provided through a research grant to the author.

## Appendix-I: References

1.  Rajaratnam, N. Erosion by Plane Turbulent Jets., Journal of Hydraulic Research., International Assoc. for Hydraulic Research, Delft, The Netherlands, Vol 19, No.4, pp339-358, 1981.

2.  Rouse, H., Criteria for Similarity in the Transportation of Sediment., Bull 20, Univ. of Iowa, Iowa City, Iowa, pp33-49, 1939.

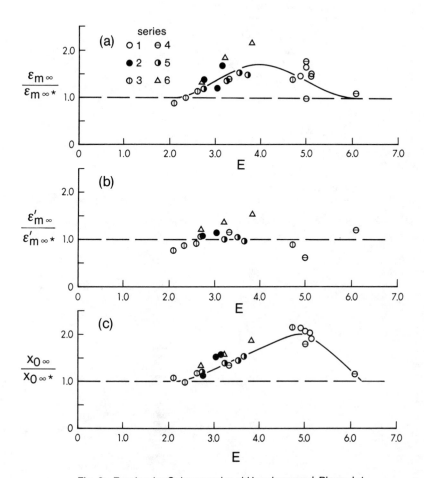

Fig. 3.  Erosion by Submerged and Unsubmerged Plane Jet.

Appendix-II: Notation

$b_0$ = half width of nozzle
$b_0'$ = half width of jet at impingement level
$D$ = mean particle size
$E$ = parameter equal to $F_0/\sqrt{H/2b_0}$
$F_0$ = $U_0/\sqrt{gD\ \Delta\rho/\rho}$
$F_0'$ = $F_0$ at impingement level
$g$ = acceleration due to gravity
$H$ = height of nozzle above sand bed
$m$ = suffix to denote maximum values
$U_0$ = jet velocity at nozzle
$U_0'$ = $U_0$ at impingement level
$x$ = lateral distance from jet axis
$x_0$ = where $\varepsilon$ first equal to zero
$\varepsilon$ = depth of erosion below original bed level
$\varepsilon_m$ = maximum value of $\varepsilon$
$\varepsilon'$ = dynamic scour
$\nu$ = kinematic viscosity of water
$\rho$ = mass density of water
$\Delta\rho$ = difference between the mass densities of sand and water
$\infty$ = suffix to denote asymptotic state
$*$ = suffix to denote the values for the corresponding submerged jet.

# PLUNGE POOL ENERGY DISSIPATOR FOR PIPE SPILLWAYS

By Clayton L. Anderson,[1] M.ASCE and Fred W. Blaisdell,[2] F.ASCE

## INTRODUCTION

Scour by a free-falling jet at pipe spillway and culvert outlets can be avoided by dissipating the destructive energy in a plunge pool. The plunge pool, Fig. 1, is a widened and deepened section of the downstream channel into which the spillway discharges. The energy in the jet is dissipated by turbulence in the pool and by shear with the surrounding water and with the boundaries of the pool. However, if the boundaries of the pool are to be stable they must be pre-excavated to a shape such that the boundary material can withstand the tractive force of the flow.

The best shape for the plunge pool is assumed to be that of a hole scoured by the jet in an erodible bed. The objective of this study is to determine the shape and location of that scour hole. Study variables are the rate of flow, the height of the pipe above the pool surface, the slope of the pipe, and the size and size distribution of the bed material. From our test results we have derived relationships to describe the geometry of the plunge pools. With this information, field design criteria will be established for shaping and riprapping plunge pools. This paper is a progress report of the results to date.

## STUDY PLAN

Our study of naturally formed scour holes in noncohesive soils was made to include a wide range of anticipated field conditions. The variables and the values of each used in the study are:

1. dimensionless discharge: $Q/(gD^5)^{1/2}$ = 0.5, 1, 2, 3, 4, and 5, where Q the discharge, g is the gravitational acceleration, and D is the pipe diameter;
2. bed material size: nominal $d_{50}$ = 0.5, 1, 2, 4, and 8 mm, with $\sigma \approx$ 1.2, where $d_{50}$ is the mean particle size, $\sigma = d_{84}/d_{50}$ is the geometric standard deviation of the log-normal size distribution, and $d_{84}$ is the particle size for which 84 percent of the bed material is finer by weight;
3. geometric standard deviation of log-normal bed material size distribution: $\sigma$ = 1.22, 1.41, and 1.60 with $d_{50}$ = 2 mm;

---

[1,2]Research Hydraulic Engineers, U.S. Dept. of Agric., Agricultural Research Service, St. Anthony Falls Hydraulic Laboratory, Third Ave. SE at Mississippi River, Minneapolis, MN 55414.

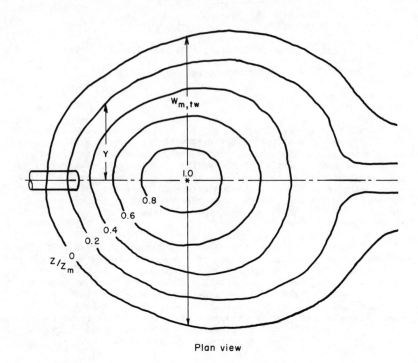

Plan view

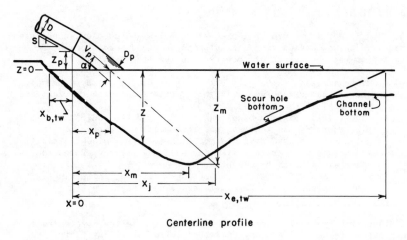

Centerline profile

Fig. 1.- Plunge Pool.

4. height of pipe invert: $Z_p/D$ = 8, 4, 2, 1, 0, -1, and -2, where $Z_p$ is the height of the pipe invert with respect to the water surface in the scour hole;

5. pipe slope: S = 0(for most tests), 0.3425, 0.428, 0.473, 0.635, and 0.782, where S is the sine of the pipe angle with the horizontal; and

6. elapsed time of scour: time-dependent measurements of scour after 10, 31, 100, 316, 1000, 3162 and 10,000 minutes; suspended-material-removed measurements after all material suspended by the jet action had been removed.

These variables were tested in combinations such that the effect of each could be evaluated with a minimum number of tests. Basic data consisted of cross-sectional profiles of the scoured region obtained at the predetermined times after the beginning of each test. The profile data were processed to yield longitudinal centerline and deepest cross-section profiles, a contour map of the scoured region, and parameters used to nondimensionalize the scour geometry.

RESULTS

The test data were used to identify forms of scour hole development, to reduce contour maps to a common dimensionless form, and to describe the position and maximum depth of the scour hole.

Scour Hole Form.--The scour holes were inverted cones for all bed material sizes when $Q/(gD^5)^{1/2}$ < 1, thus indicating that most of the energy was dissipated vertically. The scour hole was also conical when $d_{50}$ = 2 mm and $Q/(gD^5)^{1/2}$ = 2 and when $d_{50}$ = 4 and 8 mm and $Q/(gD^5)^{1/2}$ = 3, although these holes were elongated in the direction of the flow.

For the conical scour holes, the plunging jet churned and lifted material from the bottom of the hole, suspending the material in the reflection of the jet from the bottom of the hole. Part of the suspended material was carried into the downstream channel, but at a rapidly diminishing rate as the test progressed. Although very little material was leaving the scour hole when the time-dependent tests ended at 10,000 minutes, much material remained suspended in the reflected jet. The force of the reflected jet also supported the downstream slope of the scour hole at an angle steeper than the submerged angle of repose of the bed material. When the flow was diverted to measure the scour dimensions, the suspended material settled and the downstream face of the scour hole sloughed to its submerged angle of repose. This material settled in the bottom of the scour hole and masked the true depth to which the bed had been disturbed.

At dimensionless discharges exceeding those indicated above, the scour hole was elongated and a horizontal circulation in the scour hole caused flow upstream along both sides of the hole. When the shear stress of this reverse flow exceeded the critical value for the bed material on the sideslopes of the hole, particles were dislodged and carried upstream to be recirculated with the jet flow. Once started,

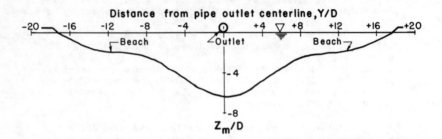

Fig. 2.-Cross section of Scour Hole with Beaching.

the circulation strengthened and caused the shallow widening at the perimeter of the scour hole illustrated in Fig. 2. This process was termed "beaching" because of the profile of the widened sections. Beaching did not occur if

$$\frac{Q}{\sqrt{gD^5}} \le 25 \frac{d_{50}}{D} + 1 \qquad (1)$$

Because the excessive width of beached scour holes is undesirable, we felt analysis of the beached scour hole data would serve no useful purpose so further analysis was limited to conditions for which beaching did not occur.

After completion of the time-dependent tests, additional tests were run to evaluate the actual depth to which the bed was disturbed. This was accomplished by removing the suspended material through a suction line extended into the reflected jet. The tests were run until the jet no longer stirred up sediment from the bottom of the hole, indicating a stable scour hole.

The scour holes from which the suspended material was removed were conical and significantly deeper than the corresponding scour holes for the time-dependent tests, particularly for the lower discharges-- $Q/(gD^5)^{1/2} \le 1$. However, as the discharge increased, the increased horizontal component of the jet swept the disturbed bed material out of the hole. As a result, the time-dependent and suspended-material-removed scour holes were quite similar for the higher discharges.

Dimensionless Scour Contours.--A general definition of the scour hole shape for all test conditions was derived by analyzing the computer-generated plots of the scour contours in terms of pertinent normalizing parameters. Dr. K. Yalamanchili, a former coworker, developed the procedure for describing the scour hole topography in terms of dimensionless contours. The dimensionless contours were similar for all five bed material sizes, all seven pipe heights, all six discharges, and all seven test times; approximately 360 contour maps were

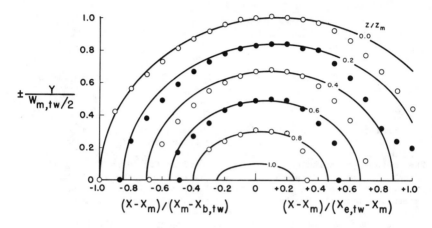

Fig. 3.—Dimensionless Contours—Observed and Computed

collapsed into a single dimensionless contour pattern.  Sample data are plotted in Fig. 3.

The dimensionless contours for all tests were enveloped by the set of elliptical contours  shown in Fig. 3.   These contours are defined by the relationships,

$$\frac{X - X_m}{X_m - X_{b,tw}} = 0_{corr} - a \left[ 1 - \left( \frac{2Y/W_{m,tw}}{b} \right)^2 \right]^{1/2} \qquad (2a)$$

and

$$\frac{X - X_m}{X_{e,tw} - X_m} = 0_{corr} + a \left[ 1 - \left( \frac{2Y/W_{m,tw}}{b} \right)^2 \right]^{1/2} \qquad (2b)$$

for the upstream and downstream portions, respectively, of the scour hole, where

$$0_{corr} = 0.15 \left( 1 - \frac{Z}{Z_m} \right) \qquad (3a)$$

$$a = 1.15 - 0.9 \frac{Z}{Z_m} \tag{3b}$$

and

$$b = 1.00 - 0.775 \frac{Z}{Z_m} - 0.125 \left( \frac{Z}{Z_m} \right)^2 \tag{3c}$$

The parameters are defined in Fig. 1. $0_{corr}$ accounts for the downstream shift of the center of the ellipses as $Z/Z_m$ decreases, and $a$ and $b$ are the semimajor and semiminor axes, respectively. The contours given by the equations plotted in Fig. 3 indicate good agreement with the sample data. Equation (2b) is perhaps a little conservative for the downstream portion of the contours for the suspended-material-removed tests, but it envelopes the test data.

In order to apply these dimensionless equations, values of the quantities $X_m - X_{b,tw}$, $X_{e,tw} - X_m$, $W_{m,tw}$, $X_m$, and $Z_m$ must be evaluated.

The ratios $(X_m - X_{b,tw})/Z_m$, $(X_{e,tw} - X_m)/Z_m$ and $W_{m,tw}/2Z_m$ were constant with time. Therefore, the ratios were averaged for all times of scour for each test condition and the averages were plotted against the dimensionless discharge. To give maximum (conservative) values to the scour hole contour coordinates, the resultant plots were represented by envelope curves, which have the equations:

$$\frac{X_m - X_{b,tw}}{-Z_m} = \frac{X_{e,tw} - X_m}{-Z_m} = \frac{3}{2} + \frac{1}{3} \frac{Q}{\sqrt{gD^5}} \geqslant 2 \tag{4}$$

and

$$\frac{W_{m,tw}}{2Z_m} = 1.75 - 0.15 \frac{Q}{\sqrt{gD^5}} \tag{5}$$

Eq. 4 represents the slopes of the ends of the scour hole, which are limited to a minimum value of 2 when $Q/(gD^5)^{1/2} < 3/2$.

Depth of Scour.--The potential ultimate depths of scour for the time-dependent data were computed after Blaisdell, Anderson, and Hebaus.[3] The ultimate scour depths computed from the time-dependent

[3]Blaisdell, F.W., Anderson, C. L., and Hebaus, G. G., "Ultimate Dimensions of Local Scour," Journal of the Hydraulics Division, ASCE, Vol. 107, No. HY3, Proc. Paper 16144, March 1981, pp. 327-337.

data plus the observed depths from the suspended-material-removed tests were used to derive a relationship for the maximum depth of scour. In Fig. 4 these maximum depths of scour are plotted against the densimetric Froude number, $F_d$, where

$$F_d = \frac{V_p}{\sqrt{gd_{50}(\rho_s - \rho)/\rho}} \tag{6}$$

and $\rho_s$ and $\rho$ are the densities of the bed material and the jet fluid, respectively. Figure 4(a) is a plot of the maximum scour depths for the suspended-material-removed tests and the ultimate scour depths for the time-dependent tests for the range of sediment sizes and size distributions tested, and Fig. 4(b) is a plot of the respective depths for the range of pipe heights and slopes tested. In Fig. 4(b) the symbols with slashes represent combinations of pipe height and slope that simulate the jet plunge angle for zero pipe slopes represented by symbols without slashes.

The curves illustrated in Fig. 4 are defined by the relationship

$$\left(\frac{Z_m}{D}\right)_{max} = A\left[1 - e^{-k(F_d-2)}\right] \tag{7}$$

where $A = 7.5$, $k = 0.6$ when $Z_p/D \lessdot 1$ and $A = 10.5$, $k = 0.35$ when $Z_p/D > 1$. The values of $A$ limit the computed scour depths to 7.5D and 10.5D because the data indicate that these depths are seldom exceeded, particularly in the range of anticipated prototype Froude numbers, which are generally less than 8.

Location of Maximum Depth of Scour.--To avoid attack of the jet on the upstream and downstream slopes of the plunge pool boundary, the pre-excavated pool must be properly located with respect to the pipe outlet. The longitudinal position of the scour hole with respect to the outlet is defined by $X_m$, the location of the maximum depth of scour. Assuming that the maximum depth of scour would occur where the jet trajectory impacts the bed, $X_m$ has been evaluated in terms of $X_j$, the location of the intersection of the jet trajectory with the minimum elevation in the scour hole as shown in Fig. 1. The distance $X_j$ is defined by

$$\frac{X_j}{D} = \frac{X_p}{D} + \frac{(Z_m/D)_{comp}\sin\alpha}{\tan\alpha} \tag{8}$$

in which $X_p/D$ is the distance from the pipe outlet to the point where the jet plunges into the pool and $(Z_m/D)_{comp}/\tan\alpha$ is the projected horizontal length of the tangent jet trajectory from the water surface to the computed depth, $(Z_m/D)_{comp}$. The sin $\alpha$ term is included because

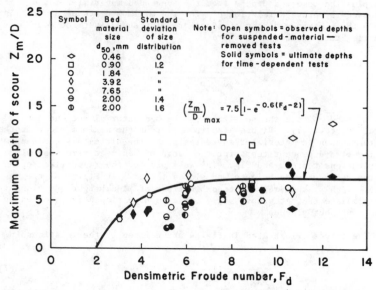

**(a) Bed material sizes and size distributions with $Z_p/D = 1$**

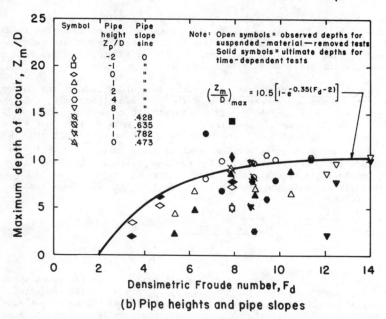

**(b) Pipe heights and pipe slopes**

Fig. 4.–Maximum Depth of Scour.

it was found to eliminate the effect of the discharge on $X_m/X_j$. $X_p$ and $\alpha$ can be computed from free fall relationships.

The jet trajectory distance to the minimum elevation, $X_j/D$, as defined above, ranged from 0.6 to 1.4 times the observed distance to the maximum scour depth $X_m/D$. Figure 5 is a plot of the ratio $X_m/X_j$ against the dimensionless discharge for the suspended-material-removed tests. Based on all tests, the average value of $X_m/X_j$ is 1.2. Therefore, the plunge pool should be placed so that the location of the maximum depth is 1.2 times the jet trajectory distance, $X_j/D$, downstream from the pipe outlet.

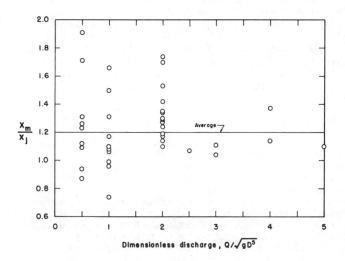

Fig. 5.-Distance to Maximum Depth of Scour

SUMMARY

The time development of natural scour holes and scour holes in which the material suspended by the jet was removed have been determined for six discharges, seven pipe heights, five mean sizes of noncohesive bed material, three standard deviations of one mean bed material size, and several pipe slopes.

The suspended-material-removed tests were made to determine the depth to which the plunging jet would be able to disturb various bed materials. Determination of this depth is necessary because the prototype plunge pool must be designed so that the boundary material will not be disturbed by the flow conditions within the pool.

The data are still being analyzed, so changes in the criteria presented herein are possible.

Prediction of Scour Depth
From Free Falling Jets

By  H.W. Coleman[1], M. ASCE

ABSTRACT

A simple method for predicting limiting scour depth in
plunge-pool energy dissipators is proposed.  The scour
depth as determined by the Veronese formula is compared
with scour hole depths observed for several model
studies.  It was found that the scour depth from the
formula, $d_s = 1.90 H^{0.225} q^{0.54}$, is a reasonable estimate
for the resulting model scour, when $d_s$ is measured in
the direction of the tangent to the jet entering the
tailwater.  The effect of size of gravel on limiting
scour depth appeared to be minimal for the range of
sizes included in these tests.

INTRODUCTION

The depth and location of scour holes from free falling
jets is of critical importance in the design of plunge
pool energy dissipators.  Where the foundation is
erodible, the proximity of the erosion to structures
such as outlet works, toe of dam, and spillway will
determine the safety of these structures from being
undermined from such hydraulic action.  For hydroelectric
dams, the flip-bucket is a common terminal structure for
the spillway.  The bucket may be placed at the end of a
chute on the abutment, at the toe of a gravity overflow,
or at the crest of an arch dam.

This paper deals with an attempt to correlate a few
model observations with the Veronese formula for
defining the magnitude and location of maximum scour
depth.

THEORY

Most theoretical treatments assume that the maximum scour
depth is a function of unit discharge, q, head, H, a
material size, such as $d_{50}$, and time of exposure to a
particular discharge.  The Veronese formula,

$$d_s = 1.90 \ H^{0.225} q^{0.54} \tag{1}$$

is an empirical attempt to identify the limiting scour

---

[1]  Assistant Head, Hydraulics Section
Harza Engineering Company, Chicago, Illinois

298

depth from vertically falling jets, where:
$d_S$ = vertical depth of scour below tailwater, in meters,
$H$ = effective energy of jet entering the tailwater
  = HW-TW-losses on the chute, in meters.
$q$ = unit discharge = Q/b, where
    Q is total discharge in $m^3$/sec, and
    b is jet width in meters.

That is, if a series of model tests were made with
progressively smaller size material, for long periods of
time, the resulting extrapolation to zero size material
would yield the limiting scour depth. In the prototype,
large rock particles are assumed to be ground to erodible
size by the action of the jet. Therefore, neither
material size nor time appear in the Veronese formula.

This formula seems clearly applicable to vertically
falling jets. However, in the case of flip-buckets at
the end of a chute spillway, the application of the
formula is not so clear. In such a case, the jet
typically enters the tailwater at a 20°-40° angle measured
from the horizontal, shown as $\alpha$ in Figure 1. Based on
model observations, the resulting scour depth is
generally considerably less than that predicted by the
Veronese formula, if $d_S$ is measured vertically.

Therefore, a very simple assumption has been made to
assist in correlation. That is, the effective scour
depth, $d_S$, predicted by the Veronese formula should be
measured along in the tangent to the jet entering the
tailwater. In other words, the expected scour depth,
$y_S$, should be as follows:
$$y_S = d_S \sin\alpha = 1.90\ H^{0.225}\ q^{\ 0.54}\ \sin\alpha, \qquad (2)$$
where $\alpha$ = jet entry angle into tailwater, as shown on
          Figure 1.

The jet trajectory is computed by the simple trajectory
formula,
$$y = x \tan\theta - x^2/4H_1 \cos^2\theta, \text{ where} \qquad (3)$$
$y$ = vertical distance above bucket lip
$x$ = horizontal distance from bucket lip
$\theta$ = exit angle of jet leaving bucket
$H_1$ = energy head leaving bucket.

The effective bucket exit angle and effective bucket head
are generally not known prior to detail computation or
model studies. The flow exit angle is generally less
than the bucket angle because the bucket radius is too
small to turn the jet completely. Therefore, for
simplicity, it is assumed that the effective jet exit
angle is equal to the bucket exit angle. With this
assumption, the effective bucket head was about 85% of

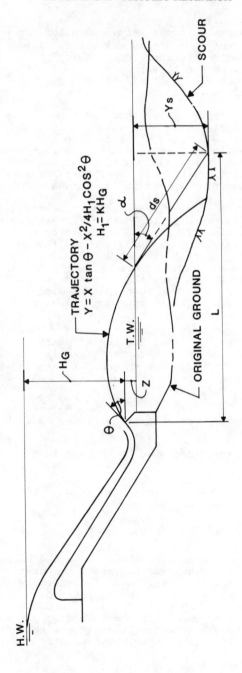

FIGURE 1    DEFINITION SCHEMATIC

the gross bucket head, based on model trajectory data.
This is not unreasonable for model scales in the range of
1:50 to 1:100, where model friction tends to be higher
than the prototype. The expected scour is then computed
based on

H = H$_1$ + Z, where                                    (4)

Z = vertical distance between bucket lip and tailwater,
and

H$_1$ = as defined above.

The location of maximum scour is computed by adding the
trajectory distance between bucket lip and entry point
into the tailwater to the x component of d$_s$,(d$_s$ cos $\alpha$).

RESULTS

The correlation of maximum scour depth is shown on Figure
2. The computed scour depths, Y$_s$, are generally slightly
higher than the quantities measured in the model. It,
therefore, appears that this prediction technique will be
slightly conservative. It is interesting that the tests
which show the greatest deviation from the predicted
depths, Mossyrock, is for a vertically falling jet from
the crest of an arch dam. The model scales used in this
correlation ranged from 1:40 to 1:100, and model material
sizes from sand to 20 mm gravel, with no apparent
relationship to results.

The correlation of location of maximum scour is shown on
Figure 3. The predicted distances are further from the
bucket lip than the observed results. It seems likely
that the jet continues to follow the trajectory curve for
some distance below tailwater, resulting in a shorter
throw than computed. In addition, the effective jet exit
angle generally is somewhat less than the bucket exit
angle. This effect would tend to shorten the throw and
result in less scour depth.

CONCLUSION

Correlation of model results with the Veronese formula
show reasonable agreement for both magnitude and location
of maximum scour, if the angle of jet entry into tailwater
can be predicted. The Veronese prediction of scour depth
should be measured in the direction of the jet entry.
This technique can be expected to apply where tailrace
material is relatively fine and uniform. In the prototype,
varying quality of geologic material and bedding planes
make prediction much more uncertain.

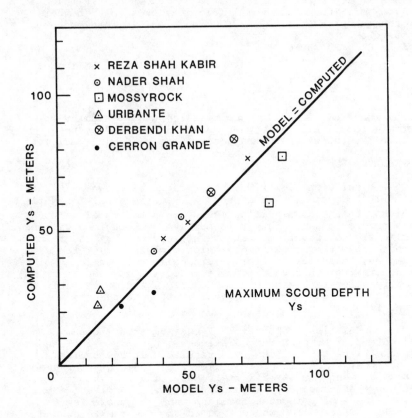

FIGURE 2     MAXIMUM SCOUR DEPTH

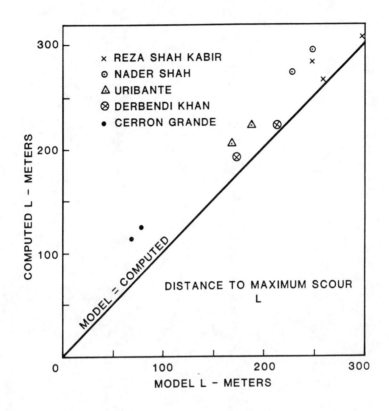

FIGURE 3     DISTANCE TO MAXIMUM SCOUR

APPENDIX - REFERENCES

1.  United States Bureau of Reclamation, "Design of Small
    Dams," 1977, p. 410.

2.  Scimemi, Ettore, "Discussion  of Paper 'Model Study
    of Brown Canyon Debris Barrier,' by Bermel and Sanks,"
    ASCE, Vol. 112, 1947, p. 1016.

EXPANSION AND REHABILITATION OF A HYDRO PROJECT

Rajeev K. Chaudhary[1], P.E., A.M. ASCE and
John E. Fisher[2], P.E., M. ASCE

ABSTRACT

This paper discusses the history of City of Sturgis, Michigan Dam and Powerhouse located on the St. Joseph River; its operation since the early twentieth century, major on site rehabilitation of turbines and generators; investigation of the dam, powerhouse and intake structures; design of minor repairs to the dam (both concrete and earthen parts); design of concrete block energy dissipation system for the apron; results of a 1:10 scale model study of the straight drop spillway system consisting of a concrete apron, impact blocks, a tailwater sill and toe sill; design of a new powerhouse to house two new standard units; results of an analytic evaluation of head losses and flow distribution of two intake systems - one with a large diameter penstock bifurcating into two units at a junction chamber and the second with two small diameter penstocks one for each turbine. At the time of this writing (April, 1982) the spillway system is 75% complete and functional and the powerhouse is under construction and due to come on line in July, 1982.

INTRODUCTION

The energy shortages and the tremendous price escalation of oil in recent years have focused renewed interests in the development of hydroelectric power in the U.S. and all over the world. Since, after hydroelectric generation, water is immediately available for reuse, hydroelectric plants are generally located in series, one downstream of another, when suitable sites are available. The Saint Joseph River in Michigan and Indiana is one such river with a total of twelve dams, ten still generating. Many of the factors which influence hydropower generation are independently variable from plant to plant and from site to site even on the same river. Most of the dams on the St. Joseph River were built in the early 1900's, about two decades after hydroelectricity was produced for the first time from flowing water on the Fox River in Appleton, Wisconsin. Lawson-Fisher Associates, in the last five years, has performed engineering work on several dams on the St. Joseph River, one of which is in the construction stage at present and will be the topic of discussion in this paper. See Figure 1 for general location of the site.

1. Project Engineer, Lawson-Fisher Associates, 525 W. Washington Street, South Bend, Indiana 46601

2. Partner, Civil Engineer, Lawson-Fisher Associates

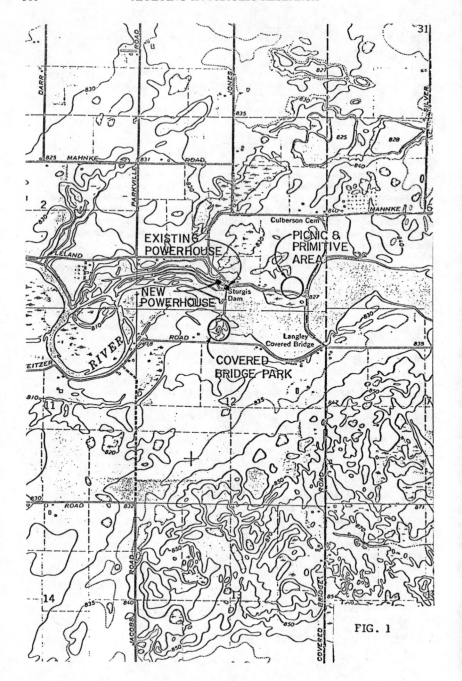

FIG. 1

## HISTORY

The dam and the powerhouse were built in 1910.  The dam is a concrete gravity arch and buttress type of structure about 300 ft. long.  The earthen part of the dam is another 500 ft. in length. The powerhouse has a reinforced concrete foundation and a masonry superstructure and houses two Allis-Chalmers reaction turbines with each generator rated at 550 kW.  The intake to each of these units is the open channel type with the trash racks upstream of the headgates.  No major repair work was done on the powerhouse, the dam or the units until 1978.  The headgates and the trash racks were replaced once in the 68 years of service between 1910 and 1978.

## DEWATERING AND INSPECTION

Dewatering and inspection of the dam and its appurtenances was done in October, 1978.  Allis-Chalmers, the original manufacturer of the turbines, was authorized to inspect the hydro equipment at the same time.  Prior to dewatering a sedimentation survey of the reservoir was made, which indicated that there was an average of about 3 ft. of sediment above the original valley floor since the reservoir impoundment in 1911.  The height of the dam above the foundation is 22 ft., with an extra 3 ft. provided by flashboards on top of the spillway.  Proper coordination with the upstream and downstream dams and property owners enabled the water to be lowered by approximately 10 ft.  All flashboards were raised to bring the water level below the top of the concrete spillway at which time the reaction turbines were removed for repairs and as a result the discharge openings in the floor of the powerhouse were used to bring the water level down another 7 to 8 ft. below the top of spillway.  An aeration system was designed and installed to prevent venturi action.

A structural inventory of all arches, walls, buttresses, aprons and related items was made to evaluate the specific condition of each.  Dye tests were made to help establish seepage through cavities.  Concrete corings (4" diameter) were made selectively to determine the probability of cavities and to allow laboratory compression tests for strength.  Coring holes were repaired with epoxy grout.  Ultrasonic tests with surface sensors were made on all major elements to establish strength and uniformity of the concrete.  A professional diver was used to visually inspect major portions of the upstream face of the Dam which remained underwater during the dewatering.

The foundation walls and the walls, crowns and bases of the intake and draft tubes were found generally to be in good condition.  The concrete corings in the foundation wall indicated generally good material with compressive strength of over 3000 psi.  Windsor probe tests indicated strengths of up to 6000 psi in the reinforced concrete powerhouse wall.  Seepage through the arches was reasonably small on all 15 arches.  The original drawings showed a baffle system on the apron, but through the years the apron had been raised to render the original energy dissipation system non-functional, which made the riprap in the downstream channel move about 150 feet downstream from its original location.  Because of the granular nature of the fill material in the embankment, the seepage was found to be quite high for an earthen structure of this type. It was estimated that the total seepage was approximately 2 cubic feet per second or 1,242,000 gallons per day.  This amount of seepage and the lack of a toe drain had caused a wet and boggy condition at the downstream toe.

## FEASIBILITY OF INCREASED GENERATION

A feasibility study was performed to investigate the possibility of increasing the output capacity at the site.  Several alternatives were evaluated and the comparison of economic analyses indicated that two new adjustable blade, propeller type units, each rated at 750 kW each, be added in a new powerhouse.  These units were selected using an in-house computer program, "POWER," which uses USGS daily flow records, flow characteristics of the units, mode of operation - peaking or run-of-river, to optimize the output at a dam.

## DESIGN OF IMPROVEMENTS AND EXPANSION

Based on the recommendation of the dam investigation report and the feasibility report, the city authorized design of the repairs to the earthen embankment part of the dam, a new energy dissipation system for the apron and a new powerhouse for the two new units.

For upstream slope protection, a 5" thick cast in place concrete mat with a filter was designed.  The concrete was placed in a fabric material known as "FABRIFORM."

To eliminate the wet and boggy condition at the downstream toe of the earthen embankment, a toe drain system was designed with perforated pipe in a trench at the toe of the embankment topped with filter fabric, filter aggregate and then backfilled.  A 12" dia. main pipe and 4" dia. laterals connected to old field tile in the original construction were found to be adequate to handle all the seepage.

The purpose of an energy dissipation system, or sometimes called a "Stilling Basin," is to slowdown the velocity of the spillway discharge to prevent the erosion of stream bed material, which if allowed to continue could undermine the dam.  The hydraulic jump is the most effective way of preventing this erosion, as it quickly reduces the velocity of water to a point where it is incapable of damaging the stream bed.  If this jump at all stages could be contained on a horizontal floor at the level of the stream bed, the erosion problem could be greatly minimized.  This was accomplished by designing an impact block basin with a tailwater sill just down-

stream from the row of blocks to create a small pool of water and a
toe sill at the downstream edge of the apron. The dissipation of
the high energy is by turbulence induced by the impingement of the
incoming flow upon the impact blocks and the still water. Fig. 2
shows this system in detail. The tailwater (TW) sill is a non-
standard feature, and is required to maintain sufficient tailwater
at low flows to contain any hydraulic jump within the length of the
apron. Except for this non-standard feature, the spillway system
was designed using the BOR method given in "Design of Small Dams" -
2nd Edition.

The Alden Research Laboratory (ARL) was retained to build a
model of one bay of the dam and spillway and evaluate the effectiveness of the TW sill in providing the required submergence over a
range of flow rates and tailwater elevations. The results of the
test observations are shown in Table 1. The main features of the
model and various tests and observations are presented below:

1. A 1 to 10 undistorted scale model of one bay of the dam and
   spillway was built in an existing 3 ft. by 3 ft. flume at
   ARL, including both the upstream face of the dam and a
   section of the riverbed downstream of the toe sill. The
   model was operated according to the Froude scaling law.
   One complete side of the model was made of plexiglass so
   that flow conditions could be easily observed. The flow
   through the model was measured by a calibrated venturi
   meter, and water levels in the model were measured by
   pressure tappings located at various points in the floor of
   the model. These tappings were connected to stilling
   wells where the level was measured with a point gauge.
   Water levels downstream of the dam are influenced by
   aeration in the free falling nappe, which is a function of
   absolute velocity and hence operation is underestimated in
   the model. The effect of aeration on the results is not
   significant and was neglected.

2. For flood conditions corresponding to flows between 7075
   (approx. $Q_{100}$) and 4775 (approx. $Q_{10}$) cfs, the flow would
   be over all fifteen bays in the prototype. These flows
   were investigated first, and these results are summarized
   in the first two lines of Table 1. Conditions in the
   plunge pool, upstream of the TW sill, were turbulent with
   considerable aeration occurring. However, because of the
   high TW elevations, the hydraulic jump which occurred just
   downstream of the TW sill was contained within the length
   of the concrete apron.

3. For flow rates less than 4775 cfs, normal operation of the
   spillway requires that the upstream elevation be maintained
   at 825.5 ft. For this elevation, the flow down one bay of
   the dam is 318 cfs. As the inflow increases, progressively
   more and more bays are opened up, increasing the flow down
   the spillway in steps of 159 cfs (since each bay has two

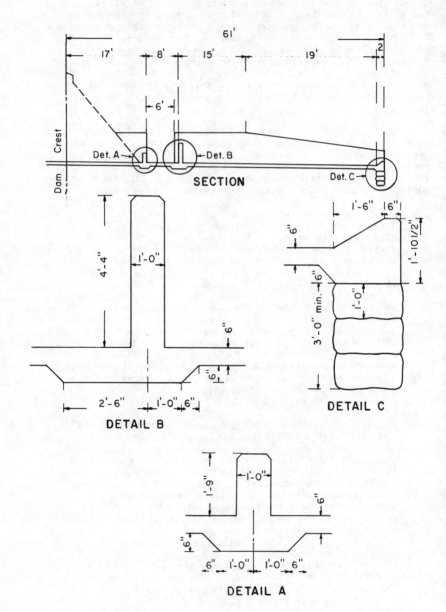

FIG. 2

flasboards so that half of the bay can be opened up at a time)
at the same time increasing the TW elevation. The flow per
unit width of the dam crest is constant (i.e., 318/20 cfs per
ft. of crest), irrespective of the number of bays open.
However, the flow per unit width of the TW sill and of the end
sill increases as the number of open bays increases. In order
to simulate prototype conditions on the one-bay model, it was
necessary to consider the spillway in two sections, the first
section being the section upstream of the TW sill, consisting
of the dam crest and plunge pool, and the second section being
the TW sill and the section downstream of the TW sill.

TABLE 1

Summary of Flow Conditions for Various Discharges Tested
(Discharge over 150' of Sill Width Assumed)

| Total Flow cfs | No. of Bays Open | Flow per Unit Width cfs per ft | TW Elevation ft MSL | Distance of Jump Downstream of TW Sill | Approx. Height of Jump | Typical Floor Velocity ft/s | Scour Patterns Observed |
|---|---|---|---|---|---|---|---|
| 7075 | 15 | 23.58 | 810.6 | 8.5' | 0.9' | 1.22 | X |
| 4775 | 15 | 15.92 | 809.3 | 5.8' | 0.7' | 0.7 | |
| 1432 | 4.5 | 9.55 | 805.85 | 8.0' | | | |
| 1273 | 4 | 8.49 | 805.6 | 12.7' | 1.9' | 4.5 | X |
| 955 | 3 | 6.37 | 805.05 | 9.3' | 1.5' | 3.1 | |
| 637 | 2 | 4.25 | 804.4 | 8.4' | 1.4 | | X |
| 318 | 1 | 2.12 | 803.7 | <2' | <1' | | X |
| 159 | 0.5 | 1.06 | 803.3 | <2' | | | X |

4.  For the first section, upstream of the TW sill with flowrates
    less than 4775 cfs, the conditions in the plunge pool of each
    fully open bay remained essentially the same corresponding to
    318 cfs flowing over each 20 ft. of crest. Thus, for all
    flowrates, conditions in the plunge pool were satisfactory.
    The TW elevation in the second section of the spillway does
    not affect the depth of the plunge pool upstream of the TW
    sill since a critical section always forms just downstream of
    the TW sill.

5.  For the second section of the spillway, the most important
    parameter to model correctly was the flow per unit width of TW
    sill and end sill corresponding to a certain number of bays
    discharging.    To avoid unnecessary testing of "safe condi-
    tions," a flow of 318 cfs per 20 ft. sill width or 15.92
    cfs per ft., was tested first and the tailwater gradually
    reduced until the hydraulic jump that formed downstream of the
    TW sill was just swept out over the end sill. This was found
    to occur at a TW elevation of EL 805.8. Thus, discharges of
    15.9 cfs per ft. of sill or less gave acceptable spillway
    conditions (because the jump was contained within the length

of the apron), provided the TW elevation was greater than EL 805.8. A TW elevation of EL 805.7 corresponds to a discharge of 1400 cfs. Assuming the discharge was spread over half the total spillway width (140 ft.), then the flow per unit width of sill would be 1400/140 = 10 cfs per ft., which is less than 15.9 cfs per ft. Thus, flows of between 1400 cfs and 2226 cfs (corresponding to 7 bays open) spread over 140 ft. of TW sill and toe sill, would be expected to give acceptable spillway conditions, and were not tested on the model. However, flows of less than 1400 cfs could give problems because the TW elevation would be lower, and these were investigated further.

6.  Flow conditions in the second section of the spillway were observed for discharges ranging from one-half bay open (159 cfs) up to 4-1/2 bays open (1432 cfs). The flow per unit width of TW sill and end sill was tested in the model on the assumption that the flow from the open bays would be uniformly distributed over 150 ft. of sill. For all flows tested, the conditions in the second section of the spillway were acceptable. The jump did not sweep out, and the maximum distance of the jump downstream of the TW sill was observed to be 12.7 ft. (see Table 1). Floor velocities in the second section were measured for a few cases, as shown in Table 1, and the highest value observed ws 4.5 ft/s. The greatest difference in water level between the TW elevation and lowest recorded level in the spillway was found to be 1.9 ft., indicating the maximum uplift pressure to be approximately 120 lbs/ft$^2$, assuming that the porous soil will allow the tailwater pressure to be sensed under the apron.

7.  Scour downstream of the toe sill was simulated using coarse sand, and running the flow for a period of about one hour. This gave a qualitative indication of likely erosion of the river bed. Initial observations with the original toe sill (6" wide x 10-1/2" high) indicated that potential scour just upstream of the toe sill could be caused by particles trapped in the backflow under the main body of water. Thus, a modified toe sill, as shown in Figure 3 was used, which forced all incoming particles to be swept out from the apron. Scour tests were carried out for a range of flow conditions as shown in Table 1, and in all cases, the flow over the end sill caused a back roller to form which washed bed material back up against the end sill, thus indicating that erosion of the toe sill foundations was unlikely to occur.

8.  Two possible refinements to the design were looked at briefly. First, the baffle blocks were removed, and a flow of 7075 cfs was tested (approximately $Q_{100}$). The position of the hydraulic jump was almost the same as with the baffle blocks in place, and the only noticeable difference was increased turbulenece in the plunge pool. Secondly, the TW sill was reduced from a height of 4.38 ft. to 3 ft., and with the baffle blocks in place, a flow of 4775 cfs (approx. $Q_{10}$)

was tested. Again, the only noticeable difference was increased turbulence in the plunge pool due to a slightly lower depth of pool.

9. In conclusion, the hydraulic performance of the spillway was found to be satisfactory for the range of flow conditions tested with the revised toe sill.

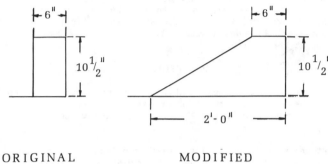

ORIGINAL                    MODIFIED

TOE  SILL  SECTION

fig. 3

## Powerhouse

The new powerhouse was located northwest of the existing (old) powerhouse about 50 ft. downstream from the discharge from the old units (See Fig. 1). The intake structure for the new powerhouse was located adjacent to the intake of the old powerhouse. On low head hydro projects, intake hydraulics losses, if improperly designed, could amount to a large percentage of the total available head. Two intake systems were considered - one involving a single large diameter pipe (penstock) leading to a bifurcation chamber, where the flow was split between the two turbines. The second one involved transporting water through two smaller diameter penstocks, one for each unit, from the intake to the turbines. Analytic evaluation of head losses in the two systems indicated extremely high losses in the system with one penstock because of the turbulence in the bifurcation chamber. This system had a lower initial cost but over the life of the project, due to extra losses, enough energy would have been lost to pay for the extra initial cost incurred in the system with two penstocks.

The evaluation of the single penstock system indicated that the flow distribution at the turbine runners would be uneven due to flow separation at the turbine intake from the junction chambers, and that the total head loss from the head pond to the turbine with two units operating at design load would be approximately 4 ft. It was estimated that maximum velocities of 17 fps would occur just upstream of the runner and backflow would exist in 20 percent of the conduit cross-section at that location, with the major separation located at the top of the conduit.

The losses in the "two penstock system" were evaluated to be less than 0.5 ft. The flow control gate originally located just upstream of the turbine was moved to the intake structure at the head pond. The reduction in penstock size from (8'x8') at the gate to 6.5' diameter was rounded with a 0.75' radius to reduce separation of flow at the intake.

CONCLUSIONS

1.  On site repairs of the old hydro units resulted in an increase in efficiency of about 14%. The pay back period for the repair cost, in 1979, was estimated to be 5 years.

2.  Earthen embankment repairs were completed in November, 1981. The toe drain system has completely eliminated the wet and boggy condition at the downstream toe of the earthen embankment and has substantially improved the factor of safety of the earthen embankment.

3.  Half of the 300 ft. apron was also completed in November, 1981. The hydraulic jump at low flows (one bay open) is within the length of the concrete apron. This apron performed extremely well in the flood of March, 1982. Flows at the downstream gages during this flood were much higher than the historic flood of record.

4.  At this writing the new powerhouse is under construction and is expected to be completed by July, 1982.

REFERENCES

1.  "Design of Small Dams," Second Edition, 1973 - Bureau
        of Reclamation

2.  "Hydraulic Model Study of Sturgis Dam Spillway,"
        July 1980, Eugene Chang & Johannes Larsen,
        Alden Research Laboratories, Holden Massachusetts

CASE HISTORY - 1.4 MW SOUTH CONSOLIDATED HYDRO PLANT
By Ken G. Laurence[1] and
William E. Shaughnessy/Member ASCE[2]

ABSTRACT

The 1.4 MW South Consolidated Canal low head hydroelectric project is a new installation located on existing irrigation canals of the Salt River Project (SRP) System near Phoenix, Arizona. This paper describes the design approach used for the hydraulic structures and power generation equipment, describes and illustrates construction details, and discusses the performance of the structures and equipment after completion.

## INTRODUCTION

The Salt River Project delivers and distributes water to the Salt River Valley canals, principally for irrigation and domestic purposes. A dependable supply of water to the irrigation canal system in the valley is ensured by the large storage reservoirs on the Salt and Verde Rivers to the northeast of Phoenix. At a point in the canal system where water flow divides into two canals, one of these canals has a 36.5-ft drop and a flow varying between 0 and 1,150 cfs, giving an average flow of 540 cfs during the irrigation season.

In 1978, a study was carried out by SRP with the assistance of Stone & Webster Engineering Corporation (SWEC) to determine the feasibility of developing this site for hydroelectric power. At that time, the federal government, through DOE, had begun promoting the development of small, low head hydro sites through demonstration projects. As the feasibility study indicated that this was a viable site, SRP decided to submit a proposal to DOE to have it considered as a demonstration project. In April 1979, DOE selected the South Consolidated site as one of seven demonstration projects, and engineering began.

## ENGINEERING ASPECTS

### General

The governing design requirement was that the primary purpose of the canal system--to provide irrigation flow demand--would be unchanged, whether the hydroturbine was in operation or not. This requirement, and economic considerations, dictated that the existing canal facilities be left intact as far as possible so that at times when the hydrogenerator was out of service or when flows exceeded the capacity of the unit, water would be passed through these facilities.

---

[1] Project Engineer, Stone & Webster Engineering Corporation
[2] Project Manager and Manager, Hydraulic Division, Denver Operations Center, Stone & Webster Engineering Corporation

To achieve this, a number of alternative schemes were investigated, the most practical and economic of which was to route all of the flow in the feeder canal into the initial stretch of the upper of the two division canals and then drop the lower canal water requirements through the hydraulic turbine. As the primary purpose of the canal system is to provide irrigation water on demand, the turbine in this instance becomes a flow control device, and power generation is a secondary, but revenue-producing and renewable energy resource. This scheme is illustrated in the aerial photograph of the completed project (Figure 1).

FIGURE 1 - COMPLETED PROJECT

The hydraulic structures required strictly for hydroelectric power generation are all new. However, modifications to existing canal structures were required to accommodate power generation while allowing the canals to continue functioning normally.

Hydraulic structures required for power generation are: 1) a gated intake channel, 2) a penstock, and 3) a tailrace channel. The hydraulic turbine and electric generator, which are controlled from a remote location, are placed in an enclosed powerhouse. Arrangement of structures is shown in Figure 2.

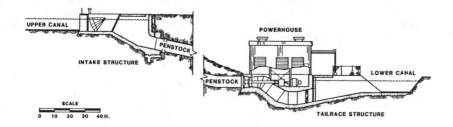

FIGURE 2 - CROSS SECTION OF HYDRAULIC STRUCTURES

Intake Channel

The 18-ft wide, 45-ft long, 21-ft deep intake channel is a reinforced concrete structure built into the first bend of the existing upper canal bank and aligned with the incoming flow to minimize hydraulic disturbance in the flow prior to its entrance to the penstock.

The intake channel was designed around a standard 18-ft wide radial control gate of the type installed throughout the canal system. The total length of the intake was determined by the requirements for the gate, stoplog slots, trashrack, and transition to the penstock entrance. It was found to be most practical, due to the narrow north canal bank, to construct the south wall of the intake channel out into the canal. This was achieved without significantly changing the maximum water velocity of that portion of the flow continuing downstream in the upper canal. The average water velocity in the intake channel is 5 ft/sec as it aproaches the trashrack.

The requirement to provide satisfactory hydraulic entrance conditions to the penstock dictated that the high point of the penstock entrance be submerged at least 8 ft below the normal upstream water surface. The floor downstream of the trashrack is sloped 2:1.

The floor and walls of the intake structure are all of reinforced concrete. When the intake gate is closed, the portion of the intake downstream of the gate empties. Due to potential seepage around the structure, the floor slab was designed for full external hydrostatic head. The walls were designed as cantilevers to resist both hydrostatic and earth pressures. Earthquake loading was also considered. To limit this potential for seepage along the exterior of the walls and floor, and prevent any potential for piping and erosion of the embankment downstream of the structure, a seepage cut-off wall was extended out from the head wall.

The top of the structure between the trashrack and the headwall is covered with standard precast bridge slabs designed to take the load of an SRP trash-raking vehicle. Access to the penstock is by means of a removable grating adjacent to the head wall.

The intake gate is operated by a wire rope hoist, with cables attached
to each side of the gate. A clutch is provided on the hoist shaft
to allow the gate to free fall on emergency closure. A friction brake
comes into operation as the clutch is released. The intake gate is the
only means of shutting off the water to the penstock and turbine.

## Penstock

The penstock is a 7-ft diameter welded steel pipe provided with one
expansion joint. It has a total length of 137 ft and is embedded in
the intake concrete and welded to the turbine inlet assembly. The
penstock diameter was selected based on a head loss assessment. The
head loss with a maximum flow of 615 cfs is less than one foot; further
increases in diameter yielded only a very small reduction in head loss.
In view of the submergence of both the intake and the turbine, and the
embankment stability requirements, it was decided that a completely
buried penstock would be the most practical and economic solution.
Bids were initially solicited for a reinforced concrete pressure pipe
penstock designed to the requirements of AWWA C302. As only one bid
was received and that was considerably higher in price than the engineers' estimate, it was decided to change the penstock to a flexible
steel design which relies upon backfill compaction for its support.
The wall thickness needed for the governing external loading in this case
was found to be 5/16 in. An acceptable bid was received for this
design. The pipe was supplied in 40-ft sections for field welding.

One flexible coupling was provided at the top end of the penstock to
allow for changes in pipe length due to temperature variations between
installation and watering up. For protection against corrosion, the
pipe was coated and wrapped on the outside and coated on the inside.
The coating used was coal tar epoxy. Anti-seepage rings were welded
to the embedded stub pipe at the intake. The penstock inlet is formed
in concrete and provided with a simple 3-ft radius bellmouth to
reduce entrance losses.

## Tailrace Channel

The 15-ft wide, 42-ft long, 16-ft deep tailrace channel is a reinforced
concrete structure built into the existing lower canal bank. The
channel is angled at 35° to the lower canal to somewhat align its
discharging flow into the canal stream.

The upstream portion of the tailrace channel takes its bottom elevation
from the turbine draft tube, which at its lowest point is 15 ft below
the lower canal water surface. The floor of the tailrace rises from
this point at a slope of 5:1 to meet the existing floor of the lower
canal. As with the intake structure, the tailrace structure was designed for external water loads, for there is a possibility that the
earth will be surcharged with tailwater pressure when the canal is
drained for maintenance. The walls were again designed as cantilevers.

The channel width was dictated by the turbine draft tube, and its length
determined by the requirements of the equipment loading bridge outside

the downstream main door of the powerhouse. The average water velocity
at the downstream end of the tailrace channel is 5 ft/sec.

The precast bridge slabs are designed for H-20 loading and will carry a
truck loaded with the heaviest piece of powerhouse equipment, the
generator.

## Turbine-Generator Equipment

The selection of appropriate power generating equipment and the determination of its expected energy output occupied a central part of the
engineering work. The starting point was to determine the expected
head and flows at the site. The gross head of 36.5 ft was obtained by
ground survey, whereas representative flows were derived from long-term
daily flows recorded at a point on the lower canal a short distance
downstream from the site. The flow duration curve developed from
these records for the mean year of the 1948 to 1966 period is presented
in Figure 3.

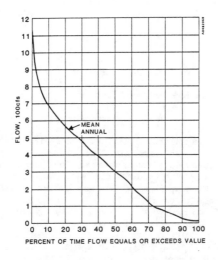

FIGURE 3 - FLOW DURATION CURVE

At the outset it was realised that the project may be viable only if
the turbine and generator could be of the package type currently
available for some applications. Most of these are of preengineered
standardized design with specific operating ranges with respect to
flow, head, and output. These packages consist of turbine, generator,
speed increaser where necessary, water passage liners and all necessary
control equipment for the main unit and auxiliaries. The principal
advantage of the stantardized turbine is the minimizing of design time
and cost, which tend to be disproportionately high for custom designed
equipment of small capacity. Some manufacturing economies are realized
also from the repeated use of casting patterns and fabrication forms.

The package unit concept results in engineering economies also because elaborate specifications do not need to be written for each major and auxiliary piece of equipment. The responsibility for matching components of the package rests with the supplier, allowing a much simplified specification to be used. The concept also reduces potential warranty problems between different suppliers, which has occurred on some conventional projects with separately purchased equipment. The purchaser is thus afforded greater protection during resolution of such problems.

The flow range shown on Figure 3 of 0 to 1,150 cfs is unusually wide for a hydro plant, and clearly could not be completely accommodated by a single machine. The choice then rested between adopting a single turbine with a permissible range of flows somewhat narrower than those available, thereby sacrificing some energy output, or going to a multiple-unit installation.

Manufacturers" responses to inquiries during the feasibility study and competitive bidding in the final dsign led to the selection of a single horizontal axis semi-Kaplan turbine and 1400 KW high speed generator for the project.

Output and Capacity

The turbine-generator output in kilowatts can be shown to equal:

P = .00085 HQN   Where H = Net head in feet
                       Q = Flow in cfs
                       N = Overall conversion efficiency (percent)

The head losses in the intake, penstock, and tailrace are estimated to be 1.5 ft, leaving a net head of 35 ft. In the turbine-generator specification, the flow-duration curve was presented in tabular form by breaking the flow range into 50 cfs steps and listing the corresponding number of hours per year for each step. Equipment manufacturers were required to complete the table by adding the guaranteed energy values in KW-hrs per year for each flow step for their particular turbine-generator. The selected equipment for the project was guaranteed to produce 6,214,500 KW-hrs per average year in the flow range between 75 cfs and 615 cfs. This represents about 90 percent of the total available energy in the water after allowing for efficiency losses in equipment and structures.

Powerhouse

The powerhouse was designed around the turbine-speed increaser-generator packaged unit, with additional floor space left for the electrical auxiliaries consisting of the air circuit breaker, station service and P. T. cubicles, the control panel, and the supervisory control rack. The powerhouse also contains the blade actuating controller, a battery rack and charger, the service water system for the turbine and speed increaser, and a sump pump.

All of this equipment is installed below grade, the generator floor elevation being dictated by the turbine "setting" requirements, that is, the necessary submergence of the turbine runner relative to the tailwater elevation as protection against cavitation damage to the runner blades. For this purpose, the runner centerline is set 2.6 ft above the minimum tailwater elevation.

The powerhouse substructure located below grade is of reinforced concrete designed for external soil and water loads; it is designed to preclude flotation and sliding. Water stops have been installed in the conrete at construction joints to prevent water from seeping into the powerhouse.

The metal siding superstructure is provided to house the 10-ton service crane. This small bridge crane has an electric motor-driven hoist and chain drives for both logitudinal and lateral traverse. The bridge rails pass through the downstream end wall of the powerhouse to service the loading deck over the tailrace. This end of the powerhouse has double-swinging doors for crane and equipment passage.

The high summer temperatures in the Salt River Valley necessitated measures to keep temperatures within the powerhouse down to an acceptable level. First of all, the metal siding and roof are insulated against the outside temperatures. Secondly, the heat produced by the fan-cooled generator is extracted by roof fans.

## Power Distribution

In addition to water storage and distribution, SRP also owns and operates an electric system which generates, purchases, and distributes electric power and energy to a 5,400-sq mi territory. The output from the generator is transmitted through a new 12 KV line for a distance of 1-1/2 mi to an existing switchyard.

## Existing Structures

Rerouting the lower canal flow through the initial stretch of the upper canal meant relocating the upper canal check structure to a position downstream of the hydroelectric facility intake. The removal of this check structure from the downstream end of the feeder canal allows the full available head at the site to be developed at the penstock intake. This increased water level, plus the freeboard requirements for emergency closure surge, necessitated that the shotcrete canal lining be extended to protect the canal banks under the new conditions. The existing lower canal check structure also needed to have a fixed crest overflow spillway added to permit the increasing water levels resulting from load rejection surge to spill safely until such time as other corrective action could be taken.

## CONSTRUCTION

The subsurface conditions at the site are typical of the Phoenix area and largely consists of well-graded alluvial deposits of silt, sand, gravel, and boulders, which are highly compact and have low permeability. Excavation for the structures was accomplished by normal overburden excavation methods.  The lightweight structures provided did not impose any more severe loadings on the foundations than the existing nearby regulating structures and canals.

The site location within the greater Phoenix area and the excellent access roads already provided for canal maintenance greatly facilitated the supply of construction materials and skilled construction labor and the movement of equipment.

The canals are normally drained for annual maintenance for one month each year usually in October or November.  The project construction work in the canals was scheduled for this time.  The constructors were successful in completing their work in the canals within a slightly extended dry up period.  Successfully completed in this period were the intake structure, the powerhouse substructure, the tailrace channel, and the relocation of the upper canal check structure.  The canal bank rebuilding and the shotcrete lining adjacent to the above structures to prevent seepage when the canals were rewatered were also accomplished.  The intake gate and draft tube stoplogs were installed after their support structures were completed to allow the canals to resume operation while the powerhouse and penstock were being constructed.

The flexible pipe design used for the penstock relies upon adequate compaction of the soil backfill around the pipe.  It was realized that compaction would be difficult in the narrow angle underneath the pipe; a concrete slurry was placed in this zone after the pipe had been assembled and welded.

Adequate access for most purposes was provided by existing canal maintenance roads adjacent to both upper and lower canals.  Access to the storage yard on the south side of the upper canal was directly off a paved highway.

Because of its nature, the bulk of the materials excavated for the structures was stockpiled and processed at the site and reused for backfill.  As the embankment containing the penstock is in essence a dike, particular attention was paid to its design and construction as an impervious structure.

The total amount of concrete required for all structures amounted to approximately 750 cu yd.  In view of the proximity of the site to concrete batching plants in the Phoenix metro area, ready mix concrete was found to be the most economic proposition.  Formwork was all custom-built at the site.

The groundbreaking took place in September 1980 and all civil work was completed by January 1981.  The installation of equipment was completed in May 1981 and the project was commissioned on September 1, 1981.

ECONOMICS

The total project cost, including engineering, equipment, construction, and project management amounted to $2.6 million or $1,857/kW.

The value of energy and capacity, when compared to the total project costs on a present worth basis, yielded a benefit/cost ratio at project completion of 1.80.

PERFORMANCE

Other than the relocated check structure and the restored canal lining, there was no opportunity to test the performance of any of the structures or equipment until construction was complete and the turbine-generator equipment ready to roll. The relocated check structure performed satisfactorily when the canals were rewatered.

After the canals had been rewatered for a few days, some seepage appeared in the embankment between canals. The upper canal was briefly unwatered and some cracks were found in the canal lining. These were repaired and the canal returned to service in a few days without further incident.

The only observed flow phenomenon were small eddies which formed downstream of the protruding intake wall. These have no adverse effect on the operation of the canals or the powerplant.

When construction and equipment installation was complete, all powerhouse auxiliary systems were checked prior to watering the turbine. The turbine was watered up by removing the draft tube bulkhead gates. Initial turbine rotation was achieved with the intake gate cracked just a few inches. The turbine rotated smoothly and was allowed to run for some time at sub-synchronous speed to check bearing temperatures, runner blade adjustment, and electrical protection circuits.

Due to the small intake gate opening, there was a great deal of turbulence at the penstock entrance; the penstock did not run full where it could be observed, but it was surmised that because the turbine ran smoothly, a portion of the penstock at the lower end probably ran full. There was a smooth discharge from the draft tube, and the tailrace was calm. Many small air bubbles were present on the tailrace.

Testing continued to obtain automatic synchronization of the generator with the line by various combinations of intake gate opening and runner blade position. The unit was difficult to synchronize repeatably and modifications were required to the synchronizer to make it dependable. It was necessary to open the intake gate about 8 inches with the runner blades in their fully closed position to achieve synchronous speed. With this gate opening, there was jet flow under the gate and much turbulence in the penstock entrance. The turbine rotation was again stable.

The unit was synchronized, brought on-line and loaded in increments. At each new load the bearing temperatures were allowed to stabilize before proceeding to the next load. A maximum output of 260 kW was generated before the unit went through an automatic trip and the intake gate closed.

For these load tests, the intake gate was only raised 20 inches. When the gate closed on emergency trip, the drum brake on the hoist shaft was not able to stop the gate from closing rapidly, and some damage occurred to the gate-bearing supports. Further testing was discontinued until a new disc brake was installed on the hoist shaft. This permitted the gate to be lowered safely upon emergency shutdown of the turbine.

The turbine-generator was subsequently tested up to full load, and operation was generally smooth and satisfactory. The turbine manufacturer had guaranteed operation down to 75 cfs flow, but with the intake gate fully open, it was not possible to pass less than 125 cfs (20 percent of maximum).

It was decided to run an unofficial performance test on the unit to verify the equipment guarantees, which had been made based on the IEC Test Code 41. Because of the particular penstock dimensions, the only acceptable means of measuring flow would be the pitot tube tranverse. Adaptors for the pitot tube had been provided in the penstock; in addition, a sonic flow-measuring instrument was installed for remote read-out during normal operation. A further check of flow rate during the test was provided by a current meter traverse in the lower canal, downstream of the tailrace.

A fair correlation was obtained between flow measurements, and it was determined that while performance was marginally lower than the level guaranteed for flows below 500 cfs, it was marginally above that guaranteed for flows above 500 cfs. In addition, while the turbine would not produce power below 125 cfs, it would produce power satisfactorily up to 650 cfs, whereas the guarantees were stated between 75 cfs and 615 cfs. All hydraulic structures performed satisfactorily. With the intake gate fully open and the turbine at maximum flow, the submergence of the penstock entrance appeared to be sufficient as there were only some small surface vortices which appeared and disappeared from time to time.

At various flows, some trashrack bars vibrated at low amplitude and frequency. The particular bars vibrating changed continually with apparently steady flow. The rack will be inspected periodically to check for cracks to bars and welds rather than attempt preventative action.

CONCLUSION

In conclusion, the design and construction of the project was carried through succcessfully, as evidenced by the satisfactory operation of the structures and equipment. Of the problems which arose during start-up, testing, and commissioning, those which took longest to resolve were associated with the turbine-generator control system. Although power generating equipment performance was not precisely as guaranteed, the shift in the performance capability of the turbine to higher flows does essentially enable the guaranteed annual energy to be produced.

# ULTRA-LOW HEAD HYDROPOWER AT AN ULTRA-LOW PRICE

By John J. Huetter, Jr.[1]

Abstract: Energy Research & Applications, Inc., in cooperation with the U.S. Dept. of Energy under a cost-shared agreement, has addressed the problem of costs at ultra-low head hydropower sites through the research, design, development and installation of a hydropower package incorporating modified marine thrusters which promises significant cost reductions for this category of sites. The first such installation has been accomplished on the main canal of the Modesto Irrigation District. Overall installed cost of the first unit, producing slightly under 500 KW at 14 ft. of head is about $1200/KW, including canal modification and an extensive test program funded by the Dept. of Energy. Civil works cost reduction of 35-60% has been obtained, even though the civil works themselves are an integral part of the downstream flow control.

## THE POTENTIAL AND THE PROBLEM

There are several thousand existing hydropower sites in the United States which will not be developed because the traditional technology costs too much. These are the sites with ultra-low hydraulic heads (defined by the DOE as 3 meters (approximately 10 ft.) and by site economics at about 15 ft.

The realistic ultra-low head hydropower potential is at least 1500 MW distributed over many sites throughout all regions of the United States. The ultra-low head hydropower dilemma suggests there are at least two ways to address the problem: increase power output per site or decrease costs.

It was our belief that economic rules are more susceptible to change than the rules of physics. Results of this project further strengthen that belief.

## A POSSIBLE SOLUTION

Energy Research & Applications' approach to this alternative energy development challenge was cost reduction through value-engineered integration of off-the-shelf components into a functional hydropower package incorporating a modified marine thruster as the prime mover.

Marine thrusters are primarily used for the maneuvering of large ships such as oil tankers, or other off-shore applications such as service barges and oil rig positioning.

[1]President, Energy Research & Applications, Inc., Santa Monica, CA

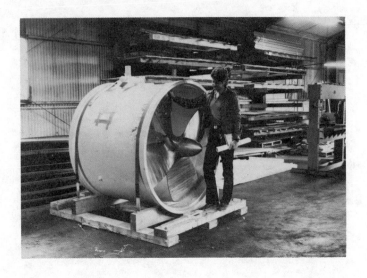

BT550 MODEL BOW THRUSTER OFF THE SHELF

These units are also referred to as tunnel thrusters or bow thrusters; manufactured by about half a dozen companies selling to the international marine market.

Thrusters have been in widespread use for only about 15 years. Both their application and hydrodynamic performance are more often observed than predicted or engineered, so there is little data.

The critical problem encountered in the Research & Development phase was the analytical determination of thruster-as-turbine performance without empirical data. Characterization of each unit of every manufacturer from which even minimal data was obtained resulted in the identification of two lines of equipment for which the calculated effective heads, in pressure equivalents, matched the 3 meter regime we were originally contracted to investigate. Fortunately, there was also complementarity in the standard unit sizes of these equipment lines over a wide power range.

Other elements of the ULHH package were selected or designed according to criteria of least cost for reliable function. This included configuring a transmission from components in the catalog, designing the turbine diffuser as part of the civil works, and even to selecting and specifying materials from a wide range of candidates.

The result of this phase of the project was the design of three hydropower package configurations: generally for horizontal, vertical or syphon applications, as determined by site-specific requirements. Based on comparison of concept designs as done by others for actual ultra-low head sites, installed cost reduction of up to 60% was calculated. The hardware package costs were, on the average, only 30% of the cost of conventional equipment.

The potential for this degree of cost reduction justified continued funding support for a field test of the concept, in DOE's opinion.

We had to arrange for the bulk of the project funding required to actually build and install a unit. Of five candidates for the test site, the Modesto, California Irrigation District was the quickest with a contract. They agreed to buy and install a single horizontal unit of 500 KW nominal power at a site on their Main Canal. ER&A is obligated to both agencies to design, fabricate, integrate, install, and test a thruster-based ultra-low head hydropower package.

## THE ORDEAL BY WATER

With the first unit installed at the Stone Drop Site in California's San Joaquin Valley, we are now in the operational test and evaluation phase.

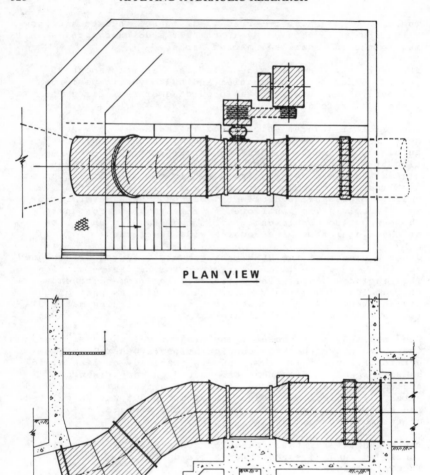

**PLAN VIEW**

**PROFILE**

ULHH PACKAGE DESIGN FOR STONE DROP

The actual fabrication of the hydraulic sections and thruster modification proceeded largely as planned.  Three and a half months elapsed from contract execution to delivery of the hydraulic section for installation at the site.

Long lead time, high cost items were the flow control gate (which was a surprise) and the generator (which was no surprise).  Both of these items defined the critical path to project completion with 5-6 month lead times.  The ULHH package had been defined as producing 480V, 3 phase power.  The utility wanted 4160V output so they could use an abandoned existing substation transformer.  This resulted in an equipment cost saving of $37,000 vs. $4000 additional for the generator and two months added to delivery time.

Ironically, while we are getting more power out of this size unit than was anticipated due to a maximum head of 14 ft. at full flow conditions, this same head increase over the 10 ft. design point uncovered a torque limitation of the standard gearbox.  All of the thruster units were characterized for a head range of 6-15 ft. in terms of equivalent pressure head.  However, at the high end of this range, the continuous duty torque rating of the gearboxes is met or exceeded in some models.  We are experimenting with different final drive ratios to see if we can work around this limitation.

One result is that the continuous duty power rating of the standard unit will be limited to a calculated maximum of 465 KW.

## APPROACHING THE BOTTOM LINE

One boundary of ER&A's project responsibility was defined as the output wires from the generator terminal box.  MID wanted to make the site modifications and additions to interface the power with their grid which was understandable and agreed to.

Cost in current 1982 dollars for site modification to accept the ULHH package and grid interface to accept the power is $266,000.  The balance of the project cost, excluding the test activity and special test equipment, is $230,000.

A feasibility study executed by another firm in 1978 had civil works costs ranging from $342,000 to $372,000.  If these costs are escalated 20% over four years (slightly less than 5% per year as a conservative figure, reflecting a depressed construction industry), they would be $410-446,000 in current dollars.  The ULHH package's required civil works cost is therefore about 60% of the conventional, which is also consistent with the cost reduction in three site-specific concept designs we developed for comparison during the previous R&D phase.

The turbine-generator equipment selected in 1978 was $690,000
for development of the site's full 900 KW potential. Consistent cost escalation of 20% over four years puts the
equipment price at $828,000. ER&A's ULHH equipment package
cost just under $95,000 in January, 1982 including hardware
engineering, fabrication and supervision.

Total project cost in 1978 was $1.4 million for 900 KW or
$1560/KW of installed capacity; in itself, not a bad price.
For the first stage, or partial, site development, we are
now looking at a cost of $1066/KW. Were the 1978 total cost
inflated to 1982 levels, the project cost reduction would be
more dramatic.

| CONVENTIONAL DEVELOPMENT (900 KW) | ER&A ULHH PACKAGE INSTALLATION (465 KW - STAGE I) | |
|---|---|---|
| 1978 | ESCALATED TO 1982 COSTS | 1982 ACTUAL COST |
| Civil Works $342-372,000 | $410-446,000 | $226,000 |
| Hydropower Equipment $690,000 | $828,000 | $ 95,000 |
| Total Project Cost $1,400,000 | $1,680,000 | $496,000 (Excl. Test) |
| $/KW Installed $1,560 | $1,866 | $1,066 |

SUMMARY OF COMPARATIVE PROJECT COSTS:  MID STONE DROP SITE

Additional cost savings could have been achieved in the
civil works, especially in powerhouse construction, but the
MID choose to build a very substantial structure, primarily
to protect against vandalism and to meet or exceed utility
standards.

CONCLUSIONS AND LESSONS

At existing avoided cost of power to the District, the
site's value is approximately $20,000 per month. This cost
saving is restricted by an eight month irrigation season,
with the compensating factor that the flows during that
period are firm and controllable by the District.

If the test program now underway confirms the power and
energy production calculated for this ULHH package, we will
have reduced the cost of hydropower development at these
former problem sites featuring very low heads by 30-50%.

More than mere percentage reduction numbers, however, the
practical effect of this innovative energy technology and
its application is to convert sites that were previously

economically infeasible into practical low-head hydropower projects producing clean, fuel-free electric power.

CONSTRUCTION PHASE OF ER&A ULTRA-LOW HEAD HYDROPOWER PROJECT AT STONE DROP:  MODESTO IRRIGATION DISTRICT.

# JACKSON MISSISSIPPI COMPUTER-AIDED FLOOD WARNING

by Thomas N. Keefer,[1] M.ASCE, and
Eric S. Clyde,[2] AM.ASCE

## ABSTRACT

As a result of a feasibility study carried out after
the April 1979 flood, Jackson, Mississippi, has purchased
an advanced flood warning system. Predicted water surface
elevations at 26 points along the Pearl River floodplain
are obtained from a linear-implicit, finite difference flow
model. Predicted elevations are used to generate real-time
flooded area maps on a color graphics terminal. Graphics
capability is backed up by lists of affected streets.
Provisions are made for easy recalibration to new flood
profiles as channel changes occur.

## INTRODUCTION

In April 1979, the City of Jackson, Mississippi, was
hit by the worst flood in its recorded history. The Pearl
River remained out of its banks for nine days, reaching
stages 10 to 12 feet above the banks. Millions of dollars
damage resulted to both residential and commercial property. The flood is known locally as the "Easter" flood.

Flooding is not unusual along the Pearl River
through Jackson; the U.S. Geological Survey (USGS) has
documented 14 major floods on the Pearl and its tributaries since 1955, four since the U.S. Army Corps of
Engineers completed the levee system in 1968. Most of the
floods are the result of heavy spring rains.

The City of Jackson extends for approximately 16
miles along the west bank of the Pearl River in west
central Mississippi. The floodplain area in the city
varies from several hundred feet to more than a half mile
wide. Flow in the river is controlled by the Ross Barnett
Reservoir immediately north of Jackson.

---

[1]Vice President, Water Resources Division, Sutron
Corporation, Fairfax, VA.
[2]Senior Water Resources Engineer, Sutron Corporation,
Fairfax, VA.

The floodplain area adjacent to the city is divided
into three sections by highway crossings. Highway 25
crosses six miles below the reservoir, and Highway 80 and
Interstate 20 cross ten miles below the reservoir. Above
Highway 25, major residential areas have encroached on the
floodplain. Between Highway 25 and Highway 80, a levee
protects a number of businesses and the city coliseum.
(This levee was overtopped by the Easter flood.) A major
tributary, Town Creek, enters the Pearl River near Highway
80. The floodplain along Town Creek has considerable
commercial development and is severly affected by the Pearl
River backwater. Below Highway 80 and Interstate 20, no
levees have been built. The floodplain development there
is not as extensive as in North Jackson, but the city's
sewage treatment plant lies in the flood-prone area and is
protected by its own small levee (also overtopped by the
Easter flood).

Traditionally, flood warning and evacuation planning
for the city has been based on forecasts of the river
stage at Highway 80 by the U.S. National Weather Service
(NWS). City officials would issue flood alerts and plan
evacuation based on lists of streets historically affected
by water at various river stages.

After the Easter flood, the city decided that its
traditional flood warning and evacuation methods based on
the Highway 80 gage were inadequate. It set out to
develop a method to accurately estimate stages in other
parts of the city and to provide more advanced warnings,
particularly for the residential areas of North Jackson.

FLOOD WARNING-SYSTEM DEVELOPMENT

A plan for development of the city's flood warning
system was completed in June 1980. The plan recommended
that the city obtain a detailed computer flow model of the
Pearl River from the Ross Barnett Reservoir to the southern city limits. It also recommended development of a
computer model of the upper Pearl River Basin and the
reservoir itself.

During the fall of 1980, a detailed description of
a warning system that could be implemented on the city's
computer was prepared. At the same time, the Pearl River
Valley Water Supply District purchased a computer model
of the upper Pearl River Basin and the Ross Barnett
Reservoir. The reservoir model is capable of predicting
water releases 36 to 48 hours in advance. The city's
model is used in conjunction with the predicted reservoir
releases to predict the flood conditions in Jackson. At
the same time, the NWS and USGS added telemetered rain
gages and increased stream gaging capabilities above the
reservoir. The increased gaging capability is used to

increase the accuracy of both the NWS and the Pearl River
Valley Development District models.

On December 31, 1980, the city issued a contract to
implement the planned flood warning system. The primary
tasks involved in developing the flood warning system were
to:

- acquire a high-resolution graphics terminal
  and hard-copy unit for predicting, displaying,
  and recording the flooded area;

- ditigize a series of maps (roughly one square
  mile each) with two-foot contour intervals for
  use in displaying flood predictions;

- develop an accurate linear-implicit unsteady-
  flow model of the Pearl River from the Ross
  Barnett Reservoir to the southern city limits,

- calibrate the flow model on three historic
  floods, including the Easter flood;

- develop computer programs that the city could
  use to (a) accurately predict flood stages all
  along the Pearl River, (b) produce maps of the
  affected streets;

- provide the capability for estimating flow and
  stage in the tributaries along the Pearl River
  in conjunction with the main Pearl River model;
  and

- install the system in the city's Emergency
  Operations Center and train city personnel
  in its use.

The flood warning system developed consists of a
graphics terminal, computer interface, hard-copy unit
(hardware), and computer programs (software). For Jackson,
very little special hardware was required because the flood
warning system is implemented on the city's Honeywell 6620
computer system.

WARNING SYSTEM COMPONENTS

The heart of the flood warning system is the computer programs. The primary ones are:

- a computer model of the Pearl River from the
  Ross Barnett Reservoir to the southern city
  limits,

- a program for use in estimating the flow in
  the tributaries from the city's rain gage
  network,

- a map retrieval and display program, and

● affected street lists retrieval and display.

   The system predicts flooded areas based on water
surface elevation predictions at 26 points along the main
stem of the Pearl River plus a similar number of points
along each of the six major tributaries that flow through
the city.  The warning system uses the predicted elevations
to determine the areas of the city that are or soon will be
flooded.  The system will reproduce the Easter flood profile with an accuracy of 0.3 foot maximum.  The Easter
flood profile was accurately established by surveys of
the U.S. Army Corps of Engineers.  The maps depicting
flooded areas are based on a 1980 series of two-foot contour interval maps provided by the Pearl River Basin
Development District.  Twenty-four individual maps of
potential flooded areas in the city are provided with the
system in addition to an overall map that gives a general
picture of the flooding.  No maps or details are provided
on the east bank of the river.

## WARNING SYSTEM OUTPUT

   The color graphics terminal in the city's Emergency
Operations Center is illustrated in Figure 1.  The overall
city map is illustrated in Figure 2.  In Figure 2, the
black indicates areas actually affected by flooding at
the indicated time and date.  The system will also indicate
predicted flooding at any other two times in the next
48-hours.

   Figure 3 illustrates a large-scale map with flooding
(black area).  Note the level of detail available.
Individual homeowners may easily identify their own residence.  The white lines in the background are the two-foot
contour markings.  The white area in which the Map B label
and the legend appear is more than five feet above the
Easter flood level and has an extremely low probability
of ever being flooded.

   These maps provide the city with extensive detail
so that it can inform citizens and business enterprises
in low lying areas.  The Emergency Operation Center's
press room is equipped with a video cassette recorder and
color camera.  The city has direct access to Cable TV
Channel 11 over which it has been proposed to televise
the series of flooded area maps so that appropriate
evacuation or prevention measures can be taken.

   In the event that the color graphics terminal mal-
functions, each detailed map is supplemented by a list of
the streets that are affected.  For each detailed map, the
city may produce a list of streets that are currently
affected as well as a list of those that will be affected

Figure 1.  COLOR GRAPHICS TERMINAL IN
JACKSON EMERGENCY OPERATIONS CENTER

Figure 2.  COMPUTER GENERATED MASTER MAP
OF FLOODED AREAS IN JACKSON

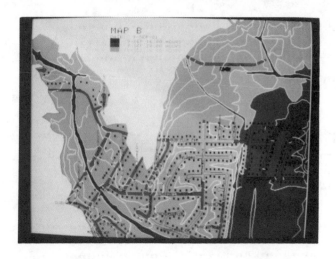

Figure 3.   DETAILED MAP OF FLOODED AREAS
IN NORTH JACKSON

at any specific time in the next 48 hours.  A typical
street list is illustrated in Figure 4.  The street lists
were developed by flooding each contour interval on each
map one at a time and are highly accurate.

ADDITIONAL SYSTEM FEATURES

In addition to the basic maps and lists of affected
streets, the warning system included the following
features:

- It has a program to estimate flow in six
  tributaries based on rainfall at city fire
  stations.

- It can plot flood water surface profiles for
  comparison with a low flood and the Easter
  flood.

- It can plot main stem and tributary cross
  sections used in the flood warning model.

- It has utility programs for easy recalibration based on up-to-date flood profiles.

```
***********************************************************************
*                                                                     *
*        LIST OF FLOODED STREETS ON MAP C AT 12 HUNDRED HOURS         *
*                                                                     *
***********************************************************************

     WESTBROOK RD. - EAST OF HARROW DR.
     HARROW DR.
     RIVERWOOD DR. - NEAR HARROW DR.
     SEDGWICK DR. - NEAR SANDLEWOOD PL.


                       - NOTICE-
        BASED ON INFORMATION AVAILABLE TO THE EMERGENCY OPERATIONS CENTER,
     THE AREAS NEAR THE ABOVE STREETS WILL BE AFFECTED AT THE INDICATED
     TIME.  THIS DOES NOT MEAN THAT WATER WILL OR WILL NOT BE IN ANY HOUSES
     OR STRUCTURES ON THESE STREETS.  IT DOES MEAN THAT RIVER LEVELS AT THE
     INDICATED TIME WILL AFFECT ACCESS EITHER BECAUSE OF WATER ON THE
     STREETS OR WATER ON ADJOINING STREETS.  BECAUSE OF DIFFERENCES IN
     ELEVATIONS, SOME STRUCTURES MAY TAKE WATER WHILE OTHERS MAY NOT.
        BECAUSE OF VARIABLES IN ELEVATIONS AND VARIATIONS IN LOCAL WATER
     LEVELS, CITIZENS SHOULD REMAIN ALERT FOR ENCROACHMENTS OF THE RIVER
     ON THEIR PROPERTY AS THE RIVER APPROACHES THE STREET ON WHICH THEY
     RESIDE.

***********************************************************************
*                                                                     *
*        LIST OF FLOODED STREETS ON MAP C AT  1 HUNDRED HOURS         *
*                                                                     *
***********************************************************************

     WESTBROOK RD. - EAST OF HARROW DR.
     HARROW DR.
     BEACHCREST DR. - EAST OF SEDGWICK DR.
     SOUTH BROOK DR. - EAST END NEAR SEDGWICK DR.
     RIVERWOOD DR. - NEAR HARROW DR. AND NEAR ROMANY DR.
     SEDGWICK DR. - NEAR SANDLEWOOD PL.
     WESTBROOK RD. - NEAR HARROW DR.
     ROMANY DR. - SOUTH OF ARGYLE ST.


                       - NOTICE-
        BASED ON INFORMATION AVAILABLE TO THE EMERGENCY OPERATIONS CENTER,
     THE AREAS NEAR THE ABOVE STREETS WILL BE AFFECTED AT THE INDICATED
     TIME.  THIS DOES NOT MEAN THAT WATER WILL OR WILL NOT BE IN ANY HOUSES
     OR STRUCTURES ON THESE STREETS.  IT DOES MEAN THAT RIVER LEVELS AT THE
     INDICATED TIME WILL AFFECT ACCESS EITHER BECAUSE OF WATER ON THE
     STREETS OR WATER ON ADJOINING STREETS.  BECAUSE OF DIFFERENCES IN
     ELEVATIONS, SOME STRUCTURES MAY TAKE WATER WHILE OTHERS MAY NOT.
        BECAUSE OF VARIABLES IN ELEVATIONS AND VARIATIONS IN LOCAL WATER
     LEVELS, CITIZENS SHOULD REMAIN ALERT FOR ENCROACHMENTS OF THE RIVER
     ON THEIR PROPERTY AS THE RIVER APPROACHES THE STREET ON WHICH THEY
     RESIDE.
     FLOOD>
     =
```

Figure 4.   LIST OF FLOODED STREETS TO
            ACCOMPANY MAP C


     The city's engineering department plans to use the
warning model to evaluate the effects of various channel
modifications and changes in floodplain management
practices.  For example, it can examine the value of
clearing the banks to reduce the resistance to flow of the
river by 10 percent.  It can also investigate the effects
of modification to restrictive highway crossings.

## POSSIBLE FUTURE APPLICATIONS

The Jackson model and graphics displays are designed for flood events occurring over periods of days. Similar methods, however, can be used in flash flood areas. More advanced graphics terminals will display maps in 1 to 2 seconds. Flooded areas can be predicted from backwater analysis if peak flow rates are obtained from rainfall/run-off models.

# Modelling Climatic Aspects of Pearl River Flooding

Charles L. Wax[1]

## Abstract

Surplus produced in parts of the upper Pearl River basin in Mississippi on 2 days, April 12 and 13, 1979, was 530% greater than normal monthly surplus, and near 100% of annual surplus for most parts of the state. Daily water balance components for April 11, 12, and 13, 1979, are estimated and mapped for the entire state to show their distribution relative to the basin. Statewide estimates of average April and average annual surpluses are modelled for comparison. Climatic models of surplus during April 1979 and December 1961 show how water balance distributions can vary yet still cause record flooding on the Pearl River in Jackson, MS. A model of the largest annual surpluses likely to occur in the state shows that the Pearl Basin is more prone to serious flooding from a climatic standpoint than other rivers in the state.

## Introduction

Flooding is a natural event, the potential for which is determined by climate. Climate delivers moisture and energy to a region, thereby setting the characteristics of the hydrologic cycle in that place. Climatological and hydrological records are used to assess flood probabilities when design or development considerations are needed. However, knowledge of the magnitude and frequency of flooding is limited by the availability and accuracy of the records and by the time period included in the records. For example, when flooding on the Pearl River in Mississippi during April 1979 reached a new, all-time record for volume of water, stage height, and economic devastation, there was nothing in the climatological or hydrological records to indicate the liklihood of such an event.

Considering the size of the Pearl River basin and its location in the humid Southeastern U.S., flooding on the river is not unusual. However, the spectacular level of flooding in April 1979 and the previously unseen volumes of water pouring into the city of Jackson prompted a public outcry and caused blame to be cast in all directions. The unusual magnitude of the flood needs to be investigated from the standpoint of the rarity of this event in terms of environmental inputs, conditions, and responses, and that is the purpose of this paper.

The flood can be placed in perspective by identifying and mapping selected components of the hydrologic cycle on a regional scale and in

[1]Assistant Professor of Geography, Department of Geology and Geography, Mississippi State University, Mississippi State, MS 39762

a historical context, thereby defining which variables or combinations of variables contributed to the production of so much floodwater in the Pearl River. This geographic/climatic explanation of the temporal and spatial characteristics of the landscape which produced the flood should be useful in assessing where to place responsibility for the record flooding and economic losses, and in assessing this event's place in the climatological record. The following analysis uses the climatic water balance technique to construct the patterns of soil moisture, surplus water, and other hydroclimatic features as they existed in the State of Mississippi during the flood and as they normally exist in the state. Comparison of regional and state-wide patterns with those inside the Pearl River basin evaluates the contribution of those particular patterns of time and space to the phenomenal flood. Comparison of the patterns associated with this specific flood to long-term average patterns within the basin evaluates the rarity of this particular hydroclimatological event.

## Geographic Setting and Background Information

The Pearl River drains approximately 3,100 square miles upstream of Jackson. Land use in the basin above the city is mostly devoted to agriculture and to forestry activities. There are no major controls on runoff north of the Ross Barnett Reservoir, which serves primarily as a control feature providing for water supply and recreation.

Major floods of record in Jackson are listed in Table 1. The table shows that river stage during the 1979 flood (43.2') was 25.2' above flood stage (18' at the time of the flood), a new record that is almost 6' higher than the previous high mark set in 1902. The flood protection levees in the city, completed in 1967, are designed to afford protection up to a 40.5' flood (Platt, 1980). Flood stage on the river in Jackson was officially changed from 18 to 28' on March 5, 1980.

Table 1: Historical Flood Data, Pearl River at Woodrow Wilson Bridge

| Date of Flood | Gage Height | Height Above Flood Stage (18') |
|---|---|---|
| 25 April 1874 | 37.0' | 19.0' |
| 5 December 1880 | 36.5' | 18.5' |
| 21 April 1900 | 36.7' | 18.7' |
| 1 April 1902 | 37.5' | 19.5' |
| 21 December 1961 | 37.2' | 19.2' |
| 17 April 1979 | 43.2' | 25.2' |

(Source: Neely, 1964, USGS, 1979)

The meteorological events that produced the precipitation for the flood are described in Figure 1. An intense center of low pressure was located over the Oklahoma panhandle at 6:00 A.M. CST on April 11, with a cold front extending through central Oklahoma and Texas. The pressure field aloft supported the surface cyclone, and upper winds, flowing around the low pressure core at the 500 millibar height (17,000-18,000'), produced a meridional (south to north) rather than a zonal (west to east) flow at that level. Acting as a steering mechanism for the surface cyclone, this circulation pattern aloft took the center of low pressure to the north-northeast and caused the cold front to slow and stall, producing a series of squall lines that moved slowly and violently over Mississippi (Figure 1a). Between 6:00 A.M. CST April 12 and 6:00 A.M. CST April 13 the storm's center moved to the north-northwest, further stalling the front's passage and prolonging the violent precipitation associated with the squall lines moving out from the front (Figure 1b).

The precipitation-producing mechanisms in the atmosphere over Mississippi did their job well. Warm, moist air from the Gulf of Mexico converged into the front and was lifted violently to condense into line after line of heavy, slow-moving thunderstorms that developed and redeveloped over many of the same areas. The result was massive amounts of rainfall in much of the state on the 12th and 13th of April. For example, Kosciusko recorded 13.0", Edinburg recorded 10.0", and Louisville recorded 19.6" during the two-day period (National Climatic Center, 1979). These locations are all in the upper part of the Pearl River basin, so these amounts are representative of the quantity of rain that fell in at least parts of the upper basin.

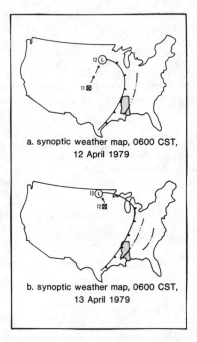

a. synoptic weather map, 0600 CST, 12 April 1979

b. synoptic weather map, 0600 CST, 13 April 1979

Figure 1:  Meteorological Setting for the Flood (NOAA, 1979)

These meteorological events and the resulting extraordinary amounts of rainfall ultimately produced the flood in Jackson, but the flood's severity was actually a consequence of other coinciding events--the simultaneous occurrence of patterns that overlapped because of timing and location. For example, antecedent conditions of rainfall and soil moisture storage coupled with seasonal low energy demand for evaporation and transpiration created the potential for a high percentage of any precipitation that occurred to be converted to surplus water, available almost immediately for surface runoff. Moreover, the drainage

divide of the upper Pearl River basin was aligned with the pattern of
rainfall over the state so as to funnel an unusually disproportionate
amount of the state's total rainfall into the Pearl River during the
two days.  Therefore, the remainder of this report will provide an enhanced understanding of the flood's origin and severity by providing a
geographical analysis of the flood-producing precipitation on the Mississippi landscape.

## Concepts and Methods

The climatic water balance methodology (Thornthwaite, 1948;
Thornthwaite and Mather, 1955) is a useful technique for evaluating
climatic variability and hydroclimatological features on a regional
scale.  It is used in this analysis to account for the disposition of
large volumes of precipitation that entered the upper Pearl River basin
on April 12th and 13th, 1979, as well as to assess the normal monthly
and annual climatic water situation in the state for comparison.  For
purposes of this report, the term "water balance" refers to the actual
short-period accounting of the moisture condition in the State of Mississippi during the first part of the year 1979 and during other specified monthly and annual periods.  Water balance technique accounts for
water as it enters an environment (precipitation), is stored there
(soil moisture), or leaves (evaporation or runoff), providing an expression of the moisture relationships, instantaneously or longer-term,
at local places or over regional watersheds.

In order to calculate the water balance components in the state for
the time period of the flood, a daily water balance model was employed.
The daily model has been adapted for use on the computer (Yoshioka,
1971), and this computerized model was used to calculate daily water
budgets for 66 locations in and around the state for the period 1 January - 30 April, 1979.  This period was chosen to allow the model to
"adjust" any erroneous assumptions before computing values for the
critical few days involved in the flood events.

The daily water balance program employed is designed to model exchanges in soil moisture storage by using a two-layer soil moisture
storage scheme.  The model allows for moisture exchanges to operate
easily within the upper soil moisture zone and for more gradual exchange to occur with depth.  The upper layer was set to hold a capacity
of 1", and the lower layer a capacity of 5" for a total soil moisture
storage capacity of 6".  All 66 locations were assumed to have these
capacities for storage and exchange, and all calculations were begun on
the assumption that the soil moisture capacity at all locations was full
on 1 January 1979.

Table 2 shows the computations for Louisville, MS, for the selected period April 7-15 to illustrate the workings of the model.  Note
particularly that soil moisture is consistently close to capacity, that
daily PE demands are low, and that surpluses are consistently a high
proportion of precipitation as a result.

Estimates like those illustrated in Table 2 were computed for 66
places in the state, forming the basis for the geographical explanation
of the flood.  To illustrate the impact of overlapping patterns, these
data were mapped using the SYMAP computer mapping program.  This

Table 2:   Daily Water Balance, Louisville, April 7-15, 1979
(Available soil moisture = 6")

| Date | PE (in) | P (in) | P-PE (in) | ST (in) | ΔST (in) | AE (in) | D (in) | S (in) |
|------|---------|--------|-----------|---------|----------|---------|--------|--------|
| 4/7  | .04 | 0    | -.04  | 5.79 | -.04 | .04 | 0 | 0     |
| 4/8  | .11 | .55  | .44   | 6.00 | .11  | .11 | 0 | .33   |
| 4/9  | .08 | 1.0  | .92   | 6.00 | 0    | .08 | 0 | .92   |
| 4/10 | .04 | 0    | -.04  | 5.96 | -.04 | .04 | 0 | 0     |
| 4/11 | .08 | 0    | -.08  | 5.78 | -.08 | .08 | 0 | 0     |
| 4/12 | .16 | 9.3  | 9.14  | 6.00 | .22  | .16 | 0 | 8.92  |
| 4/13 | .08 | 10.2 | 10.12 | 6.00 | 0    | .08 | 0 | 10.12 |
| 4/14 | .08 | .04  | -.04  | 5.96 | -.04 | .08 | 0 | 0     |
| 4/15 | .08 | 0    | -.08  | 5.78 | -.08 | .08 | 0 | 0     |

PE  = potential evapotranspiration
P   = precipitation
ST  = soil moisture storage
ΔST = change in soil moisture storage
AE  = actual evapotranspiration
D   = deficit
S   = surplus

program was prepared at the Laboratory for Computer Graphics and Spatial Analysis, Harvard University, and is useful for producing maps and diagrams which graphically depict spatially disposed quantitative information (Dudnik, 1971). Distributions of selected water balance components for specific days were thus produced, allowing spatial analysis and comparison of conditions across the entire state on those days.

Average annual statewide surpluses and average April surpluses were also computed for 85 places in the state based on 1951-1980 climatological data. Using these models it is possible to compare the surpluses produced during the two-day rainfall with long-term average surpluses, and thus to visualize how rare the event was in a normal and historical context.

Additional analyses display total monthly surpluses, statewide, for April 1979 and December 1961 to show how spatial variation in climatic components that produce flooding may occur. Distribution of the largest probable annual surpluses statewide are displayed to show the relationship of the Pearl River basin to the normal climatic aspects of the state.

## Results and Discussion

The statewide pattern of soil moisture storage conditions for the 11th of April, the day before the heavy rainfall, is shown in Figure 2. The figure shows the percent of capacity (6") to which soils were filled as a result of rainfall during the prior few days. Soils in the basin above Jackson were all between 95-100% full, so that little rainfall could be absorbed, and none of the soils could absorb more than about 0.3". In fact, the figure shows that almost all the state's soils were filled to more than 90% of capacity on the 11th of April, 1979.

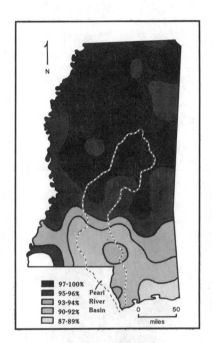

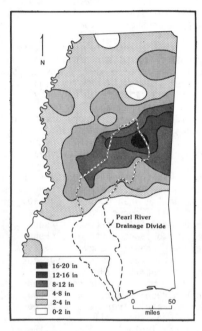

Figure 2:  Statewide Pattern of ST Showing % of Capacity to which Soils were filled April 11, 1979

Figure 3:  Statewide Pattern of Surplus Water Produced in Mississippi on April 12-13, 1979

The next two days were critical in the overlapping of patterns of time and space. As Figure 3 shows, the largest amounts of surplus water in the state on April 12th and 13th occurred over the upper Pearl River basin, overlapping the pattern of soils already filled to near capacity (Figure 2). The area of highest surplus production, shown in Figure 3 to be centered near Louisville, MS, represents about 100 square miles covered by 16-20" of water. The entire basin north of Jackson was covered with surplus water ranging in depth from almost 20" to a low of almost 4". No place in the basin above Jackson had less than 4" of water over it, all of which had no "escape" from the landscape except the channel of the Pearl River, which flowed through Jackson.

Figure 3 reveals the highly geographic character of the flood.  If the "bullseye" of surplus production had been located north, south, east, or west of the Pearl River drainage divides, the flood in Jackson would have been less severe or even minimal.  Greater flooding would have occurred elsewhere, however, for no where could the Mississippi landscape have absorbed, held back, or otherwise managed the volume of water released by the atmosphere during this event.  The unique coupling of patterns--full soils, excessive precipitation, and containment of overland flow by the Pearl River drainage divides--produced the record flood in Jackson.

Computations of average annual surplus values at 85 locations across the state for the period 1951-80 show that normal production of surplus anywhere in the state in a whole year is less than 29", with the state averaging 25.7" annually, and that average April surplus is less than 3.75" everywhere in the state (Wax, 1982).  Therefore, the 4-20" amounts of surplus produced over the upper Pearl basin on the two-day period in April of 1979 alone are seen to be between about 6% and 530% greater than that normally produced in a 30-day period anywhere in the state, and close to an entire year's production of surplus for most parts of the state.  This realization documents the abnormally rare nature of the surplus production involved in creating the April 1979 flood water on the landscape.

Figure 4 displays the estimated ratio by which precipitation was converted to surplus water, available immediately for surface runoff.

The statewide pattern again shows how a high ratio centered over the upper Pearl River basin. Most of the area north of Jackson exhibited a conversion of 93-98% of the rainfall directly to surplus.  No place north of Jackson had less than 82% conversion.

Estimates of monthly surplus for April 1979 and December 1961 are modelled for comparison in Figures 5 and 6.  Major floods on the Pearl River were produced by both of these climatic situations (Table 1).  Although the distributions and absolute values of surplus varied on the two occasions, it is evident that the upper part of the Pearl basin lay under the area of largest surplus production in both cases. Figure 7 models the largest surplus production likely to occur in a year, based on the 30-year climatic record 1951-80, and documents that the upper Pearl basin coincides with one of the areas of highest surplus production in the state.  Therefore, the Pearl basin is more prone to

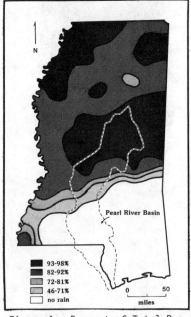

Figure 4:  Percent of Total Precipitation That Became Surplus on April 12-13, 1979

serious flooding than other rivers of comparable size in the state
of Mississippi.

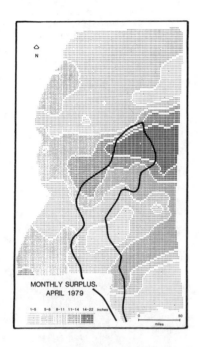

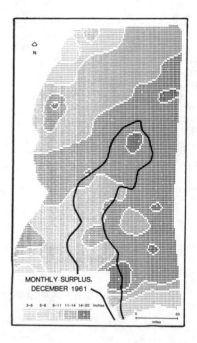

Figure 5:  Statewide Pattern of        Figure 6:  Statewide Pattern of
        Surplus, April 1979                     Surplus, December 1961


## Conclusion

     The amount of rain that fell in Mississippi on the 12th and 13th
of April, 1979, was destined to cause flooding because the rainfall
event was too rare for the landscape to treat in any other manner.
Flooding that resulted was somewhat greater than the 500-year flood
indicated in the Flood Insurance Rate Map for Jackson prepared by the
Federal Insurance Administration (Platt, 1980).  The fact that the
heaviest rain fell on saturated soils inside the drainage divides of
the Pearl River above Jackson insured where the flooding would be felt
and increased the severity of the flood.  This, in turn, caused forecasting and prediction problems because officials had no historical
meteorological or hydrological data to rely on as a basis for precise
decision making.
     The flood was beyond the limits of protection design, and it re-
mains unfeasible to design flood protection features for this level of
flooding.  This hydroclimatological evaluation of the flood points out

that it was a combination of climatic and geographic variables that was responsible for the record flooding and the resulting economic losses-- not inadequate forecasting or improper operation of the Barnett Reservoir by responsible agencies. When this unique combination of patterns and events will occur again in the Pearl basin, or any other basin, is impossible to predict. The results, however, are very predictable.

This analysis also shows that the Pearl River basin is in a preferred location in Mississippi for flooding. The alignment of the basin with climatically normal excesses of surplus water enhances the potential for flooding from a geographic standpoint (Figure 7). The super-excessive production of surplus over this area in April 1979 is now part of the climatological/hydrological record--if it happened once, it can happen again.

Figure 7:  Statewide Pattern of Largest Surpluses Likely to Occur

## References Cited

Dudnik, E. E., 1971.  "SYMAP Users Reference Manual for Synagraphic Computing Mapping."  Report No. 71-1, Dept. of Architecture, University of Illinois, Chicago, Illinois.

National Climatic Center, 1979.  Climatological Data, Mississippi, April, Vol. 84, No. 4, Asheville, N.C.

National Oceanic and Atmospheric Administration, 1979.  Daily Weather Maps, Weekly Series April 9-15. U.S. Dept. of Commerce.

Neely, B. L., Jr., 1964. "Floods of 1962 in Mississippi." State of
    Mississippi Board of Water Commissioners Bulletin 64-6. Jackson,
    Mississippi.

Platt, R. H., 1980. "The Pearl River Flood at Jackson, Mississippi,
    April 1979 - Preliminary Report." Publications of the National
    Weather Association, Vol. 5, No. 1.

Thornthwaite, C. W., 1948. "An Approach Toward a Rational Classi-
    fication of Climate." The Geographical Review, Vol. 38:  55-94.

Thornthwaite, C. W. and J. R. Mather, 1955. "The Water Balance."
    Publications in Climatology, Laboratory of Climatology, Vol. 8,
    No. 1, 104 pp. Centerton, New Jersey.

U.S. Geological Survey. 1979. Water Resources Data for Mississippi,
    Water Year 1979. U.S.G.S. Water-Data Report MS-79-1. U.S.G.S.
    Water Resources Division, Jackson, Mississippi.

Wax, C. L., 1982. Atlas of the Climatic Water Balance in Mississippi:
    Normals and Variability, 1951-1980. Water Resources Research
    Institute Technical Publication D018, Mississippi State University,
    Mississippi State, Mississippi, 82 pp.

Yoshioka, G. A., 1971. "A Computer Program for the Daily Water
    Balance." Publications in Climatology, Laboratory of Climatology,
    Vol. XXIV, No. 3, 28 pp., Centerton, New Jersey.

System Response Model for Upper Pearl River Basin

by

D. B. Simons, Fellow, R. M. Li, Member,
K. G. Eggert, Associate Member
Simons, Li & Associates, Inc.
and
C. E. Moak
Pearl River Valley Water Supply District

ABSTRACT

The purpose of the study was to allow better flood forecasting for the Pearl River in Jackson, Mississippi area. Further, the flood mitigation potential of Ross Barnett Reservoir upstream of Jackson was to be optimized under operational constraints. This was accomplished by utilizing watershed rainfall data in a multiple watershed model, which determines flow from the watershed based on physical principles governing the hydrologic cycle and hydraulics of the channel and overland flow.

Rainfall data over time from a number of locations are used as input to the watershed model, which computes flow over the land surface, into tributaries and finally into the mainstem of the river. The watershed outflow hydrograph is then used as input to the reservoir operation model. The reservoir operation model computes reservoir storage and releases so as to optimize the flood storage potential of the reservoir under operational constraints of desired pool level and minimum and maximum outflows.

INTRODUCTION

The Pearl River flows in a southwesterly direction through central Mississippi, draining a watershed of approximately 3,100 square miles above Jackson. Runoff from the upstream farmlands and forests is for the most part uncontrolled due to a lack of flood control structures. The city of Jackson is located below Ross Barnett Reservoir. The main purposes of the reservoir are public recreation and water supply.

Because Ross Barnett Reservoir was designed primarily with water supply and recreation in mind, it does not provide significant flood water storage. However, operations may mitigate the severity of the downstream flood damage to a limited degree, depending on the magnitude and duration of the storm. Therefore, the study was conducted to achieve the following objectives:

(a) Develop an integrated system response model consisting of the watershed and the reservoir. This integrated model can be used as a flood warning system that predicts the Jackson flood stage for estimated rainfall and specified reservoir operating procedures.

(b)  Based on the integrated system, provide optimal reservoir
management schemes to meet operational constraints and simultaneously
reduce downstream flood potential to the extent possible.

WATERSHED SIMULATION PROGRAM

     PRLWAT (Pearl River Watershed Model) is designed to route storm-
water runoff from watershed of complex geometry.  In order to accomplish this task, the complex watershed geometry must be simplified into
a representation suitable for computer simulation.  The geometric
approximation used in PRLWAT consists of an arbitrary numer of two-
plane and one-channel ("open book") watersheds linked together by channel segments.  All subwatershed units, planes and open books have no
upstream inflows and are routed using the method of characteristic
solution to the kinematic wave problem for overland, subsurface and
channel flow.  The channel in this case is the intersection of the two
planes of the open book.  These primary channels also have no upstream
inflow and are routed using the method of characteristics solution.
The channels that join the subwatersheds have upstream inputs and
require a numerical solution to the kinematic wave approximation.
     PRLWAT considers the physical processes of rainfall, infiltration,
overland flow surface routing, subsurface flow routing, deep percolation and routing of water in the channels.  Excess rainfall intensity
is routed through overland flow planes.  Deep percolation is computed
by subtracting a constant percentage of the infiltrated volume.
Remaining infiltration volume is routed horizontally through the soil.
This is accomplished using the Dupuit approximation.  Routing the water
in the channel segments requires a numerical solution, which collects
all the lateral inflows and routes the water down the mainstem.
     The watershed simulation is driven by rain gage data gathered
throughout the Upper Pearl River drainage.  Eight rain gages in the
upper drainage may be interrogated by telephone for rainfall data.
Figure 1 shows the distribution of gages in the basin.  Further, five
river gages (also shown in Figure 1) may also be interrogated for
essentially simultaneous information on discharges in the Pearl and
Yockanookany Rivers as well as Tuscolameta Creek.  These gages form an
unusually good data base for simulation of runoff.

Figure 1.   Upper Pearl River basin showing gage stations.

As originally formulated, infiltration was calculated using a time explicit formulation of the Green-Ampt equation. However, due to the size of the drainage and the duration of storms producing serious runoff, a constant infiltration rate approximately equal to the saturated conductivity was used. Local soil survey data served as a guide for selecting initial values of conductivity. Calibration adjusted the values used; however, the final values were within the range of reported saturated conductivities. As initially constructed, the model was chosen to predict runoff under saturated soil conditions. This calibration is now being expanded to accept a variety of initial soil moisture conditions. Initial river gage data is used to assess the base flow conditions, and storm losses to deep groundwater are coupled to these initial readings.

Overland flow slopes for the watershed basin are relatively flat, ranging from 7 percent to 0.1 percent, with the average slope being 1 percent. The watershed was subdivided into a total 175 units, including 42 channel units, 74 plane units, and 59 subwatersheds. Figure 2 is a schematic representation of the Pearl River mainstem showing the manner of representation for simulation. Slopes, lengths and areas for all the units were entered as model data. Other subdrainages such as Tuscolameta, Lobutcha, Nanawaya Creeks and the Yockanookany River are similarly modeled.

It is assumed that a negligible amount of rainfall is intercepted by ground and canopy cover. As stated before, it was also initially assumed that saturated soil conditions exist, and therefore infiltration rate is approximately equal to the saturated hydraulic conductivity. Thus the excess rainfall available for overland flow routing is equal to the input rainfall intensity minus the saturated hydraulic conductivity:

$$e_j = i_j - K_u \tag{1}$$

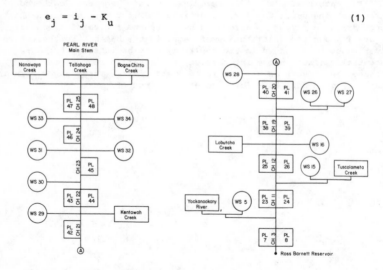

Figure 2.   Schematic representation of Pearl River mainstem.

where  e  is the excess rainfall intensity,  i  is the input rainfall
intensity,  $K_u$  is the saturated hydraulic conductivity in the wetted
zone and the subscript  j  denotes the jth time increment of a storm
hyetograph.

Deep percolation losses are subtracted from the infiltrated volume
of water and the remaining water is available for subsurface flow
routing.  Deep percolation losses are determined by subtracting the
excess rainfall intensity from the input rainfall intensity, which
represents the infiltration volume, and then multiplying the infiltration volume by a constant determined as an empirical function of stream
base flow.

$$D_j = (i_j - e_j) \, C_1 \tag{2}$$

where  D  is the deep percolation loss and  $C_1$  is the constant percentage of deep percolation loss.

Once deep percolation losses are computed, the available water for
subsurface flow routing can be computed.  This quantity is computed by
subtracting deep percolation and excess rainfall intensity from the
input rainfall intensity.

Routing schemes used in the model all utilize the kinematic wave
formulation.  Slopes in the Upper Basin are sufficiently great for use
of this method.  Basically four kinds of water routing are used here:
overland flow, subsurface flow, upland channel flow and mainstem channel flow.  The first three types of flow utilize the analytical solution to the kinematic wave problem.  That the solution technique is
identical follows from the form of the equations.

Overland flow:

Continuity --

$$\frac{\partial q}{\partial x} + \frac{\partial y}{\partial t} = e \tag{3}$$

Momentum (using Manning's equation) --

$$q = \frac{1.49}{n} S_o^{1/2} y^{5/3} \tag{4}$$

where  q  is discharge per unit width,  y  is depth of flow,  x  is the
downstream dimension,  t  is time,  n  is Manning's resistance parameter, and  $S_o$  is the bed slope.

Subsurface flow:

Continuity --

$$\frac{\partial q_s}{\partial x} + \varepsilon \frac{\partial \eta}{\partial t} = f_a \tag{5}$$

354     APPLYING HYDRAULIC RESEARCH

Momentum (Darcy's Law) --

$$q_s = \frac{kgS_o}{\nu} \qquad (6)$$

where $q_s$ is subsurface discharge per unit width, $\eta$ is height of
water table, $\varepsilon$ is porosity, $k$ is intrinsic permeability, $g$ is
acceleration due to gravity, $\nu$ is the kinematic viscosity of water,
and $f_a$ is the amount of water available for subsurface flow.

Upland channels:

Continuity --

$$\frac{\partial Q}{\partial x} + \frac{\partial A}{\partial t} = q_t \qquad (7)$$

Momentum (using Manning's Equation) --

$$Q = \alpha\, A^\beta \qquad (8)$$

where $Q$ is discharge, $A$ is the flow cross-sectional area, $q_t$ is
total of $q$ and $q_s$, and

and

$$\alpha = \frac{2.21\, S_o}{n^2\, a_1^{4/3}}^{1/2} \qquad (9)$$

$$\beta = \frac{5 - 2b_1}{3} \qquad (10)$$

Here $a_1$ and $b_1$ are the statistical parameters of a power function
fit to relate wetted perimeter $P$ and cross-sectional area.

The analytical solution for the kinematic wave problem is essentially the same derivation given by Harley, Perkins, and Eagleson
(1970), and Simons, Li and Eggert (1976). This routing procedure is
used by the plane and subwatershed units of PRLWAT to determine the
upstream and lateral inflows into the channel units. The upland
watersheds are linked together by channels as described by Simons, Li
and Spronk (1977).

In PRLWAT, the mainstem flows are routed numerically rather than
analytically. The analytical solution was used previously for its computational efficiency. However, it may present the computational difficulty known as kinematic shock (Kibler and Woolhiser, 1970) when the
upstream depth or flow area may not be assumed to equal zero. This
assumption is not possible when routing flows through river segments.
Due to the inherent averaging that occurs in finite difference schemes,
the numerical approach does not exhibit shock. The numerical routine
used was developed by Li, Simons and Stevens (1975).

CALIBRATION AND VERIFICATION OF WATERSHED MODEL

Five storms were selected for calibration and verification of
PRLWAT.  February 1979, April 1979 and November 1979 storms were used
to calibrate model parameters, and the January 1979 and April 1980
storms were used to verify the calibrated parameters.  These storms
ranged significantly in duration and intensity, from the one-in-a-
thousand-year April 1979 event to an intermittent 40-day April 1980
storm.  For calibration and verification purposes, sixteen existing
rainfall stations were used within the Pearl River Basin.

Calibrated parameters for PRLWAT included the saturated hydraulic
conductivity, deep percolation losses, Mannings' resistance coefficient
and the subsurface flow response time.  Hydraulic conductivity and deep
percolation loss values are important in determining runoff volume.
Mannings' flow resistance and subsurface flow response time are important in determining the shape of the output hydrographs.  Saturated
hydraulic conductivity values were calibrated for subwatershed and
plane units.  These fell in the range of 0.1 to 0.25 inches per hour.
Deep percolation is partly a function of soil moisture conditions.  For
example, under initially dry soil conditions, deep percolation rates
would be expected to be higher than for initially saturated soil conditions.  Base flow is often an indicator of basin moisture conditions.
One indirect way to determine initial soil moisture conditions prior to
each storm is then to measure the base flow discharge.  A low base flow
discharge indicates dry soil conditions and a high base flow discharge
indicates a saturated soil.

A simple regression was determined relating base flow discharge
and deep percolation rates for areas upstream of Walnut Grove and
Carthage.  Deep percolation above Ofahoma seemed to be independent of
streamflow.  The regressions worked well for all but the April 1980
storm, which was an intermittent 40-day storm.  Initial base flow
discharges for this storm were low, which would indicate a high deep
percolation rate; however, due to the length of the storm, saturated
soil conditions were attained within the storm duration.

Mannings' flow resistance coefficients were selected for
Tuscolameta Creek and the Yockanookany and Pearl Rivers from reported
values.  Relatively high "n" values were selected for the smaller
streams because at flood stages these waterways flow the full width of
the valley plain.  A lower resistance was selected for the Pearl River
since it is a more fully developed river and is better able to convey
higher flows.  For overland flow planes and subwatershed units, often
no defined channels exist; all,, high values of Manning's roughness
coefficient were selected for these areas.  Subsurface flow time (the
time it takes infiltrated water of the subwatershed and plane units to
travel horizontally through the soil from the top of the unit to the
channel) was calibrated to average about one day for all units.

Two storms used for calibration and for verification are discussed
here.  The April 1979 storm was one of three storms used to calibrate
the model.  Measured runoff discharges were available from the U.S.
Geological Survey gages at 15 minute intervals at Walnut Grove, Ofahoma
and Carthage.  Rainfall data was reported in daily accumulated volumes
for sixteen existing precipiation gages within the watershed.  Results
between measured and predicted results agree remarkably well for this
storm at all three of the gage sites.  Figure 3 shows the agreement of

calibrated and measured flows at Carthage.   Similar plots were obtained
for the Walnut Grove and Ofahoma gages.

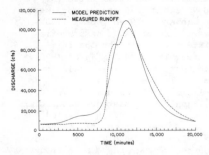

Figure 3.   Calibration comparison        Figure 4.   Verification comparison
    for storm of April, 1979.                 for storm of January, 1979.

        The January 1979 storm was one of two verification runs for pro-
gram PRLWAT using the calibrated parameters.   The storm occurred from
January 17 to January 27 with the majority of rainfall occurring January 20th and 21st.   Only a drizzle occurred preceding and after these
two days.   Rainfall and runoff were reported on a daily basis.   A comparison between computed and measured hydrographs is shown in Figure 4
for flows at Carthage.   Results are good for Ofahoma, Carthage and
estimated flows into Ross Barnett Reservoir.   At Walnut Grove it
appears that the gage record missed the first peak that occurred around
January 22nd due to the procedure of taking daily readings.
        All in all the initial calibration and verification runs seem to
provide very good agreement.   However, subsequent tests at the model
last spring have indicated that the calibration may require some modification for drier conditions.   This work will be performed in the
summer of 1982.

RESERVOIR OPTIMIZATION

        Analysis has shown that the flood mitigation capacity of the Ross
Barnett Reservoir is very limited.   The reservoir was not designed for
this purpose.   However, in some cases it is possible to operate the
reservoir in a manner that will minimize flood impacts.   The runoff
forecasts from the watershed routing model are used to provide input to
the reservoir optimization scheme.   This program optimizes reservoir
releases subject to constraints of minimum storage, desired release and
initial pool levels.   Optimization techniques are employed to determine
operational plans to minimize flood damage below the reservoir by properly regulating the gate openings as dictated by physical constraints.
As can be seen, the optimization problem for the operation of the Ross
Barnett Reservoir is complex because of limited flood storage capacity
and extremely large flood events, as evidenced by the April 1979 flood.
The constraints on the inflow-storage-outflow relation, the maximum
pool level of 300 feet in extreme floods, and the required reservoir
level maintained for its design purposes all add to the complexity of
the problem.   As mentioned earlier, the main purpose of the Ross

Barnett Reservoir is water supply, water quality control and recreation. Reservoir operation can only provide limited flood storage capacity. The following optimization analysis considers the concept of
mitigating the severity of flooding within the physical constraints of
reservoir design.

The optimization scheme was developed by first assuming that a
storage-stage relationship has been derived for the reservoir based on
historical records. For Ross Barnett this expression is

$$\log S_i = -17.9784 + 0.1034\ h_i \tag{11}$$

where S is storage in the ith time increment and h is the pool elevation corresponding to S .

The water balance equation for the reservoir can be written as

$$S_{i+1} = S_i + C_2(u_i - R_i) \tag{12}$$

where u is the upstream inflow volume, R is the release volume and
$C_2$ is a proportionality constant whose value depends on the time-step
and units of measure used in the computation.

The constraints on the reservoir water surface elevation can be
expressed by the following inequalities:

$$h_{min} < h_i < h_{max} \tag{13}$$

Constraints on the amount of water to be released from the reservoir
can be separated into two groups:

(1)  for downstream water conservation purposes:

$$R_i > R_{min} \tag{14}$$

(2)  for avoiding overbank flow in downstream:

$$R_i + \delta^-_{R_i} - \delta^+_{R_i} = R_{max} \tag{15}$$

where one seeks to minimize $\delta^+_{R_i}$ .

For water supply and recreational purposes, it is desired to maintain the water surface elevation as close as possible to a certain
desired stage $h_d$ . Thus, this "goal constraint" can be expressed by
the following equation:

$$h_i + \delta^-_{h_i} - \delta^+_{h_i} = h_d \tag{16}$$

where one seeks to minimize the "over-achievement" $\delta^+_{h_i}$ and the
"under-achievement" $\delta^-_{h_i}$ to attain the desired objective, or goal,
$h_d$ .

The objective of the reservoir regulation for this case can be stated as follows:

"Find a set of releases $R_i$ , i=1,2,...,N to achieve the goals (15) and (16) and satisfy the constraints (12), (13), and (14), when a set of inflows $u_i$ , i=1,2,...,N and the initial and final values (i.e., at time 1 and N+1) of the water in storage are given."

Expressing all of the above constraints in terms of $R_i$ , $u_i$ , $S_1$ , and $S_{N+1}$ . From Equation (12), the following relation is obtained:

$$\sum_{i=1}^{N} R_i = \sum_{i=1}^{N} u_i + \frac{1}{C_2}(S_1 - S_{N+1}) \qquad (17)$$

where $S_1 = \exp[a + b \times h_1]$ (i.e., initial storage)

$S_{N+1} = S_d = \exp[a + b \times h_d]$ (i.e., desired storage at the final time)

Let $S_{min} = \exp[a + b \times h_{min}]$ (i.e., the minimum amount of water required to be stored in the reservoir at any time)

$S_{max} = \exp[a + b \times h_{max}]$ (i.e., the maximum amount of water that can be stored in the reservoir at any time).

Using these expressions, the set of inequalities (16) can be replaced by the following constraints:

$$\frac{1}{C}(S_1 - S_{max}) + \sum_{i=1}^{k} u_i < \sum_{i=1}^{k} R_i < \frac{1}{C_2}(S_1 - S_{min}) + \sum_{i=1}^{k} u_i$$

$$k = 1,2,...,N \qquad (18)$$

Also, the water supply and recreation constraints can be replaced by the following equations:

$$\sum_{i=1}^{k} R_i + \delta^-_{SR_k} - \delta^+_{SR_k} = \frac{S_1 - S_d}{C_2} = \sum_{i=1}^{k} u_i$$

$$k=1,2,...,N \qquad (19)$$

The decision variables in this study are

$$(R_1, R_2,...,R_N; \delta^-_{SR_1};...; \delta^-_{SR_N}; \delta^+_{SR_1},...,\delta^+_{SR_N}; \delta^-_{R_1},...,$$

$$\delta^-_{R_N}; \delta^+_{R_1},...,\delta^+_{R_N})$$

and the sequence of optimal releases $R_1, R_2, \ldots, R_N$ are obtained by solving the following mathematical programming problem:

$$\min \sum_{i=1}^{N} W_F \, \delta_{R_i}^{+} + \sum_{i=1}^{N} W_D (\delta_{SR_i}^{-} + \delta_{SR_i}^{+}) \qquad (20)$$

total flooding - total deviation from the desired reservoir water surface level $h_d$

where $W_F$ and $W_D$ are weighting coefficients set by the reservoir manager according to "goal priorities" (i.e., the more important goal will have a bigger value for the weighting coefficient) subject to the constraints (15), (16), (17), (18) and (19).

Finally, if the predicted runoff volume from the storm is less than the readily available average, no optimization is performed. In this case only continuity routing through the reservoir is performed.

CONCLUSIONS

The optimization scheme may be used to examine alternative reservoir operation schemes resulting from storm inflows. In combination with the watershed model and the interrogatable rainfall information network the system model provides a real time system for predicting stormwater flows and to the degree possible minimizing downstream flood stages. The system response model represents a practical example of the application of how sophisticated hydrologic simulation and optimization techniques may be used for the direct benefit of society.

ACKNOWLEDGMENTS

The authors also wish to acknowledge Mr. Michael Ballantine, Dr. Nguyen Duong, Ms. Jan Kimzey, and Ms. Lan-Yin Li for their valuable constributions to this work.

REFERENCES

Harley, B. M., F. E. Perkins and P. S. Eagleson, "A Modular Distributed Model of Catchment Dynamics," MIT, Department of Civil Engineering, Hydrodynamics Laboratory Report No. 133, 1970.

Simons, D. B., R. M. Li, and K. G. Eggert, "Storm Water and Sediment Runoff Simulation for Upland Watersheds Using Analytical Routing Technique, Volume I - Water Routing and Yield," Colorado State University, Engineering Report CER77-78DBS-RML-KGE16.

Simons, D. B., R. M. Li and B. E. Spronk, "Storm Water and Sediment Runoff Simulation for a System of Multiple Watersheds, Volume I - Water Routing and Yield," Colorado State University Engineering Report CER77-78DBS-RML-BES47, 1977.

Kibler, David F., and D. A. Woolhiser, The Kinematic Cascade as a Hydrologic Model, Hydrology Paper #39, Colorado State University, Fort Collins, Colorado, 1970.

Li, R. M., D. B. Simons, and M. A. Stevens, "Nonlinear Kinematic Wave Approximations for Water Routing," Water Resources Research, Vol. 11, No. 2, 1975.

# MODAL TESTING OF TRASHRACK AT HIWASSEE DAM

By Gerald A. Schohl and Patrick A. March, M.ASCE[1]

## ABSTRACT

During pump-turbine commissioning tests for the Tennessee Valley Authority's Hiwassee Project, the lower pool (draft tube) trashracks fell off, apparently due to anchor bolt failure caused by static or dynamic forces. The lower pool trashracks were recently repaired and replaced using epoxy-type anchor bolts. Vibration tests were conducted to determine whether the trashrack failure was caused by excessive static forces or by flow-induced vibrations. An instrumented impact hammer and a fixed accelerometer were used to perform a modal analysis. By determining transfer functions between impact forces and the corresponding accelerations for a variety of locations on the trashrack, natural frequencies, damping values, and mode shapes were determined. A large number of lightly damped modes were identified, and the modal results guided the selection of locations for plunge (parallel-to-flow) and lateral (transverse-to-flow) accelerometers. A strain gage was placed on a trashrack bar at the point of maximum expected strain.

Outputs from the strain gage and the accelerometers were monitored at five percent wicket gate opening intervals during generating operation. Buffeting response was observed in plunge and lateral modes previously identified during impact tests. Maximum stress levels less than 4000 $lbf/in^2$ were recorded, and "locked-in" trashrack response was not observed. It was concluded that the previous trashrack failure was due to inadequately strong anchorage, not flow-induced vibrations, and that the current performance was adequate without additional modifications.

## INTRODUCTION

Hiwassee Dam, located on the Hiwassee River near Murphy, North Carolina, contains a 2-unit hydropower plant with a combined generating capacity of 117 MW. Unit 1 is a conventional Francis-type hydroturbine. Unit 2 is a reversible pump-turbine which requires trashracks on the upper pool and lower pool intakes to protect the pump-turbine during both generating and pumping operation. The lower pool trashracks failed during commissioning tests in 1956, but the racks were not replaced until 1980 because of other power system priorities.

[1]Mechanical Engineers, Water Systems Development Branch, Engineering Laboratory, Tennessee Valley Authority, Norris, Tennessee.

Several other pumped-storage plants have also experienced trashrack failures (for example, Crandall, et al., 1975; Vigander and March, 1974; Fiske and Daly, 1971; Louie and Allen, 1971) due to flow reversal and high flow velocities. The cause of the trashrack failure at Hiwassee had not been determined, so it was decided to perform onsite testing to determine the dynamic characteristics of the rein-stalled racks and to monitor rack vibrations during startup and normal operation. Modal testing, using impact-response techniques, was performed on the draft tube trashrack in air to determine the natural frequencies and vibration modes of the structure. Two accelerometers and a strain gage were attached to the trashrack and subsequently monitored during pump-turbine operation. This paper presents the results and conclusions from these field tests.

## DESCRIPTION OF TRASHRACK

Figure 1a shows the three draft tube trashrack sections after reinstallation to the three exit bays for Unit 2. From left to right in Figure 1a, the exit bays are designated as the west, center, and east bays. Figure 1b shows a schematization of the east bay trashrack section used for the modal analysis. Every fourth vertical bar is included.

Figure 2a gives the dimensions of the east and west bay trashrack sections, shows the locations of the instruments that were installed on the east bay section, and shows the measurement points (1 through 10) used in determining the plunge vibration modes (rack motion in the direction of the flow) of the east bay section. The two horizontal bars at the center of the trashrack are not joined together, so relative motion between the upper and lower sections is possible. As shown in Figure 1, a 2-foot, 9-inch tall rack section with curved bars is anchored to the wall above the 20-foot tall vertical section. The center trashrack is 18 feet, 4 inches wide with one less vertical bar, but otherwise identical to the east and west sections. Figure 2b shows the measurement points (11 through 19) used in determining the lateral vibration modes (horizontal motion perpendicular to the direction of flow) of a single vertical bar.

## INSTRUMENTATION AND PROCEDURES

### Modal Testing of Draft Tube Trashrack

Instrumentation for the modal testing of the trashrack included a dynamic signal analyzer consisting of a data sampler, an analog-to-digital converter, a Fast Fourier transform processor, and a cathode-ray tube display; an accelerometer, fastened to a magnet for attachment to the trashrack; and a hammer, with an attached force gage to measure impact force. The dynamic signal analyzer contains algorithms to determine natural frequencies and damping values by curve fitting the transfer functions, which relate response acceleration to impact force. Additional algorithms are used to determine and to animate mode shapes. Measurement data are stored on tape cassettes.

Modal testing by the impact-response technique involves measuring transfer functions between impact forces and response accel-

a. PHOTOGRAPH                                    b. SCHEMATIZATION

## Figure I: Unit 2 Draft Tube Trashrack

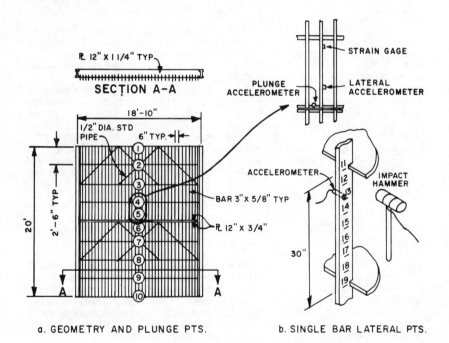

a. GEOMETRY AND PLUNGE PTS.          b. SINGLE BAR LATERAL PTS.

## Figure 2: Dimensions, Instrumentation, and Measurement Points

erations at enough points on a structure to define its deformation shape. It is sufficient to measure transfer functions relating the impact at all points to the response at one point, or to measure transfer functions relating the response at all points to the impact at one point. Because of the limited time available for the tests, transfer functions were measured using only the nineteen selected points illustrated in Figure 2.

For the plunge mode test, transfer functions were measured between the impact force at points 1 through 9 in succession and the plunge response measured by the accelerometer mounted at point 3. (Because it was submerged, point 10 was not struck by the hammer; it was assumed to be stationary.) Note that point 5 is on the top rack section and point 6 is on the lower section.

For the single bar modal test, transfer functions were measured between the lateral impact force at points 11 through 19 in succession and the lateral response measured by the accelerometer mounted at point 13. Lateral motion of a single bar was tested because relatively massive and stiff end supports inhibit lateral motion of the trashrack as a whole.

## Vibration Monitoring During Pump-Turbine Operation

The accelerometers and the strain gage mounted on the east bay trashrack (see Figure 2) were monitored during startup operation (as the lower reservoir was filled), during normal power generation (with the water surface in the lower reservoir at a normal level), and during pumping operation. During the startup and normal generation tests, wicket gate position was varied in 5 percent increments from 20 to 95 percent open, and the accelerometer and strain gage signals were recorded on analog tape and a strip-chart recorder. During steady operation of Unit 2 as a pump, the signals from the accelerometers were recorded on the strip-chart recorders.

## RESULTS AND DISCUSSION

## Modal Testing of the Draft Tube Trashrack

Figure 3 illustrates typical impact force and response acceleration signals obtained during the plunge mode testing of the east bay trashrack section. The frequency spectrum of the impulsive force (Figure 3a) is approximately uniform; the impulsive force delivers energy to the trashrack at all frequencies. The response acceleration (Figure 3b) is a complex waveform representing the superposition of the trashrack vibrations at discrete frequencies in various modes. Fourier analysis is used by the dynamic signal analyzer to decompose a complex waveform into its constituent frequencies. Response accelerations measured during the plunge mode and lateral mode tests were decomposed into their constituent frequencies and normalized by the impact force to obtain the transfer functions illustrated in Figure 4. The ordinate of these transfer functions represent acceleration per unit force. Because they are normalized, transfer functions obtained from impact forces and accelerations at various points on the trashrack indicate relative vibration amplitudes from which the mode shapes can be determined. The peaks in the transfer functions indicate the natural frequencies (vibration modes) of the trashrack.

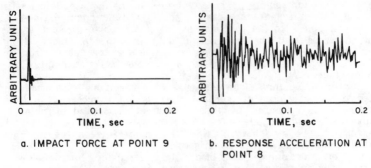

a. IMPACT FORCE AT POINT 9

b. RESPONSE ACCELERATION AT POINT 8

**Figure 3: Typical Impact and Response Signals**

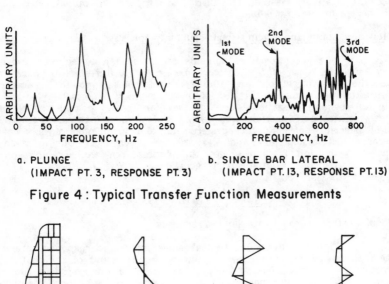

a. PLUNGE
(IMPACT PT. 3, RESPONSE PT. 3)

b. SINGLE BAR LATERAL
(IMPACT PT. 13, RESPONSE PT. 13)

**Figure 4: Typical Transfer Function Measurements**

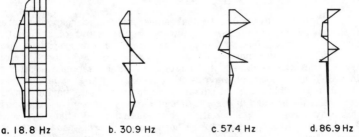

a. 18.8 Hz     b. 30.9 Hz     c. 57.4 Hz     d. 86.9 Hz

**Figure 5: Plunge Mode Shapes (Side View of Trashrack)**

The frequency and damping values associated with the eight transfer function peaks in Figure 4a, which represent plunge modes, are listed in Table 1. The natural frequency of the first plunge mode was calculated by assuming pinned end connections to be 21.5 Hz, which closely agrees with the measured value of 18.8 Hz.

TABLE 1:   Trashrack Plunge Modes

| Freq., Hz | 18.8 | 30.9 | 57.4 | 86.9 | 108 | 147 | 187 | 221 |
|---|---|---|---|---|---|---|---|---|
| Damping, % of critical | 2.6 | 1.8 | 0.5 | 0.8 | 0.9 | 0.6 | 0.7 | 0.2 |

In an edge view of the trashrack, Figure 5 illustrates the four plunge modes that correspond to the four lowest natural frequencies indicated in Figure 4a. For visualization, the vibration amplitudes are greatly exaggerated in Figure 5, and the top and side sections of the trashrack are not shown in b, c, and d. Because data was collected only from points 1 through 9 (see Figure 2) for the plunge mode, only the center vertical bar is deflected in Figure 5. All four mode shapes exhibit a discontinuity at the horizontal centerline of the rack because the upper and lower sections are not joined. At 19 Hz (Figure 5a) the upper and lower sections vibrate in phase as cantilever beams in the first bending mode. At 31 Hz, the upper and lower sections vibrate 180 degrees out of phase as cantilever beams in the second bending mode. At 57 Hz, the upper section vibrates as a cantilever in the third mode while the lower section vibrates as a beam supported at both ends in the second mode. At 87 Hz, the upper section vibrates as a cantilever in the fourth mode while the lower section apparently vibrates as a cantilever in the first mode.

The frequency and damping values associated with the three transfer function peaks in Figure 4b that agree most closely with computed values for the first three lateral bending modes of a single bar are listed in Table 2. In an edge view of a single vertical bar, Figure 6 illustrates the first three bending modes determined from the measured data. The transfer function peak near 700 Hz in Figure 4b probably corresponds to a twisting mode vibration of a single bar. The mode shape determined at 700 Hz did not look reasonable, apparently because too few measurements were taken to adequately describe the motion.

TABLE 2:   Single Vertical Bar Lateral Bending Modes

| Mode No. | Computed Freq. Hz | Measured Freq. Hz | Measured Damping % of Critical |
|---|---|---|---|
| 1 | 144 | 135 | 0.6 |
| 2 | 398 | 369 | 0.1 |
| 3 | 780 | 774 | 0.4 |

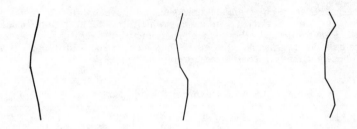

a. 1st MODE, 135 Hz    b. 2nd MODE, 369 Hz    c. 3rd MODE, 774 Hz

Figure 6 : Lateral Bending Modes of Single
Vertical Bar

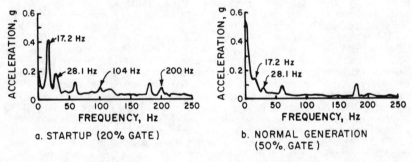

Figure 7 : Plunge Accelerometer Data

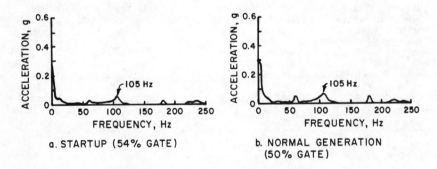

Figure 8 : Lateral Accelerometer Data

Vibration Monitoring During Operation

The vibrations measured during the operating tests were all relatively small, with no locked-in vibrations observed. Figure 7 compares plunge accelerometer spectra obtained during the startup test and during the normal generation test. The spikes at 17.2 and 28.1 Hz correspond to the lowest two plunge modes at 18.8 and 30.9 Hz, with the frequencies slightly reduced because of the added mass effect of the water on the submerged trashrack. The spikes near 100 and 200 Hz in Figure 7a probably correspond to the fifth and eighth plunge modes at 108 and 221 Hz. The spikes at 60 and 180 Hz are due to electrical noise. The spectra presented in Figure 7 show the largest accelerations at 17.2 and 28.1 Hz that were measured during the startup and normal generation tests. Not shown in Figure 7 was an acceleration of 0.31 g at 53.6 Hz which corresponds to the third plunge mode at 57.4 Hz and was measured during the startup test with the wicket gates set at 63 percent open.

Figure 8 compares lateral accelerometer spectra obtained during the startup and normal generation tests. The 104 Hz spike presumably corresponds to the first lateral mode of a single vertical bar, and is in reasonable agreement with a theoretical calculation of 115 Hz which includes the added mass effect of the water. The accelerations of about .06 g were nearly the same in both the startup and normal generation tests.

Table 3 summarizes the data obtained from the accelerometers during operation of Unit 2 as a generator. This table includes displacement values obtained by twice integrating the acceleration spectra and bending stress values estimated from the displacement values and assumed bar deformation shapes. A conservative value of maximum bending stress in the vertical bars, obtained by adding the maximum RMS values from the startup test for the first three plunge modes and multiplying by 1.4 to obtain a peak value, is 3900 lbf/in$^2$ which is far below the fatigue limit of typical steel.

The strains measured with the strain gage were small during all of the tests. The largest individual strain value from all the recorded data corresponded to a lateral bending stress of 1200 lbf/in$^2$ in the vertical bar.

Strip-chart recordings of the signals from the two accelerometers during steady pumping showed no measurable vibration.

Trashrack Anchorage Loads

The maximum drag force on a trashrack section was estimated to be approximately 600 lbf. The total dynamic streamwise load on a trashrack section due to first and second mode plunge vibration was estimated from the measured acceleration data and the plunge mode shapes. Using the measured acceleration, the acceleration at the center of each horizontal bar was scaled from the mode shapes. Pinned end connections were assumed in deriving the bending profiles for the horizontal bars. Newton's second law was applied to each horizontal bar individually and the computed forces were added, to obtain the total dynamic force. The maximum peak total dynamic force derived from startup test data was estimated to be 3600 lbf. The maximum peak total dynamic force derived from normal generation test data was estimated to be 1300 lbf. Although the vibrations measured

TABLE 3:  Vibration Monitoring Data

| Accelerometer | Freq. Hz | Accel. g | Maximum RMS Values Disp. inches | Bending Stress, lbf/in$^2$ Hor. Bar | Ver. Bar |
|---|---|---|---|---|---|
| **STARTUP TEST:** | | | | | |
| Plunge | 17.2 | 0.42 | .015 | 650 | 1700 |
| Plunge | 28.1 | 0.18 | .002 | 100 | 550 |
| Plunge | 53.6 | 0.31 | .001 | 50 | 550 |
| Lateral | 104 | 0.059 | .00005 | -- | 25 |
| **NORMAL GENERATION TEST:** | | | | | |
| Plunge | 17.2 | 0.14 | .005 | 200 | 550 |
| Plunge | 28.1 | 0.072 | .0008 | 50 | 200 |
| Lateral | 106 | 0.065 | .00006 | -- | 30 |

during the generation tests were relatively small, the resulting dynamic load estimates were approximately six times the steady drag force during the startup test and two times the steady drag force during the normal generation test. Taking the sum of the drag force and dynamic force, the estimated total maximum streamwise loads on the east bay trashrack were 4200 lbf during startup and 1900 lbf during normal generation.

## Vortex-Shedding Frequencies

The vortex-shedding frequencies of the 5/8-inch by 3-inch vertical bars were estimated using 0.22 as the Strouhal number (Hoerner, 1965). The Strouhal number is defined as follows:

$$\$ = \frac{fh}{V}$$

where f is the vortex-shedding frequency, h is the bar dimension perpendicular to the flow, and V is the approach velocity. The approach velocity necessary for vortex-shedding excitation of the first lateral mode of a single vertical bar is approximately 24 ft/sec which is six times the maximum average approach velocity of 4 ft/sec. The lateral vibrations recorded at this frequency must have been caused by random buffeting. The approach velocities necessary for vortex-shedding excitation of the first two plunge modes are both less than 4 ft/sec, implying that locked-in response of these two modes is possible although no locked-in response was observed. The damping values of these two modes (Table 1) are nearly identical to those measured for the first two plunge modes of the Raccoon Mountain trashrack (Vigander and March, 1974). In that study, first mode locked-in vibrations were observed but were inconsequential in magnitude, which may indicate that damping equal to two percent of critical is relatively large for a trashrack. Apparently, relatively large damping and the observed non-uniformity in space and time of the velocity distribution through the Hiwassee trashrack prevented the locked-in condition from being obtained.

## CONCLUSIONS

The results suggest that the previous trashrack failure at Hiwassee Dam was due to a combination of steady drag force, and plunge mode dynamic force caused by turbulent buffeting. The dynamic forces measured were not large in absolute terms, but they were large relative to the steady drag force. If the original trashrack anchorage was designed solely on the basis of expected drag force, it may have been inadequate. Also, because the tested trashrack had been modified for increased lateral stability, it is possible that the original trashrack experienced more severe vibrations.

The reinstalled trashrack survived the low tailwater startup tests that were shown to subject the trashrack to streamwise forces more than twice as large as those experienced during normal power generation. On this basis, additional modifications to strengthen the trashrack supports and anchorage were considered to be unnecessary.

## REFERENCES

Crandall, S., S. Vigander, and P. March, "Destructive Vibration of Trashracks Due to Fluid-Structure Interaction," ASME Journal of Engineering for Industry, November 1975, pp. 1359-1365.

Fiske, R., and C. Daly, "Trashrack Problems and Remedies at Muddy Run Pumped Storage Generating Plant," ASCE National Water Resources Engineering Meeting, Phoenix, Arizona, January 11-15, 1971.

Hoerner, S., Fluid-Dynamic Drag, Midland, New Jersey: (published by author), 1965.

Louie, D., and A. Allen, "Trash Rack Studies - Seneca Pumped-Storage Plant," ASCE National Water Resources Engineering Meeting, Phoenix, Arizona, January 11-15, 1971.

Vigander, S., and P. March, "Trashrack Vibration Studies, Raccoon Mountain Pumped Storage Project," TVA Engineering Laboratory Report No. 43-47, October 1974 (Revised March 1979).

# MODEL/PROTOTYPE COMPARISONS OF
# TRASHRACK VIBRATIONS

by

Patrick March[1]
Member, American Society of Civil Engineers
and
Svein Vigander[2]
Member, American Society of Civil Engineers

This paper compares results from vibration tests on a half-scale trash-rack model to results from vibration tests in the field on a full-scale trashrack. Modal frequencies and relative damping values were in good agreement between model and prototype.

## INTRODUCTION

Trashracks for the Tennessee Valley Authority's Raccoon Mountain Pumped Storage Plant are placed at the lower pool discharge structure for protection of the hydraulic machinery during pumping. Because of flow reversals and relatively high velocities in the proto-type and reports of trashrack failures in other pumped storage systems, including TVA's Hiwassee Unit No. 2, a half-scale model study of a typical trashrack unit was conducted. This paper reviews the results of the half-scale model study and reports the results of preliminary vibration measurements on the Raccoon Mountain trashracks during pumping operation.

## DESCRIPTION OF PROTOTYPE

The Raccoon Mountain Pumped Storage Plant is located near Chattanooga, Tennessee. Four 385 MW reversible pump-turbines operate between Nickajack Reservoir and an upper pool which is about 1000 feet above, as shown in Figure 1. The trashrack structure covers an area 74 feet wide by 55 feet high and consists of 20 individual racks stacked five high in each of four columns. The racks are bolted into slots and supported by streamlined concrete piers, as shown in Figure 2.

[1]Mechanical Engineer, Water Systems Development Branch, Tennessee Valley Authority, Norris, Tennessee 37828.
[2]Group Head, Water Systems Development Branch, Tennessee Valley Authority, Norris, Tennessee 37828.

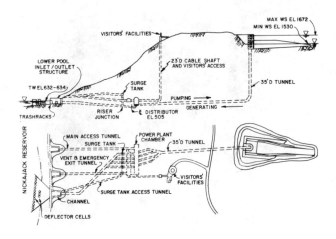

Figure 1:   Overall Plant Layout

        The individual racks are 15 feet wide by 11 feet high and
are fabricated from stainless steel.   Typical bar dimensions are given
in Figure 3.   Vibration damping is provided by chloroprene rubber
bearing pads between the racks and by hydraulic baffle plates welded
onto the racks (see Figures 2 and 3).   Flow deflectors were constructed in the discharge channel to distribute the flow uniformly
through the trashracks during generating operation.   Results from a
1:40 scale model study, supported by qualitative observations in the
field, indicated a uniform velocity distribution at the trashracks during
both pumping and generating.


## HALF-SCALE MODEL STUDY

        A half-scale model of a typical trashrack unit was con-
structed and tested during the design phase of the Raccoon Mountain
Project [Vigander and March, 1979].   A half-scale model of the original
trashrack design, which used circular trashrack bars, was clamped
into place within a steel support frame.   Mode shapes, frequencies,
and damping values were measured in air using accelerometers and an
electromagnetic shaker.   A lock-filling culvert at TVA's Melton Hill
Project was used as a water tunnel for hydrodynamic tests with the
half-scale trashrack.   "Locked-in" vibrations were observed over a
wide range of velocities, and fractures occurred in 17 bars after less

Figure 2: Prototype Trashracks and Support Piers

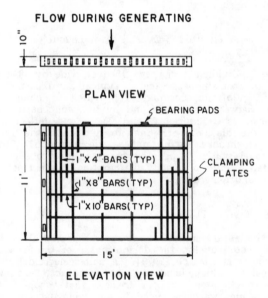

Figure 3: Design of Individual Prototype Trashrack Unit

than 30 minutes of testing. The "locked-in" vibrations were in line with the flow, in modes with very low damping, and at frequencies corresponding to second harmonic excitation from the drag component of the unsteady hydrodynamic force [Crandall, et al., 1975; Vigander and March, 1979; March and Vigander, 1980].

The trashrack design was modified to reduce the destructive vibrations experienced during the flow tests. Rectangular bars were used to stiffen the rack in the flow direction and to reduce the amount of energy transferred from the water by vortex-shedding. A half-scale model of the modified design was constructed, clamped into the support frame, and shaker-tested to determine mode shapes, frequencies, and damping values. The rectangular-bar trashrack was also tested in the lock-filling culvert over a velocity range from 0 ft/s to 18 ft/s. Results from these tests are discussed along with proto-type results in a subsequent section.

## FIELD TESTS OF TRASHRACK VIBRATIONS

The four top-most trashracks protrude about one foot above the water surface at normal operating levels. Two accelerometers were cemented in the center of the top beam for one of the racks to measure heave (vertical) vibrations and plunge (parallel-to-flow) vibrations. Two types of vibration measurements were taken. An instrumented impact hammer and 2-channel dynamic signal anaylyzer were used to determine transfer functions for the trashracks in still water. Frequencies and damping values for heave modes and plunge modes were obtained from the transfer functions. The heave and plunge accelerations were also monitored during 1-unit and 2-unit pumping operations, and the acceleration spectra were recorded. Pumping flow rates were based on differential head measurements from the Winter-Kennedy taps.

## RESULTS AND DISCUSSION

Mode shapes for heave modes of the half-scale trashrack model are presented in Figure 4, and plunge mode shapes are presented in Figure 5. (A detailed explanation of the nomenclature used to categorize mode shapes appears in Vigander and March [1979]). Transfer functions for impact tests on the prototype trash-racks are presented in Figure 6. Modal frequencies and damping values in the prototype were obtained by curve-fitting the measured transfer functions. The model and prototype results are compared for the heave modes in Table 1 and for the plunge modes in Table 2. Note that the model results correspond to measurements taken in air, before installation of damping devices, and the prototype results correspond to measurements taken in water, with damping pads and hydraulic damping plates installed. Frequencies measured for the half-scale model have been converted to prototype frequencies by dividing by two, due to the geometric and material similarity [Vigander and March, 1979].

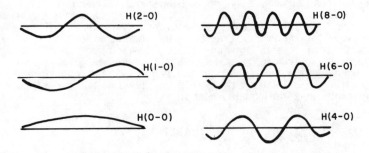

Figure 4:  Heave Modes for Half-Scale Trashrack

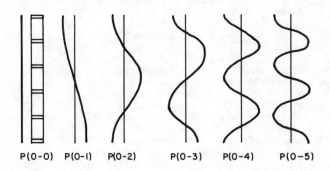

Figure 5:  Plunge Modes for Half-Scale Trashrack

TABLE 1

HEAVE MODE RESULTS

| Mode Designation | Model Results Frequency (Hz) | Damping (%) | Prototype Results Frequency (Hz) | Damping (%) |
|---|---|---|---|---|
| H(0-0) | 16.7 | 3.8 | 12.9 | 1.4 |
| H(1-0) | 39.9 | 2.5 | 28.6 | 1.2 |
| H(2-0) | 59.8 | 0.2 | 60.9 | 1.0 |
| H(4-0) | 118.0 | 0.1 | 118.8 | 0.9 |
| H(6-0) | 171.0 | 0.2 | 176.5 | 2.8 |
| H(8-0) | 230.5 | 0.1 | 249.7 | 2.6 |

TABLE 2

PLUNGE MODE RESULTS

| Mode Designation | Model Results Frequency (Hz) | Damping (%) | Prototype Results Frequency (Hz) | Damping (%) |
|---|---|---|---|---|
| P(0-2) | 59.5 | 0.10 | 59.4 | 0.2 |
| P(0-3) | 136.5 | 0.05 | 133.8 | 2.3 |
| P(0-4) | 251.0 | 0.04 | 245.3 | 1.5 |
| P(0-5) | 385.0 | 0.03 | 376.7 | 0.9 |

A variety of heave modes and plunge modes were observed in the half-scale trashrack model and in the prototype. Damping values measured for the prototype trashrack were greater than the damping values measured for corresponding modes in the half-scale model, as would be expected because of the additional damping from damping pads, damping plates, and hydraulic damping in the prototype. Only the even-numbered heave modes are readily discernible from the transfer function due to the location of the accelerometer on a nodal point for odd-numbered heave modes (see Figure 4). Modal frequencies predicted from the half-scale trashrack model correspond closely to measured frequencies for the plunge modes and higher-order heave modes. Modal frequencies for the first two heave modes are lower in the prototype than in the model, presumably due to added-mass effects [Blevins, 1979].

During flow tests, the half-scale trashrack's response corresponded to low-level random buffeting over much of the velocity range. "Locked-in" vibrations at 30 Hz (model), indicating the (0-0) heave model, were observed for approach velocities from 5.3 ft/s to 6.3 ft/s. "Locked-in" vibrations at 116 Hz (model), indicating the (0-2) plunge mode, were observed at an approach velocity of 11.2 ft/s, and "locked-in" vibrations at 267 Hz (model), indicating the (0-3) plunge mode, were observed for approach velocities from 16.0 ft/s to 16.8 ft/s. With either the damping pads or damping plates installed in the model, as subsequently incorporated in the prototype trashracks, no "locked-in" vibrations were observed.

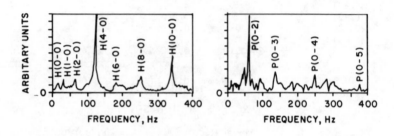

A. Heave Mode Results      B. Plunge Mode Results

Figure 6: Transfer Functions from Impact Tests
on Prototype Trashrack

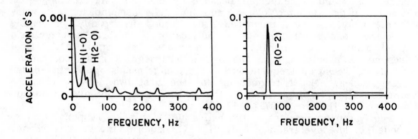

A. Heave Mode Results      B. Plunge Mode Results

Figure 7: Trashrack Acceleration Spectra for
Two Unit Pumping Operation

Acceleration spectra for the prototype trashracks were measured for pumping flow rates of 3,700 ft³/s (one unit) and 7,400 ft³/s (two units). The measured vibrations were negligible when only one unit was pumping. Trashrack acceleration spectra with two units pumping, corresponding to an approach velocity of 2.5 ft/s, are presented in Figure 7. The spectra include small peaks due to low-level buffeting response in the (1-0) and (2-0) heave modes and in the (0-2) plunge mode. The acceleration due to "locked-in" (0-2) plunge mode vibrations in the half-scale trashrack model, measured during flow tests before the installation of damping pads or plates, were about 60 times larger than the prototype (0-2) plunge mode acceleration shown in Figure 7B.

## CONCLUSIONS

There was excellent agreement between the modal frequencies and relative damping values measured in the half-scale trashrack model and in the prototype trashrack. This suggests that useful design information could have been obtained using a smaller scale model, which would have reduced the model construction costs and the flow requirements for hydrodynamic tests. Also, modal analysis using impact tests, as described by Schohl and March [1981], greatly reduces the time required for vibration testing compared to the swept-sine technique used with the half-scale trashrack. Additional vibration monitoring tests will be conducted at a variety of operating conditions in pumping and generating to document the performance of the Raccoon Mountain trashracks.

## REFERENCES

Blevins, R. D., Flow-Induced Vibration, New York: Van Nostrand Reinhold Company, 1977.

Crandall, S., S. Vigander, and P. March, "Destructive Vibration of Trashracks Due to Fluid-Structure Interaction," ASME Journal of Engineering for Industry, November 1975, pp. 1359-1365.

March, P., and S. Vigander, "Some TVA Experiences with Flow-Induced Vibrations," in Naudascher and Rockwell, eds., Practical Experiences with Flow-Induced Vibrations, Berlin: Springer-Verlag, 1980.

Schohl, G., and P. March, "Modal Analysis and Vibration Monitoring of Trashrack Structures, Hiwassee Project, Unit No. 2," Tennessee Valley Authority, Water Systems Development Branch, Report No. WR28-1-5-100, June 1981.

Vigander, S., and P. March, "Trashrack Vibration Studies, Raccoon Mountain Pumped Storage Project," TVA Engineering Laboratory Report No. 43-47, October 1974 (revised, March 1979).

# GARVINS FALLS HYDROELECTRIC EXPANSION PROJECT

R. P. Brecknock, P.E.*
J. E. Lyons, P.E.*

## ABSTRACT

The Garvins Falls Hydroelectric Replacement Project was a fast tracked installation of two 3.375 MW turbine/generators. These 2.75 meters Allis-Chalmers tube turbines (Units 1 & 2) replaced two obsolete units retired in 1971. Including generation from two existing units, annual station production will approximate 51,000 MWH, representing an increase of 20,000 MWH.

Prior to 1980 the Garvins Falls Station consisted of a granite block gravity dam with an ogee spillway, a power canal with head gates, two fixed blade vertical propeller turbines driving generators of 2.4 MW (Unit #3) and 3.2 MW (Unit #4), an access deck and vacant powerhouse adjacent to the operating structure, and a tailrace. The gross head fluctuated from 23 to 33 feet.

Preliminary investigations determined that only the access deck and empty structure needed alterations to allow installation of new turbine/generators. Economic analysis of various alternates indicated that the twin 2.75 meter units were the optimum size. Based on these investigations and a site feasibility study, this project was awarded funding from the Department of Energy for 15 percent of its estimated costs.

Cofferdam construction began in August, 1980. Both the upstream and downstream coffers were required to allow generation from Units #3 and #4. The design that accomplished this was a series of braced steel soldier beams with steel panels stacked in between. Reinforced concrete foundations were fixed to ledge with prestressed rock anchors. The upstream coffer provided access during construction to the operating powerhouse while the downstream structure supported temporary power and control cables from the existing units. Additionally, the steel panels could be reused in lieu of stop logs for all four units.

After both cofferdams were placed, the superstructure of the empty powerhouse was razed. Blasting and excavation of the unreinforced granite masonry foundation and underlying ledge followed and proceeded throughout the winter. Approximately 4,000 cubic yards of material was removed by April, 1981.

*Robert P. Brecknock - Staff Engineer - Public Service Co. of N.H.
*John E. Lyons - Manager, Supplementary Energy Sources - Public Service Co. of N.H.

Concreting began in May, 1981, with placement of the intake transition foundations and was completed in November, 1981 after the final
roof section was placed. Approximately 3,700 cubic yards were used.
Both cofferdams were removed by October, as soon as the structure
could withstand full hydrostatic load.

Equipment installation had proceeded coincident with powerhouse
construction to provide as compact a schedule as possible. Unit #1
first generated energy on December 18, 1981; Unit #2, twelve days
later. Both units were placed in service before January, 1982.

## INTRODUCTION

Since 1977, several factors have combined to stimulate development
or redevelopment of small hydroelectric projects. Among these, which
include renewed emphasis on using non-consumable resources, escalating
costs and increasingly vulnerable supply of foreign oil, and environmental legislation, improved economics is paramount. Recent tax credit legislation and increased costs of all alternate sources of
electricity result in capital expenditures in small hydroelectric
facilities today providing a relatively greater return compared to
past investments to develop this resource. Simply stated, all energy
has become more valuable.

Considering this background, Garvins Falls Hydroelectric Station
on the Merrimack River in Bow, New Hampshire, which previously had
more than 5.6 MW in capacity, became a prime target in a search for
redevelopment possibilities. Public Service Company of New Hampshire,
the owner, made a decision in 1979 to study the feasibility of a capacity addition to this site.

The site was first utilized for navigation in 1813 when a dam,
canal, and four locks were constructed. Mechanical power was inciden-
tally provided to a grist and saw mill. The present granite block
gravity dam was constructed in 1901 as part of an extensive redevelopment of the site. By 1906 a total of 6 horizontal 650 KW turbine
generators were in place at the end of an enlarged canal. An indoor
substation was constructed and interconnected to the station in 1915.

In 1925, two thirds of the generating facilities and powerhouse
were removed and replaced with two vertical fixed propeller units of
2.4 MW and 3.2 MW. The remaining two horizontal units lasted until
their retirement in 1971.

Prior to 1980-81, the station consisted of the following structures and equipment. A granite block dam having a rubble masonry core
and an ogee shaped overflow spillway, consists of a 475 feet long low
crest section, USGS el. 216.85, plus a high crest section, USGS el.
218.85, 75 feet long. Removable wood flashboards provide a maximum
pond elevation of 219.85 when flow permits. A headgate house that is

situated adjacent to the dam allows water through six 10 feet wide by
12 feet high wood gates into the canal.

A rubble masonry lined canal approximately 500 feet long, 18 feet
deep, and 65 feet wide serves as an intake channel. A small waste
gate is located at its downstream end, adjacent to the 1925
powerhouse. This powerhouse has a concrete substructure and brick
masonry/structural steel frame superstructure. Plan dimensions are 68
feet by 31 feet. The remaining portion of the 1901 powerhouse was 65
feet by 30 feet, and located between the indoor substation on the west
river bank and the 1925 facility. It had a granite masonry substructure below a brick superstructure. An upstream bridge deck provided
access to the operating units.

Before redevelopment, energy produced at the station was considered "run-of-river" although the 250 acre pond could facilitate
some daily ponding. River flows greater than 2,863 cfs, the combined
maximum discharge of both units, were spilled. Flowage at the site
did fall below the minimum discharge of one unit (500 cfs), but con-
denser cooling water requirements for a coal fired steam station
downstream require a minimum continuous release of 485 cfs.

Drainage area of the Merrimack River behind Garvins Falls Dam is
approximately 2,427 square miles. As is characteristic of southerly
flowing rivers in New Hampshire, flows are generally highest from
mid-March to mid-June and moderately high in early autumn. Power
developments upstream provide no flow regulation. Flood control dams
regulate only during relatively high flowage.

Under these conditions, historical maximum, median and mean flows
at the site were estimated to be 122,000 cfs, 2,650 cfs, and 4,190
cfs, respectively. Average annual generation was 30,600 MWH.

## EVALUATION

Public Service Company hired Tippetts-Abett-McCarthy-Stratton
Engineers and Architects, New York (TAMS) to study the feasibility of
expanding existing facilities, and if warranted, to prepare a proposal
for submission to the Department of Energy requesting a construction
grant under the Small Hydroelectric Demonstration Projects Program.

Orientation of any new unit(s) was constrained by both the vacant
powerhouse location and the requirement that existing facilities in
use remain operational. Horizontal shaft tube turbines with
adjustable blades were chosen for initial study, and subsequently
used. This choice was based on the variable head (23 to 33 feet) at
the site, less civil work associated with horizontal versus vertical
units, and the apparent domestic availability of "standard" units.

The new powerhouse was located where the remaining portion of the
1901 structure had been. Centerline elevation of the new units was

considerably lower than that for the old equipment, requiring removal
of the existing floor and excavation below the existing substructure.
Also, the tube turbines required more length.  These facts resulted in
the decision to completely remove the existing 1901 powerhouse and
deck, except for the substructure on either side, a concrete and granite masonry structure adjacent to the 1925 powerhouse and a rubble
masonry retaining wall below the indoor substation.

Intake walls for all four units were in the same vertical plane.
Since the new powerhouse is approximately 80 feet long, it extended
farther downstream than the old structure.  Excavation was required
upstream of the intake, within the powerhouse area and in the
tailrace.  Otherwise, all existing facilities, including the headgate
house and canal, remained unchanged.

All preliminary project estimates were based on reinforced
concrete construction.  A description of the actual design is provided
later in sections headed Phases I and II.

Preliminary economic analysis by TAMS indicated the optimum capacity addition to be two 2.75 meter turbine/generators, each rated at
2.95 MW at 30 feet net head.  Estimated construction costs (without
grants) were used with the estimated annual incremental energy of
20,000 MWH produced by these units to establish or disprove viability.
Two different analytical methods were used.

First, additional energy produced at Garvins Falls was assumed to
replace an equal amount of oil fired generation from a PSNH
fossil-fueled station.  Since this would occur throughout each year, a
current average fuel cost was escalated through time to obtain a levelized value for saved fuel.  This was added to annual fixed and O&M
costs for a 5.9 MW thermal generational addition and compared to estimated annual fixed and O&M costs for the Garvins Falls project.  All
costs were in levelized values over an expected 50 year project life.
A positive annual savings resulted, indicating the project's economic
feasibility when compared to the alternate new source.

Second, incremental energy from the project was assumed to replace
generation from portions of all components in PSNH's system.  Avoided
costs were therefore comprised of various percentages of costs per KWH
for nuclear, coal, oil, gas and purchased generation.  The estimated
project capital expenditure was used with PSNH's current cost of
capital, all pertinent taxes, the proper method of depreciation, and a
50 year expected life to estimate the actual annual revenue required
to support the investment.  This value was then compared on a cumulative present worth basis for each year to the total avoided energy
costs (which change each year depending on the planned energy mix)
plus any yearly capacity credits due PSNH.  The year that cumulative
present worth (cpw) benefits (avoided costs plus capacity credits)
equal cpw costs (revenue requirements) is when the project has paid
for itself.  The 50 year benefit cost ratio of 1.44 for this project
indicated economic feasibility when compared to estimated savings.

   Two equal size units were selected because: 1) equipment costs
for two small horizontal units equalled costs for one larger vertical
one, but civil costs were less; and, 2) water utilization at a run-of-
river site can be efficiently accomplished with four units of similar
capacities. Therefore, any savings in fabricating identical equipment
should be realized.

   Environmental impacts of the project were considered extremely
small, having negligible effect on costs. There would be some short
term turbidity and sedimentation from cofferdam construction, some
additional downstream migrant fish mortality, although tube turbines
are more efficient at safely passing fish, and some increase in noise
during construction. Operational impacts were considered non-existant.

   Based on the above considerations, the project was judged as
viable and a decision to submit a proposal to DOE was made. A
cooperative agreement between PSNH and DOE was signed in June, 1980
providing the Company a $924,750 construction grant in return for
collecting and disseminating data concerning construction, operation,
and maintenance costs, engineering characteristics, and maintenance
requirements of the new facility during construction and for two years
of operation. This grant further enhanced project economics.

View from roof of Indoor Substation looking upstream showing completed
project.

The following sections describe design and construction aspects of the Garvins Falls Project. Phase I defines the period from FERC license requirements to excavation complete. Phase II starts with initial concreting and finishes with the units becoming operational.

## PHASE I

Since 1972, Garvins Falls Station has operated under F.E.R.C. regulations as a 30 year licensed project. Consequently, the major revisions proposed required submission and acceptance of an Application for Amendment to the license. Fortunately, since project changes utilized an existing dam, streamlined procedures for applications applied. The application was submitted on March 31, 1980. The amendment was received shortly before construction started five months later.

Other pre-construction requirements included N.H. Public Utility Commission approval, obtaining State and Federal dredge and fill permits, obtaining a local building permit, and consulting with Federal and State environmental agencies. Meeting regulatory requirements for this project was a relatively simple process.

Design by TAMS commenced in January, 1980. Hydrologic considerations required that, in order to be minimized, cofferdam work be started and completed during mid-summer to mid-autumn. Both phases of construction would last well beyond this period, requiring the tailrace cofferdam to withstand seasonly high flows. The original water control scheme was to place a relatively large earth cofferdam across the tailrace and to close the canal headgates. This provided the most economical construction method but resulted in almost $1½ million of lost energy since both existing units would not generate during the nine month construction period.

A more economical solution with respect to both construction and operating costs was to place an earth dike across the tailrace and close the canal headgates while a long-term cofferdam was constructed. The earth cofferdam was then removed, allowing generation by existing units throughout normal and high water periods.

Early in the project's life a management decision was made to expedite construction, thereby reducing overheads and interest during this period, and starting the payback period as soon as possible. PROJECT/2*, a computer software system for integrated project schedule and cost control, was used as a planning and control technique to support this decision. The system uses critical path method techniques to monitor schedules and an earned value approach to monitor costs. It proved useful in identifying items or areas in both phases of the project requiring additional effort to realize the scheduled completion date.

Design of both upstream and downstream long-term cofferdams consisted of an L-shaped series of braced steel soldier beams spanned by steel panels. Reinforced concrete pads and foundation sills supported the braces and the beams and panels, respectively. Reinforced concrete end closures provided torsional strength and water seals. Steel wide flange soldier beams had sealing angles on their webs and were braced by either single or double diagonal pipe braces. These were either compression or tension braces depending on site constraints. Steel panels, made from wide flange sections welded to steel plate, were 5½ feet high. Perimeter sealing surfaces allowed them to be stacked to heights corresponding to maximum canal and tailwater elevations. All panels were to be reused in lieu of stop logs for either the new or existing units. Their lengths were governed by this constraint, but the soldier beams could be flexibly spaced to accomodate varying dimensions.

The major economical consideration of this scheme was panel reuse. However, the downstream structure provided support for temporarily relocated power and control cables from the existing units. These had been located within and above the 1901 powerhouse floor and had to be moved and reconnected prior to demolition but within the period that the earth dike was in place and the units were down. Additionally, the upstream cofferdam provided access for operation and maintenance personnel to the existing equipment.

Concurrent with project construction, PSNH took advantage of the dewatered tailrace area and constructed piers with stop log slots behind the existing units. This will greatly facilitate future maintenance and inspection of these units since the tailrace will, upon project completion, hopefully not be dewatered for another 50 years.

Cofferdam construction, including some minor ledge excavation , lasted three months, with demolition commencing thereafter. After the superstructure had been razed and before mass excavation began, existing remaining walls were shored using horizontal walers and inclined rock anchors. Also, the set of compression braces on the upstream soldier beam nearest the forebay wall was removed and replaced with a 12" round pipe, braced against the intake canal wall 46 feet away. This was done at the contractor's suggestion to facilitate continuous excavation and was accomplished while the beam was loaded.

Precautions taken to reduce potentially harmful blast effects included extensive line drilling to isolate existing structures plus continuous seismographic monitoring of peak particle velocities at various strategic points throughout the site. In addition to mitigating structural damage, a primary concern was how the existing unit would be affected. If one was down during blasting, damage potential on the unlubricated face of the Kingsbury thrust bearing was a concern.

The contractor used explosives consisting of 40% dynamite, with each hole loaded using 1/8 to 5 pounds depending on the proximity to

critical structures or equipment. Throughout ledge and masonry excavation in the intake and powerhouse areas, no significant damage
occurred to buildings or equipment (some as close as 10 feet to the
cut). Leakage through the cofferdam did increase however, but was
stopped by a combination of welding horizontal joints and bolting
panels to vertical sealing surfaces.

The most difficult blasting operations involved destroying masonry
walls and arches, which were too well constructed to economically use
mechanical demolition methods. Charges had to be placed in locations
that prevented dissipation of the explosive forces through mortar
joints. Large charges could not be used since the majority of this
type of excavation occurred in the rubble masonry forebay wall, with
existing walls to remain on either side of the 30 foot deep cut.

Approximately 4000 cy of ledge and masonry was removed in a three
and a half month period during the winter.

Interior view of powerhouse showing speed increasers in foreground,
turbine covers beyond and control panel on upper level.

## PHASE II

By March, 1981, near the end of the fixed price excavation
contract, the decision was made to overlap design and construction,
then proceed on a cost-plus contractual basis. This provided the
best method for attaining the scheduled completion date of November
13, 1981. On-site supervision was augmented to provide additional
coordination between PSNH, as construction manager, and the
contractor. Construction then proceeded as design drawings became
available.

Equipment for each unit supplied by Allis-Chalmers consisted of
a hydraulically controlled vertical roller intake gate, a steel water
passage (intake transition, intake ring, and draft tube), to which the
gate guides are connected, an adjustable four blade propeller turbine,
a speed increaser, a blade positioner, hydraulic power supply system,
control panel, generator and miscellaneous support equipment. (Units
are rated @ 3.25 MW each at 30 feet head.) Equipment from other
suppliers included the switchgear, transformer, and bridge crane.

Concreting began in May, 1981, with placement of the intake transition support structures. Foundation work continued until receipt of
embedded parts from Allis-Chalmers. The ten disassembled pieces per
unit (2 intake transition halves, four upper and four lower draft tube
sections) were preassembled into three sections, then lowered from the
staging area along with the intake ring onto their foundations.
Roughly one thousand manhours of welding and grinding were required to
install each water passage before and during concrete embedment.

Intake gate guide slots were aligned then tack welded to the
intake transition sides. Twelve hours later both sets were out of
alignment because of relative temperature induced movement between the
top and bottom of the plate steel. The guides were cut, realigned,
and embedded independently from the transition. Joints were then
welded as required and air spaces left behind them for concrete protection were grouted. Otherwise, embedment proceeded as designed.

As previously mentioned, centerline elevation of the 1901 units
was considerably higher than that for the tube units. Consequently,
the new draft tube floor level is 35 feet below maximum tailwater
elevation. (Generator floor level is 4 feet below minimum tailwater.)
If both draft tubes are dewatered during high tailwater, hydrostatic
uplift forces are not balanced by the building's weight. Three rows
of hollow-core prestressed rock anchors were used to assure stability.

The lower turbine setting allowed the new powerhouse roof level to
equal the 1901 access bridge elevation. It also provided access to
the existing powerhouse tailrace area, which had previously been economically unfeasible. The resulting roof design supported both vertical crane loads with T-beams and horizontal hydrostatic loads
through diaphram action of the slab.

Construction materials and methods for the roof were specified as cast-in-place reinforced concrete, posing a scheduling restraint. To meet the completion date, equipment installation and civil construction had to occur concurrently. Extensive shoring to support roof beams and slabs would have delayed equipment installation, and thus project completion, by at least two months. The support system devised to overcome this included high early strength concrete, corrugated steel decking used as permanent slab forms, and 7 feet wide beam shoring platforms supported by two center columns and the crane rails on either end. Two such structures were used to support two 2 feet by 6 feet by 56 feet beams until concrete had attained sufficient strength to withstand dead and construction loading. The platforms were reused for other beams as the roof deck and slab were placed. Shorter beams over non-critical areas were shored from the floor. Turbine, speed increasers and generators were placed and aligned during this period since the only obstructions on the generator floor were two support columns.

The 10 ton station maintenance crane arrived at the site while roof construction was proceeding. Rather than dismantling and reassembling the bridge, the entire structure was rotated 90° and lowered through an opening between reinforcement of two beams, then set on the crane rails. The flexible shoring system described above facilitated this operation.

After completion of the new intake wall and upstream portion of the roof, installation of trash racks and support structures, and placement of upstream and downstream stop panels, the construction area was rewatered. Prior to this an earth coffer was placed across the tailrace (again during low water) while upstream and downstream steel cofferdams were dismantled and ledge was excavated in the tailrace behind all four units.

Approximately 2,000 cy was removed to create a 12° slope in this area. This increased the tailrace efficiency behind the 1925 units, providing a total additional capability of 200 KW. Blasting was accomplished using primer cord because of stray ground currents created by an overhead 34.5 kV transmission line. This is the primary circuit into the State capital and had to remain in service. Stand-by crews remained available during the blasting period.

The project first generated energy on December 18, 1981. Both units were placed in-service before January, 1982.

Although a variety of union and non-union crews worked at various times throughout the project, no labor problems surfaced. Also, no serious accidents to personnel or structures occurred, even though the work was at times dangerous. Completion of the project only slightly behind schedule was a credit to all concerned.

*PROJECT/2 - software developed by Project Software and Development, Inc.

# SHAWMUT REDEVELOPMENT PROJECT

By Gerald C. Poulin[1], Member ASCE

- Abstract -

Central Maine Power Company's Shawmut Hydro Project is
located on the Kennebec River in the towns of Fairfield
and Benton, in Central Maine. The existing project
facilities were constructed early in this century and
they remain nearly unchanged today. The construction of
major upriver storage reservoirs during the 1950's
substantially improved river flows at the site. However,
a plant expansion could not be economically justified at
that time. Not until the dramatic fossil fuel price
increases of the 1970's did the redevelopment of Shawmut
become attractive. CMP has been joined in this project
by the Department of Energy. DOE has provided financial
assistance to CMP in exchange for project data that will
be used to promote future low head hydro developments.

The Shawmut Hydroelectric Project is located on Maine's
largest river, the Kennebec. The site is approximately
25 miles north of Augusta, the state capital. At
Shawmut, the Kennebec River has a drainage area of nearly
4200 square miles and is well regulated with an upriver
storage capacity of over 1 million acre-feet. The site
has had some form of development since the mid-nineteenth
century. Saw mills and grist mills flourished on the
site throughout the latter 1800's as falling water was
harnessed to provide mechanical power.

In 1912 the Shawmut Manufacturing Company undertook
redevelopment of the site. The present dam, headworks,
and powerhouse were constructed and 4 turbines were put
into service. Three of the wheels powered electric
generators, while the fourth provided mechanical power
for a pulp grinder. Expansion of the project facilities
continued until 1921 when the current arrangement of six
turbine/generators was established.

Central Maine Power acquired the site in 1924 and the
facilities have remained virtually unchanged since that
time.

[1]Mr. Poulin is currently Manager of Engineering for CMP.

The six existing units have a nameplate capacity of 4650 KW while utilizing 4200 cfs under 23 feet of gross head. The station currently produces over 41 million KWH annually.

CMP presently has a hydro electric generating system with a dependable capacity of over 300 MW. The 24 generating stations that make up this system are the result of continual site review and re-evaluation. Potential projects are periodically reviewed to assess the impact of new technological developments or changing economic conditions.

Spiraling oil costs which occurred during the mid 1970's prompted CMP to re-examine expansion of its Shawmut Station. Increased capacity at Shawmut had been envisioned since 1941. At that time the forebay was reconstructed and provisions were made for a future vertical unit. However, economic conditions were never such that a new unit could be economically justified.

Late in 1977 the Department of Energy announced a grant program to stimulate interest in low head hydro projects. The Program Research and Development Announcement (PRDA) made grants of up to $100,000 available on a cost sharing basis to conduct feasibility studies for small hydro projects.

CMP, in conjunction with Stone & Webster Engineering, submitted a proposal for a Shawmut feasibility study under the PRDA guidelines. On September 25th, 1978 the proposal was accepted and the Shawmut feasibility study was begun.

The study by Stone & Webster focused on two alternatives for redeveloping the Shawmut site; the so-called East-side and West-side schemes. A third option which called for replacing the existing units was quickly dismissed after preliminary examination. The East-side expansion called for a new powerhouse and tailrace to be constructed at the east end of the dam across the river from the existing facilities (see Figure 1). This location had no space restrictions but it did involve more complicated and expensive access problems. On the other hand, the West-side scheme was restricted by the existing provisions for redevelopment but was much more accessible to the utility grid and construction personnel (see Figure 2).

In order to evaluate the relative economics of each scheme, a detailed cost estimate was developed for each proposal. Table 1 is a breakdown of the estimates developed by the Stone & Webster feasibility study.

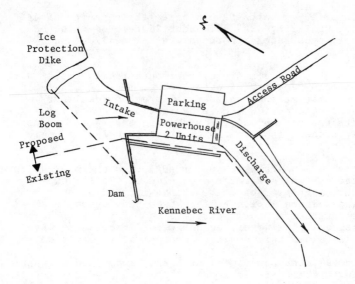

FIGURE 1 - EAST SIDE LAYOUT

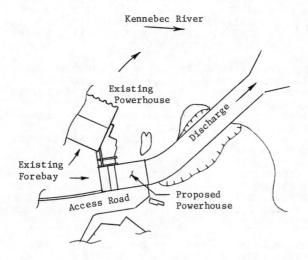

FIGURE 2 - WEST SIDE LAYOUT

- TABLE 1 -
SHAWMUT EXPANSION
PRELIMINARY COST ESTIMATE SUMMARY
(1979 $1000's)

| Item | West Side | East Side |
|---|---|---|
| Land | -- | -- |
| Powerhouse | 492 | 796 |
| Waterways | 543 | 1,231 |
| Turbine/Generator | 1,535 | 2,277 |
| Acc. Elec. Equipment | 313 | 318 |
| Aux. Plant Equipment | 42 | 53 |
| Roads | 10 | 81 |
| Substation | 45 | 68 |
| Transmission | 2 | 34 |
| Sales Tax, where applicable | 18 | 42 |
| DIRECT CONSTRUCTION COST | 3,000 | 4,900 |
| Omissions & Contingencies | 390 | 600 |
| Engring & Supervision | 650 | 1,000 |
| Legal-Environmental Administration & Overheads | 250 | 250 |
| Interest (AFUDC) | 200 | 310 |
| GROSS PLANT INVESTMENT | 4,490 | 7,060 |

Per unit costs were developed from the figures in Table
1.  Table 2 contains the per unit economic analysis for
the two Shawmut schemes.

- TABLE 2 -
SHAWMUT POWER PLANT EXPANSION
ECONOMIC COMPARISON

| | West Side | East Side |
|---|---|---|
| **Hydro** | | |
| Rated Capacity (KW) | 3,440 | 5,000 |
| Ave.Annual Energy (MWH) | 18,000 | 21,000 |
| Gross Plant Investment ($1000's) | 4,490 | 7,060 |
| Investment/Installed KW ($/KW) | 1,305 | 1,412 |
| Annual Cost ($1000's) | 898 | 1,412 |
| Annual Energy Cost (Mills/KWH) | 49.9 | 67.2 |
| **Fossil Alternative** | | |
| Gross Plant Invest.($1000's) | 798 | 1,160 |
| Annual Fixed Cost ($1000's) | 160 | 232 |
| *Annual Fuel Cost ($1000's) | 936 | 1,092 |
| Total Annual Fossil($1000's) | 1,096 | 1,324 |

Comparison
Annual Savings with
  Hydro ($1000's)                    198              (88)
Benefit/Cost Ratio,
  Hydro/Fossil                       1.22              .94

*Levelized cost (52 mills) over 50 years assuming 6%
annual increases and a present day cost of 24 mills/KWH.

The final Shawmut Report for the PRDA program was
submitted on May 4th, 1979.  As indicated in Table 2
the study showed that a West-side redevelopment
alternative was economically feasible.

Having identified an economical approach for Shawmut CMP
retained Kleinschmidt & Dutting of Pittsfield, Maine to
act as Architect/Engineer for the Shawmut Redevelopment
Project.  The redevelopment project can be divided into
four distinct phases:  licensing, design, construction,
and start-up.

The licensing phase of the project involved securing the
necessary local, state, and federal authorizations for
the project.  This portion of the project spanned a two
year period from the spring of 1979 until the start of
construction in June of 1981.  Table 3 contains a
compilation of the various permits and licenses that were
required for the Shawmut Project.

- TABLE 3 -

| Authorization | Agency | Application Date | Approval Date |
|---|---|---|---|
| FERC License | FERC | 8/04/80 | 1/05/81 |
| Certificate of Public Necessity | MPUC | 5/02/80 | 8/28/80 |
| Dredging Permit | COE | 1/14/81 | 6/03/81 |
| Water Quality Certificate | MeDEP | 10/09/80 | 10/14/80 |
| Building Permit | Town of Fairfield | 3/31/80 | 3/31/80 |
| Stream Alteration Permit | Me. Fish & Wildlife | 7/24/80 | 8/29/80 |

In addition to the normal licensing work, CMP also
prepared an application to DOE under the Program
Opportunity Notice II (PON II) program.  Unlike the PRDA
program which provided grant money to conduct feasibility
studies, the PON II program offered federal dollars for
the actual construction of low head hydro projects.
Twenty developments from across the country were selected
to serve as demonstration projects in exchange for
partial federal funding.

Information collected by the developers would be gathered by DOE and distributed to the public. DOE is providing $850,000 (approximately 15% of the projected $5.8 million Shawmut cost) in exchange for information regarding project design, material specification and procurement, construction, and operating experience.

The next phase of the project, project design, was actually begun at the same time as the licensing effort. Preliminary engineering developed conceptual plans to support the license applications. However, detailed design work did not begin until December of 1979.

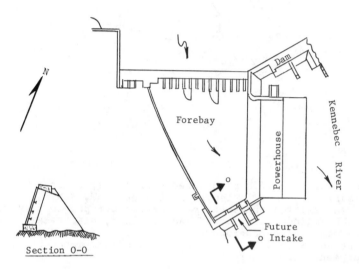

FIGURE 3 - EXISTING FACILITIES

As previously stated, the forebay of the existing project facilities contained provisions for expansion of the project (see Figure 3). Two unequal openings were provided in the south wall of the forebay during a 1941 rebuild. A 2400 KW vertical unit had been planned with the two openings to lead to a standard spiral scroll case. The Stone & Webster feasibility study proposed a pair of 2500 mm (1720 KW) tube type generating units to utilize the two openings. The compact tube turbine layout would most efficiently take advantage of the expansion provisions. Detailed examination by K & D revealed that slightly larger units (2750 mm, 2000 KW) could be accommodated by the existing forebay openings. These larger units became the basis for the project design.

The final design of the project consists of a powerhouse
structure to enclose the turbine/generators, a new
tailrace and tailrace training wall, new intake
facilities, an access bridge to the old powerhouse, and a
substation addition (see Figure 4).

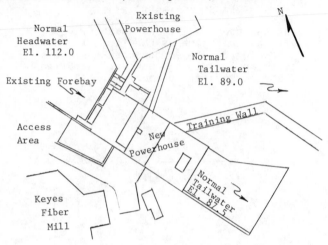

FIGURE 4 - SHAWMUT REDEVELOPMENT FACILITIES

The powerhouse is a reinforced concrete structure
measuring 70 feet in length by 50 feet in width. A
mezzanine level above the generator floor provides space
for the main control board and auxiliary electrical
equipment. The powerhouse also includes an overhead
crane and a roof hatch for handling major pieces of
equipment (see Figure 5).

The new intake facilities include a gatehouse, an intake
deck, and intake trashracks. The gatehouse encloses two
hydraulically operated sliding gates which serve as the
unit control gates. Since the turbines are fixed blade
and have no wicket gates, the hydraulic headgates are the
only means of flow control. However, they are designed
strictly as on or off type devices. The gatehouse also
houses an emergency gate and stop log slots which serve
as a back-up system to the main headgates. The emergency
gate is constructed in two sections. An electric chain
hoist suspends one section of the gate over the stop log
slot just upstream of each headgate. A shallow trench
interconnects the stop logs slots for both units and
allows the gate sections to be piggy-backed in a single
slot to form a full closure of the intake. In addition
to emergency protection, these gates also provide a means
for dewatering the headgates for maintenance functions.

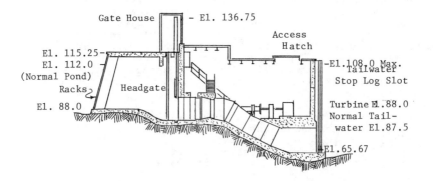

FIGURE 5 - LONGITUDINAL SECTION OF POWERHOUSE

Upstream from the gatehouse an intake deck provides a
small work area and accommodates a mechanical trash
rake.  The Berry trash rake is a hydraulically operated
clam shell type device that is manufactured in Berlin,
New Hampshire by the Cross Machine Company.  The rake is
designed for one man operation and it will clean the
entire rack surface, top to bottom.  The rake will even
clean debris from the racks while the units are operating.

An access bridge walkway, a trussed structure, spans some
60 feet between the old and new powerhouses.  This bridge
is highly insulated and will serve as a heating duct as
well as provide worker access to the old powerhouse.
Waste heat from the new generators will be circulated to
the old powerhouse as needed to maintain building
temperature when the old units are not operating.

Project design was sufficiently completed by the spring
of 1981 to allow a civil specification to be prepared.
Bids were requested for the civil portion of the project
in early May.  The civil contract was awarded to the
Cianbro Corporation of Pittsfield, Maine on May 27th,
1981.  Contractor mobilization and site clearing began
immediately.  On June 3rd construction of a cofferdam
started.

Following the site clearing and cofferdam work, the focus
of the construction work shifted to excavation activity.
At Shawmut bedrock was exposed in several locations in
the construction area. Unfortunately, the ledge at
Shawmut is relatively soft, fissile bedrock. The
somewhat fragile nature of the rock along with the close
proximity of the existing structures dictated cautious
rock removal. During presplitting and blasting particle
velocities were kept below 2 inches per second. In
addition, mechanical removal means were employed in the
immediate vicinity of existing structures. In all, some
5000 cubic yards of rock were removed from the powerhouse
and tailrace areas.

The nature of the bedrock at Shawmut also required
precautions to insure the stability of the excavation as
well as the permanent structures. Approximately 5000
linear feet of Dywidag rock bolts were installed at
Shawmut. This included both vertical and horizontal
bolts. The vertical bolts provided uplift resistance to
oppose hydrostatic forces on the powerhouse while the
horizontal bolts stabilized the rock cut.

The powerhouse excavation was completed by mid-August and
construction emphasis shifted to the powerhouse
substructure and turbine/generator installation. The
turbine/generators selected for Shawmut were
Allis-Chalmers, standardized tube units. The units have
2750 mm diameter runners and are rated 2000 KW while
passing 1200 cfs at a 22.5 net head. These units consist
of two major types of components; embedded parts and
rotating parts. The embedded parts consist of reinforced
steel plate sections that form the penstock and draft
tube liners. These liners act as a large concrete form.
The powerhouse substructure, including reinforcing and
concrete, is placed around the liners in vertical lifts
of about 3 feet. This vertical limit was imposed by the
manufacturer because of possible liner distortion during
embedment. In addition to the deformation concern,
embedment of the liners also presented another problem.
The reinforcing ribs which encircle the liner sections
create pockets which would trap air and prevent full
concrete flow. To remedy this a super plasticizer
concrete additive was used to temporarily increase the
normal 2-3 inch slump to approximately 9 inches. In
addition, relief holes were cut into each pocket to allow
trapped air to escape. The resulting embedment was
completely satisfactory.

The construction of the powerhouse substructure also
included the installation of anchoring provisions for the
generator and speed increaser, electrical duct lines, the
powerhouse drainage system, and grounding system.

Concrete placement for the substructure continued through the summer and fall along with the construction of the powerhouse walls and the tailrace retaining wall. The intake structure took shape as the powerhouse construction progressed. Spalled concrete was removed from the existing forebay structure and the new intake was securely doweled to sound concrete in the existing piers. Equipment foundations and duct banks for the substation addition were also installed during the late summer and fall.

In the fall, weather conditions deteriorated and a period of above average rainfall was experienced. The Kennebec River rose to unexpected levels and considerable time was spent raising and reinforcing the cofferdam. Nevertheless, on September 24th and again on October 24th CMP was forced to intentionally flood the construction site. This action prevented a sudden failure of the cofferdam and avoided the massive clean-up problem that a sudden breaching would have caused.

The powerhouse was enclosed early in January and the interior mechanical and electrical work was begun. CMP's own electrical maintenance crews handled the electrical equipment installation while the general contractor provided millwrights to accomplish the mechanical erection work. An Allis-Chalmers field representative was present throughout the installation of both the embedded and rotating parts to provide technical assistance and advice.

On March 17th the first turbine/generator was placed on line (Unit 7). Preliminary tests indicate that output from the unit is consistent with the manufacturer's claims. However, unit efficiency can not be determined until detailed performance tests are completed. The second Shawmut Unit (Unit 8) became operational on March 24th. Once again, preliminary output results are satisfactory. But formal acceptance of the turbine/generators will await the efficiency tests scheduled for May.

The Shawmut Hydro Project is important for two reasons. First, Shawmut demonstrates the economic viability of small, low head hydro developments in today's environment of costly fossil fuel. The Shawmut Project was justified on the basis of oil displacement alone, completely disregarding the capacity benefit of the installation. Second, the new, standardized tube type turbines being installed at Shawmut may prove to be an acceptable approach to reducing first costs and speeding project construction. Only experience will determine whether or not this concept is a viable alternative for future development plans.

# PENACOOK LOWER FALLS HYDROELECTRIC PROJECT
by

Harold Turner, Jr., P.E. M.ASCE[1] and

John R. Lavigne, Jr., P.E. M.ASCE[2]

## ABSTRACT

This case study concerns the design of a new concrete hydroelectric facility situated on the Contoocook River approximately 1,000 feet upstream from its confluence with the Merrimack River in the communities of Penacook and Boscawen, New Hampshire. Design features of the power-house, spillways, channels and gates are presented in addition to discussion of river hydrology and facility operation. The new run-of-river project generating salable electric power replaces an old concrete and timber crib mill dam previously used to drive machinery. Private developers will own and operate the facility as a semiunmanned station. Computer-based hydraulic river models from existing Federal studies were used to optimize channel excavation and design.

## PROJECT DATA

| | |
|---|---|
| Capacity: | 4110 Kilowatts |
| Turbine Type: | (1) Allis-Chalmers 3,000 MM horizontal tube-type full Kaplan |
| Plant Operation: | Automatic run-of-river |
| Avg. Annual Energy Output: | 15.4 million kwh |
| Estimated Avg. Head: | 28 feet |
| Reservoir Surface Area: | 8.4 acres |
| Gross Storage Capacity: | 54 acre-feet |
| Max. Plant Hydraulic Cap.: | 2200 cfs |
| Estimate Avg. River Flow: | 1255 cfs |
| Estimate Project Cost: | $7.5 million |
| Project Timetable: | Construction Start-Oct. 1981 Commercial Operation - Spring 1983 |
| Project Developer: | New Hampshire Hydro Assoc. Concord, NH |

---

1 A-E Division Manager, Chief Structural Engineer, Anderson-Nichols & Co., Inc., Concord, NH.
2 Chief Water Resource Engineer, Anderson-Nichols & Co., Inc., Concord, NH.

Prime Contractor:          Perini Corp., Framingham, MA

Design Engineer:           Anderson-Nichols & Co., Inc.
                           Concord, NH

## HYDROLOGY

The project site is located near the mouth of the Contoocook River watershed in the Merrimack River Basin. See
Figure #1. The river has a drainage area of about 766 sq.
miles and total fall of 760 feet, 92 feet of which occurs
in the lower three miles through Penacook. Gage records
just upstream of the site indicate an average annual flow
of 1255 cfs. These same records were used in developing
the annual flow duration for estimating power production,
shown in Figure #2. The climate of the watershed is typical of the basin, with warm summers and cool winters.
Temperatures during the month of July range from an average
high of 80°F to a low of 58°F. Temperatures in January
range from an average high of 32°F down to 11°F. Average
annual precipitation is 40 inches, of which approximately
20 percent is in the form of snow.

The flood of record which occurred in 1936 was 46,800 cfs.
Since that flood, the US Army Corps of Engineers has built
four upstream dams to control floodwaters. As a result,
the probable 100-year and 10-year flood events have been
reduced by 44% to 23,300 cfs, and 51% to 10,500 cfs,
respectively.

## POWER POOL

Normal pool elevation for the project is 278.0. This
elevation is 6' higher than originally proposed. Hydraulic analyses indicated that the project could support the
additional head with only minor impacts on upstream properties and facilities. This level will be stabilized by
maintaining the powerplant operation as a run-of-river
mode; outflows equal inflows. The pond, extending 1500
feet upstream of the powerhouse, will have an average depth
of 10 feet; a maximum of 41.5 feet at the powerhouse forebay.

## FACILITIES DESIGN

### Powerhouse

A concrete, bedrock-based powerhouse will be constructed
near the centerline of the river profile. See Plate #1.
Final determination of the powerhouse siting and adjacent spillway orientation will take advantage of the
high and low profiles of the natural river channel to
minimize rock excavation and economize the size of these
structures. The overall length of the powerhouse will
be 97.5 feet and the width perpendicular to the profile
will be 35 feet. The upstream face will rise from
elevation 236.52 at the invert of the intake to elevation 283.00 at the top of the wall. Exit level at the
downstream end of the draft tube is at elevation 224.90.

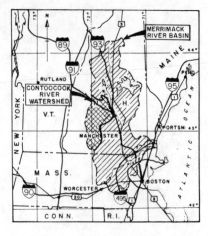

WATERSHED  MAP

FIGURE-1

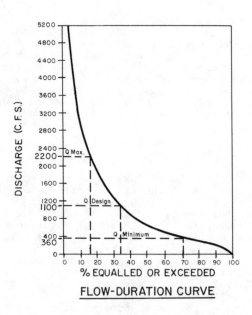

FLOW-DURATION CURVE

FIGURE-2

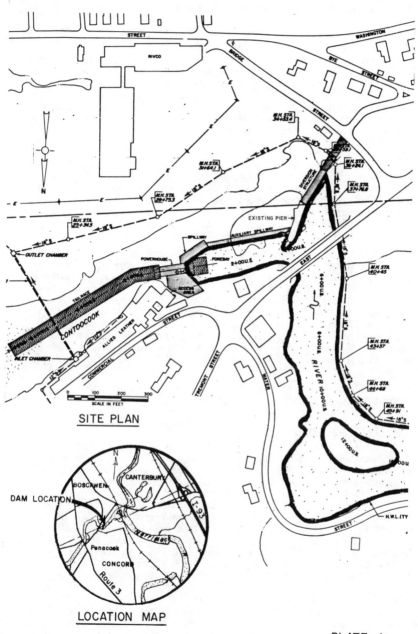

SITE PLAN

LOCATION MAP

PLATE-1

The powerhouse roof level steps down from elevation 283.00 at the upstream end to elevation 268.00 at the downstream end. Permanent roof hatches are provided to give crane access for equipment installation or removal.

A 55 foot wide rock filled access area ties the north side of the powerhouse to the north river bank at elevation 281.00. Upstream and downstream sides will be formed by concrete retaining walls to bedrock and serve as part of the power pool containment. Provisions have been designed into the access area to incorporate a fishway passage at a future date.

## Spillway

From the southwest corner of the powerhouse, at an angle of 25° south from its axis, a concrete, gated spillway will extend upstream for 106 feet. Contained within the spillway will be four 9.5 foot wide by 15.5 foot high fully operable timber sluice gates. Four timber stoplog bays of similar size, and one 12 foot wide by 3 foot high operable timber ice gate are also included.

The operable spillway gates in conjunction with the operable diversion spillway gates are proportioned to safely discharge the 10-year flood event. All operable gate guides contain heating elements for de-icing. When all gates are lifted and all stop logs released in both spillways, the 100-year flood event can be discharged while maintaining the permanent pool elevation of 278.00. All operable gates utilize manual lifting mechanisms except for one fully automated gate at the spillway structure. Stoplogs are released via the pin release of a center steel stanchion.

The spillway base is keyed to bedrock with an invert elevation of 262.50. Transverse concrete walls rise above the base to a walkway elevation of 283.00.

## Auxiliary Spillway

A concrete gravity auxiliary spillway emanates from the upstream end of the spillway at an angle of 15° north from the spillway axis. The auxiliary spillway will be constructed and keyed in bedrock. The height will vary up to 14 feet at the downstream end and step in relation to the rock profile. The maximum base width of 18 feet will extend upstream for 316 feet and terminate at the existing granite roadway pier. The concrete cross section is proportioned to absorb full static ice pressures and potential ice jamb forces. Hydraulic capacity of the auxiliary spillway is utilized during flood flows greater than the 100-year discharge.

## Existing Granite Pier

The existing abandoned granite roadway pier extends southward for a distance of 115 feet. It is approximately 35 feet wide by 20 feet high, and is situated on

bedrock. All faces of the existing granite exposed to
the power pool will be refaced with a new concrete wall.
See Plate #1.

## Diversion Spillway

A concrete, gated diversion spillway, 134 feet in length,
will be constructed from the south end of the existing
granite pier across to an existing granite abutment on
the south river bank.  The structure will be situated
between supports once spanned by a previously removed
roadway bridge, and forms the final link in retaining
the new power pool.  Contained within the diversion
spillway will be three 9.5 foot wide by 10 foot high
operable timber sluice gates.  Seven timber stoplog bays
of similar size and two 9.5 foot wide by 3 foot high
operable timber ice gates are also included.  Gates and
stoplogs are operated as described in the spillway
section.

The diversion spillway is keyed to bedrock with an
invert elevation of 268.00.  Transverse concrete walls
rise above the base to a walkway elevation of 283.00.

## UPSTREAM CHANNEL IMPROVEMENTS

The spillway structure adjacent to the powerhouse and the
auxiliary spillway structure combine to form the south
bank of the upstream channel for almost 400 feet until
the existing granite roadway pier is met.  Thus, a unique
river closure will impound the forebay and power pool of
the dam in conjunction with the diversion structure.  Once
impounded, the river will be diverted to the powerhouse by
the granite pier and diversion spillway at nearly a 90°
angle to the upstream flow line.

Design of channel improvements upstream of the powerhouse
incorporate a variety of considerations.  Of primary
importance was the removal of a rock constriction in the
channel at the north end of the granite pier.  Rock
removal in this area to enlarge the channel cross section
will eliminate an existing head loss condition and provide
a uniform flow transition.

Selective rock removal, extending some 500 feet upstream
from the powerhouse, was designed to decrease turbulence
and keep channel velocities below 2 feet per second.  The
combination of low velocities and smooth channel flow will
reduce head loss and insure an ice sheet cover to prevent
active frazil ice production in this reach.  Having a
solid ice cover over the initial 500 to 1,000 feet upstream of the intake will also turn incoming active frazil
ice into a non-sticky consistency, thereby minimizing the
potential for clogging the intake or equipment with frazil
ice or reducing the channel cross-section with anchor ice.

The hydraulic model for the Contoocook River from the
Concord and Boscawen flood insurance studies was used to

determine the upstream extent of the pond and the resulting backwater effect resulting from influent floodflows.
The model was used as the design tool to determine the
pre- and post-conditions of the pond and upstream areas.
By using the computer model, the optimum rock excavation
limits were determined.

FOREBAY DESIGN

Because of the turbine design, the powerhouse and intake
will be excavated 24 feet below river channel.  In order
to achieve a smooth-flowing intake having a minimum of
turbulence, a semi-rectangular forebay will be excavated
just ahead of the intake.  Side and front sloping of the
forebay sidewalls will transition this area back to existing grade.  The forebay is designed such that its area is
sufficiently large, relative to the upstream river channel,
to efficiently transfer the approaching flow to the intake.

DRAFT-TUBE DESIGN

Because of the sensitivity of geometry on turbine
efficiencies, the hydro-machinery vendor is normally responsible for the design of the draft tube.  However, it
was felt that a cost-effective draft-tube shape could be
achieved, without jeopardizing turbine efficiencies, by
having the project's engineer submit to the equipment
vendor alternatives to the original draft-tube layout.
These alternatives were then checked by the manufacturer
to insure that the equipment efficiency specifications
were not being violated.  This has resulted in a draft-
tube design that will require a minimum of concrete forming to achieve the necessary transition from the turbine
exit to the end of the powerhouse.

TAILRACE DESIGN

The tailrace channel is 700 feet long with its inlet level
at elevation 224.90, the draft-tube exiting elevation.
The bottom profile climbs at a rate of 10.% for 125 feet
downstream.  At that point, the slope decreases significantly; 0.7% with the bottom of the tailrace running out
at this slope until it meets the existing channel grade
at elevation 241.00.  The profile and cross-section of
the tailrace are shown on Figure #3.

The computer-based hydraulic river model was used to
establish pre- and post-construction conditions to determine optimum rock excavation limits.  Computer modeling
also afforded the opportunity to review the tailrace
design under flow conditions above and below the design
discharge and to interject the added variability of
fluctuating backwater influences from the confluence with
the Merrimack River.  The end product of this hydraulic
analysis was a simplistic graphical relationship of tailrace elevation at the powerhouse versus main stem,
Merrimack River elevations for various Contoocook River
discharges.  See Figure #4.

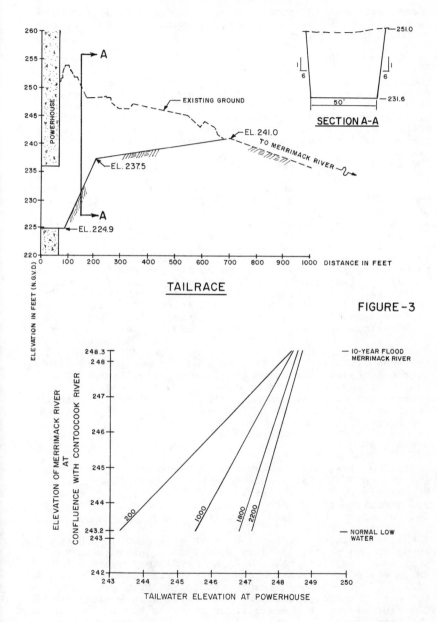

SECTION A-A

TAILRACE

FIGURE-3

FIGURE-4

These tailwater relationships were used by the hydro-
machinery vendor as an aid in setting the optimum elevation
of the turbine runner.

ICE PASSAGE DESIGN

Located in central New Hampshire, the Contoocook River
experiences a winter ice regime of considerable importance
in the design of a hydroelectric facility.  A steep river
gradient immediately upstream of the site provides open
water throughout the winter period, and ideal conditions
for frazil ice production.  Conversely, much of the upper
reaches of the Contoocook are placid and result in a sheet
ice cover of thicknesses up to 2 feet.

The run-of-river nature of the site and small storage area
provided by the river gradient necessitated the inclusion
of features to accommodate the passage of frazil, block
and anchor ice.  At the upstream diversion spillway, two
9.5 foot by 3 foot ice passage gates were located adjacent
to the west bank to allow for accessibility and to match
existing river orientation.  A 12 foot by 3 foot ice
passage gate was located in the downstream spillway
immediately adjacent to the powerhouse.  All gate guides
are heated to allow easy operation throughout the winter
months for frazil and anchor ice passage.  Gate depths
were kept shallow to minimize lost river discharge but
still allow adequate ice discharge capacity at pool level.

Under an extreme condition of spring flow and break-up of
the sheet ice, the auxiliary spillway is designed to discharge excessive ice flows and thereby alleviate any ice
jam conditions which might possibly occur.  Under normal
break-up conditions, discharges through the ice and other
spillway gates would be sufficient to handle the ice flows
and river discharge.

OPERATIONS

Operating flows and gross heads are expected to range from
360 cfs to 2200 cfs, and 34 feet to 30 feet respectively.
The single full-Kaplan turbine unit is expected to attain
93.8% maximum efficiency, generating 2,700 kilowatts at
30 feet of net head, and 1,100 cfs of discharge.

A Philadelphia Gear speed increaser will provide 600 rpms
input to a 3 phase, 60 Hertz McGraw-Edison synchronous
generator.  The generator will be rated to produce 4,000
KW at 0.9 power factor, and 4,160 volts.  Power output
will feed via overhead transmission line from the powerhouse to the main transformer and switchgear located some
220 feet away on the south river bank.  A tie to Concord
Electric Company's transmission lines 120 feet away will
then wheel the power output to the ultimate purchaser,

Public Service Company of New Hampshire.

The plant will be under an automatic float control system
with manual overrides.  A microprocessor will integrate
the operation of the wicket gates, runner blades, pond
level controller and automatic spillway gate to maintain
a constant pool elevation of 278.00 and produce electricity at optimum efficiency.  At incoming flows below
360 cfs, the turbine will shut down and flows will pass
through the automatic spillway gate.  During flows in
excess of 2,200 cfs, discharge will again begin through
the spillway gate in order to maintain a constant pool and
maximum turbine discharge.  A reactivated USGS gaging
station up stream of the project site will be incorporated
and aid in the automatic regulation of the system.

The station is designed to function as a semi-remote
operation.  An operator will be assigned for routine
maintenance purposes.  All systems and output information
will be monitored and controlled at the developer's central
operations in Lawrence, Massachusetts some 50 miles away.
Automatic shutdowns due to system malfunction or transmission line failure will require manual restart at the
station.  Scheduled or otherwise usual shutdowns can be
initiated and reversed by Lawrence Control.

A hydraulic trash rake will be used to keep the trash
racks clear of debris and ice, thereby minimizing downtime
and loss of generation.  This feature becomes important in
the removal of anchor ice formations which find their way
to the powerhouse intake.  Electrically motorized intake
and draft-tube slide gates provide means for the dewatering of the turbine and powerhouse passages for either
emergency repairs or scheduled maintenance.

# HYBRID MODELING OF ESTUARINE SEDIMENTATION

W. H. McAnally, Jr.[1], M. ASCE, and J. P. Stewart[1]

## Abstract

A hybrid modeling method using physical and numerical models in an
integrated solution method has been developed for use in solving estuarine sedimentation problems.  The method was applied to the Columbia
River estuary with a large physical model, finite element numerical
models RMA-2V and STUDH, a finite difference wave propagation model and
several analytical techniques.

## Introduction

This paper presents the hybrid modeling method developed at the
U. S. Army Engineer Waterways Experiment Station (WES).  The U. S. Army
Corps of Engineers maintains a 48-ft-deep by 2600-ft-wide (15-m by 792-m)
navigation channel through the Columbia River estuary entrance (Fig. 1)
which is protected by jetties.  Annual maintenance dredging of 4.8 million cubic yards (3.67 million cubic metres) in the entrance has led the
U. S. Army Engineer District, Portland (NPP) to seek means of reducing
shoaling in the entrance channel.  Accordingly, in 1976 NPP asked WES to
determine if newly developed numerical modeling techniques could be used
to address the shoaling problems in the estuary.  It was concluded from
a pilot study that a hybrid modeling method, combining physical and numerical models, offered the potential to provide better results than
heretofore possible.

Field (prototype) data collection and analysis serves both as an
important aspect of the other solution methods and as an independent
method.  It is an indispensable element in verification of numerical and
physical models.  And, to a limited extent, field data can be used to
estimate the estuary's response to different conditions of tide and river
discharge.  Obtaining sufficient temporal and spatial data coverage in
the field, however, is a formidable and difficult task.  Field testing of
structures costing millions of dollars is generally too risky.

Analytic solutions are those in which answers are obtained by use of
mathematical expressions.  Analytical models usually combine complex phenomena into coefficients that are determined empirically.  The usefulness
of analytical solutions declines with increasing complexity of geometry
or increasing detail of results desired.

---

[1]Estuaries Division, U. S. Army Engineer Waterways Experiment Station,
Vicksburg, MS.

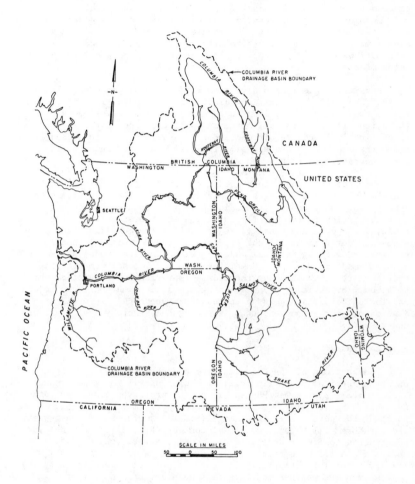

Figure 1.   Site map

Numerical models employ special computational methods, such as
iteration and approximation, to solve mathematical expressions.  In
coastal hydraulics problems, they are of two principal types--finite
difference and finite element.  They are capable of simulating some
processes that cannot be handled any other way.  Numerical models provide much more detailed results than analytical methods and may be more
accurate, but they do so at the expense of time and money.  They are
also limited by the modeler's ability to formulate and accurately solve
mathematical expressions that truly represent the physical processes
being modeled.

Physical scale models have been used for many years to solve
coastal hydraulic problems.  Physical models of estuaries can reproduce
tides, freshwater flows, long shore currents, and three-dimensional
variations in currents, salinity, and pollutant concentration.  Con-
flicts in similitude requirements for the various phenomena usually
force the modeler to neglect similitude of some phenomena in order to
more accurately reproduce the more dominant processes.

The preceeding have described the four principal solution methods
and some of their advantages and disadvantages.  Common practice has
been to use two or more methods jointly, with each method being applied
to that portion of the problem for which it is best suited.  For example, field data are usually used to define the most important processes and verify a model that predicts hydrodynamic conditions in an
estuary.  Combining physical modeling with numerical modeling is termed
a hybrid modeling method.  Combining them in a closely coupled fashion
that permits feedback among the models is termed an integrated hybrid
solution.  By devising means to integrate several methods, the modeler
can include effects of many phenomena that previously were either neglected or poorly modeled, thus improving the accuracy and detail of
the results.

## Physical model

The keystone of the Columbia hybrid method studies was the Corps'
existing physical hydraulic model of the estuary.  The model reproduces
approximately 350 sq mi (906 sq km) of the prototype area, including the
Columbia River to mile 52, the Pacific Ocean from 9 miles (14.5 km)
north of the north jetty to 6 miles (9.7 km) south of the south jetty,
and offshore well beyond the 120-ft contour (Fig. 2).  The model was
constructed to linear scale ratios, model to prototype, of 1:500 horizontally and 1:100 vertically.  The salinity ratio for the model was
1:1.  The model is completely inclosed to protect it and its appurtenances from the weather and to permit uninterrupted operation.

The model is equipped with the necessary appurtenances to reproduce
and measure all pertinent phenomena such as tidal elevations, saltwater
intrusion, current velocities, freshwater inflow, dispersion characteristics, and shoaling distribution patterns.  This equipment is described
in detail by Herrmann (1968).

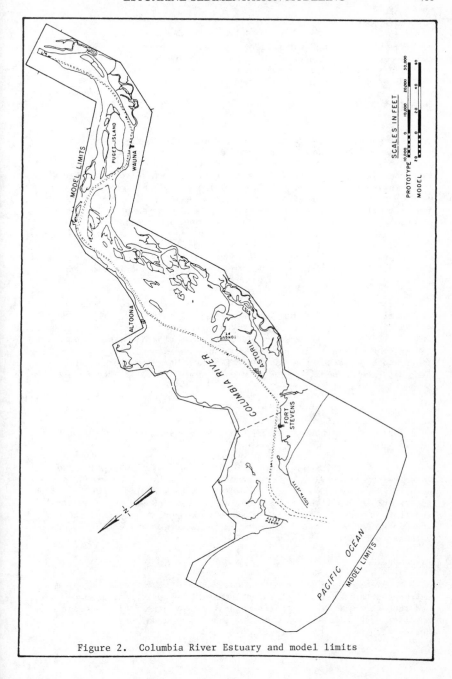

Figure 2. Columbia River Estuary and model limits

Numerical models

In the Columbia River hybrid model study, three large numerical
models were used--RMA-2V, RFAC, and STUDH.  These programs modeled
hydrodynamics, wave conditions, and sediment transport, respectively.
The following describes two of these models.

RMA-2V

The generalized program, RMA-2V, solves the depth-integrated
equations of fluid mass and momentum conservation in two horizontal
directions.  The code RMA-2V is based on the earlier version RMA-2
(Norton and King, 1977) but differs from it in several ways.  First, it
is formulated in terms of velocity (v) instead of unit discharge (vh),
which improves some aspects of the code's behavior; it permits drying
and wetting of areas within the grid; and it permits specification of
turbulent exchange coefficients in directions other than along the  x
and  z  axes.  The Chezy roughness formulation of the original model was
modified in the input portion so that Manning's  n  roughness coefficients are specified by nodes.  A short input routine computes elemental
Chezy coefficients from input Manning's  n  values and initial water
depth.

The equations of mass and momentum conservation are solved by the
finite element method using Galerkin weighted residuals.  The elements
may be either quadrilaterals or triangles and may have curved (parabolic)
sides.  The shape functions are quadratic for flow and linear for depth.
Integration in space is performed by Gaussian integration.  Derivatives
in time are replaced by a nonlinear finite difference approximation.
Variables are assumed to vary over each time interval in the form

$$f(t) = f_o + at + bt^c \qquad\qquad t_o \leq t < t_1 \qquad\qquad (1)$$

which is differentiated with respect to time and cast in finite difference form.  It has been found by experiment that the best value for  c
is 1.5.

The solution is fully implicit and the set of simultaneous equations
is solved by Newton-Raphson iteration.  The computer code executes the
solution by means of a front-type solver that assembles a portion of the
matrix and solves it before assembling the next portion of the matrix.
The front solver's efficiency is largely independent of bandwidth and
thus does not require as much care in formation of the computational
mesh as do traditional solvers.

Wave model

A wave model, RFAC, provided estimates of wave conditions over the
entrance area by refracting and diffracting deepwater waves shoreward
and through the entrance to the upper limits of the finite element mesh
used for hydrodynamics and sedimentation.  The model was developed by
Resio and Vincent (1977) of WES.

## STUDH

The generalized computer program, STUDH, solves the depth-integrated convection-dispersion equation in two horizontal dimensions for a single sediment constituent. The form of the solved equation is

$$\frac{\delta C}{\delta t} + u \frac{\delta C}{\delta x} + w \frac{\delta C}{\delta z} = \frac{\delta}{\delta x}\left(D_x \frac{\delta C}{\delta x}\right) + \frac{\delta}{\delta z}\left(D_z \frac{\delta C}{\delta z}\right) + \alpha_1 C + \alpha_2 \qquad (2)$$

where

C = concentration of sediment

u = velocity in  x  direction

w = velocity in  z  direction

$D_x$ = dispersion coefficient in  x  direction

$D_z$ = dispersion coefficient in  z  direction

$\alpha_1$ = coefficient of concentration dependent source/sink term

$\alpha_2$ = coefficient of source/sink term

STUDH is related to the model SEDIMENT II (Ariathurai, et al., 1976) developed at the University of California, Davis, under the direction of R. B. Krone. It is the product of joint efforts of WES personnel (under direction of W. A. Thomas) and R. Ariathurai, now a member of Resource Management Associates.

The source/sink terms in equation 2 are computed in routines that treat the interaction of flow and the bed. Separate sections of the model handle computations for clay bed and sand bed problems. In the tests described here, only sand beds were considered. The source/sink terms were evaluated by first computing a potential sand transport capacity for the specified flow conditions, comparing that capacity with the amount of sand actually being transported, and then eroding from or depositing to the bed at a rate that would approach the equilibrium value after sufficient elapsed time.

The potential sand transport capacity in these tests was computed by the method of Ackers and White (1973), which uses a transport power (work rate) approach. It has been shown to provide superior results for transport under steady flow conditions (White et al., 1975) and for combined waves and currents (Swart, 1976). WES flume tests have shown that the concept is valid for transport by estuarine currents.

The Acker-White equations result in a computed potential sediment concentration, $G_p$ . This value is the depth-averaged concentration of sediment that will occur if an equilibrium transport rate is reached with a nonlimited supply of sediment. The rate of sand deposition (or erosion) is then computed as

$$R = \frac{G_p \, C}{t_c} \tag{3}$$

where

      $C$ = the present sediment concentration

      $t_c$ = a time constant

For deposition, the time constant is

$$t_c = \text{the larger of } \Delta t \text{ or } \frac{h}{V_s} \tag{4}$$

and for erosion it is

$$t_c = \text{the larger of } \Delta t \text{ or } \frac{C_L h}{U} \tag{5}$$

where

      $\Delta t$ = computational time step

      $C_L$ = a response time coefficient

      $h$ = water depth

      $V_s$ = sediment settling velocity

      $U$ = current speed

Equation 2 is solved by the finite element method using Galerkin weighted residuals. Like RMA-2V, which uses the same general solution technique, elements are quadrilateral and may have parabolic sides. Shape functions are quadratic. Integration in space is Gaussian. Time stepping is performed by a Crank-Nicholson approach with a weighting factor (theta) of 0.66. The solution is fully implicit and a front-type solver is used similar to that in RMA-2V.

## Organization of the hybrid method

The preceeding paragraphs have described the complexity of the physical processes contributing to the shoaling problems in the Columbia River estuary and introduced a new approach to addressing these processes. We have termed this approach the hybrid solution method and have described the principal models involved.

The following paragraphs describe the actual operations involved in a general application of the method. Figure 3 shows the general sequence of steps performed in applying the hybrid solution method as used in the Columbia River entrance studies.

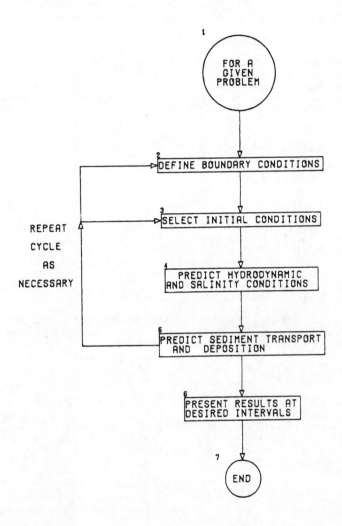

Figure 3.   Steps in the hybrid solution process

Step 1:  For a given problem

Each problem is first tested with existing conditions and then one or more plans representing channel or structural changes is tested.

Step 2:  Define boundary conditions

The numerical solution methods require that the user supply proper boundary condition information.  In estuary sedimentation problems this includes freshwater inflows at the upstream boundary, water surface elevations, wave conditions, salinities, and sediment concentrations at each time step.  Normally, the hydraulic data would be obtained from a physical model.

Step 3:  Select initial conditions

The initial conditions are defined, including the initial bathymetry of the study area.

Step 4:  Predict hydrodynamic and salinity conditions

Driven by the physical model current and water surface elevation measurements at selected points, the hydrodynamic model, RMA-2V, interpolates the data and calculates currents and water surface elevations at each computation point.  By integrating the physical model current measurements over depth, the effect of density currents are indirectly incorporated into the two-dimensional numerical model, provided that physical model measurements are made at appropriate points.

Wave conditions in the entrance are computed by the RFAC model, and longshore currents are predicted analytically by the Longuet-Higgins technique (US Army, 1973).  These currents are linearly superposed on tidal currents predicted by the physical and numerical models.

Step 5:  Predict sediment transport and deposition

Using the hydrodynamic data from Step 4, the numerical model, STUDH, predicts sediment transport, erosion, and deposition.  Depth changes are monitored and computations halted when changes become large enough to change the hydrodynamic response of the system.  If this occurs, the solution process returns to Step 3, and the bathymetry is updated and new hydrodynamics are computed before resuming sediment modeling.

When a period of modeling is complete for a given combination of hydrodynamic events (waves and currents), the solution process returns to the boundary condition step.

Summary

This paper has presented a general description of a solution method for complex coastal hydraulic problems.  The hybrid modeling solution method combines the output from various models, each designed for a particular aspect of the overall problem.

Acknowledgments

Original development of the hybrid modeling method and application to the Columbia River entrance were funded by the U. S. Army Engineer District, Portland. Subsequent development of RMA-2V and STUDH were funded by the Office, Chief of Engineers research program Improvement of Operation and Maintenance Techniques. We wish to thank Emmanuel Partheniades, who first suggested the hybrid concept to the senior author, and the many personnel of the WES who contributed to this effort. Permission to publish this paper was granted by the Chief of Engineers.

## REFERENCES

Ackers, P. and White, W. R.  1973.  "Sediment Transport:  New Approach and Analysis," Journal of the Hydraulics Division, HY11, American Society of Civil Engineers.

Ariathurai, R., MacArthur, R. C., and Krone, R. B.  1977.  "Mathematical Model of Estuarial Sediment Transport," Technical Report D-77-12, U. S. Army Engineer Waterways Experiment Station, Vicksburg, MS.

Herrmann, F. A.  1968.  "Model Studies of Navigation Improvements, Columbia River Estuary, Report 1, Hydraulic and Salinity Verification," Technical Report No. 2-735, U. S. Army Engineer Waterways Experiment Station, Vicksburg, MS.

Herrmann, F. A.  1970.  "Tidal Prism Measurements at Mouth of Columbia River," Miscellaneous Paper H-70-3, U. S. Army Engineer Waterways Experiment Station, Vicksburg, MS.

Norton, W. R. and King, I. P.  1977.  "Operating Instructions for the Computer Program RMA-2," Resource Management Associates, Lafayette, CA.

U. S. Army Coastal Engineering Research Center.  1973.  "Shore Protection Manual," Washington, D. C.

White, W. R., Milli, H., and Crabbe, A. D.  1975.  "Sediment Transport Theories: An Appraisal of Available Methods," Vol 1 and 2; Hydraulics Research Station, Report Int 119, Wallingford, England.

# A NUMERICAL METHOD TO DESCRIBE SEEPAGE FLOW INTO A ROW OF DRAIN TILE

By
A.Swain[1] and S.N.Prasad[2]

ABSTRACT: This paper presents an exact solution of an axisymmetric seepage flow through drain tiles embedded in homogeneous isotropic soil by means of a singular integral equation of the first kind with a Cauchy Kernel. In addition it is shown that a similar integral equation with a Cauchy type Kernel can be obtain for the seepage flows into drain tiles embedded in homogeneous anisotropic soil. The numerically calculated results are compared with other published numerical solutions.

## INTRODUCTION

In modern design of flood water retarding structures, such as an earth dam, numerical solutions of the differential equation for one dimensional drainage of soil columns have received considerable attention in the literature. An initially saturated column of soil at a given location of an earth dam that is allowed to drain the water by means of drain tiles is typical of the mixed initial and boundary value problem of the potential theory. The differential equation describing this process is usually obtained by combining Darcy's law with the equation of continuty. However, a gap exists between the theoretical approaches and the practices used in the field for design and instalation of subsurface drainage system. This is due to the fact that the initial and boundary conditions needed for the theoretical and/or mathematical solutions are usually hard to attain in the field. In order to have a better understanding between theory and practice, this paper presents an exact solution of an axisymmetric seepage flow through drain tiles embedded in homogeneous isotropic/anisotropic soil by means of a singular integral equation of the first kind with a Cauchy Kernel. For design purposes, this paper discusses the effect of drain gap, pipe radius, depth of burial, and hydraulic conductivity on flow through homogeneous isotropic and anisotropic soil and presents dimensionless plots.

## FORMULATION OF THE PROBLEM

Homogeneous and isotropic soil: Consider cylindrical polar coordinate$(r, \theta, z)$ in which r=vertical coordinate; z = horizontal coordinate coinsides with the drain center line; and $\theta$ represents the azimuthal coordinate. The unit of length is h as shown in Fig.1. The axisymmetric

---

1.Research Hydraulic Engineer, Wave Dynamics Division,USAE,WES, P.O.Box 631, Vicksburg, MS 39180; Formarly Graduate Student,University of Mississippi,Oxford,MS.
2. Professor of Civil Engineering, University of Mississippi, Oxford, MS 38677.

flow condition can be obtained by solving the following mixed boundary value problem (1) of the potential theory:

$$\nabla^2 \phi = 0 \qquad \lambda_1 \le r \le \lambda_2 \ , \ |Z| \le 1 \qquad \text{------ (1)}$$

$$\phi(\lambda_2, Z) = V_s \ , \quad |Z| \le 1 \qquad \text{------ (2)}$$

$$\phi(\lambda_1, Z) = V_0 \ , \quad |Z| \le d \qquad \text{------ (3)}$$

$$\frac{d\phi}{dz}(r \pm 1) = 0 \ , \quad \lambda_1 \le r \le \lambda_2 \qquad \text{------ (4)}$$

$$\frac{d\phi}{dr}(\lambda_1, Z) = 0 \ , \quad d < Z \le 1 \qquad \text{------ (5)}$$

in which $\phi(r,z)$ = potential; and $\lambda_1, \lambda_2, r, z, d$ are dimensionless parameters (nondimensionlized with respect to $k$).

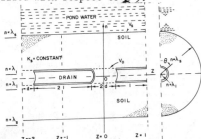

FIG. 1
FIG.- SECTION VIEW OF DRAIN EMBADED IN ISOTROPIC MEDIA

The following separation of variable solution satisfies Eqs. 1, 2, and 4.

$$\phi_s(r,Z) = \frac{1}{2} B_{os} \log(\lambda_2 / r) + \sum_{n=1}^{\infty} a_{ns}(r) \cos n\pi Z + V_s \quad \text{--(6)}$$

where $\quad a_{ns}(r) = B_{ns} \left[ K_0(n\pi r) I_0(n\pi \lambda_2) - I_0(n\pi r) K_0(n\pi \lambda_2) \right] \quad \text{----(7)}$
$$n = 1, 2 \cdots$$

in which $B_{ns}$ = constant; and $I_0$ and $K_0$ are modified Bessel functions of first and second kind respectively (order zero). The remaining boundary conditions, Eqs. 3 and 5, lead to the following dual series:

$$\frac{1}{2} B_{os} \log(\lambda_2 / \lambda_1) + \sum_{n=1}^{\infty} a_{ns}(\lambda_1) \cos n\pi Z = V_0 \ , \ |Z| \le d \qquad \text{-------- (8)}$$

$$-\frac{1}{2\lambda_1} B_{os} + \sum_{n=1}^{\infty} a'_{ns}(\lambda_1) \cos n\pi Z = -f(z) \ , \ d < Z \le 1 \qquad \text{-------- (9)}$$

in which $f(z) = -\dfrac{d\phi}{dr}$ and $f(z) = 0, \ d < |z| \le 1$

Multiply both sides of Eq. 9 by $\cos n\pi z$ and integrate between limits $(-1,1)$
Results: $\quad a'_{ns}(\lambda_1) = -\int_1 f(t) \cos n\pi t \ dt \ , \ n = 0, 1 \cdots \quad \text{-------(10)}$

Note that the orthogonal property of $\cos m\pi t$ and $\sin n\pi t$ or $\cos m\pi t$ and $\cos n\pi t$ is used in Eq.10 and variable z is replaced by t which is also dimensionless. Likewise the constant $B_{os}$ can be obtained by integrating Eq.9 between the limits$(-1,1 )$. This gives:

$$B_{os} = \lambda_1 \int_{-1}^{1} f(t)\,dt \quad\text{------------- (11)}$$

Substituting Eqs.7,10 and 11 into Eq.6 and interchanging the order of integration and summation yields:

$$\phi_s(r,z) = \frac{\lambda_1}{2}\log\left(\frac{\lambda_2}{r}\right)\int_{-1}^{1} f(t)\,dt + \sum_{n=1}^{\infty}\frac{1}{n\pi}\left(\frac{K_0(n\pi r_1)I_0(n\pi\lambda_2)-I_0(n\pi r_1)K_0(n\pi\lambda_2)}{K_1(n\pi\lambda_1)I_0(n\pi\lambda_2)+I_1(n\pi\lambda_1)K_0(n\pi\lambda_2)}\right)*$$
$$* \int_{-1}^{1}\cos n\pi z \cos n\pi t\, f(t)\,dt + V_s \quad\text{-----(12)}$$

Denote the terms under ( ) as $F_n(r)$. Eqs.3,5 and 12 indicate that:

$$V_0 = \int_{-d}^{d}\left[\frac{\lambda_1}{2}\log\left(\frac{\lambda_2}{\lambda_1}\right)+\sum_{n=1}^{\infty}F_n(\lambda_1)\cos n\pi z \cos n\pi t\right]f(t)\,dt + V_s, \; |z|\leq d \quad\text{----(13)}$$

Introduce the relation

$$\sum_{n=1}^{\infty}\frac{1}{n}\cos n\pi z \cos n\pi t = -\frac{1}{4}\log\left[4\left(\cos\pi t - \cos\pi z\right)^2\right] \quad\text{-----(14)}$$

and differentiate with respect to z. Eq. 13 becomes

$$\int_{-d}^{d}\left[\frac{\sin\pi z}{\cos\pi t - \cos\pi z} + 2\sum_{n=1}^{\infty}(F_n(\lambda_1)-1)\cos n\pi t \sin n\pi z\right]f(t)\,dt \quad\text{---(15)}$$

Note $V_0$ is constant in the range $|z|\leq d$
Expand $\cos\pi t - \cos\pi z$ in a Taylor series about the point t=z and retain only the leading term of the series. Eq. 15 becomes

$$\frac{1}{\pi}\int_{-1}^{1}\frac{f(xd)\,dx}{x-y} + \int_{-1}^{1}K(xd,y)f(xd)\,dx = 0 \quad\text{------- (16)}$$

Where

$$K(xd,y) = d\left[\sum_{n=1}^{\infty}2(F_n(\lambda_1)-1)\sin n\pi yd \cos n\pi xd + \frac{\sin\pi yd}{\cos\pi xd - \cos\pi yd} - \frac{1}{\pi(x-y)}\right] \quad\text{---(17)}$$

in which the transformations t=xd and z=yd are used. The Kernel in the first integral of Eq.16 is a Cauchy type and the second integral is the Fredholm Kernel. In view of the convergence of $(F_n(\lambda_1)-1)$, consider the asymptotic expansion of $F_n(\lambda_1)$. It is seen that

$$F_n(\lambda_1)-1 \approx -\frac{1}{2n\pi\lambda_1} \quad\text{------------ (18)}$$

The expression $1/2n\pi\lambda_1$ is added to $F_n(\lambda_1)-1$ to nullify the effect of convergence of $F_n(\lambda_1)$.

The total flow rate, $Q_d$, is given by

$$Q_d = 2\pi\lambda_1 d \int_{-1}^{1} f(xd)\,dx \quad\text{------------(19)}$$

The system of Eqs.16 and 19 posses unique solution. If the theory given in Ref.2 is used, this system of Eqs. can be replaced by

$$\sum_{k=1}^{N}\frac{1}{N}g(x_k)\left[\frac{1}{x_k - y_m} + \pi K_1(x_k, y_m)\right] = 0, \; M = 1, \cdots(N-1) \quad\text{----(20)}$$

and

$$2 \sum_{k=1}^{N} \frac{\pi}{N} g(x_K) \Big/ \Big(Q_d/2\pi\lambda_1 d\Big) = 1 \quad ------ (21)$$

(factor 2 in Eq.21 is taken due to symmetry about the r axis)
Where

$$x_K = \cos\frac{\pi}{2N}(2K-1) \quad --------- (22)$$

$$Y_M = \cos\pi M \quad ------- (23)$$

and

$$f(xd) = \frac{g(x)}{\sqrt{1-x^2}} \quad ----- (24)$$

$$K_1(x_K, Y_M) = \frac{1}{\pi}\Bigg[ 2\sum_{n=1}^{\infty}\Big\{ F_n(\lambda_1) - 1 + \frac{1}{2n\pi\lambda_1} \Big\} \cos n\pi x_K d \sin n\pi y_M d$$

$$- 2\sum_{n=1}^{\infty}\frac{1}{2n\pi\lambda_1} \cos n\pi x_K d \sin n\pi y_M d + \frac{\sin\pi y_M d}{\cos\pi x_K d - \cos\pi y_M d} - \frac{1}{x_K - Y_M} \cdots (25)$$

Eqs.20 and 21 form a system of N linear algebric equations whose solution provide extremely accurate results for $g(x_K)$.

FLOW RATIO:

The ratio of the flow rate, $Q_d$, through the drain and the flow rate $Q_h$ through an open drain without pipes where L = 0 and d = 1, defines the flow ratio R = $Q_d/Q_h$. The flow rate through an open drain without pipes is

$$Q_h = \frac{4\pi(V_0 - V_s)}{\log(\lambda_2/\lambda_1)} \quad --------- (26)$$

An expression for $V_o-V_s$ can be obtained by substituting Eqs.14,18,19, and 21 into Eq.13. That is

$$R = 1\Bigg/\Bigg[1 + \frac{4}{\lambda_1 \log\frac{\lambda_2}{\lambda_1}}\Bigg\{\Big(\sum_{n=1}^{\infty}F_n(\lambda_1) - i + \frac{1}{2n\pi\lambda_1}\Big)\sum_{k=1}^{N}\frac{g(x_K)\cos n\pi x_K d}{N}\Big) - \frac{1}{2\pi}\sum_{n=1}^{\infty}\frac{1}{n^2}\sum_{K=1}^{N}\frac{g(x_K)}{N}*$$

$$* \cos n\pi x_K d - \frac{1}{4}\sum_{K=1}^{N}\frac{g(x_K)}{N}\Big(\log[4(1-\cos\pi x_K d)^2]\Big)\Big\}\Bigg] \quad -------(27)$$

Eq. 27 is plotted in Fig.2 for typical values of $\lambda_1$, $\lambda_2$ and d(note that d is dimensionless = d/h).

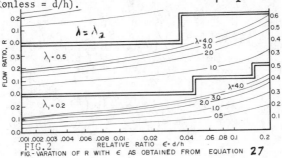

FIG.2
FIG.- VARATION OF R WITH $\epsilon$ AS OBTAINED FROM EQUATION 27

HOMOGENEOUS ANISOTROPIC SOIL:

It is known that when drain tiles are either partially or completely covered with gravel, the spacing between the drains would increase significantly as compared with drains without gravel surrounds.

Consider the following expression for the potential of the gravel layer:

$$\phi_g = \frac{1}{2} B_{og} \log\left(\frac{\lambda_g}{r}\right) + \sum_{n=1}^{\infty} a_{ng}(r) \cos n\pi z + V_g \quad \cdots \cdots (28)$$

in which

$$a_{ng}(r) = B_{ng} K_o(n\pi r) + C_{ng} I_o(n\pi r) \quad - - - - - - - - (29)$$

where $B_{ng}$ and $C_{ng}$ are constants. $\lambda_g$ = dimensionless radial distance to the interface between the gravel and the soil; and $V_g$ is the potential at the interface. Eq.6 represents the expression for potential for the soil layer.

The following interfacial conditions must satisfy:

$$\phi_g(\lambda_g, Z) = \phi_s(\lambda_g, Z) \quad - - \cdots - - - - \cdots (30)$$

$$-K_g \frac{\partial \phi}{\partial r}(\lambda_g, Z) = -K_s \frac{\partial \phi}{\partial r}(\lambda_g, Z) \quad - - - - - - - (31)$$

Follow an analysis similar to the homogeneous and isotropic soil and incorporate Eqs.30 and 31 in it. This will result in the following results:

$$B_{og} = \frac{\lambda_1}{K_g} \int_{-1}^{1} f(t) dt$$

$$a_{ng}(r) = \frac{1}{n\pi} F_{ng}(r) \int_{-1}^{1} f(t) \cos n\pi t \, dt$$

$$\phi_g(\lambda, g) = \frac{\lambda_1}{2K_g} \log\left(\frac{\lambda_g}{\lambda_1}\right) \int_{-1}^{1} f(t) dt + \frac{1}{K_g} \sum_{n=1}^{\infty} \frac{1}{n\pi} F_{ng}(\lambda_1) \cos n\pi z \int_{-1}^{1} f(t) \cos n\pi z \, dt + V_g \cdots (34)$$

in which $F_{ng}(\lambda_1)$ is a function of $\lambda_1, \lambda_g, K_s, K_g, I_o, K_o, I_1$ and $K_1$. In addition Eq.25 is the same for this case when

$$F_n(\lambda_1) = F_{ng}(\lambda_1)$$

FLOW RATIO:

Consider open drain without pipes. Define the potentials for the soil layer as $\phi_s$ and for the gravel layer $\phi_g$. Where

$$\phi_s = \frac{1}{2} A_s \log\left(\frac{\lambda_s}{r}\right) + V_s \quad - - - - - - - - - (35)$$

$$\phi_g = \frac{1}{2} B_g \log\left(\frac{r}{\lambda_1}\right) + V_o \quad - - - - - - - - - (36)$$

Use Eqs.30 and 31 to solve Eqs.35 and 36. This gives the value of $Q_h$. Where

$$Q_h = 4\pi K_s(V_o - V_s) / \left[\log\left(\frac{\lambda_s}{\lambda_g}\right) + \frac{K_s}{K_g} \log\left(\frac{\lambda_g}{\lambda_1}\right)\right] \quad - \cdots (37)$$

In order to evaluate the value of $V_o - V_s$, use Eqs.6 and 28. Introduce interfacial conditions given by Eqs.30 and 31 and define

This gives

$$-K_g \frac{\partial \phi_g}{\partial r}(\lambda_1, Z) = f(t) \quad - - - - - - - - - - \cdots (38)$$

$$V_g = V_s + \left\{\frac{\lambda_1}{2K_s} \int_{-1}^{1} f(t) dt\right\} \log\left(\frac{\lambda_s}{\lambda_g}\right) \quad - - - - - - \cdots (39)$$

For $r = \lambda_1$ and $-d \leq z \leq d$ Eq. 28 becomes

$$V_g = V_0 - \left\{ \frac{\lambda_1}{2K_g} \int_{-1}^{1} f(t)\,dt \right\} \log\left(\frac{\lambda_g}{\lambda_1}\right) - \sum_{n=1}^{\infty} a_n(\lambda_1) \cdots (40)$$

Eqs. 19, 21, 39 and 40 can now be solved to give

$$V_0 - V_s = \frac{Q_d}{4\pi K_s} \left\{ \log\frac{\lambda_s}{\lambda_g} + \frac{K_s}{K_g}\left[ \log\frac{\lambda_g}{\lambda_1} + \frac{4}{\lambda_1}\left(\sum_{n=1}^{\infty}\frac{F_{ng}(\lambda_1)}{n} \sum_{K=1}^{N}\frac{g(x_K)}{N}\cos n\pi x_K d\right)\right] \right\} \cdots (41)$$

Substitute Eq. 41 into Eq. 37 and consider Eqs. 14 and 18. (use the relation $F_{ng_1}(\lambda_1) - 1 \approx -1/2n\pi\lambda_1$). Then:

$$R = \cfrac{1}{1 + \cfrac{4K_s}{\lambda_1 K_g\left(\log\frac{\lambda_s}{\lambda_g} + \frac{K_s}{K_g}\log\frac{\lambda_g}{\lambda_1}\right)}\left\{ \begin{array}{l} \text{the same expression under } \{\ \} \\ \text{in Eq. 27, except change} \\ F_n(\lambda_1) \text{ to } F_{ng}(\lambda_1) \end{array} \right\}} \cdots (42)$$

Figs. 3 and 4 show typical variation of R as obtained from Eq. 42. Fig. 5 shows a variation of R for constant $\lambda_1, \lambda_2, \lambda_s$ and $\lambda_g$ and varied $K_s/K_g$.

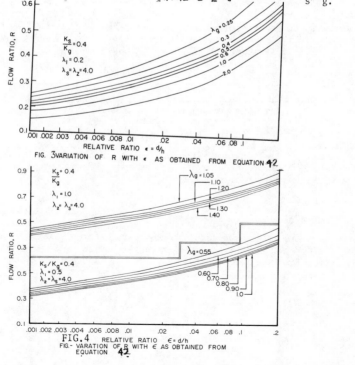

FIG. 3 VARIATION OF R WITH $\epsilon$ AS OBTAINED FROM EQUATION 42

FIG. 4 RELATIVE RATIO $\epsilon = d/h$
FIG.- VARATION OF R WITH $\epsilon$ AS OBTAINED FROM EQUATION 42

DRAIN TILES EMBEDDED IN MULTI-LAYERED ANISOTROPIC SOILS:
Consider three layered soil stratum. The expressions of the potential for the top and bottom layers are given by Eqs.6 and 28 respectively. Consider the potential for the middle layer as:

$$\Phi_f = \frac{1}{2} B_{of} \log\left(\frac{\lambda_f}{r}\right) + \sum_{n=1}^{\infty} a_{nf}(r) \cos n\pi z + V_f \quad ------(43)$$

in which $B_{of}$ = constant; and

$$a_{nf}(r) = B_{nf}\left[K_0(n\pi r) I_0(n\pi\lambda_f) - I_0(n\pi r) K_0(n\pi\lambda_f)\right] \quad ---(44)$$

The potentials for the top and bottom layers when the drain is open are given by Eqs.35 and 36. Let the potential for the middle layer is:

$$\Phi_f = \frac{1}{2} A_f \log\left(\frac{\lambda_s}{r}\right) + V_f \quad --------(45)$$

Following a procedure similar to the two layered soils and using the interfacial conditions analogous to Eqs.30 and 31.Eq. 42 reduces to:

$$R = \frac{1}{1 + \frac{4K_s}{\lambda_1 K_g\left(\log\frac{\lambda_s}{\lambda_f} + \frac{K_s}{K_f}\log\frac{\lambda_f}{\lambda_g} + \frac{K_s}{K_g}\log\frac{\lambda_g}{\lambda_1}\right)}\left\{\begin{array}{l}\text{the same expression under}\\ \{\ \} \text{ in Eq. 27, except}\\ \text{change } F_n(\lambda_i) \text{ to } F_{ns}(\lambda_i)\end{array}\right\}}(46)$$

Likewise,if there are five layered stratum in which a drain is buried whose permeability vary from the bottom to the top layer as $K_g, K_f, K_a,$ $K_c, K_s$ and the corresponding radial distances to the top of a layer measured from the center of the drain be $\lambda_g, \lambda_f, \lambda_a, \lambda_c, \lambda_s$ .The expression for R can be written by just examining Eqs.42 and 46. This is possible due to the harmonic continuation of the potential and the velocity in the soil stratum. Thus

$$R = \frac{1}{1 + \frac{4K_s}{\lambda_1 K_g\left(\log\frac{\lambda_s}{\lambda_c} + \frac{K_s}{K_c}\log\frac{\lambda_c}{\lambda_a} + \frac{K_s}{K_a}\log\frac{\lambda_a}{\lambda_f} + \frac{K_s}{K_f}\log\frac{\lambda_f}{\lambda_g} + \frac{K_s}{K_g}\log\frac{\lambda_g}{\lambda_1}\right)}\left\{\begin{array}{l}\text{the same}\\\text{expression under } \{\ \} \text{ in Eq. 27 except}\\\text{change } F_n(\lambda_i) \text{ to } F_{nf}(\lambda_i)\end{array}\right\}} \cdots (47)$$

in which $F_{nf}(\lambda_i)$ = constant,which can be obtained by following a procedure similar to the case for the homogeneous isotropic soil and using the interfacial boundary conditions.

CONCLUSIONS

This paper has examined the mixed boundary value problem of flow through drain tiles embedded in homogeneous isotropic/anisotropic soils. It is shown that the problem can be solved exactly by means of a singular integral equation of the first kind with a Cauchy Kernel

In addition it is shown that a similar integral equation with a Cauchy Kernel can be obtained for the seepage flows into drain tiles embedded in homogeneous anisotropic soils.

APPENDIX.-REFERENCES

1. Sneyd,A.D;and Hosking,R.J.,"Seepage flow through homogeneous soil into a row of drain pipes",Journal of Hydrology,Vol.30.,1976,pp127-146.
2.Erdogan,F.,and Gupta,G.D.,"On the Numerical solution of Singular Integral Equations",Quarterly of Applied Mathematics,January,1972.

ACKNOWLEDGMENT

The research reported herein was developed at the University of Mississippi while the first author was a graduate student. The patience and financial assistance received through Dr. Samuel Deleeuw,Chairman,Department of Civil Engineering,University of Mississippi is highly appreciated and unforgetable.

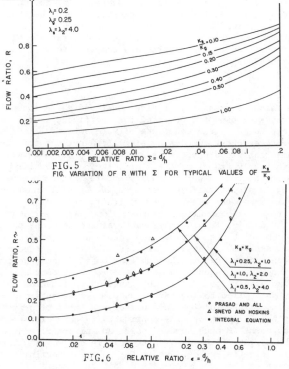

FIG.5
FIG. VARIATION OF R WITH $\Sigma$ FOR TYPICAL VALUES OF $\frac{K_s}{K_g}$

FIG.6    RELATIVE RATIO $\epsilon = \frac{d}{h}$

MONOCLINAL-WAVE SOLUTION OF THE DAM-BREAK PROBLEM

by  Nikolaos D. Katopodes[1], Member A.S.C.E.

ABSTRACT

The adoption of kinematic-wave theory in the computation of Dam-Break flood waves leads almost exclusively to formation of kinematic-shock fronts in the solution, which limits seriously the predictive ability of kinematic-wave based models. Although mathematically correct, the kinematic shock is physically unrealistic and should be replaced by a monoclinal wave. A simple and inexpensive scheme is presented, which permits calculation of the kinematic-shock profile. The kinematic-shock front trajectory is drawn by accurately determining the envelope of intersections of kinematic-wave characteristics and then the sharp discontinuity is eliminated by introducing a second order dissipative term, which accounts properly for the significant free-surface slope in the vicinity of the front. The results are compared to more accurate models and experimental data.

INTRODUCTION

The unbalanced pressure, gravitational and shear forces immediately following the collapse of a dam lead to significant inertia and free-surface slope terms in the equations of wave motion, which for long discouraged researchers from using simplified methods for the computation of Dam-Break flood waves. It is now accepted, however, that following the early stages of the flood, the causative forces approach an equilibrium state, in which gravity and resistance become dominant and the flow is almost uniform locally although still unsteady. Under these circumstances the kinematic-wave theory provides an inexpensive and robust means of computing Dam-Break flood waves at some distance away from the dam (3,7). In many practical applications the main task is indeed to predict the maximum flood depth and time of arrival at long distances downstream of the dam. If, in addition, the technical difficulties often encountered in the application of dynamic-wave models to natural channels are taken into account (2), the use of kinematic-wave models become highly desirable. The range of application of kinematic-wave theory as applied to the Dam-Break problem is determined by the relative value of the local Froude number and ratio of depth to local uniform depth (7). When general applicability criteria are desired, however, the normal Froude number, based on uniform flow at a depth equal to the height of the dam, is the single controlling parameter. Quantitative criteria are derived in Ref. 4 for determining the range of application of the kinematic-wave model, by comparison of dimension-

---

[1]Asst. Professor, Dept. of Civil Engineering, University of Michigan, Ann Arbor, MI

less dynamic and kinematic-wave solutions for various values of the
normal Froude number.  It is interesting that many of the limitations
of the kinematic-wave model stem from the fact that the latter leads
almost exclusively to formation of a kinematic-shock front when applied
to the dam-break problem.  Although this shock front is mathematically
correct (5), it has very little resemblance to physical reality, which
shows a rather long-wave front for dam-break at low normal Froude
numbers.  Strelkoff et als. (7) suggested that such a wave could be
satisfactorily approximated by expansion of the kinematic-shock.  In
fact, when such an expansion is implemented the wave is no longer
kinematic but possesses all the features of the so-called monoclinal
wave (1, pp. 365-373).

KINEMATIC-WAVE SOLUTION

   A long distance downstream of a breached dam the flow is adequately described by the equation of conservation of mass

$$\frac{\partial y}{\partial t} + \frac{1}{B} \frac{\partial Q}{\partial x} = 0 \tag{1}$$

coupled with the statement of locally uniform flow, i.e.,

$$Q = \frac{A}{n} R^{2/3} S_0^{1/2} \tag{2}$$

In Eqs. (1) and (2) $y$ = depth of flow; $Q$ = volumetric flow rate; $B$ =
channel width; $t$ = time; $x$ = distance along the channel; $A$ = channel
cross-sectional area; $R$ = hydraulic radius; $n$ = Manning coefficient of
roughness; and $S_0$ = bottom slope.

   When the discharge hydrograph at the breached dam is known, an
almost exact solution of Eqs. (1) and (2) can be found using the method
of characteristics.  Without loss of generality, the channel may be
assumed prismatic and of rectangular cross section.  The partial
derivative in $Q$ may be eliminated from Eq. 1, resulting in

$$\frac{\partial y}{\partial t} + c \frac{\partial y}{\partial x} = 0 \tag{3}$$

in which

$$c = \frac{1}{B} \frac{\partial Q}{\partial y} = \frac{Q}{3B} \left( \frac{5}{y} - \frac{4}{B + 2y} \right) \tag{4}$$

is the kinematic-wave speed.  Equation 3 suggests that the depth and
discharge remain constant along the so-called characteristic curves
whose inverse slope is equal to $c$.  Furthermore, because of the
invariance of depth and discharge along the characteristics, the latter
are straight lines, which suggests a simple and inexpensive solution
of the dam-break problem.  In fact, the solution would be exact if
Eq. 3 were not nonlinear.  The consequences of nonlinearity become
obvious in the presence of a rapidly rising inflow hydrograph.  Since
each characteristic carries a larger value of $Q$ as time increases,
their inverse slope increases too until the characteristic lines intersect.  The intersection implies a discontinuity in depth and discharge
which is termed a kinematic shock (5).  It is easy to show that the
coordinates of the first intersection are

$$t_s = \frac{5}{2} Q_0 (0) \left[ \frac{dQ_0(0)}{dt} \right]^{-1} \tag{5}$$

and

$$X_s = \frac{25}{6} S_0^{3/10} n^{-3/5} Q_0(0)^{7/5} \left[ \frac{dQ_0(0)}{dt} \right]^{-1} \tag{6}$$

in which $Q_0(t)$ is the discharge at the breached dam. Equations 5 and 6 lead obviously to immediate shock formation with any abrupt change in discharge or if the initial flow rate is zero. This means that a kinematic-shock front will develop instantaneously in all cases of sudden dam failure.

The presence of the kinematic-shock front does not present major numerical difficulties (6) and a simple scheme for carrying out the computation is shown in Fig. 1

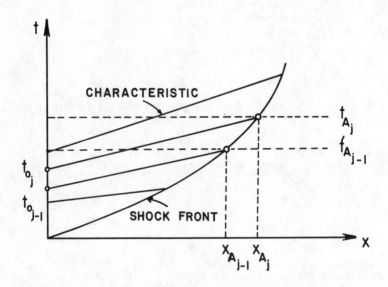

Fig. 1 Kinematic-wave computational scheme.

For an inflow hydrograph typical of instantaneous dam failure, the sudden increase in discharge is followed by a gradual decrease and therefore, once the shock front is formed, the characteristics behind it carry constant values of depth and discharge that decrease with time. Their inverse slope decreases too although it always remains greater than the speed of the shock W. The coordinates of points on the shock trajectory are given by (c.f. Fig. 1)

$$
t_{A_j} = t_{0_j} + \frac{\ddot{x}_{A_{j-1}} - \bar{w}\left(t_{A_{j-1}} - t_{0_j}\right)}{c\left(t_{0_j}\right)}
\tag{7}
$$

and

$$
x_{A_j} = x_{A_{j-1}} + \bar{w}\left(t_{A_j} - t_{A_{j-1}}\right)
\tag{8}
$$

in which

$$
\bar{w} = \frac{1}{2}\left(\frac{Q_{U_j} - Q_{D_j}}{A_{U_j} - A_{D_j}} + \frac{Q_{U_{j-1}} - Q_{D_{j-1}}}{A_{U_{j-1}} - A_{D_{j-1}}}\right)
\tag{9}
$$

In Eqs. 7 - 9, subscript 0 indicates conditions along the t axis, subscript A indicates points on the shock, and U and D points on the upstream and downstream sides of the shock, respectively. With the shock
location determined at all times, the solution at specified times is
easily found by extending characteristic lines from the t axis until
they intersect the desired time level. An important observation can be
made at this point with respect to the nature of the solution. Because
the depth and discharge decrease with time at the origin, and because
of the presence of the shock front, the kinematic wave exhibits a strong
attenuation in amplitude as it propagates down the channel. This is in
contradiction with the well-known property of kinematic-waves of propagation without subsidence and is totally due because of the presence
of a shock front, which intersects the characteristics terminating thus
their role of transferring constant values of depth and discharge along
the channel. Therefore the depth behind the front decreases with time
and so does the entire free-surface profile. Nevertheless there exists
an abrupt discontinuity at the wave front at all times, which is very
different from a dynamic-shock wave, or bore. Indeed, since the
kinematic-wave solution is expected to be meaningful at low values of
the Froude number, a bore is an entirely unrealistic form of wave front.
Instead what is mathematically termed as a shock-front should indeed be
in reality a long wave known as the monoclinal wave (1, p.370).

MONOCLINAL-WAVE SOLUTION

In most cases of modeling long waves in rivers, the kinematic-shock
never appears in the computations. The reasons vary from slowly rising
inflow hydrographs combined with channel reaches that are not long
enough to permit intersection of characteristics within their length,
to use of diffusing finite-difference of finite-element numerical
schemes, which smear the sharp discontinuity over a long distance. The
fact that these models provide usually satisfactory results, makes it
clear that a second-order dissipative form is appropriate for the elimination of the kinematic-shock front. In fact, the very presence of the
discontinuity should be interpreted as a consequence of neglecting the
free-surface slope term in the equation of motion near the wave front.
The computational scheme based on the monoclinal wave is centered around
this simple fact. In other words, although the flow may be kinematic

over most of the channel length, it cannot be so in the vicinity of the
front.  There the kinematic-wave equation must be modified as follows

$$\frac{\partial y}{\partial t} + c \frac{\partial y}{\partial x} = \varepsilon \frac{\partial^2 y}{\partial x^2} \tag{10}$$

in which $\varepsilon$ is a function of the free surface slope (1, p.384).

When Eq. (10) is solved over the entire flow domain, the solution
obtained is termed the zero-inertia model, which is described in detail
in Refs. 4 and 7.  The monoclinal-wave solution results by maintaining
the kinematic-wave solution but simply introducing the effects of free-
surface slope in the vicinity of the front, so that the discontinuity
is eliminated.  Figure 2 shows a computed kinematic-wave profile terminated by a chock-front on a dry bed.  The equation of motion near the
front can be more accurately stated as

$$\frac{\partial y}{\partial x} = S_0 - \frac{n^2 \, W^2(t)}{R^{4/3}} \tag{11}$$

in which $W(t)$ is the kinematic-wave speed.  Based on the assumption of
uniform velocity behind the front, Eq. 11 can be easily integrated to
generate the free-surface profile shown by the dashed line in Fig. 2

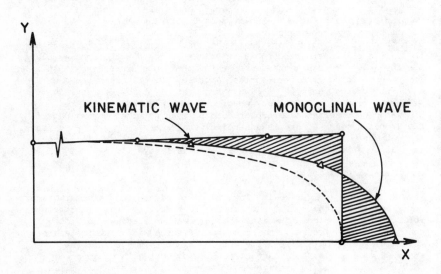

Fig 2.   Computation of kinematic-shock profile.

The singularity at zero depth is avoided by observing that $S_0$ becomes
negligible as $y \to 0$ in Eq. 11 and so a short distance $\xi$ behind the front

$$y(\xi,t) = \frac{7}{3} \left[ n^2 \, W^2(t) \xi \right]^{3/7} \tag{12}$$

The computation of the free-surface profile indicated by Eq. 11 continues until the kinematic-wave profile is intersected. At this time the entire shock profile is stretched to the right until the volume under the kinematic and monoclinal waves is exactly the same. This very simple technique guarantees that the continuity equation is not violated and provides a physically realistic wave front. There is no doubt that the computed profile is a crude approximation to the true solution. However, any possible errors are not propagated in the solution since at all times the almost exact kinematic-wave solution is used for the computation of the current shock profile, independently of any previously computed profiles.

The scheme is very simple to implement and inexpensive to execute with one exception. If the computed monoclinal wave reaches the origin before intersecting the kinematic-wave profile, the above technique fails. This obviously is the case at the early stages of dam-break when in fact the wave is not long enough in order for it to be kinematic away from the front. An alternative scheme is suggested in this case, which permits computation of the monoclinal profile independently. Referring to Fig. 3, the continuity equation is integrated over a control volume JMRL

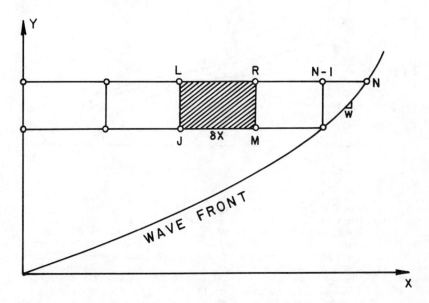

Fig. 3 Integrated difference scheme.

as follows

$$\left[(1-\theta)(Q_M - Q_J) + \theta(Q_R - Q_L)\right]\delta t + B\left(y_R - y_M + y_L - y_J\right)\frac{\delta x}{2} = 0 \qquad (13)$$

in which $\theta$ is a weighting factor. At any time at which a solution is

required, the speed of the monoclinal wave is given a trial value and
an initial value of the depth near the front is computed from Eq. 12.
Then Eq. 13 is solved simultaneously with Eq. 11 in which now the shock
speed is replaced by the fluid velocity. The profile is computed until
the dam is reached, at which time the computed discharge $Q_L$ is compared
with $Q_0(t)$, known from the inflow hydrograph. A shooting procedure is
initiated as shown in Fig. 4., and is continued until $Q_L$ agrees with
$Q_0(t)$ within a specified tolerance. Only few iterations are usually
required since the method is only used for the short lengths of
computation associated with the early stages of flow.

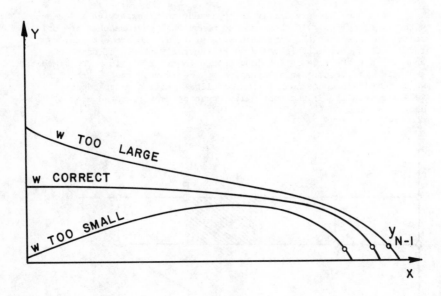

Fig. 4.  Shooting Procedure.

The number of iterations required for the shooting scheme to converge increases significantly with the length of the wave. In fact,
the procedure becomes very sensitive for long distances and eventually
double precision arithmetic is required for convergence. This is mainly
due to the fact that although the governing equation is of the so-called
parabolic type the natural direction of computation is downstream. The
problem is then similar to a convection-diffusion problem with dominant
convective parameters. Berturbations at the downstream boundary are
amplified exponentially if the solution proceeds against the direction
of flow and the scheme becomes very inefficient. In most practical
computations, however, the flow at the upstream boundary approaches
normal conditions long before the computational scheme becomes ill-
conditioned. Then, the procedure is abandoned and the monoclinal wave
is approximated be expanding the kinematic shock, as explained earlier.

RESULTS OF COMPUTATIONAL EXPERIMENTS

The performance of the monoclinal-wave model was tested by comparison with the experimental results obtained by the Waterways Experiment Station (8). The tests were made in a plastic-coated plywood flume of rectangular cross section. The flume was 400 feet long, 4 feet wide, set on a 1/2% slope and terminated in a free overfall. The model dam was 1 foot high and was located in the middle of the flume. Two series of experiments were performed identified in this work as smooth and

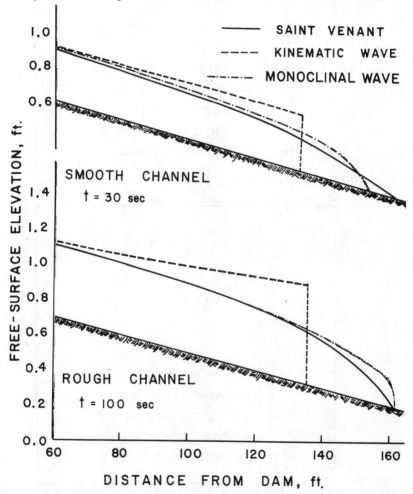

Fig. 5  Computed Free-Surface Profiles.

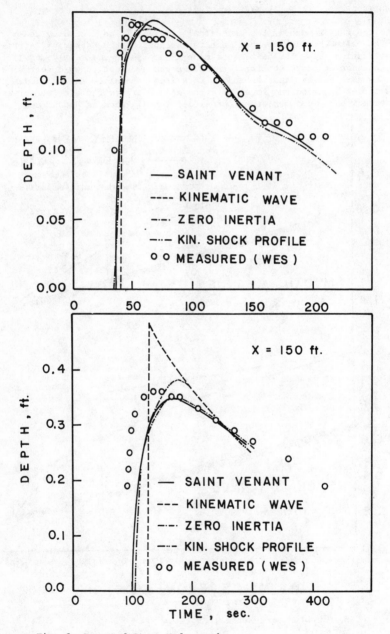

Fig. 6  Computed Stage Hydrographs

rough channels, respectively. In the smooth channel a value of the Manning n equal to 0.009 is used. In the rough channel the Manning n is taken equal to 0.05.

Figure 5 shows the computed free surface profiles for the two series under full breach conditions. The solid curves represent the full dynamic equations (7). The dashed lines correspond to the kinematic-wave solution, in which the profile is terminated by the kinematic shock. Finally the monoclinal wave is shown to provide satisfactory results with respect to front location and depth.

Figure 6 shows the computed stage hydrograph 150 ft downstream of the dam for the smooth and rough channels under 30% breach conditions. The kinematic-wave solution is surprisingly accurate in the case of the smooth channel, but, as explained in Ref. 4, this is simply a coincidence due to the fact that the smooth channel was set almost at critical slope. In the rough channel the kinematic-wave results are disappointing, but the fitted kinematic-shock profile provides a very satisfactory estimate of the observed hydrograph. The agreement with the more accurate models is good, especially when computational cost is considered. The monoclinal wave or expanded kinematic-shock profile model appears to yield satisfactory results in most cases tested with very little additional expense and complexity over the kinematic-wave solution.

APPENDIX I - REFERENCES

1. Henderson, F.M., "Open Channel Flow," McMillan Publishing Co., New York, 1966.

2. Franz, D.D., "Dam-Break Flood Wave Analysis: Problems, Pitfalls and Partial Solutions," Proceedings of Dam-Break Flood Routing Model Workshop, held in Bethesda, Md., Oct. 1977.

3. Hunt, B., "Asymptotic Solution for Dam-Break Problem," Journal of the Hydraulics Division, ASCE, Jan. 1982, pp. 115-126.

4. Katopodes, N.D., and Schamber, D.R., "Applicability of Dam-Break Flood Wave Models," Journal of the Hydraulics Division,ASCE, Submitted for publication.

5. Lighthill, M.J., and Whitham, G.B., "On Kinematic Waves: I - Flood Movement in Long Rivers," Proceedings of the Royal Society of London, Series A, Vol. 22, No. 1178, 1955, pp. 281-316.

6, Smith, R.E., "Border Irrigation Advance and Ephemeral Flood Waves," Journal of the Irrigation and Drainage Division, ASCE,June 1972, pp. 289-307.

7. Strelkoff, T., Schamber, D.R., and Katopodes, N.D., "Comparative Analysis of Routing Techniques for the Floodwave from a Ruptured Dam," Proceedings of Dam-Break Flood Routing Model Workshop, held in Bethesda, Md. Oct. 1977.

8. Waterways Experiment Station, "Floods Resulting from Suddenly Breached Dams," Miscellaneous Paper No. 2-374, U.S.Army Corps of Engineers, Vicksburg, MS, 1961.

# VELOCITIES AND PERIODIC FORCES FOR TRASHRACKS

By Wallis S. Hamilton,[1]L.M. ASCE and J. Paul Tullis,[2]M. ASCE

## ABSTRACT

Trashrack design for pumped-storage hydro plants requires an estimate of the periodic forces that may act on the bars and panels. Forces caused by eddies shed from trashrack members and forces caused by pressure pulsations from the pump/turbine runner were used as part of the load that the Bath County trashracks would need to endure. To help estimate these loads, velocities and pressures during generation were measured downstream from the draft tube exits of a 1:12.76–scale model of a Bath County pump/turbine which included the water passages downstream to the lower reservoir. For some operating conditions, velocities were badly distributed across the trashrack plane, and a time-averaged peak magnitude of 10.4 m/s (prototype) was recorded. The largest average velocity through an area about 2 meters square was 8.8 m/s prototype.

A sample calculation shows that the amplitude of the periodic force acting on a 25.4 mm-thick bar--the force caused by eddy-shedding when the approach velocity is 8.8 m/s--could be about 240 N/m of length. During resonant vibration, the equivalent static load would be 50 to 100 times this much.

The Bath County runner can generate, in the water, sound waves having a frequency 21 times the runner speed in rps, or 90 Hz. At a resonant frequency of 90 Hz, an elastic bar driven by such a wave could respond with a displacement amplitude of about one millimeter.

## INTRODUCTION

Reference [1] cites eight trashrack failures caused by flow-induced vibration. All but one occurred when the racks were located near the draft tube exits of pumped-storage plants. Draft-tube trashracks of pumped-storage schemes are especially vulnerable during generating, when they are subjected to high velocities and periodic forces. Force amplitudes may be small, but the probability of resonance of a lightly-damped structure can be large.

The lower trashracks of the Bath County Pumped Storage Project are located 24.4 m downstream from the turbine centerline. Each machine discharges up to 153 m³/s through a pair of draft tube passages. At the trashrack plane most of the flow can be expected to pass through a cross-sectional gross area of 60 m², giving an average velocity of about 2.5 m/s and suggesting local maxima of perhaps 10 m/s.

A wide range of local velocities produces a correspondingly large variation in the frequency of periodic forces caused by eddy-shedding. Consequently, the natural frequencies of many modes of vibration of a trashrack panel are likely to be matched by equal driving frequencies. The panel must be designed to withstand stresses produced by resonant vibrations. When resonant vibrations are anticipated, the first design requirement is a reasonably accurate knowledge of the velocity distribution in the flow cross section. For the Bath County Project, velocity traverses were made in a hydraulic model

---

1 Senior Hydraulic Specialist, Harza Engineering Co., 150 S. Wacker Drive, Chicago, IL 60606.

2 Prof. of Civil Engineering, Utah State Univ., Water Research Lab., Logan, UT 84322.

and also an attempt was made to measure pressure pulses emanating from the runner. Such pulses have identifiable frequency components caused by the runner blade tips passing close to the wickets.

The purposes of this discussion are to describe the tests, give an example of velocity distribution in a passage downstream from the draft tube, present a sample calculation of the eddy-shedding forces and frequency that may drive the vibration of a trashrack panel, and outline, by an example, how sound waves generated by pressure pulsations in the runner could cause appreciable strains in a trashrack.

## MODEL AND INSTRUMENTS

The Allis-Chalmers Corporation provided a 1:12.76–scale model of a Bath County pump/turbine, which was set up for testing at the Water Research Laboratory at Utah State University. The elbow draft tube, the two water passages downstream from it, and a portion of the tailwater pool were reproduced. Suitable blockage was installed to represent the obstruction to be caused by the trashracks. Velocity and pressure measurements were made just upstream from the trashrack plane.

A laboratory pump supplied up to 0.28 m$^3$/s at 34 m head to run the turbine. The turbine output was absorbed by a second pump and throttling valve. Because part of the water supply came from a mountain stream and was not recirculated, no significant temperature rise resulted from a continuous input to the system of up to 90 kW.

The water velocity in the model was not based upon Froude scaling but upon the peripheral velocity of the runner, which was somewhat arbitrary. For example, when the model was run at 1003 rpm, which is 3.90 times the prototype speed, the velocity scale ratio of the runner periphery was 3.90/12.76, or 1:3.27, model to prototype. To make the water velocity scale ratio equal to 1:3.27 the head scale ratio was set at 1:(3.27)$^2$. More formally, the speed coefficient $\phi$ in the model always was made equal to the prototype $\phi$, where

$$\phi = \frac{r\omega}{(2gH)^{1/2}}$$

r = throat radius of the runner
$\omega$ = angular velocity, radians/second
g = acceleration of gravity
H = net head on the turbine.

The throat radius was measured at the exit cross section--exit during generation. During ordinary operation, values of the speed coefficient lie between 0.52 and 0.56.

Time-averaged velocities were measured with a five-hole cylindrical Pitot-type tube manufactured by United Sensor. Three differential manometers and the manufacturer's calibration curves allowed one to determine the magnitude and direction of the vector at each traverse point. The probe slid through a packing-gland that fit four receptacles built into the vertical outside wall of each of the two water passages just upstream from the trashrack plane. Thus the tip of the instrument could be moved horizontally across a flow passage at four different elevations.

To save time, measurements were not made above the plane (extended) of the draft-tube ceiling. The highest velocities usually appeared within one meter (prototype) of the floor.

Instantaneous pressures and pressure differences were measured with Entran semiconductor sensors flush-mounted in a streamlined strut. The strut, which was about 16 mm thick and 60 mm deep, also slid through a packing-gland that fit the wall receptacles. The strut partially spanned a flow passage horizontally, and the pressure sensors were located about 22 mm from the free end. The sensors were arranged like the openings in a two-dimensional Pitot tube--one on the leading edge of the strut, one on the top about halfway between the leading and trailing edges, and one on the bottom. The thought was that instantaneous pressure differences would represent fluctuating lift and drag components on the strut--a body that would represent a horizontal girder about 0.2 m thick and 0.8 m deep in the prototype.

Amplified sensor readings were recorded on magnetic tape and later analyzed to get pressure and pressure-difference spectra.

## VELOCITIES AND STROUHAL FORCES

The highest velocity measured was 10.4 m/s prototype. It occurred during a runaway test at 95% gate opening, $\phi = 0.70$, and was located near the lower outside corner of the right-hand passage of the prototype, looking downstream. (In a plan view, the prototype runner rotates counter-clockwise.)

The magnitudes of velocities on which a distribution for the trashrack design was based are shown in Fig. 1. Pitot readings were taken at the location of the decimal points in the figure. The measurements were made at 95% gate and $\phi = 0.52$, a possible full load operating condition. Velocity vectors for this condition and two others are presented in Reference [2] as three-dimensional plots. Reference [2] shows directions in the model which are opposite to those for the prototype because the model runner rotated clockwise.

One result of the direction-of-velocity measurements was that the transverse members of the trashrack panels (laterals and girders) were tilted upward, looking downstream, 6.6° from the floor to reduce head loss.

Eddies shed from a trashrack member exert periodic forces on the members in both the upstream-downstream direction and a direction perpendicular to the flow. These forces may be thought of as periodic drag and lift. They are called Strouhal forces because their frequencies are proportional to an experimental number called the Strouhal number. Thus, the first harmonic of the Strouhal frequency is given by

$$f_1 = \frac{SV}{D}$$

and the second harmonic is

$$f_2 = \frac{2SV}{D}$$

and so on. In these expressions

$$
\begin{aligned}
f &= \text{frequency, cycles/s (Hz)} \\
S &= \text{Strouhal number, dimensionless} \\
V &= \text{velocity approaching the member} \\
D &= \text{thickness or diameter of the member.}
\end{aligned}
$$

The Strouhal number depends on the shape of the member, its angle of attack and, in the case of a trashrack or grid, the ratio of the blocked area to the total flow area just upstream from the grid. For deep rectangular bars, small angles of attack, and a blockage up to 15%, the value of the Strouhal number is about 0.22.

The first harmonic component exerts the major force in the direction perpendicular to the flow (lift or heave). The second harmonic also exerts lift, but it provides the major force in the upstream-downstream direction (plunge direction in vibrating trashrack usage).

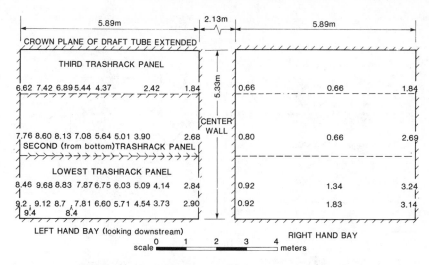

FIG. 1. VELOCITY MAGNITUDES FOR PROTOTYPE, m/s, 95% GATE OPENING, HEAD=384m
DISCHARGE RATE = 153m³/s, $\emptyset$ = 0.52

One must understand that the eddies shed by a stationary bar may not be synchronized along its length. If the bar vibrates near a Strouhal frequency $f_1$, $f_2$, $f_3$, however, the shedding pattern becomes organized and the resulting periodic forces act in phase throughout much of the span of the bar. The vibration is sustained or amplified.

Thus, if one attempts to estimate a resonant deflection (due to eddy shedding) by calculating the small static deflection that a Strouhal force would cause, he must consider the local Strouhal force per unit length, the probable length along which the local forces are correlated, and a resonant amplification factor. The amplification factor depends upon damping; an estimate that the resonant deflection of a bar or panel may be 50 to 100 times the static deflection that would be caused by applying Strouhal or other periodic forces is reasonable. A simplified example to illustrate how to estimate Strouhal forces follows.

Suppose a trashrack panel composed of horizontal and vertical bars one inch (0.0254 m) thick were subjected to the high velocities shown for the lowest panel on the left of Fig. 1. Because of diagonal stiffeners, the panel acts as a truss and is estimated to have a natural frequency of 80 cps (Hz) when it oscillates in a second heave mode as shown in Fig. 2a. This mode could be sustained by periodic forces that acted upward on the left-hand half of the panel and downward on the right at one instant and then reversed direction a half period later. The problem is to estimate the force amplitude on each half of one of the horizontal members.

We first set $f_1$ and $f_2$ equal to 80 Hz and calculate the corresponding required velocities. Thus

$$V_1 = \frac{f_1 D}{S} = \frac{(80)(0.0254)}{0.22} = 9.24 \text{ m/s}$$

$$V_2 = \frac{f_2 D}{2S} = 4.64 \text{ m/s}$$

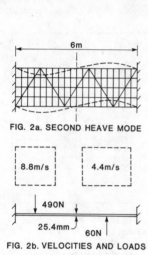

FIG. 2a. SECOND HEAVE MODE

FIG. 2b. VELOCITIES AND LOADS

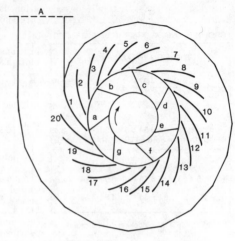

a1, b4, c7, d10, e13, f16, g19, a2,
FIG.3. BLADE PASSING SEQUENCE

Next, a study of Fig. 1 shows average approach velocities for the left-hand third and right-hand half of the lower panel of 8.8 and 4.4 m/s respectively, as shown in Fig. 2b. These numbers then are compared with the required velocities $V_1$ and $V_2$. An approach velocity of 8.8 m/s is close enough to the required first harmonic velocity of 9.24 m/s to produce eddies that would lock in with an 80 Hz oscillation. Hence, we conclude that eddy-shedding will be synchronized along a two-meter stretch near the left end of a horizontal bar. Further, this length will be driven by a first harmonic Strouhal force calculated as follows:

$$F_1 = C_1 \rho \ell D \frac{V_1^2}{2} = 0.25(1000)(2)(.0254)\frac{8.8^2}{2}$$

$$= 490 \text{ N}$$

The lift coefficient $C_1 = 0.25$ comes from a number of sources; the density $\rho$ is 1000 kg/m$^3$, and the length $\ell$ is the two-meter correlation length.

Similarly, the average velocity of 4.4 m/s is close enough to the required second harmonic velocity of 4.62 m/s to produce eddies whose second harmonic would lock in with 80 cps. The correlation length is estimated to be 2 m in the center of the right-hand half of a horizontal bar. A calculation using a second harmonic lift coefficient of 0.125 and an approach velocity of 4.4 m/s yields

$$F_2 = 60 \text{ N}$$

Note that the lift-force amplitude calculation uses bar thickness, not chord length (depth). The force amplitudes $F_1$ and $F_2$ are shown acting in opposite directions in Fig. 2b.

To arrive at stresses caused by resonant vibrations, one needs to calculate all the forces like $F_1$ and $F_2$ that could contribute to the motion. When more than one size of member appears in a proposed rack design, the velocity that will trigger vibration at some natural frequency depends directly on the member thickness and, of course, upon the harmonic of the Strouhal frequency that one chooses. Finally, when a set of Strouhal forces has been calculated, each force must be multiplied by the resonant amplification factor already mentioned.

## RUNNER SPEED AND BLADE PULSES

A pump/turbine runner generates a pressure pulse each time a blade passes a wicket. The pulses travel through the runner passages and combine to produce "noise" in the water flowing out the draft tube. The frequencies of the blade-passing noise components are higher than the frequency associated with the wobble or oscillation of the vortex that develops under certain loads in the draft-tube cone.

The pulses also travel around the spiral case and broadcast sound waves that travel up the penstock. References [3] and [4] describe difficulties caused by such waves.

The pressure sensor records from the Bath County model tests yielded several frequency components that were related to runner speed. One of these, 350 Hz at 1000 rpm, can be explained by referring to Fig. 3. Let us think in terms of the pumping mode and visualize pressure pulses that originate at wicket stations (1) through (20) as the seven blades (a) through (g) pass these stations. When a runner exerts a torque on the water passing through it, the relative velocity along the back of a blade must differ from that along the front. The result is a sudden change in the pattern of absolute velocities as a blade tip passes close to a fixed object such as a wicket. The resulting change in pressure on the wicket "nose" can be expected to be quite large when, for example, the blade tip speed is about 85 m/s, as it is for the Bath County prototype. A conservative guess is a local change of 30 or 40 m of head.

Fig. 3 is drawn with Blade (a) opposite Wicket (1). Let this be the source of the first pulse. An instant later Blade (b) passes Wicket (4) and then Blade (c) passes Wicket (7). We see that the source of the pulses steps around the circle 3 wickets at a time. After seven steps, the eighth pulse occurs as Blade (a) passes Wicket (2). Now the clue to the observed frequency is that the pulse location traveled from Wicket (1) around the circle to Wicket (2), i.e., 21/20 of the circumference, while Blade (a) traveled 1/20 of the circumference. Therefore, the pulse locus makes 21 trips around the wicket ring while Blade (a) makes one trip. The trip frequency is 21 times the runner frequency measured in revolutions per second. When the model is running at 1000/60 rps, the trip frequency is exactly 350 circuits per second. When the prototype is running at synchronous speed, i.e., 257.14 rpm, the trip frequency is 90 Hz.

But why should a Fourier component of the measured noise have the same frequency as the trip frequency? For ease of visualization, suppose a microphone were located at A, the exit of the spiral case during pumping. During the first trip, pulses from Wickets (1), (4), (7),. . . (19) travel with the velocity of sound from their points of origin to Section A. Assume that the most viable route is through the water between the wickets and in the spiral case. Each pulse will have its own travel time and attenuation; upon its arrival at Section A it will have its own shape also. Depending on the size and speed of the machine, however, the sum of the seven pulses generated during the first trip will add up to a pressure fluctuation at A having a period about equal to the time of one trip. Similarly, during the second trip pulses from Wickets (2), (5), (8),. . . (20) will produce a pressure fluctuation having the same period but a slightly different shape. The third trip generates a pressure wave at A originating at Wickets (3) to (18), 6 pulses instead of 7, but the fourth trip duplicates the wave produced by the first trip. Hence, we envision a pattern of three somewhat different trip waves having an average frequency of 21 cycles per revolution—a pattern that repeats 7 times per revolution.

The pattern repeat frequency of 7 per revolution is the fundamental frequency associated with seven blades. No Fourier component having this frequency was found in the draft tube model of the Bath County Project. Unaccountably, a strong component having a frequency of 6 per revolution sometimes appeared in the model.

The concept of adding pulses generated at various places and traveling as elastic waves for various distances to a chosen measuring point was applied to turbine vibrations at least 50 years ago. In connection with the present discussion, it shows that the transfer of model results to a prototype cannot be taken casually. Because elastic pulses travel about as fast in a model as they do in the prototype, they may arrive at a model measuring station in the wrong sequence and consequently produce an improper wave form and amplitude.

## ELASTIC WAVES AND FORCES

To approach the question of how an elastic wave in water may act on a trashrack panel, we first consider the motion of the water particles during the passage of a plane wave and then calculate the possible effect on a solid object, such as a trashrack member suspended in the water. Forces can be significant because the accelerations of water particles are large even though amplitudes are small.

Suppose the pulses from a large pump/turbine runner add up to a pressure wave that travels through the penstock toward the upper reservoir. Let the wave frequency be 90 Hz and the amplitude of the pressure, assuming a sine wave form, be 69 kN/m$^2$ (10 psi). For a wave velocity of 1300 m/s the wavelength will be 14.44 m. As the pressure wave passes a section of the penstock, the alternate compression and rarifaction of the water will cause the water particles in a cross-sectional plane to move back and forth with an amplitude slightly less than 0.09 mm. Because the circular frequency is 180 $\pi$ radians/s, however, the corresponding amplitude of acceleration is nearly 3g.

A bar having the same density and elastic modulus as water and freely floating in the liquid would oscillate with an amplitude of only 0.09 mm, but the force amplitude exerted on it by the water to make it perform the oscillation would be about 28 N/kg.

The final step is to increase the density of the bar to that of steel and suspend it from the ceiling of the passage with two of the proverbial cantilever springs that have neither mass nor volume. With proper tuning it then becomes the analog of a trashrack member having a natural frequency in water of 90 Hz. The picture now has changed to that of a damped vibration forced by the pressure wave. Calculation shows that for a damping factor of 1/80, the resonant amplitude of the bar motion will be 12 to 15 times the amplitude of the original water motion--say 1 mm. When the bar thickness and depth are small compared to the wavelength, the multiplication factor depends upon the bar density and somewhat on its shape, but not upon its size.

Thus, if the amplitude and frequency of a sound wave can be determined, the resonant deflection caused by the wave can be estimated for a particular panel design and mode of vibration. Accompanying stresses would add to those caused by Strouhal forces of the same frequency.

## ACKNOWLEDGMENT

The work described was funded by the Virginia Electric and Power Company as part of design studies for the Bath County Pumped Storage Project.

**REFERENCES**

[1] Behring, A. G. and Yeh, C. H., "Flow-Induced Trashrack Vibration," Pump Turbine Schemes Planning. . ., Joint ASME-CSME Applied Mech. Conf., Niagara Falls, NY, June 1979, published by ASME, New York, NY 10017.

[2] Hamilton, W. S. and Tullis, J. P., "Velocities and Pressure Forces for Design of Lower Reservoir Trashracks, Bath County Pumped Storage Project," Water Power and Dam Construction, Sutton, Surrey, England, August, 1982.

[3] Den Hartog, J. P., "Mechanical Vibrations in Penstocks of Hydraulic Turbine Installations," Transactions, ASME, Vol. 51, HYD 51-13, 1929.

[4] Alming, Knut, "Damping of Pressure Pulsations from a Turbine by an Energy-Absorber in the Pipeline," Paper No. 1, Symposium on Vibrations in Hydraulic Pumps and Turbines, Manchester, England, Sept. 1966, published by the Inst. of Mech. Engineers, London.

# MT. ELBERT TRASHRACK VIBRATION STUDIES

Robert V. Todd 1/, Brent W. Mefford 2/, and Thomas J. Isbester 3/

## Abstract

A finite element analysis was performed on a prototype trashrack to determine if the analysis could be used in the design stage to predict frequencies and mode shapes of vibration. Both laboratory and field tests were performed on the trashrack to provide comparative data of the vibrations in air and in water. The effects of added mass were apparent for heave modes resulting in reductions in frequencies of about 20 percent in the field tests as compared to the laboratory tests in air. Plunge modes were not affected by added mass. Due to the random nature of flow velocity field on the trashrack, no lock-in vibrations were observed in the field tests. The presence of an intermittent draft tube surge with vortex breakdown was found to be the most severe condition the trashrack was subjected to. The periodic surge did not produce adverse conditions on the trashrack. A finite element analysis of a proposed trashrack design will give sufficiently accurate frequencies and mode shapes for evaluation purposes.

## Introduction

The Bureau of Reclamation's experience to date with trashracks has been with low flow velocities, usually less than 0.6 m/s. Under these conditions, there is insufficient energy in the flowing fluid to cause damage to the trashrack. Because of their suitability to provide power system regulation, reserve, and peaking power, pumped-storage facilities are becoming popular. Due to economic considerations, trashracks in these installations are being designed for higher velocities than 0.6 m/s. Depending on the overall operational parameters of the facility, there is a possibility of failure due to vibration. The source of vibration may arise from pulsations from the pump-generator in the form of intermittent or periodic surging, and water velocities resulting in vortex shedding from the trashbars.

Depending on the structural arrangement of the trashrack, failure may arise from the individual bars being excited at their fundamental frequency or the trashrack as a whole may be forced to vibrate at one of its natural frequencies. A number of papers have been written covering case histories of trashrack vibration failures with explanations of the

---

1/ Mechanical Engineer, Division of Design, Bureau of Reclamation, Denver, Colorado.
2/ Hydraulic Engineer, Division of Research, Bureau of Reclamation, Denver, Colorado.
3/ Hydraulic Engineer, Division of Research, Bureau of Reclamation, Denver, Colorado.

exciting mechanisms and the structures' response (see Appendix - Bibliography).

The Bureau needed a new design approach to take into account possible vibration of a trashrack under operational conditions. This investigation was undertaken to establish that design approach.

## Proposed Design Approach

Initially the trashrack would be sized by the conventional approach based on a static head loading based on trash accumulation. Then a vibration analysis would be performed which would consider forcing frequencies; individual bar fundamental transverse frequencies; and natural frequencies and mode shapes of the trashrack under its proposed field conditions.

The determination of the trashrack's response to operating conditions is the most complex part of the analysis. It is dependent upon a number of factors which are difficult to predict at the design stage. These factors are the trashrack's boundary conditions or fixity, the velocity profile at the trashrack location, and the damping of the trashrack.

A program was proposed to obtain the natural frequencies and mode shapes of a prototype trashrack in air using both analytical and laboratory testing methods, and then comparing the results. The trahsrack would then be field tested to determine its response to actual operating conditions. Comparison could be made between the laboratory and the field results to determine the impact of the various factors mentioned previously.

## Description of Trashrack Chosen for Tests

A trashrack from Mt. Elbert Pumped-Storage Powerplant was selected for the tests because it was economically feasible to utilize an existing trashrack located a reasonable distance from the laboratory. The trashracks - two bays of three trashracks each, one on top of the other - are located at the draft tube bellmouth, 40.84 m from the centerline of the pump-generator (see fig. 1). The draft tube is split by a center pier along its length; during generation, when operating at best efficiency the flow distribution at the trashracks is unsymmetrical with the higher velocities occuring on the left side (looking downstream.) The draft tube centerline is positioned 10° to the horizontal. The trashracks are located vertically in their guides; consequently, flow during generation will impinge on the transverse bars at an angle of attack of 10°. The upper trashrack in the left bay was selected because it required minimal handling compared with the lower trashracks.

The trashrack had been designed by the conventional method with no consideration given to fluid-structure interaction. The trashrack is 6.78 m wide by 2.5 m high made of seven 32- by 254-mm transverse load-carrying bars and forty 16- by 51-mm vertical bars. Loading is transferred to the side members, 152- by 102- by 16-mm angles, which bear on the steel guides.

The vertical bars have an aspect ratio of 3.2 which results in the flow forming a separation zone without reattachment taking place. The transverse bar spacing of 406 mm makes the fundamental transverse frequency of the vertical bars between welds equal to 334 Hz; this value is far in excess of any probable forcing frequency.

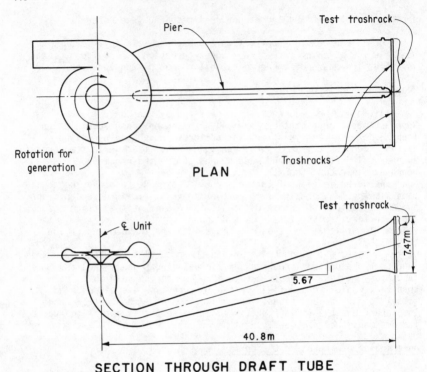

PLAN

SECTION THROUGH DRAFT TUBE

FIG. I - General Arrangement of Pump - Turbine and Trashracks

## Choice of Boundary Conditions

The trashacks at Mt. Elbert are located in guides with no clamping devices securing them. The total transverse clearance between the trashrack and the guides is 32 mm laterally and 10 to 13 mm in the fore and aft directions. When the trashracks are subjected to flow, the drag will cause them to seat against the guides. Unevenness of the trashrack and the guides may result in a nonuniform bearing. The effect of possible uneveneness on the frequencies cannot be determined in advance.

Considering two sets of boundary conditions, it was planned to cover the actual response of the trashrack in its field location and have two different cases with which to compare the finite element model to the laboratory setup. The two cases were designated setup A, short sides of trashrack clamped; and setup B, the four corners clamped.

## Finite Element Analysis

An in-house finite element program, STR5, was used to determine the natural frequencies, mode shapes, and moments of the trashrack for two sets of boundary conditions. STR5 is a structural analysis program for static and dynamic analysis of linearly elastic systems. The dynamic analysis part of STR5 provides a means of determining the natural frequencies and mode shapes for the structural system being modeled.

Initially a two-dimensional model was used. The complete trashrack was modeled using beam elements to represent the actual rectangular bars. The nodes were modeled at the intersection of the actual members. There were 294 nodes, each with 6 degrees of freedom resulting in 1764 equations to be solved. The computer can handle this size of problem, but it uses an excessive amount of time if more than three frequencies are required. The model was first run with the corners restrained in the x, y, and z directions to obtain the first three frequencies. The results indicated that there was no coupling between the heave and plunge modes. The terms heave and plunge are used to describe the motion of the vibrating trashrack. A heave mode takes place in the x, y plane, and plunge mode in the y, z plane. The axes relative to a trashrack are defined as: x parallel to the width; y parallel to the height, and z normal to the surface. Nodal lines are used to define the mode shape; for example, a (1-2) mode would have one vertical nodal line and two transverse nodal lines.

For the plunge modes, the rotation about the z axis and the displacements in the x and y directions were insignificant. For the heave mode, the rotations about the x and y axes and the displacement in the z direction were insignificant. By fixing three redundant degrees of freedom at each node, one type of mode would be obtained with half the number of equations to be solved. With the reduction in unknowns, a greater number of frequencies could be obtained from the same computer time. Running the program twice with the appropriate fixities, the first 10 frequencies for the plunge and heave modes could be obtained for one set of boundary conditions at an economical computer usage rate.

Two different sets of boundary conditions were required. The first, "test setup A," had the z displacements fixed along the short sides of the model. This simulated one of the laboratory setups where the short sides were clamped. It also approximated the actual installation conditions of the trashrack at the end of the draft tube. The second set of boundary conditions, "test setup B," had the x, y, and z displacements at the four corners fixed and simulated the second setup in the laboratory.

While running the laboratory tests, there appeared to be coupling between the heave and plunge modes. This was due, however, to the offset structural configuration. The vertical, 51- by 16-mm trashbars are welded to the short face of the 254- by 32-mm transverse beams. When the vertical bars deflect in a plunge mode configuration, the transverse bars are forced to rotate. These transverse bars have negligible torsional stiffness because of their rectangular shape, and due to the rotation, a vertical component of displacement is obtained which could be interpreted as a heave mode. This effect was simulated by utilizing a three-dimensional model.

Because of the node limitation in the STR5 program, only one-fourth of the trashrack was modeled. All of the mode shapes could be obtained by modifying the boundary conditions. Apart from the boundary nodes, the remaining nodes were allowed 6 degrees of freedom resulting in a model with 168 nodes having an average of 970 equations to be solved. Eight different sets of boundary conditions were required to obtain the plunge and heave modes for the two laboratory setups.

Laboratory Tests

    The laboratory testing of the trashrack unit was accomplished in the vibration test building operated by the Concrete and Structural Branch of the Bureau of Reclamation Engineering and Research Center in Denver, Colorado. The trashrack was placed horizontally and supported on its short sides on 457- by 298-mm-wide flange beams to allow for access to the underside with electromagnetic shakers.

    Six accelerometers and six strain gages were attached to the trashrack as shown in figure 2. In addition, a roving accelerometer was used to help define the mode shapes.

    Two electromagnetic shakers were available to excite the trashrack. Some modes required two shakers either in phase or 180° out of phase.

    In addition to data from the STR5 program, the transfer function capability of a spectrum analyzer was used to locate excitation frequencies for the laboratory study. The transfer function measurement yielded a ratio of cross power spectrum to autopower spectrum, or roughly the ratio of output to input at the selected bandwidth interval. Pseudorandom noise fed to an electromagnetic shaker was used to excite all frequencies within the total bandwidth. The transfer function output was also used as an indicator of the torsion present at each

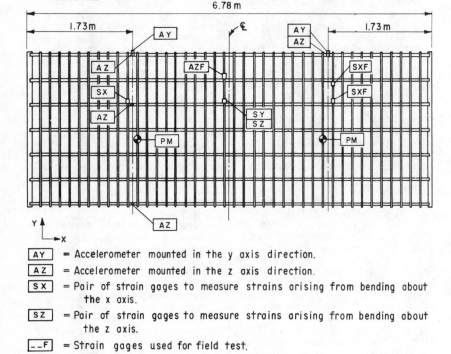

AY   = Accelerometer mounted in the y axis direction.

AZ   = Accelerometer mounted in the z axis direction.

SX   = Pair of strain gages to measure strains arising from bending about the x axis.

SZ   = Pair of strain gages to measure strains arising from bending about the z axis.

--F   = Strain gages used for field test.

PM   = Propeller meter.

FIG. 2  Trashrack Instrumentation

excited mode by comparing transfer functions of accelerometers located with their sensitive axis at 90° from the trashrack excitation direction. This allowed for comparison and adjustment of trashrack boundary conditions on a real-time basis.

The remainder of the tests utilized discrete finely tunable sinusoidal excitation to drive the electromagnetic shakers and vibrate the trashrack.

Plunge modes were tested for setups A and B. Heave modes were tested for setup B only. Some predicted modes were well defined and easy to attain, while others were difficult to achieve or did not conform to the predicted shape.

Individual strain gage and accelerometer data were obtained with a measurement and control processor and stored on floppy disk for later FFT (Fast Fourier Transform) analysis. In addition, data were analyzed on an FFT analyzer, channel by channel during the tests.

The trashrack damping factor was determined for both the plunge and heave modes. After the steady-state data were obtained for the test frequency, the circuit to the electromagnetic shaker was opened and rack vibration allowed to decay. A retentive oscilloscope equipped with a Polaroid camera was used to record the rate of decay of the vibration. The traces were then analyzed using the logarithmic decrement method to obtain the damping factor.

## Comparison Between Two- and Three-dimensional Finite Element Model Values and Laboratory Test Results

The frequency values obtained from the finite element models and the laboratory tests were tabulated and compared (tables 1, 2, and 3). There was good agreement in the case of test setup A except for the (1-3) plunge. In the case of test setup B, good agreement was obtained except for the (0-3) plunge with the three-dimensional model. Three heave modes obtained in the laboratory tests which correlated well with the two-dimensional finite element model were not obtained with the three-dimensional model.

## Field Tests

The trashrack was installed on a new pump-turbine unit, and testing was accomplished during the startup phase which included loads of 25, 50, 60, 70, 75, 80, 90, 100, and 115 percent.

The instruments, consisting of six accelerometers, six strain gages, and two propeller meters, were installed on the trashrack according to details on figure 2. All lead wires were fixed to the rack bars with plastic wire ties and fiberglass tape to eliminate slack and funneled to the top centerline of the trashrack. From this point, the wires were placed within a single polyethelene plastic tube which extended to the top of the afterbay wall and then to the instrumentation shack.

A 1:1 coal-tar epoxy was used to coat the accelerometers, strain gages, and lead wires. This coating fixed the position of the lead wires to the trashrack bars and prevented their vibration.

Frequency analysis of field data was performed using a combination of real time output from the spectrum analyzer and computer-controlled analog-to-digital sweeps of all transducer channels.

The spectrum analyzer was used to obtain accurate amplitude response. Time domain records were windowed using a flattop passband function before the FFT. Amplitude spectrums were obtained using long-term power spectrum averaging of successive FFT's.

Continuity between trashrack transducers was established using multiple-channel computer data acquisition. Time domain data sweeps of all transducers were taken at a 454-Hz rate. To minimize aliasing, analog filters were placed on each channel. Filter cutoff points were set to 110 Hz, dropping 32 dB per octave. The FFT's run on the time domain data yielded near simultaneous amplitude and phase data for all transducers.

Table 1. - Plunge mode frequencies of the trashrack for test setup A

| Mode shape (1) | Two-dimensional finite element model, natural frequency in Hz (2) | Three-dimensional finite element model, natural frequency in Hz (3) | Laboratory results, natural frequency in Hz (4) |
|---|---|---|---|
| 0-0 | 11.47 | 11.61 | 11.60 |
| 0-1 | 12.57 | 12.70 | 13.10 |
| 0-2 | 21.51 | 20.22 | 21.42 |
| 1-0 | 45.57 | 46.34 | 44.20 |
| 0-3 | 46.48 | 41.06 | 42.43 |
| 1-1 | 47.46 | 47.39 | 46.09 |
| 1-2 | 52.48 | 51.16 | 53.40 |
| 1-3 | 68.43 | 61.00 | 61.82 |

Table 2. - Plunge mode frequencies of the trashrack for test setup B

| Mode shape (1) | Two-dimensional finite element model, natural frequency in Hz (2) | Three-dimensional finite element model, natural frequency in Hz (3) | Laboratory results, natural frequency in Hz (4) |
|---|---|---|---|
| 0-0 | 10.42 | 10.50 | 11.00 |
| 0-1 | 12.54 | 12.46 | 12.94 |
| 0-2 | 20.81 | 20.57 | 21.20 |
| 1-0 | 27.80 | 28.42 | 32.20 |
| 2-0 | 43.96 | - | 42.17 |
| 0-3 | 46.34 | 32.88 | 42.13 |
| 1-1 | 46.72 | 46.43 | 45.66 |
| 1-2 | 51.08 | 50.30 | 50.41 |
| 1-3 | 67.22 | 60.58 | 64.08 |

Table 3. - Heave mode frequencies of the trashrack for test setup B

| Mode shape (1) | Two-dimensional finite element model, natural frequency in Hz (2) | Three-dimensional finite element model, natural frequency in Hz (3) | Laboratory results, natural frequency in Hz (4) |
|---|---|---|---|
| 0-0 | 7.11 | 6.90 | 6.92 |
| 1-0 | 15.21 | 14.94 | 14.43 |
| 2-0 | 25.11 | - | 24.78 |
| 3-0 | 37.29 | 36.18 | 36.81 |
| 4-0 | 52.03 | - | 51.94 |
| 6-0 | 89.07 | - | 88.41 |

Table 4. - Comparison between the trashracks' natural frequencies under laboratory and field conditions

| Mode shape (1) | Laboratory results, natural frequency in Hz (2) | Field results, natural frequency in Hz (3) | Percentage difference between field and laboratory frequencies. (4) |
|---|---|---|---|
| Heave, 0-0 | 6.92 | 5.60 | -19.10 |
| Heave, 1-0 | 14.43 | 11.50 | -19.60 |
| Heave, 2-0 | 24.78 | 19.60 | -20.10 |
| Plunge, 0-2 | a/ 21.42<br>b/ 20.81 | 20.90 | a/ -2.43<br>b/ 0.43 |

a/ Test setup A.
b/ Test setup B.

The location of the installed trashrack made the measurement of instantaneous velocity impractical. Current meters which produce an electrical pulse based on propeller rotation were utilized. The two meters used were calibrated for a range of velocities of 0.18 to 5.49 m/s, producing one pulse for every 20 revolutions of the propeller. The meters were placed on the trashrack as shown on figure 2. The large number of revolutions per pulse prevents measurement of instantaneous velocities but was considered adequate for these field tests. The current meters are unable to determine when flow might be moving in a reverse direction.

The pulse data were recorded on a two-channel oscillograph. The

452       APPLYING HYDRAULIC RESEARCH

records were then analyzed for average velocity by taking a long record
and counting pulses, and converting to average velocity.  In addition,
the maximum and minimum values were obtained from single pulses which
covered the shortest and greatest time periods, respectively.
The velocity data confirmed the unsteady nature of flow in the
draft tube.  The percent deviation between the extremes and the average
was considerably less at or near the 100-percent load condition.

While operating the turbine at 80-percent load and 63-percent
wicket gate opening, some rather large accelerometer and strain gage
readings were encountered.  At the time draft tube surging was
suspected, however, records of draft tube pressures were not being
obtained.  After reviewing the data from the pump-turbine model test
report, it was determined that an intermittent surge with associated
vortex breakdown was probably occurring in the draft tube.  During
December 1981, additional data were taken of draft tube pressure in the
suspected range of operation and the presence of the intermittent surge
confirmed.  Intermittent surges in the draft tube occur during the
breakdown of a helical vapor core vortex.  The formation of the vortex
and subsequent breakdown are random occurrences and produce large
pressure pulsations during the collapse of the vapor cavity.  During
this phenomenon, rapid changes in pressure and velocity occur within
the draft tube and are felt by the trashrack at the downstream end.

Comparison Between Laboratory and Field Results

Table 4 lists the modes and frequencies obtained from the field
test and the corresponding laboratory results.  Only four frequencies
were encountered in the field tests; they occurred over the range of
power settings from 60- to 115-percent load.  A maximum average velocity of 2.29 m/s occurred on the left side of the trashrack at 90-percent
load with a corresponding velocity of 1.83 m/s on the right side.  The
flow velocity fluctuated appreciably during one run.  From the results,
there appeared to be no lock-in at any frequency in the forcing
functions.  Resonant conditions were not present, the forcing spectrum
being provided by the turbulent nature of the velocity profile.

The highest g loadings occurred for the 80-percent load case, which
further testing indicated an intermittent draft tube surge was present.
The heave mode frequencies from the field test were approximately
20 percent lower than those obtained in the laboratory.  Field and
laboratory frequencies for the plunge mode were comparable.  The
reduced frequency when the trashrack is submerged may be interpreted as
being caused by an increase in the mass of the trashrack or an added
virtual mass.  The virtual mass is a function of the mode shape; the
heave modes which displace more water than the plunge modes are subject
to a greater reduction in their frequencies when submerged.  In the
case of the heave modes, to obtain the reduction in frequency from air
to water, the ratio of added mass to the structural mass was approximately 58-percent.

The stresses recorded in the field were extremely low; the highest
values obtained were for the (1-0) heave mode with 80-percent load.
The highest bending stress measured resulted in a maximum bending
stress in the vertical bar of 1865 kN/m$^2$, which is well below the
fatigue endurance limit.  Good correlation was obtained between the
stresses from the finite element models and the field test results for
the (1-0) heave mode.

It was not possible to determine the degree of fixity of the
trashrack from the field results.  The vertical heave mode frequencies
from the finite element analysis were found to be practically independent of the side fixity.  There were differences for the plunge modes,
but the plunge mode obtained in the field was found to have practically
the same frequency as the two fixities tested in the laboratory.

## Conclusions

1.  The finite element analysis gave mode shapes, frequencies, and
stresses that were in good agreement with those obtained from laboratory testing of a full-sized trashrack.
2.  A two-dimensional model was sufficiently accurate compared with a
more complicated three-dimensional model.
3.  A finite element analysis of proposed trashrack designs will give
sufficiently accurate frequencies and mode shapes for evaluation
purposes.
4.  More field testing is recommended specifically to cover the following situations:  uniform velocity profiles equal to and greater than
3 m/s, and whether the rectangular bar with a minimum aspect ratio of
4 is susceptible to forming periodic vortices with sufficient energy to
cause fatigue damage to a trashrack structure.
5.  The random nature of the flow velocity field near the trashrack
prevented the lock-in of vibration on any of the trashrack's natural
frequencies.

## Reference - Bibliography

1.  Behring, A. G. and C. H. Yeh, "Flow-Induced Trashrack Vibration,"
    Pump Turbine Schemes, Planning, Design and Operation, the Joint
    ASME-CSME Applied Mechanics, Fluids Engineering and Bioengineering
    Conference, Niagara Falls, New York, June 1979, pp. 125-134.
2.  Crandall, S. H., S. Vigander, and P. A. March, "Destructive
    Vibration of Trashracks due to Fluid-Structure Interaction," Journal
    of Engineering for Industry, Transactions of the ASME, November 1975.
3.  Fortey, J. W. and R. F. Tirey, "Flow-Induced Transverse Vibrations
    of Trashrack Bars," Civil Engineering, ASCE, May 1972, pp. 44-45.
4.  Jones, G. V., "Vibration Study of Wallace Dam Trashracks," presented
    at the August 8-11, 1971 Proceedings of the Specialty Conference on
    Conservation and Utilization of Water and Energy Resources held at
    San Francisco, California, pp. 349-356.
5.  Louie, D. S. and A. E. Allen, "Trash Rack Studies - Seneca Pumped-
    Storage Plant," ASCE National Water Resources Engineering Meeting,
    Phoenix, Arizona, January 1971, pp. 1-26.
6.  Sell, E. E., "Hydroelectric Power Plant Trashack Design," Journal
    of the Power Division, Proceedings of the ASCE, POI, January 1971,
    pp. 115-121.
7.  Vigander, Svein, "Trashrack Vibration Studies," Raccoon Mountain
    Pumped-Storage Project, Laboratory Report No. 1, Tennessee Valley
    Authority, Division of Water Control Planning Engineering Laboratory,
    October 1974.

MODELLING RIVER-CHANNEL CHANGES USING ENERGY APPROACH

by Howard H. Chang,[1] Joseph C. Hill[2], M. ASCE

ABSTRACT

A computer-based flood- and sediment-routing model which simulates river-channel changes is described. As the change in channel width reflects the river's adjustment in resistance--that is, in energy expenditure; it is therefore simulated using the energy approach. This model has been applied to study river-channel changes of the San Dieguito River; predicted results using this model are supported by field measurements. Changes in channel widths, while closely interrelated to the changes in channel-bed elevation and other variables, reflect the river's tendency to seek equal power expenditure along the channel.

INTRODUCTION

River channel changes generally include channel-bed aggradation and degradation, width variation, and lateral migration in channel bends. These changes may occur naturally or as a result of a change in the environment. Man is also regarded as a geomorphic agent with certain activities such as sand and gravel mining, bridge construction, river control schemes, etc. having contributed to river channel changes. Langbein and Leopold maintain that the equilibrium channel represents a state of balance with a minimum rate of energy expenditure or an equal rate of energy expenditure along the channel (10). Changes induced by nature or men's activities distort the channel equilibrium and therefore result in river-channel changes.

The three types of river channel changes in channel-bed elevation, channel width, and lateral migration are closely interrelated to each other and may occur concurrently. Changes in channel-bed elevation are often inseparable from width variation because a channel tends to become narrower during degradation, and it tends to widen during aggradation. Earlier versions of the FLUVIAL model have considered channel-bed aggradation and degradation (2,3,4) and width variation (3,4). The current version FLUVIAL-11 which simulates all three types of changes has been formulated, developed and applied in a case study. This paper describes this model and its application

_____

[1] Professor of Civil Engineering, San Diego State University, San Diego, CA
[2] Principal Civil Engineering, County of San Diego

in the case study.  Special attention is given to the lateral
migration and to the nature of energy (or power) transformation
in alluvial rivers associated with river-channel changes.

ANALYTICAL BACKGROUND

This mathematical model has five major components:  1) water routing;
2) sediment routing; 3) changes in channel width; 4) changes in
channel-bed profile; and 5) lateral migration of the channel.  This
model employs a space-time domain in which the space domain is represented by the discrete cross-sections along the reach and the time
domain is represented by discrete time steps.  In water routing, the
time and spatial variations of the discharge, stage, velocity, energy
gradient, etc., along the reach are obtained by an iterative procedure.  At each time step, sediment discharge at each cross-section is
computed; changes in channel width, channel-bed profile and lateral
migration are obtained and applied to each cross-section.  The bed-
material composition is updated at each time step.  Since changes in
channel geometry and bed-material composition are slow in comparison
to water routing, corrections for them are made separately for each
time step.  The component on width changes is related to the energy
expenditure as described below:

Changes in Channel Width--Simulation of width variation is based upon
the concept of minimum stream power.  At a time step, width corrections for all cross-sections are such that the total stream power
(or rate of energy expenditure) for the reach is minimized; these
corrections are subject to the physical constraint of rigid banks
and limited by the amount of sediment removal or deposition along
the banks within the time step.  Total stream power of a channel
reach is

$$P = \int_L \gamma Q S dx \qquad (1)$$

in which P = total stream power of the reach: L = length of the
reach; Q = discharge; S = energy gradient; $\gamma$ = specific weight of
water and sediment mixture; and x = distance in the flow direction.
Written in finite difference form, this equation becomes

$$P = \sum_{i=1}^{N-1} \frac{1}{2} \gamma (Q_i S_i + Q_{i+1} S_{i+1}) \Delta x_i \qquad (2)$$

in which N = total number of cross-sections for the reach; i = number
of a cross-section; and $\Delta x_i$ = distance between Sections i and i
+ 1.  Previous studies have established that minimum stream power for
an alluvial river is equivalent to equal power expenditure per unit
channel length, that is, constant $\gamma Q S$ along the reach (4,9,10).  A
river channel undergoing changes usually has uneven spatial distribution in power expenditure or $\gamma Q S$.  Usually the spatial variation
in Q is small but that in S is pronounced.  Total stream power of a
reach is reduced with the reduction in spatial variation in QS or S
along the reach.  Adjustments in channel widths are made in such a
way that the spatial variation of QS is minimized subject to the
constraints and limitations stated above.  An adjustment in width

reflects the river's adjustment in flow resistance--that is, in power expenditure. A reduction in width at a cross-section is usually associated with a decrease in energy gradient for the section whereas an increase in width is accompanied by an increase in energy gradient. Using these guidelines, a technique for width correction has been developed as described in a previous publication (4).

Width changes in alluvial rivers are characterized by the formation of small widths at degrading reaches and widening at aggrading reaches (3,4,7,9,10,11). This type of width formation represents the river's adjustment in resistance to seek equal power expenditure along its course. A degrading reach usually has a higher channel-bed elevation and energy gradient than its adjacent sections. Formation of a narrower and deeper channel at the degrading reach decreases its energy gradient due to reduced boundary resistance. On the other hand, an aggrading reach is usually lower in channel-bed elevation and energy gradient. Widening at the aggrading reach increases its energy gradient due to increased boundary resistance. These adjustments in channel width reduce the spatial variation in energy gradient and total power expenditure of the channel. Since the sediment discharge is proportional to the stream power (1), these adjustments favor the establishment of channel's equilibrium.

A CASE HISTORY OF RIVER CHANNEL CHANGES

The San Dieguito River at Rancho Santa Fe, California, experienced significant changes in a two-mile reach (see Figure 1) during recent floods. Documentation of river channel changes and flood hydrographs were made by the County of San Diego (5,12), providing a valuable set of field data for river studies.

Physical Conditions -- The study reach is about four miles from the ocean and about five miles below Lake Hodges Dam. The channel has a wide and flat natural configuration; the natural slope and bed-material size decrease significantly in the downstream direction. Bed material of the study reach varies from coarse sand ($d_{50}$ = 0.85 mm) at the upstream end to find sand ($d_{50}$ = 0.24 mm) downstream.

The natural channel configuration was distorted prior to recent flood events by men's activities including sand mining and construction of the Via de Santa Fe Road and bridge as shown in Figure 1. As a result of sand mining, several large borrow pits with a depth as great as 25 feet were created. The natural wide channel is encroached upon by the road embankment on each side of the bridge (Section 51). While the river channel has an erodible bed and banks, the banks, however, are constrained by the hills at the south bank of Section 51 and along the north banks of Section 60 to 63 and by bank protections at the north banks of Sections 51 and 58.

Two floods passed through the river, one in March, 1978 (peak flow = 4,400 cfs), and another in February, 1980 (peak flow = 22,000 cfs), when Lake Hodges spilled. Hydrographs of these floods are shown in Figure 2. Prior to these events, Lake Hodges had not spilled for 26 years.

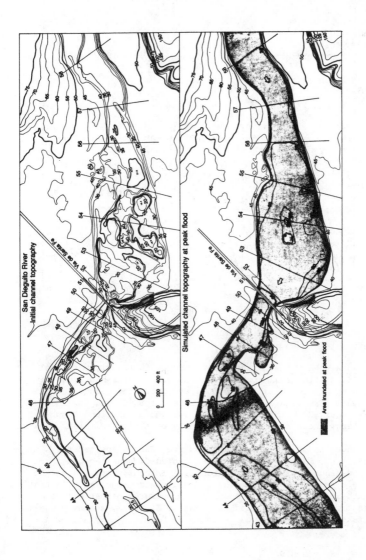

Figure 1.  Topographies and cross-section locations

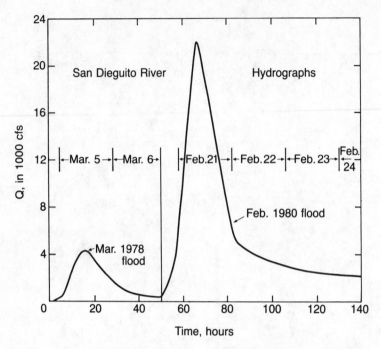

Figure 2.  Flood hydrographs

River Channel Changes -- Significant changes in the river channel
were observed after the March, 1978, flood.  Channel-bed scour
occurred near borrow pits and notably at the bridge crossing where
measurements were made as shown in Figure 3 for Section 51.  Deposition was observed in the borrow pits.  With limited flood discharge
and duration, these borrow pits were only partially refilled.

Major changes in the river channel occurred during the greater
February, 1980, flood.  These changes included channel-bed aggradation and degradatin, width variation, and lateral migration of the
channel are described below.  These changes were recorded by a high
water mark at Section 52, by channel-bed measurements at selected
cross-sections after the flood as shown in Figure 3, and by photographs taken during the flood as shown in Figures 4, 5, and 6.

SIMULATION AND RESULTS

The mathematical model FLUVIAL-11 was used to simulate river channel
changes in the San Dieguito River during the 1978 and 1980 floods.
Graf's equation (6) for bed-material load was used in computing the
sediment movement.  Channel roughness in terms of Manning's n was

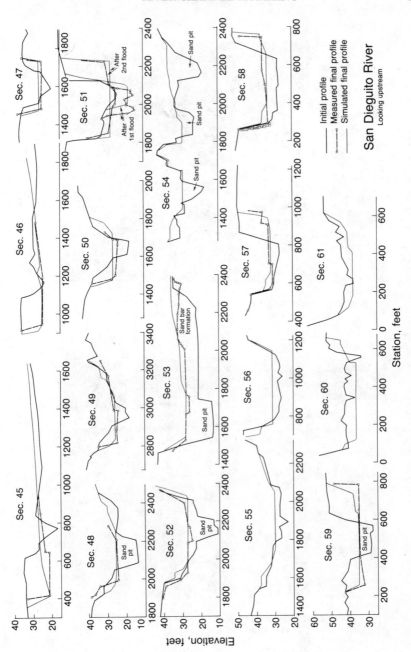

Figure 3.  Simulated and measured cross-sectional changes

Figure 4.   San Dieguito River on February 21, 1980
            (Looking toward south)

Figure 5.   Sand bar formation near Section 53
            (Looking toward south)

Figure 6. Channel bank scour near Sections 45 and 46
(Looking toward north)

selected to be 0.035 in consideration of the channel irregularity
and minor vegetation growth; it was estimated to be 0.04 at the
bridge crossing. The combined duration of 140 hours for these two
floods was computed using 2,000 time steps.

Simulated results as presented in Figures 1, 3, and 7 are described
below.

Changes in River Channel Configuration -- River channel changes,
including those in channel-bed profile, channel width, and lateral
migration, as simulated by the computer model, are described herein.
Changes in the longitudinal channel-bed profile (see Figure 7) are
characterized by aggradation in the borrow pits, erosion at higher
grounds, and the gradual formation of a more or less smooth channel-
bed profile at the end of the flood. In that process, considerable
variation in the longitudinal channel-bed elevation through the
downstream portion of the river reach is predicted at the peak flood
as shown in Figures 1 and 7. The higher channel-bed elevations at
Sections 45, 46, and 48 are associated with large channel widths
while the lower elevations at Sections 47 and 50 are due to their
small widths.

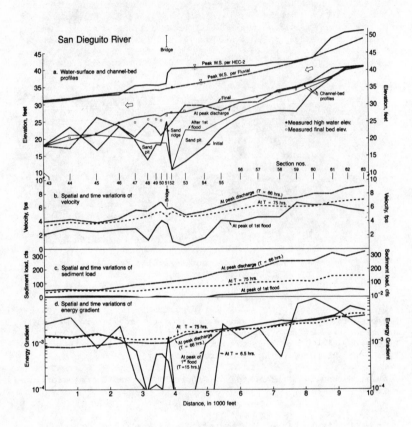

Figure 7.  Simulated and measured results

Changes in channel width which occur concurrently with variations in channel-bed elevation and lateral migration are simulated. Width changes are characterized by the gradual widening at those initially narrow sections, notably at Sections 47, 49, 50, 51, 57, 58 and 59 and reductions in width at initially wide sections, notably at Sections 53 and 54. Initial and simulated final cross-sectional profiles of these sections are shown in Figure 3 together with some measured profiles. Simulated channel width at the peak flood (shown in Figure 1) is highly uneven in its spatial variation along the river. This variation is gradually reduced during the flood as reflected by the simulated final cross-sectional profiles in Figure 3. By comparing the initial and final channel-bed profiles, one finds that widening at a section is through bank erosion and that the reduction in width is usually through sand-bar formation along the bank(s). Simulated variation of the cross-sectional profile for Section 53 illustrates the sand-bar formation and the associated reduction in channel width. A picture of this sand bar taken at the end of the flood is shown in Figure 5.

That changes in channel width and channel-bed elevation are closely related may be illustrated by the simulated time variation of the cross-sectional profile at Section 51 (see Figure 9). Initially Section 51 is on a sand ridge with borrow pits existing on both sides. Gully formation through this sand ridge during the first flood is simulated followed by gradual widening and lessening of the gully depth during the second flood. The maximum scour depth is predicted to occur in the initial gully. The simulated results correlate well with measurements at this section shown in Figure 3, in which the uneven final channel-bed profile as measured is related to the removal of several piers during the flood (see Figure 4).

Lateral migration of the channel at Sections 44, 45, and 46 in the river bend as simulated is illustrated by the variation of the cross-sectional profile at Section 46. Lateral migration is through gradual erosion of the concave bank and deposition on the convex bank. A picture of the eroded concave bank is shown in Figure 6.

RIVER-CHANNEL CHANGES IN RELATION TO POWER EXPENDITURE

Changes in river-channel configuration are accompanied by changes in flow resistance and hence the rate of energy (or power) expenditure. The $\gamma QS$ product represents the rate of energy expenditure per unit channel length. Since the spatial variation of $Q$ is small, the spatial variation of $\gamma QS$ may be represented by the spatial variation of the energy gradient $S$ shown in Figure 7.

Simulated river channel changes are associated with the gradual reduction of the spatial variation of energy gradient along the channel subject to the physical constraint of rigid banks. That the adjustment in river-channel configuration is closely related to the change in power expenditure can be illustrated by the sequential changes of cross-sectional profile at Section 51 as shown in Figure 3. Because it is initial sand ridge (see Figure 7), the energy

gradient at this section is initially much greater than those of
its adjacent sections. This pronounced spatial variation in energy
gradient is reduced through gully formation at this section and
deposition at the adjacent sections. The gully which is small in
width and has a low channel-bed elevation provides the least possible
flow resistance and hence lowest energy gradient at this section; it
also reduces the back-water effect on the upstream section and
thereby increases its energy gradient. At subsequent time intervals,
the energy gradient at Section 51 becomes less than its adjacent
sections. Cross-sectional changes at this section then include
widening in channel width and aggradation in the gully. These
changes are accompanied by increases in boundary resistance and
energy gradient at this section, favoring the establishment of equal
energy gradient along the reach. This pattern of river channel
changes characterized by the formation of narrow channel width during
channel-bed degradation and widening during aggradation, is evident
in nature and has been reported in the literature (3,4,7,8,9,10,11).


SUMMARY AND CONCLUSIONS

The mathematical model FLUVIAL-11 has been formulated and developed;
it has been employed to simulate flood- and sediment-routing and
associated river channel changes in the San Dieguito River near the
Via de Santa Fe bridge. Simulated results using this model are
supported by field observations and measurements.

An alluvial river is the author of its own geometry; therefore it
will respond to any change imposed upon by nature or by men through
self adjustments. River channel changes may include channel-bed
aggradation and degradation, width variation, and lateral migration
in channel bends. These changes are interretated as they may occur
concurrently. Therefore, a mathematical model for erodible channels
must include these variables.

River channel changes in the San Dieguito River are characterized
by the trend toward a more uniform configuration from the initially
distorted configuration. In this process, the river channel tends
to become narrower during channel-bed degradation and it tends to
widen during aggradation. This pattern of adjustment reflects the
river's tendency to seek equal power expenditure along the channel.
River channel changes provide a mechanism with which the river seeks
to establish the dynamic equilibrium in sediment transport, that is
equal sediment load along the reach.

In erodible channels, flood-level computation using a fixed-bed model
may be limited in accuracy. Improved accuracy for flood-level determination in such channels may be provided by an erodible-bed model.

APPENDIX I.  REFERENCES

1.  Bagnold, R. A., "An Approach to Sediment Transport Problem from General Physics," U.S. Geological Survey Professional Paper 422-I, Washington, D.C., 1966, 137 pp.

2.  Chang, H. H. and Hill, J. C., "Computer Modeling of Erodible Flood Channels and Deltas," Journal of the Hydraulics Division, ASCE, Vol. 102, No. HY10, October 1976, pp. 1461-75.

3.  Chang, H. H. and Hill, J. C., "Minimum Stream Power for Rivers and Deltas," Journal of the Hydraulics Division, ASCE, Vol. 103, No. HY12, December 1977, pp. 1375-89.

4.  Chang, H. H., "Mathematical Model for Erodible Channels," Journal of the Hydraulics Division, ASCE, Vol. 108, No. HY5, Proc. Paper 17062, May 1982.

5.  "Flood Plain Changes During Major Floods," Department of Sanitation and Flood Control, Community Services Agency, County of San Diego, 1979.

6.  Graf, W. H., Hydraulics of Sediment Transport, McGraw-Hill Book Co., 1971, 509 pp.

7.  Gregory, K. J., Editor, River Channel Changes, John Wiley and Sons, 1977, 448 pp.

8.  Lane, E. W., "A Study of the Shape of Channels Formed by Natural Streams Flowing in Erodible Material," Missouri River Division Sediment Series No. 9, U.S. Army Engineer Division, Missouri River Corps of Engineers, Omaha, Nebraska, 1957.

9.  Langbein, W. B. and Leopold, L. B., "River Meanders, Theory of Minimum Variance," U.S. Geol. Survey Prof. Paper 282-B, 1957.

10. Leopold, L. B., Wolman, M. G., and Miller, J. P., Fluvial Processes in Geomorphology, W. H. Freeman and Co., San Francisco, 1964, 522 pp.

11. Schumm, S. A., The Fluvial System, Wiley-Interscience, 1977, 338 pp.

12. "Storm Report, February 1980," San Diego County Flood Control District, 26 pp.

# RATE OF ENERGY DISSIPATION AND SEDIMENT TRANSPORT

by Chih Ted Yang [1], M. ASCE and Albert Molinas [2], A.M. ASCE

## INTRODUCTION

Laboratory data are used in this paper to illustrate some of the basic weakness of conventional assumptions used in the development of sediment transport equations. The same data will be used to demonstrate the superiority of the assumption that sediment discharge or concentration should be related to the rate of energy dissipation of the flow. The use of unit stream power for the determination of sediment concentration will be justified from general physics and basic turbulence theories. Field data from six river stations will be used to support sediment transport equations derived directly or indirectly from the rate of energy dissipation approach.

## CONVENTIONAL APPROACHES

Comparisons between measured and computed sediment discharges from different transport equations were made by many investigators. The comparisons made in the Sedimentation Engineering (2) and by Vanoni, et al., (11) as shown in Figs. 1 and 2, respectively, are among the notable ones. The computed results from different equations as shown in Figs. 1 and 2 differ drastically from each other and from the measurements. It is prudent to examine the generality and validity of the assumptions used in these equations to determine the basic reason of disagreements shown in Figs. 1 and 2.

The conventional approach in the development of most sediment transport equations is to assume that sediment discharge can be expressed as a function of a dominant variable as shown in one of the following equations, i.e.:

$$q_t = A_1 (Q - Q_{cr})^{B1} \qquad (1)$$

$$q_t = A_2 (V - V_{cr})^{B2} \qquad (2)$$

$$q_t = A_3 (S - S_{cr})^{B3} \qquad (3)$$

$$q_t = A_4 (\tau - \tau_{cr})^{B4} \qquad (4)$$

---

[1] Civil Engineer, U.S. Bureau of Reclamation, Engineering and Research Center, Denver Federal Center, Denver, Colorado.

[2] Research Associate, Department of Civil Engineering, Colorado State University, Fort Collins, Colorado.

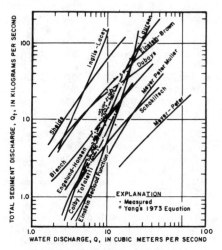

Fig 1 – Comparison between measured total sediment
discharge of the Niobrara River near Cody, Nebraska
and computed results of various equations.

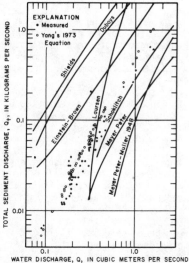

Fig. 2 – Comparison between measured total sediment discharge
of the Mountain Creek at Greenville, South Carolina and
computed results of the various equations.

Where $q_t$ = sediment discharge per unit channel width; Q = water
discharge; V = average flow velocity; S = water surface or energy
slope; $\tau$ = shear stress; cr = subscript denoting the critical value
required at incipient motion; and $A_1$, $A_2$, $A_3$, $A_4$, $B_1$, $B_2$, $B_3$, $B_4$ =
coefficients. The 0.93 mm sand laboratory data collected by Guy,
et al., (8) as shown in Figs. 3 to 6, will be used to demonstrate
the weakness of assumptions used in Eqs. (1) through (4).

Fig. 3 shows that total sediment discharge cannot be uniquely
determined by water discharge. More than one sediment discharge can
be obtained by a given water discharge, and vice versa. Fig. 4
indicates that for approximately the same velocity, the sediment
discharge can differ considerably due to the steepness of the curve.
Some of Gilbert's data (7) also indicate that the correlations
between sediment discharge and velocity are very weak. Fig. 5
indicates that there is no one-to-one relationship between sediment
discharge and slope. Fig. 6 shows that there is a one-to-one
correlation between sediment discharge and shear stress in the
middle range of the curve. For either higher or lower sediment
discharge, the curve becomes vertical, which means for the same
shear stress, numerous values of sediment discharge can be obtained.
Thus, the generality and applicability of a sediment transport
equation which is based on one of the basic assumptions shown in
Eqs. (1) through (4) is questionable. More detailed analyses
on the validity of different transport equations are given by Yang (14).

POWER APPROACH

Based on general physics, the rate of energy used in transporting a
given material should be related to the rate of material being
transported. Bagnold (4) defined the stream power as the product of
shear stress and average velocity. The stream power thus defined
has the dimension of power per unit bed area.

To compare the validity of the stream power approach to conventional
approaches, data used in Figs. 3 to 6 are replotted by using stream
power as the dominant variable. With the exception of two sets of
data collected with dune bed, Fig. 7 shows a closer and more definite
correlation between sediment discharge and stream power.

Engelund and Hansen (6) developed their transport function indirectly
from Bagnold's stream power concept and by using the similarity
principle. Fig. 8 shows fairly good agreements between computed
results from Engelund and Hansen's equation and measured results by
different investigators from six river stations.

Based on Bagnold's stream power concept, Ackers and White (1) applied
dimensional analysis technique to express the mobility and sediment
transport rate in terms of dimensionless parameters. Fig. 9 shows
fairly good agreements between computed results from Ackers and
White's equation and measurements from six river stations.

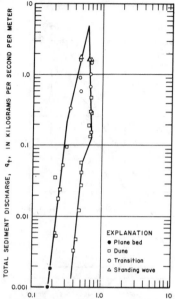

Fig. 3-Relationship between total sediment discharge and water discharge for 0.93mm sand.

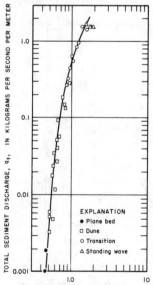

Fig. 4-Relationship between total sediment discharge and average flow velocity for 0.93mm sand.

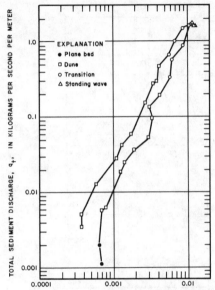

Fig. 5-Relationship between total sediment discharge and water surface slope for 0.93mm sand.

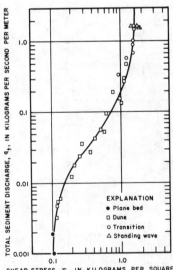

Fig. 6-Relationship between total sediment discharge and shear stress for 0.93 mm sand.

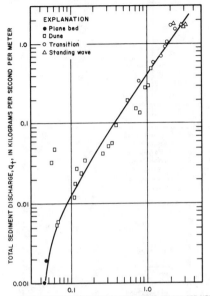

Fig. 7—Relationship between total sediment discharge and stream power for 0.93 mm sand.

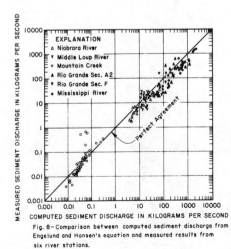

Fig. 8—Comparison between computed sediment discharge from Engelund and Hansen's equation and measured results from six river stations.

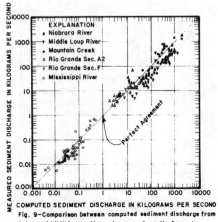

Fig. 9-Comparison between computed sediment discharge from
Acker and White's equation and measured results from six
river stations.

From general physics, sediment concentration being transported by water should be related to the rate of energy dissipation per unit weight of water. Yang (12) defined unit stream power as the rate of potential energy dissipation per unit weight of water. The unit stream power can be written as the product of velocity and slope, i.e.,

$$\frac{dY}{dt} = \frac{dx}{dt}\frac{dY}{dx} = VS \tag{5}$$

Where Y = potential energy per unit weight of water, t = time, x = longitudinal distances, and VS = unit stream power. Based on the unit stream power concept, the relationship between total sediment concentration and unit stream power can be written as:

$$C_t = M (VS - VS_{cr})^N \tag{6}$$

Where $C_t$ = total sediment concentration; $VS_{cr}$ = unit stream power required at incipient motion; and M, N = parameters related to flow and sediment characteristics. To compare the validity of Eq. (6) with conventional approaches, data used in Figs. 3 to 6 are replotted in Fig. 10 by using the unit stream power as the dominant variable. Fig. 10 shows a stronger and more definite correlation than those shown in Fig. 3 to 6.

The basic form of Eq. (6) can also be obtained from basic theories in fluid mechanics and turbulence (16). The energy equation for turbulent flows can be written as (9):

$$\frac{d}{dt}\frac{q^2}{2} = -\frac{\partial}{\partial x_i}\overline{u_i\left(\frac{p}{\rho}+\frac{q^2}{2}\right)} - \overline{u_iu_j}\frac{\partial U_j}{\partial x_i} + \nu\frac{\partial}{\partial x_i}\overline{u_j\left(\frac{\partial u_i}{\partial x_j}+\frac{\partial u_j}{\partial x_i}\right)}$$

$$\qquad\quad\text{(I)}\qquad\qquad\text{(II)}\qquad\qquad\text{(III)}\qquad\qquad\text{(IV)}$$

$$-\overline{\left(\frac{\partial u_i}{\partial x_j}+\frac{\partial u_j}{\partial x_i}\right)\frac{\partial u_j}{\partial x_i}} \tag{7}$$

$$\text{(V)}$$

Where $q^2 = u_iu_i$; $U_j$, $u_i$ = time averaged and fluctuating components of velocity in i directions, respectively; p = fluctuating component of pressure; $\rho$= density; and $\nu$ = kinematic viscosity. Hinze (9) explained the physical meaning of Eq. (7) as follows:

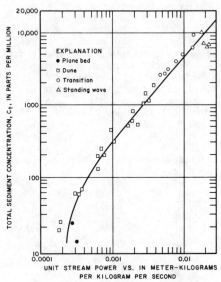

Fig. 10-Relationship between total sediment concentration and unit stream power for 0.93 mm sand.

"The change (I) in kinetic energy of turbulence per unit of mass of the fluid is equal to (II) the convective diffusion by turbulence of the total turbulence energy, plus (III) the energy transferred from the mean motion through the turbulence shear stresses, or the production of turbulence energy, plus (IV) the work done per unit of mass and of time by the viscous shear stresses of the turbulent motion, plus (V) the dissipation per unit of mass by turbulent motion."

It should be noted that the term "energy" just quoted is actually the rate of energy which is apparent by the dimension of (I).

The second term on the right-hand side of Eq. (7) is of particular interest to the writers. The writers (16) have shown that the rate of turbulence energy production at a given depth can be directly related to the sediment concentration at that depth by using the logarithmic velocity distribution and Rouse's (10) equation for sediment concentration distribution, i.e.,

$$
\frac{\overline{C}}{\overline{C}_a} = \left[ \frac{\tau_{xy} \dfrac{d\overline{U}_x}{dy}}{\left( \tau_{xy} \dfrac{d\overline{U}_x}{dy} \right)_{y\,=\,a}} \right]^{Z_1} \tag{8}
$$

Where $\overline{C}$ and $\overline{C}_a$ = time-averaged sediment concentration at distance of $y$ and $a$ above the bed, respectively; $\tau_{xy}$ = shear stress due to viscosity; $d\overline{U}_x/dy$ = velocity gradient; $Z_1$ = coefficient; and

$$
\frac{\tau_{xy}}{\rho} \frac{d\overline{U}_x}{dy} = - \overline{u_x u_y} \frac{d\overline{U}_x}{dy} \tag{9}
$$

which is item (III) of Eq. (7).

By integrating the concentration through the depth of the flow, total suspended sediment concentration can be expressed as a function of unit stream power which has the basic form shown in Eq. (6). Assuming a continuity of sediment concentration at the interface between suspended and bed load, and following similar procedures used by Einstein (5), the bed load as well as the total load concentrations can also be written in the basic form shown in Eq. (6). Thus, the basic form of unit stream power equation can be justified not only from measured data, but also from general physics as well as from basic theories in fluid mechanics and turbulence.

There are two dimensionless unit stream power equations proposed by Yang (13, 15). The one with incipient motion criteria is (13)

$$\log C_t = 5.435 - 0.286 \log\frac{\omega d}{\nu} - 0.457 \log \frac{U_*}{\omega}$$

$$+\left(1.799 - 0.409 \log \frac{\omega d}{\nu} - 0.314 \log \frac{U_*}{\omega}\right) \log\left(\frac{VS}{\omega} - \frac{V_{cr}S}{\omega}\right) \quad (10)$$

in which the dimensionless critical velocity required at incipient motion can be expressed by:

$$\frac{V_{cr}}{\omega} = \frac{2.5}{\log\left(\frac{U_* d}{\nu}\right) - 0.06} + 0.66, \quad 1.2 < \frac{U_* d}{\nu} < 70 \quad (11)$$

and

$$\frac{V_{cr}}{\omega} = 2.05, \quad 70 \leq \frac{U_* d}{\nu} \quad (12)$$

Where $C_t$ = total sediment concentration in parts per million by weight; $\omega$ = average fall velocity of sediment particles; d = median diameter of sediment; $\nu$ = kinetic viscosity of water; $U_*$ = shear velocity; V = average flow velocity, S = energy slope; VS = unit stream power; $V_{cr}S$ = critical unit stream power; and VS - $V_{cr}S$ = effective unit stream power. Another equation without any criteria for incipient motion is (15).

$$\log C_t = 5.165 - 0.153 \log \frac{\omega d}{\nu} - 0.297 \log\frac{U_*}{\omega}$$

$$+ \left(1.780 - 0.360 \log \frac{\omega d}{\nu} - 0.480 \log \frac{U_*}{\omega}\right) \log \frac{VS}{\omega} \quad (13)$$

Comparisons between measurements from six river stations and computed results from Eqs. (10) and (13) are shown in Fig. 11 and 12, respectively. Results in Fig. 11 and 12 indicate that Eq. (10) and (13) are about equally accurate in the prediction of total sediment load in natural rivers under diversified flow and sediment conditions. The use of effective unit stream power instead of unit stream power will change the curve in Fig. 10 to a straight line. Detailed analyses and comparison (15) indicate that the inclusion of incipient motion criteria in Eq. (10) is helpful to improve the accuracy of predicting sediment concentration when the concentration is low. Thus, the writers prefer the use of Eq. (10) under all conditions when the sediment is in the sand size range. Fig. 1 and 2 indicate that Eq. (10) is more accurate than other equations. Detailed comparisons made by Alonso, et al., (3) of the accuracy of different

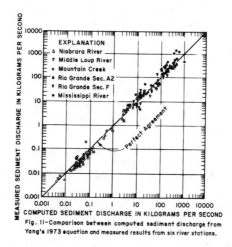

Fig. 11-Comparison between computed sediment discharge from Yang's 1973 equation and measured results from six river stations.

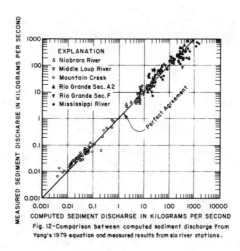

Fig. 12-Comparison between computed sediment discharge from Yang's 1979 equation and measured results from six river stations.

transport equations in the sand size range also concluded that
Eq. (10) is the most accurate one for both laboratory flumes and
natural rivers under laminar as well as turbulent flow conditions.

SUMMARY AND CONCLUSIONS

The basic assumptions used in different sediment transport equations
are reviewed to determine their validities and generalities. Laboratory data are used to compare the validity of these assumptions.
River data are used to compare the accuracies of different sediment
transport equations. This study has reached the following conclusions:

1. The assumption that sediment discharge can be determined from
   either water discharge, average flow velocity, energy slope,
   or shear stress as a general rule is questionable.

2. The assumption that sediment discharge should be related to
   stream power can be supported by general physics and
   laboratory data.

3. The assumption that sediment concentration should be related
   to unit stream power can be supported by laboratory data,
   general physics and basic theories in fluid mechanics
   and turbulence.

4. Comparisons between computed and measured results from rivers
   indicate that equations derived directly or indirectly from
   the concept that sediment transport rate should be related
   to the rate of energy dissipation of the flow are more
   accurate than other equations.

REFERENCES

1. Ackers, P., and White, W.R., "Sediment Transport: New Approach
   and Analysis," Journal of the Hydraulics Division, ASCE,
   Vol. 99, No. HY11. Proceeding Paper 10167, 1973, pp. 2041-2060

2. American Society of Civil Engineers Task Committee for the
   Preparation of the Manual on Sedimentation of the Sedimentation
   Committee of the Hydraulics Division, "Sedimentation Engineering,"
   Vanoni, V.A., Editor, 1975, Reprinted 1977, p. 222

3. Alonso, C. V., Neibling, W. H., and Foster, G. R., "Estimating
   Sediment Transport Capacity in Watershed Modeling," Transaction
   of the American Society of Agriculture Engineering, Vol. 24,
   No. 5, September-October 1981, pp. 1211-1220 and 1226

4. Bagnold, R. A., "An Approach to the Sediment Transport Problem
   from General Physics," U.S. Geological Survey Professional
   Paper 422-I, 1966

5. Einstein, H. A., "The Bed-load Function for Sediment Transport in Open Channel Flows," U.S. Department of Agriculture Soil Conservation Service, Technical Bulletin No. 1026, 1950

6. Engelund, F., and Hansen, E., "A Monograph on Sediment Transport in Alluvial Streams," Teknish Forlag, Denmark, 1967

7. Gilbert, K. G., "The Transportation of Debris by Running Waters," U.S. Geological Survey Professional Paper 86, 1914

8. Guy, H. P., Simons, D. B., and Richardson, E. V., "Summary of Alluvial Channel Data from Flume Experiment, 1956-1961," U.S. Geological Survey Professional Paper 462-I, 1966

9. Hinze, J. O., "Turbulence," 1st ed., McGraw-Hill Book Company, Inc., New York, N.Y. 1959, pp. 64-66

10. Rouse, H., "Modern Concepts of the Mechanics of Turbulence," Transactions of ASCE, Vol. 102, 1937

11. Vanoni, V. A., Brooks, N. H., and Kennedy, J. F., "Lecture Notes of Sediment Transport and Channel Stability," Report KH-RI, California Institute of Technology, Pasadena, California, 1960

12. Yang, C. T., "Unit Stream Power and Sediment Transport," Journal of the Hydraulics Division, ASCE, Vol. 98, No. HY10, Proceeding Paper 9295, 1972, pp. 1805-1826

13. Yang, C. T., "Incipient Motion and Sediment Transport," Journal of the Hydraulics Division, ASCE, Vol. 99, No. HY10, Proceeding Paper 10067, 1973, pp. 1679-1704

14. Yang, C. T., "The Movement of Sediment in Rivers," Geophysical Survey 3, D. Reidel Publishing Company, Dordrecht, Holland, 1977, pp. 39-68

15. Yang, C.T., "Unit Stream Power Equations for Total Load," Journal of Hydrology, Vol. 40, No. 1/2, 1979, pp. 123-128

16. Yang, C. T. and Molinas, A., "Sediment Transport and Unit Stream Power Function," Journal of the Hydraulics Division, ASCE, Vol. 108, No. HY6, 1982

Application of Variation Principle

to River Flow

by

Charles C. S. Song,[1] M. ASCE and Chih Ted Yang,[2] M. ASCE

ABSTRACT

The variational principle is used to explain phenomena related to laminar-turbulent flow transition, velocity distribution, stability of moving particles, self generation of bed roughness, meandering and braiding, and equilibrium channel geometry.

INTRODUCTION

The most challenging aspect of fluid mechanics is the problem of indeterminacies: the possibility of a nonunique solution for a given set of initial and boundary conditions. A classical example is the existence of laminar flow and turbulent flow. Other examples are the occurrences of boundary layer separation, secondary current, and vortex, which may be either steady or unsteady. In river mechanics the complexity and the degree of uncertainty are compounded due to the fact that the boundary is moveable and multicomponent flow is involved. consider a case of a river's response to a change of water discharge. Because the water surface is the most flexible part of the boundary, the flow depth and velocity distribution will change first. This change will cause the sediment transport capacity to be altered and, hence, affect the bed roughness. The bed roughness induced by ripples and dunes causes the well-known analytical indeterminacy. The questions related to the indeterminacy of river dynamics are: Should the sediment transport be in the form of bedload or suspended load and should there be a transport of air in the form of entrained air bubbles?

A somewhat slower but equally important response of a river to the change in discharge occurs in the form of modified cross-sectional

---

[1]Professor, St. Anthony Falls Hydraulic Laboratory, Department of Civil and Mineral Engineering, University of Minnesota, Minneapolis, Minnesota 55414.

[2]Civil Engineer, U. S. Bureau of Reclamation, Engineering and Research Center, Denver, Colorado 80225.

shape, bed slope, and sinuosity.  A river in its natural state must be
very flexible to accommodate the highly variable hydrological condition.  This further adds to the dimension of the indeterminacy.
Traditionally, the problems of indeterminacy or multiplicity are
regarded as the stability problem.  The purpose of this paper is to
show how the variational principle, the principle of minimum energy,
and energy dissipation rate can be used to help solve the problem of
indeterminacy.

A variational principle introduced by Song and Yang (6) applies to
a closed and dissipative mechanical system.  A mechanical system herein
refers to a system which depends only on the mechanical energy, i.e.
potential energy and kinetic energy, but not on any other form of
energy.  A closed system is a system insulated from its surroundings in
terms of mass and mechanical energy.  A system is called dissipative
when its total mechanical energy can only decrease due to friction or
viscosity if there is relative motion.  Only when all relative motion
stops does the total mechanical energy remain constant.  The
variational principle may be divided into four parts:

1.  A state of stable static equilibrium is a state of minimum
mechanical (potential) energy.

2.  If there is more than one stable static equilibrium condition
for a given system, then, one with smaller potential energy is more
stable (more likely to occur) than one with larger potential energy.

3.  A state of stable dynamic equilibrium is a state of minimum
energy dissipation rate.

4.  If there is more than one stable dynamic equilibrium condition
for a given system, then, one with smaller energy dissipation rate is
more stable (more likely to occur) than one with greater energy dissipation rate.

## APPLICATIONS OF THE VARIATIONAL PRINCIPLE

### Laminar-Turbulent Flow Transition

Flow in a river or a pipe can be either laminar or turbulent
depending mostly on the Reynolds number.  Since there is no essential
difference between river flow and pipe flow as far as the question of
laminar-turbulent transition is concerned, a better known pipe flow
problem will be considered herein.

Consider a smooth pipe of diameter  D  and length  L  attached to
a large reservoir at each end.  Water is placed in this system in such
a way that the two reservoir levels differ by H.  Because of this differential head there will be a discharge of  Q  flowing from one reservoir to the other.  Clearly the system is closed and dissipative.  If
two reservoirs are very large, then, the flow will change slowly enough
to be regarded as practically in a dynamic equilibrium condition and
the variational principle is applicable.

The energy dissipation rate of the system  $\Phi$  is

$$\Phi = \gamma Q H \tag{1}$$

in which  $\gamma$ = the specific weight of water.  The energy equation is

$$H = (K_e + 1 + f \frac{L}{D}) \frac{Q^2}{2gA^2} \tag{2}$$

in which $K_e$ = entrance loss coefficient, $f$ = Darcy-Weisbach friction factor, and $A$ = cross-sectional area. Since $\gamma$ and $H$ are given constants, minimization of $\Phi$ is equivalent to minimization of $Q$. According to Eq. 2 minimization of $Q$ is equivalent to maximization of $f$. According to the last part of the variational principle, the flow with greater $f$ should prevail over the flow with smaller $f$.

The friction factor for laminar flow in a pipe is

$$f = 64/R \tag{3}$$

in which $R$ = Reynolds number. The Blasius empirical formula for $f$ for smooth pipe is

$$f = 0.316/R^{0.25} \tag{4}$$

Equations 3 and 4 intersect at $R = 1200$. The friction factor for laminar flow is greater than that of turbulent flow when $R < 1200$ but the opposite is true when $R > 1200$. According to the variational principle, then, the critical Reynolds number for a pipe flow should be 1200. It should be pointed out that the turbulent flow is also an equilibrium flow because $\Phi$ of the system is stationary even though the flow may be locally unsteady.

It is interesting to note that the theoretical critical Reynolds number for pipe flow according to the classical stability theory for small perturbation is infinity (1, 3). Clearly the present theory yielded a much better result than the classic stability theory in this instance. Because the actual transition depends also on the existence of disturbances, the real critical Reynolds number may not be exactly the same as the value computed by the theory.

Laminar and Turbulent Velocity Distribution

As the system considered in the previous section selects the preferred flow regime, laminar or turbulent, it must also select its preferred velocity distribution at the same time. In an open channel, the depth of the flow may also be a free parameter to be selected.

Velocity distribution in a very wide open channel, one-dimensional problem, was considered by Song and Yang (4). An exact solution for laminar flow and an approximate solution for turbulent flow were obtained using the variational principle for an open system derived from the Navier-Stokes equation by Yang and Song (9) and Song and Yang (5). This variational principle for an open system is much more restrictive than the variational principle for closed systems proposed by Song and Yang (6).

It was shown analytically that the velocity distribution for a laminar flow satisfying the equation of continuity and minimizing the energy dissipation rate is a parabola. For a turbulent flow all possible combinations of a logarithmic distribution and parabolic

distribution were compared. It was found that a combination of
logarithmic distribution in the inner half and parabolic distribution
in the outer half of the depth results in least energy dissipation
rate. This conclusion is more or less supported by available experimental data.

## Stability of Moving Particles

The most important parameter characterizing a sediment in sediment
transport analysis is the fall velocity of a particle. For a particle
with very small fall velocity in stokes range, the fall velocity is
practically independent of the shape and the orientation of the particle. The fall velocity of larger particles, however, is sensitive to
its shape and orientation. Thus, the shape of the particle and its
effect on the stability of motion becomes important as the particle
size exceeds the Stokes range.

Stringham et al. (7) noted that a circular disk may fall in one of
the following four patterns.

1. Steady-flat fall - When the Reynolds number was less than 100,
the disk fell at a constant velocity with the maximum projected area
perpendicular to the direction of fall without any tendency to
oscillate.

2. Regular oscillation - When the Reynolds number was slightly
greater than 100, the disk rotationally oscillated about a diameter and
also oscillated in a horizontal direction.

3. Glide-tumble - At somewhat greater Reynolds number, the motion
was a combination of oscillation, gliding, and tumbling.

4. Tumble - In this fourth fall pattern, the disk rotated through
360 degrees at a nearly constant angular velocity, and the path of travel was nearly a straight line inclined at an angle with the horizontal.

A particle falling in a large body of liquid is a closed and
dissipative mechanical system. The rate of energy dissipation of the
system is

$$\Phi = WV \qquad (5)$$

in which W = submerged weight and V = the vertical component of fall
velocity. The drag coefficient $C_D$ is given by the following
equation.

$$W = C_D A_o \ 1/2 \rho V^2 \qquad (6)$$

in which $A_o$ = a representative cross-sectional area and $\rho$ = density
of liquid.

Since W is fixed, minimization of $\Phi$ is equivalent to the minimization of V. According to Eq. 6, minimization of V is equivalent to
maximization of $C_D A_o$. If $A_o$ is taken as a nominal cross-sectional
area which is fixed for a disk, then minimization of $\Phi$ is also
equivalent to maximization of $C_D$. The recalculated data for $C_D$ and
R were replotted by Song and Yang (6). The plot appeared very much
similar to that of f ~ R curve for flow in pipe and the 4th part of
the variational principle is strictly applicable using the following
analogy.

    a.  Steady-flat fall is analogous to laminar flow.

    b.  Tumble fall is analogous to turbulent flow.

    c.  Regular oscillation and glide-tumble are analogous to transition flows.

Maximization of drag coefficient by adjusting particle orientation and fall pattern can be achieved only if the particle is given sufficient time to establish a dynamic equilibrium condition. When a sediment particle is suspended in a turbulent flow field, the relative velocity between the particle and the fluid may vary so rapidly that equilibrium condition is not possible. Under this condition the experimentally determined fall velocity may not be a good representative particle characteristic.

Another evidence that sediment particles are likely to move in such a way that the drag coefficient is maximized is found in the orientation of pebbles resting on river beds. Rust (2) measured the orientations of elongated pebbles in dry channels on river braided bars. He found that for isolated large pebbles, the orientation of the long axis is overwhelmingly normal to the direction of flow. This directional preference was weaker in the case of concentrated large pebbles or isolated small pebbles. Considering the uncertainty in the flow direction due to turbulence, the strong orientational preference of large isolated pebbles is remarkable.

## Bed Roughness

After adjusting for the depth and velocity distribution, a river may also adjust the roughness if the stream is powerful enough to transport sediment. Two types of bed roughness generating mechanisms, the generation of bed forms and coarsening of bed material on the bed surface will be considered here.

Song and Yang (6) tried to explain the growth and damping of ripples and dunes in a channel transporting uniform size sediment using the variational principle. In order to isolate the variation due to roughness from other geometrical factors, a very wide open channel with slope S and drop H was considered. To make the system closed, a very large reservoir with unlimited supply of water and sediment was assumed to exist at each end. By dividing the sediment transport into bed load and suspended load, the total energy dissipation rate of the system is written as

$$\Phi = (\gamma q + \gamma_s q_b + \gamma_s q_s)H \tag{7}$$

in which $q$ = water discharge per unit width, $q_b$ = bed load per unit width, $q_s$ = suspended load per unit width, and $\gamma_s$ = specific weight of sediment.

For a closed system, the specific energy

$$E = \frac{q^2}{2g\,y^2} + y \tag{8}$$

is fixed.  In the above equation, E = specific energy, g = gravitational acceleration, and y = flow depth.  A generalized resistance equation may be written as

$$q + q_s = \sqrt{\frac{8g}{f}}\ y^{3/2}\ s^{1/2} \tag{9}$$

The reason why $q_s$ appears in Eq. 9 is that y is the gross depth shared by water and suspended sediment.

Equation 7 may be made dimensionless and written as

$$\tilde{\phi} = \tilde{\phi}_w + \tilde{\phi}_b + \tilde{\phi}_s \tag{10}$$

$$\tilde{\phi} = \phi/\sqrt{2g}\ \gamma\ HE^{3/2} \tag{11}$$

$$\tilde{\phi}_w = 2\tilde{f}(1 + 2\tilde{f})^{-3/2} \tag{12}$$

$$\tilde{f} = f/8S = F^{-2} \tag{13}$$

in which F is the Froude number.

The energy dissipation rate due to water alone, $\tilde{\phi}_w$ , is plotted as the function of $\tilde{f}$ in Fig. 1.  Note that $\tilde{\phi}_w$ takes a maximum value at the critical flow, $\tilde{f}=1$, and decreases monotonically away from the critical point.  If there is no sediment transport, then Eq. 8 and Eq. 9 without $q_s$ together are sufficient to solve for the two unknowns, q and y, because f is fixed and cannot be changed by the flow.  If there is bed load but without suspended load, then $q_b$ or $\tilde{\phi}_b$ is usually so small that $\tilde{\phi}$ is not significantly different from $\tilde{\phi}_w$ .

Consider a subcritical flow initially at point A shown in Fig. 1.  Since the flow is capable of changing the bed roughness, it will try to minimize $\tilde{\phi}$ by increasing $\tilde{f}$ or f .  If the sediment is uniform, then, the increase in roughness can be achieved only by generating bed forms.  This is believed to be the mechanism of ripple and dune generation.  On the other hand, if the bed material consists of nonuniform sediment, then the size distribution of the surface layer may change in such a way that $\tilde{f}$ or f is increased.  Since large f requires large sediment size, coarser material is sorted and tends to stay on the bed surface.

Now consider a supercritical flow initially represented by point B in Fig. 1.  If the bed material is cause enough, the sediment transport rate may still be so small that $\tilde{\phi}$ is not substantially different from $\tilde{\phi}_w$ .  The variational principle suggests that the roughness should be decreased if it is possible.  It is a well-known fact that ripples and dunes are absent in steep gravel channels.

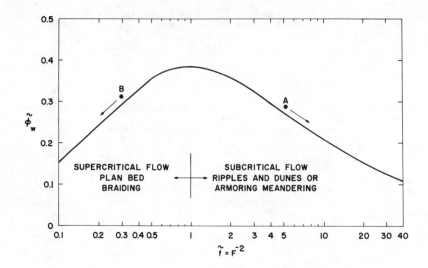

Fig. 1.  Dimensionless Energy Dissipation Rate

## Meandering and Braiding

Meandering and braiding of rivers have attracted much research
interests but still must be regarded as largely unsolved problems.
Needless to say, most previous analytical works were based on the traditional vector mechanics approach.  Yang (8) attempted to explain the
river's tendency to meander using the theory of minimum unit stream
power.

Qualitative explanation of meandering and braiding will be given
here based on the variational principle.  Quantitative analysis of the
problems is deferred to future works.  Consider again, the hypothetical
closed river-reservoir system described in the last section.  If the
valley slope and sediment size are both relatively small, the flow is
subcritical and there is modest amount of sediment transport.  The initially nonequilibrium condition is represented by point $\underset{\sim}{A}$ in Fig. 1.
The river will then adjust itself in order to increase $\tilde{f}$ until
further increase is not possible, i.e., until an equilibrium condition
is obtained.  According to Eq. 13, increase in $\tilde{f}$ implies increase in
f  or decrease in  S .  This last case is the mechanism of meandering
to be discussed next.

In a watershed where the total drop  H  varies very slowly in comparison with the time required to develop meandering, a reduction in
river slope accompanied by increased river length is possible.  That
is, after maximizing its roughness, if the river still possesses sufficient power to erode its bank, meandering would be the next available
means to minimize the overall energy dissipastion rate.

Now consider a case when both the slope and the sediment size are large so that point B in Fig. 1 is obtained. The variational principle requires that $\tilde{f}$ be decreased. This can be achieved by decreasing f or increasing S . Reduction of f has been discussed in the previous section but increase in S implies a straight river. It appears that a straight river can exist in two different ways. If, relatively speaking, the stream power is small or the boundary is rigid, then the stream geometry is permanent and a straight channel remains straight. On the other extreme, if the stream is powerful and supercritical, and if the bank is erodible, then the natural tendency is to generate a straight but braided river.

## Equilibrium Channel Geometry

Recently Yang et al (10) and Yang and Song (11) studied the optimum width to depth ratio of a trapezoidal channel using the variational principle. A trapezoid was described by its surface width W, depth D, and an inverse to the side slope z. Rectangles and triangles were included as special cases of a trapezoid (11).

The total energy dissipation rate was expressed as a function of two variables, i.e., W and z. Different equilibrium solutions are possible depending on other constraint conditions. Following are three particularly interesting solutions:

1. Side slope constrained solution - When the river bank consists of noncohesive material, the bank slope cannot be greater than the angle of repose of the sediment. Thus, a side slope constraint applies if the uncontrained solution results in greater slope than the angle of repose. By setting the partial derivative of $\Phi$ with respect to W equal to zero, the z-constrained solution is obtained as

$$M = 2 \sqrt{1 + z^2} \tag{14}$$

in which M is the width-depth ratio W/D. Equation 14 clearly shows that the width-depth ratio of an equilibrium trapezoidal channel increases monotonically as the angle of repose decreases.

2. Width constrained solution - The width constrained solution may be appropriate if bank erodibility is significantly restricted by the existence of cohesion, vegetation, or other protective measures. By setting the partial derivative of $\Phi$ with respect to z equal to zero, it follows that,

$$M = \frac{2(z^2 - 1 - z \sqrt{1+z^2})}{4z - 3 \sqrt{1 + z^2}} \tag{15}$$

3. Unconstrained solution - The unconstrained solution is the solution of the two simultaneous equations. By solving Eqs. 14 and 15, simultaneously, it follows that

$$z = \frac{1}{\sqrt{3}} \, , \quad M = \frac{4}{\sqrt{3}} \tag{16}$$

Equation 16 indicates that the unconstrained equilibrium solution is half of a hexagon which is also known as the most hydraulically efficient trapezoid.

It is interesting to note that an equilibrium geometry obtained by the variational principle herein coincide with the most efficient hydraulic cross section. For example, by setting z=0 in Eq. 14, the equilibrium rectangle is found to be half of a square.

In order for the unconstrained equilibrium to occur, the angle of repose of noncohesive sediment has to be at least 60 degrees. Since the angle of repose rarely exceeds 40 degrees, the unconstrained equilibrium condition is not expected to occur in practice. Clearly, it is even more difficult to find a self generated rectangular channel with M=2. Available laboratory and field data agrees fairly well with the theoretical result.

## CONCLUSION

It is hoped that the paper has demonstrated quite clearly that the variational principle is a useful tool in solving difficult problems in hydraulics and sediment transport.

REFERENCES

1. Davey, A. and Drazin, P. G., "The stability of Poiseuille flow in a pipe," Journal of Fluid Mechanics, Vol. 36, Pt. 2, 1969, pp. 209-218.

2. Rust, B. R., "Pebble orientation in fluvial sediment," Journal of Sedimentary Petrology, Vol. 42, No. 2, June, 1972, pp. 384-388.

3. Salwen, H. and Grosch, E., "Stability of Poiseuille flow in a pipe of circular cross section," Journal of Fluid Mechanics, Vol. 54, Pt. 2, 1972, p. 93.

4. Song, C.C.S. and Yang, C. T., "Velocity profiles and minimum stream power," Journal of the Hydraulics Division, ASCE, Vol. 105, No. HY8, Paper No. 14780, Aug., 1979.

5. Song, C.C.S. and Yang, C. T., "Minimum stream power: theory," Journal of the Hydraulics Division, ASCE, Vol. 106, No. HY9, Paper No. 15691, Sept., 1980.

6. Song, C.C.S. and Yang, C. T., "Minimum energy and energy dissipation rate," Journal of the Hydraulics Division, ASCE, Vol. 108, No. HY5, May, 1982.

7. Stringham, G. E., Simons, D. B., and Guy, H. P., "The behavior of large particles falling in quiescent liquids," U. S. Geological Survey Prof. Paper 562-C, 1969.

8. Yang, C. T., "On river meanders," Journal of Hydrology, Vol. 13, No. 3, 1971, pp. 231-253.

9. Yang, C. T. and Song, C.C.S., "Theory of minimum rate of energy dissipation," Journal of the Hydraulics Division, ASCE, Vol. 105, No. HY7, Paper No. 14677, July, 1979.

10. Yang, C. T., Song, C.C.S., and Woldenberg, M. J., "Hydraulic geometry and minimum rate of energy dissipation," Water Resources Research, Vol. 17, No. 4, Aug., 1981, pp. 1014-1018.

11. Yang, C. T. and Song, C.C.S., "Optimum channel geometry and minimum rate of energy dissipation," unpublished.

# METRICATION IN THE UNITED STATES

By A. Ivan Johnson, F., ASCE[1]

## ABSTRACT

Although the U.S. Congress made the metric system legal in 1866 for the transaction of business, metrication in the U.S. did not really start moving for over 100 years. On December 23, 1975 Public Law 94-168, "The Metric Conversion Act," was signed by President Ford and the U.S. was on its way to joining the rest of the world in use of the metric system. The act declared that the National policy of the U.S. shall be to coordinate and plan the increasing use of the SI metric system in the United States and to establish a U.S. Metric Board to coordinate the voluntary conversion to the metric system. Although progress for much of the time since 1975 admittedly has been millimeter by millimeter rather than in giant strides, this paper summarizes policies and progress leading towards metrication in various engineering activities.

Following a discussion of the main elements of the U.S. Metric Conversion Act, the paper describes the organization and objectives of the American National Metric Council and the U.S. Metric Board. The policies and progress regarding metrication in industry and various Federal agencies is summarized and the activities of the Interagency Committee on Metric Policy and the Office of Water Data Coordination are described. Progress towards metrication in various technical and standardization societies, such as ASCE, ASTM, and ANSI, is discussed in some detail. The coordinating efforts of the ANMC Water Resources Sector Committee and the Joint Committee of National Societies for Metrication in Water Resources show how 11 national societies, as well as representatives from industry and the public sector, are tackling the problem of developing a widely accepted list of recommended metric units for water-resources measurements. The paper concludes by presenting the ASCE Policy Statements regarding use of the metric system—two in 1876, and one each in 1959 and 1970—and in 1977, the support given by the ASCE Committee on Metrication. Recent action of ASCE regarding the Committee on Metrication further strengthens the role of ASCE in metrication in the United States.

## INTRODUCTION

In the past few years several U.S. Presidents have publicly stated that they support and encourage the use of the international metric system of measurement in the United States. With all the other industrialized nations of the world already using the metric system—or soon to complete their conversion—the Presidents pointed out that we

---

[1] Water Resources Consultant, Woodward-Clyde Consultants, Denver, CO, and Chairman, ASCE Committee on Metrication

will find we are at a serious disadvantage in selling products to
other countries if we do not make progress in our metric conversion.
Even though President Ronald Reagan had to cut government funded
metrication activities as part of his over-all program to reduce
government spending, he has publicly stated his support for the policy
of voluntary metrication expressed in the Metric Conversion Act of 1975.
    Although progress towards metric conversion in the U.S. admittedly
has been millimeter by millimeter rather than by giant strides, progress is being made more rapidly than most people are aware. It is
the purpose of this paper to demonstrate that, in addition to the
Presidents and other high officials, there are many other individuals
actively moving metrication forward today in the United States.

HISTORICAL BACKGROUND

    For nearly two centuries the U.S. Government has been considering
the pros and cons of a single measurement system to serve the varied
needs of all Americans. Our nation's first President, George Washington, requested that Congress study alternatives and provide the nation
with a simple and uniform system of weights and measures. Following
investigation of this matter, Secretary of State Thomas Jefferson proposed to Congress a scientific measurement system based on units of
ten, like the new American currency. At about the same time, the
leaders emerging from the French Revolution were establishing the first
metric system. In 1790, the French National Assembly requested the
French Academy of Sciences to work out a system of units suitable for
use by the entire world. The new system, based on the meter and gram,
was a decimal system of measurement tied to scientific standards.
Though the French made the new metric system compulsory in 1795, the
law was ignored and in 1812, Napoleon legalized returning to the pre-
metric system units. In 1840, France re-established the metric system
as its sole system of measurement and an international movement toward
metrication began. By 1900, 40 nations had gone metric.
    In 1866, the U.S. Congress made the metric system legal for the
transaction of business, but its use was entirely voluntary. In 1875,
the United States joined 16 other nations in signing the Treaty of the
Meter. This treaty created the International Bureau of Weights and
Measures, made the meter the accepted international unit of measurement, and set worldwide standards for length and mass. A General Conference on Weights and Measures was also constituted and controls the
International Bureau of Weights and Measures. Through the years, the
National Bureau of Standards has represented the U.S. in General Conference on Weights and Measures activities, which mainly are related
to preservation of the metric standards, comparison of national standards with the metric standards, and research needed to develop new
standards. In 1893, the United States became officially metric and the
accuracy of every customary measuring device used in America since
then has been defined in relation to metric standards. In succeeding
decades, all of the industrialized nations adopted the metric system
to take advantage of its usefulness for international trade. However,
the inch-pound system continued to be used in domestic commerce and
industry in the U.S. and several other countries throughout the 19th
and first half of the 20th Century.
    In 1960, the 11th General Conference on Weights and Measures
adopted a new rationalized and coherent system of units based on the

meter, kilogram, second, and ampere units. The new system was known as
SI, which is an abbreviation for "Le Système International d'Unités"
or "International System of Units." Following adoption of this new SI
system, there was a strong move among the scientific and technical
communities to increase the use of the metric system in the United
States.

Public Law 90-472, which authorized a U.S. metric study by the U.S.
Department of Commerce, was signed into law by President Johnson. In
1971, the report, entitled "A Metric America: A Decision Whose Time
Has Come" was sent to Congress. The report recognized that by this
time only Liberia, Yemen, Brunei, Burma, and the United States were
non-metric countries. The report also pointed out that the conversion
movement was underway voluntarily in the U.S. and recommended that it
was in the best interest of the United States to undertake a planned
program of voluntary metric conversion as soon as possible. The U.S.
metric study report stated "A metric America would seem desirable in
terms of our stake in world trade, the development of international
standards, relations with our neighbors and other countries, and
national security."

## U.S. METRIC LAW OF 1975

Following release of the 1971 report by the Department of Commerce,
a number of metric bills were introduced in Congress in order to implement the report's recommendations. For a number of reasons, none of
the laws were enacted until December 23, 1975 when Public Law 94-168,
"The Metric Conversion Act," was signed by President Ford. When he
signed the bill, President Ford stated, "This legislation establishes
a national policy of coordinating and planning for the increased use
of the metric measurement system in the United States." He stated
further "...the real impetus for the bill came from the private sector,
people in the business of buying and selling American products here
and overseas." The Act had passed without objection in the Senate and
300 to 63 in the House so the conversion process was well supported by
Congress as well as by the President.

That Act declared that the national policy of the United States
"shall be to coordinate and plan the increasing use of the SI metric
system in the United States and to establish a U.S. Metric Board to
coordinate the voluntary conversion to the metric system." Thus, it
is emphasized that the change to the metric system is not to be
achieved by the force of law or under any time constraints, but is to
be by the voluntary, coordinated decisions of the members of each segment and sector of our economy.

## U.S. METRIC BOARD

The Conversion Act mandated that a U.S. Metric Board shall carry
out such activities as (1) consult with and take into account the
interests, views, and conversion costs of U.S. commerce and industry,
(2) provide for appropriate procedures whereby various groups may
formulate specific programs for coordinating conversion, (3) publicize
proposed programs, (4) encourage standardization organizations to develop or revise engineering standards on a metric measurement basis,
(5) assist the public, through information and education programs, to

become familiar with the meaning and applicability of metric terms and measures, (6) collect, analyze, and publish information about the extent of usage of metric measurements, evaluate the costs and benefits, and make efforts to minimize any adverse effects resulting from increased usage of the metric system, (7) consult and cooperate with foreign governments, intergovernmental, and international organizations, (8) conduct research and recommend to Congress and to the President such action as appropriate to deal with any unresolved problems, (9) submit annually to the Congress and to the President a report on its activities which may include recommendations for legislation needed to implement programs of conversion and (10) submit to the Congress and to the President a report on the need to provide an effective structural mechanism for converting customary units to metric units in statutes, regulations, and other laws at all levels of Government.

Although the Metric Conversion Law was signed by President Ford in 1975 and his nominees were submitted in 1976, it was not until October 29, 1977 that President Carter submitted his first nominees for membership on the Board. The Senate Committee on Commerce, Science, and Transportation held confirmation hearings on two sets of nominees and the full 17-member Board was confirmed finally on June 20, 1979. The Board is made up of members representing many different parts of the economy and social structure, including industry, science, engineering, standards, construction, small business, labor, state or local Government, retailing, and education. Terms of office run from 2 to 6 years. Dr. Louis F. Polk, former chairman of the U.S. Department of Commerce's National Metric Advisory Council and author of the 1971 report on a metric America, was appointed by President Carter to be the first chairman of the U.S. Metric Board. The Board, supported by a staff of around 40, operated during fiscal year 1981 with a budget of over two million dollars. Following some reductions in budget and staff during fiscal year 1982, the Board is being discontinued on September 30, 1982 by Presidential order. In a letter of April 6 to the American National Metric Council, President Reagan stated that his Administration supports the policy of voluntary metrication. He noted that as the representative of the private sector he looked forward to the Administration working together with ANMC to make the U.S. more competitive in world markets. Some of the duties of the U.S. Metric Board are being continued on a very limited scale under the Department of Commerce's Office of Productivity, Technology, and Innovation and that organization is already working closely with ANMC to develop an active, coordinated approach to metrication in the U.S.

AMERICAN NATIONAL METRIC COUNCIL

During the 70's, the metric conversion process accelerated voluntarily, primarily in response to the need for competition with foreign firms. The final passage of the Metric Conversion Act of 1975 accelerated the move of organizations voluntarily towards metric conversion. Today, 62 percent of Fortune Magazine's list of the top 1,000 companies produce at least one metric product and 32 percent of all sales are metric. Such giants of business as GM, IBM, and Caterpillar were among the first major manufacturers to convert to metric measurements. For example, GM's conversion began with the 1976 Chevette and Caterpillar is now 100 percent metric. In general, industry found that

total costs for conversion were much lower than their original estimates.

To serve as a representative of industry and others in the private
sector in coordinating public/private approach to conversion to the
SI system of metric units, the American National Metric Council (ANMC)
was established in 1973 under the auspices of the American National
Standards Institute (ANSI). In July of 1976, ANMC became incorporated
as a separate organization and formally cut its corporate ties with
ANSI.

The ANMC is organized into nine "Coordinating Groups" that represent
various segments of the U.S. economy. (See figure 1.) At present,
these cover the subjects of (1) Materials, (2) Engineering and Transportation, (3) Construction Industries, (4) Consumer Products, (5)
Education and Training, (6) Food and Grocery Products, (7) Medical and
Health, (8) Power and Energy, and (9) Service Industries.

Each of the Coordinating Groups are subdivided into five to twelve
"Sector Committees," which are the main working groups of the ANMC
structure. The membership of a Sector Committee typically is representative of industry, trade and professional groups, government, small
business, labor, and consumers. The Sector Committee's purpose is to
work towards a planned conversion program within a specific industry
for the guidance of all interested parties who desire to voluntarily
convert to metric measurements wherever that makes economic sense. The
number of sector committees at ANMC has increased from 19 in 1973 to
over 54 today with nearly 2,000 major companies and other organizations
supporting financially, and participating in, the work of ANMC. ANMC's
major information tool is the "Metric Reporter," a bi-weekly news-
letter with over 12,000 readers. In addition, many special reports are
produced by ANMC staff or sector and operations committees. One of
these reports, the "Metric Editorial Guide" is recommended by many
organizations, including ASCE, as one of three primary references for
metric conversion.

The Construction Industries Coordinating Group (CICB) can be used
as an example of the working structure within ANMC. Presently there
are eight Sector Committees: 01-Design, 02-Codes and Standards, 03-
Products Manufacturers, 04-Contractors, 05-Information, 06-Real Estate,
07-Users, and 08-Surveying and Mapping. The American Society of Civil
Engineers serves as the Secretariat for the Coordinating Group. The
efforts of the CICB and its sector committees has resulted in a coordinated Construction Industries Metric Conversion Timetable, which
proposes a network of step-by-step schedules that propose complete
conversion by 1985. At a recent meeting, the ASCE Committee on Metrication endorsed the proposed target date. The ASCE/COM members are
also members of the CICB Design Sector Committee, but other members
represent architectural firms, consumers, manufacturers and other
sectors.

PROGRESS AMONG FEDERAL AGENCIES

Shortly after signing of the 1975 Metric Conversion Act, the National Bureau of Standards established a Metric Information Office and
organized a number of Federal Metric Panels related to various disci-
plines such as surveying and mapping, construction, hydrology, etc.,
with all interested Federal agencies invited to provide representation

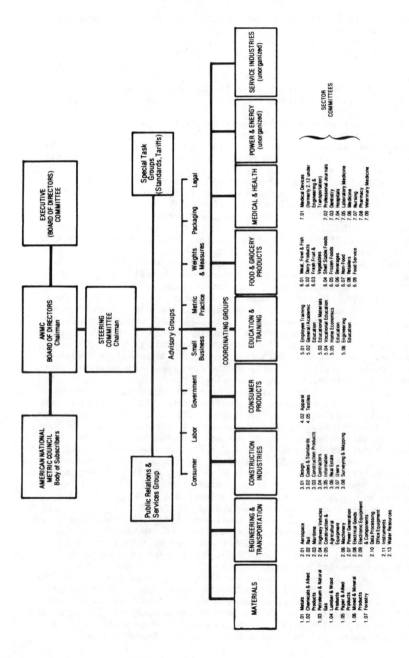

Figure 1.—Organization of the American National Metric Council (Metric Reporter, 1981-12-21)

to each panel of interest to the agency. Once the U.S. Metric Board
was organized the NBS stepped out of its early role as metric coordin-
ator for government agencies. However, NBS still has a very important
role in the metric process. The Metric Conversion Act of 1975 stated
that official interpretation of the SI metric system for application
within the United States shall be by the Department of Commerce and NBS
has been given that responsibility within that Department. In implementation of this authority, tables and associated materials were published in the Federal Register of December 10, 1976 and a refinement
of that earlier interpretation and modification was published in the
Federal Register of October 26, 1977. The latter notice, entitled "The
Metric System of Measurement (SI)," is recommended by many organizations, including ASCE, as one of three primary references for metric
conversion.

One of the responsibilities assigned to the U.S. Metric Board is to
plan and coordinate metric conversion in the public, or Federal Government, sector. Thus, the Interagency Committee on Metric Policy (ICMP)
was formed in 1978 to coordinate Federal efforts to implement metric
conversion. The ICMP is composed of high-level policy representatives
from 37 Federal departments and agencies. The ICMP serves as vehicle
for (1) coordination of metric policy and activities of Federal
agencies, (2) mutual reinforcement of metrication activities of Federal
agencies, (3) encouragement of top management commitment to Federal
metrication activities, including participation in private sector con-
sensus planning, and (4) information exchange.

Federal agencies also are represented on the Metrication Operating
Committee (MOC), a subcommittee of the ICMP that conducts the day-to-
day business of metric conversion planning within the Federal government. The MOC is made up of the metric coordinators from 45 Federal
agencies. These coordinators function at the technical or operating
level and are involved directly in the development of metric conversion plans of their respective agencies. The MOC has established nine
subcommittees of related interests: (1) Construction, (2) Consumer
Affairs, (3) Transportation, (4) Fuel and Power, (5) Procurement and
Supply, (6) Legislation and Regulations, (7) Federal Employee Training,
(8) Public Awareness and Education, and (9) Metric Practices and Preferred Units. From these subcommittees have come two documents to
date: "Metric Practice Guidelines for Federal Government Agencies" and
"Preferred Metric Units for General Use by the Federal Government." In
addition, the ICMP adopted the "Federal Agency Guidelines for Implementation of Metric Conversion Policy" drafted by the MOC. This letter
document was published in the Federal Register of January 8, 1980.

Over a dozen agencies have now issued policy statements on metric
usage and about an equal number are developing policy statements. The
U.S. Department of Commerce issued its plan to convert by 1985 and all
industries, farms, and homes will be affected by this agency's plans.
Because of America's role in NATO, the Department of Defense has undertaken an aggressive conversion program. By using a common system of
measurement to design and produce defense equipment, the U.S. and our
NATO allies can purchase, maintain, and operate major weapons systems
at a lower cost. In a recent memorandum to all Military Departments
and Defense Agencies the DOD has established January 1, 1990 as a target date for availability of a complete set of metric standards and
specifications. DOD components also have been instructed to

participate with national standardization activities of the private
sector in preparation of metric documents and to assume a fair share of
the work load. However, the components also were requested to begin
scheduling preparation of metric specifications and standards on an
accelerated basis wherever the private sector cannot or will not prepare the documents.

In 1964, Office of Management and Budget Circular A-67 prescribed
guidelines for coordinating Federal activities related to acquisition of
water data from streams, lakes, reservoirs, estuaries, and ground
waters. To implement Circular A-67 policies and guidelines, the Office
of Water Data Coordination (OWDC) was established administratively
within the Geological Survey, U.S. Department of the Interior. One
activity started by OWDC was development of the "National Handbook of
Recommended Methods for Water-Data Acquisition." This handbook is designed to document the methodologies that the collectors and users of
hydrologic data believe to be most suitable and thereby provide the
mechanism to assure greater comparability, compatibility, and usability
of water data. This activity is guided by a coordinating council, with
representation from 19 Federal agencies, and by a non-Federal working
group, with representation from technical societies, state water
agencies, universities, and consulting firms. To prepare the handbook,
nearly 200 technical personnel from around 30 Federal agencies are
assigned to 12 working groups, each responsible for one chapter covering a specific phase of the hydrologic cycle. From the beginning, the
working groups concluded that the handbook should emphasize use of the
SI metric system. An appendix of the handbook will contain recommendations related to metric units and conversion factors, precision of
metric measurements, and metric conversion of equipment for the broad
field of hydrology. The metric activities are being coordinated closely with representatives of Canada and Mexico, as well as with ANMC,
ASTM, ASCE, and other technical organizations.

PROGRESS AMONG TECHNICAL SOCIETIES

The American Society for Testing and Materials issued its Metric
Practice Guide for the first time in 1964 and a revised version was
adopted as an ASTM standard in 1968. The edition of 1976 was published
as ANSI.Z210.1-1976 "American National Standard for Metric Practice"
with the joint sponsorship of IEEE and ASTM. Since then IEEE has
issued a separate metric practice guide and ASTM has published its own
edition as Standard E380-79 "Standard for Metric Practice." Attempts
are now being made to get these two organizations together to produce
a single metric guide for ANSI approval. The ANSI standard is the
third main reference recommended by ASCE and many other organizations
for metric conversion activities. It also should be noted that the
Interagency Committee on Metric Policy has indicated that the authoritative sources for questions related to SI units in the U.S. shall be
(1) the NBS item "The Metric System of Measurement (SI)" published in
the October 26, 1977 Federal Register, (2) ANSI Z210.1-1976 (ASTM 380),
and (3) the ANMC Metric Editorial Guide, with authoritative priority in
that order.

In recent years, a metric coordinating group known as the Joint
Committee of National Societies for Metrication in Water Resources
was organized. The committee consists of representatives of 11

national societies:  American Society for Civil Engineers, American
Water Works Association, Water Pollution Control Federation, American
Geophysical Union, American Public Works Association, American Society
of Agricultural Engineers, National Water Well Association, Universi-
ties Council on Water Resources, American Water Resources Association,
American Society of Mechanical Engineers, and Soil Science Society of
America.  Its main purpose has been to coordinate the standard metric
units that will be used by all societies involved in the field of water
resources.  The Joint Committee now forms the nucleus of ANMC's new
"Sector Committee on Water Resources," located organizationally under
ANMC's "Coordinating Committee on Engineering and Transportation"
(figure 1).

The American Society of Civil Engineers' concern with metrication
goes back to August 1876 when the following resolution was approved by
letter ballot of the membership and adopted by the ASCE Board of Directors:  "Resolved that the American Society of Civil Engineers will
further, by all legitimate means, the adoption of the Metric Standards
in the Office of Weights and Measures at Washington, as the sole
authorized standard of weights and measures in the United States."
This resolution was reaffirmed by the Board of Directors on December 6,
1876 and again on October 19, 1959.  In October 1970, the ASCE Board
provided additional support for the metric conversion process in the
U.S. by adopting the following resolution:  "The American Society of
Civil Engineers will carry forward the following program with respect
to metrication:  (1) Support actively the conversion to the SI (System
International; (2) Recommend that no customary or old metric units be
incorporated into the SI; (3) Start immediately the mandatory use of
dual units in all ASCE publications.  Customary units are to be given,
followed by SI units in parentheses until six months after the presumed
enactment of federal metrication law.  For a suitable period thereafter, the order of unit presentation should be reversed; (4) Adopt the
revised edition of ASTM's Metric Practice Guide for ASCE use; (5)
Recommend the use of SI units in all cartographic and geodetic projects.
Conversion of existing property titles and deeds should take place at
the time of property transfer; (6) Recommend the revision of all pertinent books, handbooks, catalogs, etc., using SI units; (7) Recommend
that schools teaching civil engineering initiate immediately instruc-
tion in the SI units."  In 1976, the ASCE  Committee on Metrication
(COM) was organized and in January 1977 the COM recommended the following up-dated and abbreviated form of the 1970 Board resolution:  "(1)
Support actively in the engineering profession and in the public domain a progressive change-over to the International System of Measuring
Units, officially known as SI; (2) Promote the preferred usage of SI
as determined by NBS, ANSI, ANMC, and ISO, especially in the technical
councils and divisions of ASCE; (3) In all publication activities of
ASCE, to encourage henceforth the primary statement of units in SI,
giving customary units in parentheses whenever appropriate; and (4)
Adopt the current American National Standard Metric Practice Guide as
the basic reference document for authors and editors."  Recently, the
position of COM was strengthened when it was moved out of the Technical Council on Codes and Standards and placed administratively under
ASCE Management Group A.

A number of other technical societies in the United States have
adopted policies requiring use of the SI metric system in their publications and several societies also have published detailed metric

practice guides for use by their members. One of the reasons for organization of the Joint Committee of National Societies for Metrication in Water Resources, and later of the ANMC Water Resources Sector Committee, was to recommend the particular SI metric units, including specific prefixes, that could be used by all technical societies in expressing the many different types of water-resources measurements. As the U.S. moves toward the metric world, such coordination is extremely important in order to prevent the confusing multiplicity of units now common for many engineering measurements.

SUMMARY

It has not been the purpose of this paper to promote metric conversion through a detailed discussion of the legal implications of the U.S. metric law or by pointing out the many advantages of the metric system. The fact that the metric conversion movement is moving steadily and quietly forward on a completely voluntary basis is admirable but also is the reason that so many people think that very little is happening in response to the U.S. metric law. However, this paper demonstrates that there really are many individuals and organizations that are convinced of the value of the metric system and of the inevitability of metric conversion in the United States.

APPENDIX.—BIBLIOGRAPHY

1. American National Metric Council, "ANMC Metric Editorial Guide," ANMC Pub. 1, 3rd. Ed., Amer. National Metric Council, Bethesda, MD, 1981.
2. American Society for Testing and Materials, "Standard for Metric Practice," ASTM Standard E380-79, Amer. Soc. Testing and Materials, Philadelphia, 1970.
3. Goldman, David T. and Bell, R. J., Ed., "The International System of Units (SI)," NBS Pub. 330, National Bureau of Standards, Washington, D. C., 1981.
4. Interagency Committee on Metric Policy, "Preferred Metric Units for General Use by the Federal Government," Federal Standard 376, National Bureau of Standards, Washington, D. C., 1981.
5. National Bureau of Standards, "The Metric System of Measurement (SI)," Federal Register Notice of October 26, 1977.
6. Pedde, Lawrence D., and others, "Metric Manual," U.S. Dept. of the Interior, Bureau of Reclamation, Denver, CO, 1978.

# INTERFACING WITH METRICATION ON THE RIO GRANDE

Oscar M. Saldana [1] and Cruz Ito [1]

## ABSTRACT

Because of the international character of the
International Boundary and Water Commission and the need
for daily exchange of flow and storage data with Mexico
required for the joint operation of several projects and
activities, the United States Section in 1980 converted
most of its stream gaging stations on the Rio Grande and
its tributaries to the metric system.  Since then the basic
data and flow records are obtained and computed in metric
units, eliminating conversion of units from one system
to the other as was done in the past.

Exchange of data with Mexico is necessary for
the following activities:  (1) daily joint operation of
international reservoirs and diversion dams, (2) preliminary weekly and final monthly water accounting to determine international ownership of stored waters in Amistad
and Falcon Reservoirs, (3) flood operations to determine
the rate of releases from Amistad and Falcon Reservoirs,
and (4) publication of water bulletins.

Many of the stream gaging instruments were involved
in the conversion, and the component parts for some of them
were changed by means of conversion kits already available.
Other instruments required the use of standard gear pulleys,
coupled with some that were manufactured to meet the requirements.

The International Boundary Commission was created pursuant to
the Treaty of March 1, 1889, and its jurisdiction was extended by subsequent treaties.  It was reconstituted as the International Boundary and
Water Commission, United States and Mexico, by the Water Treaty of 1944
with expanded responsibilities and functions under the policy direction
of the Department of State and the Mexican Secretariat of Foreign Relations, respectively.  The United States Section, a federal agency, also
operates under various congressional acts.  The headquarters of the
United States Section is in El Paso, Texas, and those of the Mexican
Section in Ciudad Juarez, Chihuahua, Mexico.

The Commission, consisting of the United States Section and
the Mexican Section, is charged with implementing the provisions of
existing treaties dealing with boundary and water matters affecting the

---

[1] International Boundary and Water Commission, United States and Mexico

two countries, to include preservation of the international boundary; distribution between the two countries of the waters of the boundary rivers; control of floods on the boundary rivers; river regulation by joint storage works to enable utilization of the waters in the two countries; improvement of quality of waters of the boundary rivers; sanitation measures; and use of waters in the boundary section of the Rio Grande to jointly develop hydroelectric power.

Since 1931, when the Hydrographic activities of the International Boundary Commission were established, the joint operation of the Rio Grande network of gaging stations by the United States and Mexican Sections calls for close cooperation in day-to-day exchange of data and interface of certain operational phases of the mission. Each section operates a proportional number of gaging stations on the main stem of the Rio Grande and all the stations on each of their tributaries. The United States Section, previous to 1980, operated in English units while Mexico has always used the metric system. All data obtained is exchanged, compiled, checked, agreed to and published separately by each country in Water Bulletins issued yearly in metric units by Mexico and in English units by the United States. All these data are utilized in maintaining a joint water accounting of all international waters and the determination of the ownership of such waters by each country on a weekly and monthly basis.

With the introduction in the summer of 1975 of a Bill in Congress for eventual U.S. conversion to the metric system, all IBWC offices were instructed to begin preparation for the total conversion to the metric system, and as part of that preparation, both the English and metric units were to be used in all correspondence, memoranda, and engineering plans. The signing of the President of the <u>Metric Conversion Act of 1975</u> (Public Law 94-168; 15 VSC 205a) on December 23, 1975 made it official policy of the United States to coordinate and plan for the increasing use of the metric system in this country.

After a period in which the use of metric units by conversion were coupled with the regular English units in all figures of our station descriptions and in yearly and period summaries published in our Water Bulletin, it was decided that the Commission should initiate the process for eventual total conversion to the metric system within a time frame consistent with the actions of other federal agencies and the private sector. A work order was issued September 30, 1977, authorizing the work and expenditures required to convert all gaging station equipment to the metric system. The investigation, design, purchase, and implementation of the program was assigned to the Laredo Field Office which directs all field hydrographic activities along the Rio Grande for the United States Section and whose experience in metric units is based on over 40 years of cooperation and interchange of data with the Mexican Section who operates in the metric system. The joint activities of the Commission facilities our conversion to the metric system.

Of the total gaging stations operated by the United States Section it was decided to defer action on those sites where other entities are involved in ownership, joint effort, or cooperation, such as irrigation districts diversion stations and National Weather Service GOES satellite stations. Likewise, action on all observation well and rainfall

station instrumentation was deferred until further notice. This decision left a total of 66 gaging stations on the main stem of the Rio
Grande, its tributaries, floodways, and reservoirs that required conversion to the metric system. Involved in the changeover were 119
instruments including water stage recorders, remote transmitters,
manometers, and shaft encoders.

The conversion of the Leupold and Stevens instruments is fairly
routine, with the component parts being furnished by the manfacturer and
the work on individual instruments such as the type A and the ADR recorders
being performed in the field with a minimum of lost record. Remote transmitters and receivers, especially if dating back many years, require coordination with the manufacturer as to required components and in cases calls
for shipment to their shop facilities for conversion. The cost of conversion kits for L&S Type A and ADR recorders is less than $100 each. For
coupling these recorders to a manometer, Leupold and Stevens also furnishes
a set of sprockets and a bead chain drive for about $80.00.

For the other instruments and recorders the conversion was achieved
by the use of a positive drive "no-slip" belt drive of molded polyurethane
with plastic drive pins on "no-slip" gear pulleys. These components are
produced by PIC Design-Benrus Division of Ridgefield Ct. We found that
some of their standard gear pulleys, coupled with some manufactured by
the company to our specifications, would fulfill our needs. We were able
to position our coupled instruments to the length of their standard drive
belts. The cost of components required for conversion of a float-operated
Fischer & Porter ADR was $100, while the cost of a manometer-driven ADR was
$76.00. All conversion components were assembled in our shop and the change
made in the field with a loss of record of less than one hour. The "PIC"
type components were also used to couple shaft-encoders driven by either
a Type A-35 or ADR recorder.

At the present time all of our gaging stations on the Rio Grande
and tributaries are metric. The change-over was accomplished in steps
as components and assembly became available. The stations above Fort
Quitman came on line in February 1980; the Presidio District converted in
April 1980 and the Del Rio District later that same month; Eagle Pass
District converted in May 1980 and the Lower Rio Grande Valley recorders
and encoders were likewise converted in May 1980. Leupold & Stevens is
working on the components for the remote Transmitting/Receiving recorders.
At the present time, there remains outstanding the remote registering
instrumentation at Falcon Reservoir and power plant, similar units at the
control room of Anzalduas Dam, and the recorders at four GOES satellite
stations operated in cooperation with NOAA. The first two cases will be
converted as soon as Leupold & Stevens is ready to furnish the needed
components. This is estimated to be within the next six months. the
conversion of the GOES satellite stations can be performed within 10
days of notification to proceed from NOAA since these components are in
our possession. To date a total of $15,00 has been expended for components
plus an estimated $6,500 for labor by hydrographic personnel. The balance
of expenditures to complete the program is estimated at $5,000 to $6,000
for a total outlay of $27,000.

Our smallest published metric unit for stage elevation or gage-height is the centimeter. Punched tape records read to the nearest unit, while strip chart records can, if desired, be estimated to smaller increments by visual inspection. We judge that estimates to one-half centimeter can be done with accuracy, and our reservoir storage tables are kept at the half-centimeter unit. Our regular streamflow rating tables are computed to the nearest centimeter, and in practice we round recorded chart gage-heights up or down to the nearest whole unit for processing. For rainfall and evaporation records the unit used is the millimeter. For streamflow, the smallest published unit is .01 $m^3$/sec. Larger amounts are rounded to four significant figures. Smallest unit published for volume flow is one-hundred cubic meters for quantities below 100,000 cubic meters, and one thousand meters for greater volumes. Reservoir storage is published to the nearest one-hundred thousand cubic meters. River mileages are published to the nearest meter. Dimensions, when shown, are to the smallest unit required in each case.

Our experience in data gathering and processing in the metric mode has been very satisfactory during our almost two years of operation. We currently published the Water Bulletin in English units with summaries also shown in metric units and are now prepared to publish our Water Bulletins in all metric data although no changeover date has been set. We may, for a period of time, couple significant portions with English figures in parenthesis until the use of the metric system becomes more wide-spread and familiar to the general public. The operating personnel, both in the field and in the office, have found no difficulty in adapting to the decimal system which is basic to the metric mode.

In summary the exchange of data with Mexico is necessary for the following activities: (1) daily joint operation of international reservoirs and diversion dams, (2) preliminary weekly and final monthly water accounting to determine international ownership of stored waters in Amistad and Falcon Reservoirs, (3) flood operations to determine the rate of releases from Amistad and Falcon Reservoirs, and (4) publication of Water Bulletins.

Advantages: (1) exchange of basic data with Mexico is done directly in metric units; (2) joint projects are expressed in the same system of units; (3) water accounting results are compared easily.

Disadvantages: (1) lack of readily available, economical conversion kits for some recorders; (2) public users of data not receptive to metric units; (3) not all equipment which is used in acquisition of basic data has been converted.

Copies of a summary and of technical notes and listings of the instrument components involved in the metric changeover are available on request.

# CANADA'S METRICATION EXPERIENCE
# IN THE FIELD OF HYDRAULICS

## S.R.M. Gardiner, P.Eng.[1]

### ABSTRACT

In 1970, the Canadian Government expressed its intention to adopt the Systeme International Units of Measurement as the predominant system of weights and measures in the country. Conversion in the private sector was to be voluntary, however, with government encouragement and due to events in the United States, most large enterprises decided to also proceed with metrication.

Components of Hydraulic engineering fall into a wide cross-section of sectors of the Canadian economy and thus hydraulic engineering did not have a single conversion plan. Metrication, however, proceeded smoothly in the field of hydraulic engineering with most engineers now quite benign to the system. The reasons advanced for the smooth conversion include the nature of hydraulic engineering, the commitment of the profession, good planning, and the large number of European trained engineers in Canada.

### INTRODUCTION

In today's style of management, managers are stressing milestones as a means of achieving objectives. But what is a milestone? A quick consultation with the dictionary indicates that a mile is a unit of measurement equivalent to 5,280 feet and a stone is an English measurement equivalent to 14 pounds. Thus:

$$1 \text{ milestone} = 7.39 \times 10^4 \text{ ft - lb}$$
$$= 0.270 \text{ Horsepower - hour}$$
$$= 95.0 \text{ BTU}$$
$$= 1.0 \times 10^5 \text{ Joule}$$
$$= 0.02779 \text{ Kw - hours}$$

or, in other words, a unit of work. Are managers really stressing work? Perhaps, in a disguised fashion they are stressing hard work toward objectives. Certainly, in the old imperial (inch-pound) system of measurement, the milestone could be legitimately used in that context. The fact is that many other imperial units, none of which appear to have any rational relationship to one another, also signify work. This illustrates one of the most frequently used criticisms of measurement systems based upon the old English Engineering Units - the conversion factors between one unit and another are arbitrary and often confusing.

---

1.  Head, Special Projects and Hydraulic Structures, Western Canada Hydraulics Laboratories Ltd., Port Coquitlam, British Columbia.

The exception is the metric system, first adopted in France in the eighteenth century and by much of the rest of the world including Canada in the twentieth century. The decision of the Canadian government to convert to a metric based system of measurement had repurcussions on all facets of Canadian life. The subject of this discussion is its effect on engineering in the field of Hydraulics. The following sections include a background to metric conversion in Canada, its implementation, and its impact on hydraulic engineering.

## BACKGROUND

By a queer twist of fate, Canada's first official measurement system was metric. The initial Weights and Measurements Act passed in 1871 used the metric system because it was the only system based on a single, international standard. However, in 1873, Great Britain signed the Treaty of the Metre which based the imperial system on the metric standard and thus Canada amended the Weights and Measurements Act to include the imperial version of the English system. Since then, the imperial system has prevailed.

In 1970, the Canadian government published a white paper (1) recommending that Canada adopt as the dominant system of measurement the Systeme International (SI), the modern integrated metric system. The reason stated for this recommendation was not directly related to the rational basis of the metric system, although that was recognized, but rather:

> "adoption of the metric system of measurement is ultimately inevitable -and desirable - for Canada. It (the government) would view with concern North America remaining as an inch-pound island in an otherwise metric world - a position which would be in conflict with Canadian industrial and trade interests and commercial policy objectives." (1)

The interests and objectives to which the white paper refers is the diversification of markets for Canadian goods and services beyond its traditional market in the United States. Offering goods and services in metric would improve Canada's position in those markets.

At the time the white paper was published, only the United States, Canada, New Zealand, and Australia amongst all the industrially advanced countries had not embarked on conversion to metric. New Zealand and Australia have since gone ahead of Canada in metrication. By the mid 1980's the U.S. will be alone in its use of English based units.

The implications to Canada of converting to metric without a similar policy in the U.S. to soon follow suite are immense. The Canadian economy is heavily dependent on trade with and investment from our neighbours to the south. Converting to metric unilaterally was a "gutsy" move on the part of the Canadian government to say the least.

While some U.S. policies were a hindrance to Canadian conversion, some were a catalyst. Conversion was to be voluntary and thus it fell to the government to convince industry to follow its lead. This voluntary system would probably have failed if several major companies in the U.S., including General Motors, had not decided to convert their production to metric. The potential spillover into Canada due to the interrelated economies convinced many Canadian companies that conversion was in their best interest.

Upon publication of the white paper, the government commenced assembling the bureaucratic machinery required to implement the conversion. A Metric Commission was appointed with a mandate to co-ordinate and review metrication, and advise the government concerning any changes to standards legislation requested from segments of industry. The economic community was split into groups of interrelated organizations with a steering committee to co-ordinate conversion of each group. Further, the groups were divided into sectors and each sector had a conversion plan (2) drawn up by a sector committee. The sector plans laid out a logical and gradual course of events leading to SI Day, the day that a particular sector would be completely converted to metric.

Each company or organization was encouraged to set up metric committees to oversee conversion and to provide a liaison with the government appointed bodies. It must be said that conversion could not have been attempted without a high degree of committment on the part of the economic community.

## METRIC CONVERSION IN THE FIELD OF HYDRAULIC ENGINEERING

Hydraulic engineering encompasses a wide range of engineering activities and, as a result, does not fall into any one sector as defined by the Metric Commission. Sectors such as agriculture, electric power, mining, and many others include some component of hydraulic engineering. Any report on metrication in this diverse field must, therefore, involve some generalities.

The diversity of the field also implies that there is no one source of material for such a report. To overcome this difficulty, solicitations were made to various individuals across Canada to obtain a cross section of experience and opinion. Again, these contributions are blended such that they appear to apply quite generally across the hydraulics field.

For the most part, conversion commenced with an investigation phase in January of 1972. Actual implementation of conversion was started some three years later in April of 1975. The intervening years were used to draw up the various sector policies and plans. SI Day for most sectors containing some component of hydraulic engineering was year end 1980. Generally, conversion is two years behind schedule even though some sectors have met the schedule.

Implementation was phased approximately in the following order:

a)    define policy and strategy
b)    identify areas affected
c)    specify new metric equivalents
d)    consult with organizations
e)    develop schedule
f)    start conversion monitoring (continuous)
g)    change codes and standards
h)    convert engineering design and computer programs
i)    convert instrumentation

j)     convert operational records
k)     SI Day

The above smears a great number of activities under the same heading and seems to indicate a chronological order which may not be true for all sectors. In the main though it is a fair representation of the events and their order.

While the sector plan and schedule were drawn up by the sector committee, the true implementation rested with the sector companies and organizations themselves. The implementation varied somewhat but a consistent thread can be identified:

a)     Metrication began with a certain design project even though other projects within the organization were continuing under the old system. An example of this is the huge James Bay Hydroelectric Project in the province of Quebec. La Grande 3 and previous projects were designed in the old imperial system. La Grande 4 was completely designed in metric.

b)     Instrumentation and metering systems were converted where possible during regular maintenance.

c)     Activities which take place on a continuing basis were phased in according to the requirements of each organization.

## HOW WELL DID CONVERSION PROCEED?

Since SI Day was late 1980 for most sectors involving hydraulic engineering, one can look back in retrospect and see how smoothly the conversion took place. Even though some sectors are behind schedule, the message from across the country is that engineering converted with very few hitches. In fact, most engineers are now quite benign to the metric system. The analogy is often made that it is similar to being bilingual, equally comfortable in both the old and new systems.

Some minor annoyances were voiced:

a)     The Human Factor

Basically, mature people do not like to change their set ways and deferrence must be given to the psychologists to explain the reasons. At the very least, most people were initially ambivalent to metric if for no other reason than they no longer recognized or had a feeling for the magnitude of measured quantities.

b)     Phase Lag

Conversion was phased in a series of steps. As a result, some segments of companies and even sectors were metric while the remainder were still using the imperial system of units. This gave rise to problems when the metric and imperial groups had to interface. For instance, engineers were writing specifications in metric before suppliers were ready to fill these orders in metric.

c)     Design Formulae

There is a large volume of empirical design formulae which is unit specific. All such formulae had to be converted and individuals had to refamiliarize themselves with them. There were instances where mistakes were made using incorrect formulae.

c)     Poor Planning

While the sector committees on the whole did a credible job in planning conversion, there were some oversights which caused problems. One of the most dramatic in this regard occurred outside of hydraulic engineering but is worthy of note. In the steel industry, the construction sector and the manufacturing sector had converted but the suppliers (middlemen) had been left out of the sector plan. Construction complained that they couldn't purchase metric materials and the mills complained that there was no demand for metric materials. This oversight set SI Day in that sector back by two years.

d)     Politics

Metrication was a political decision and from time to time the politicians have stuck their fingers into the workings to change the course of events. In general, political interference has had a negative effect on conversion.

## REASONS FOR SMOOTH CONVERSION

The exact reasons for the smoothness of conversion to date are hard to pinpoint but the following have been suggested:

a)     Due to the nature of hydraulic engineering, there was no formidable obstacles or large capital investments in the way of smooth conversion. In fact, many companies had been doing offshore work which required use of metric and their employees had become familiar with the system.

b)     Good planning and co-ordination on the part of the sector committees.

c)     Professionals had a high level of commitment to conversion and, through technical publications, had been using the system for several years prior to conversion.

d)     A large number of Canadian engineers were educated in Europe either as Canadians doing post graduate studies abroad or as foreign undergraduates who immigrated to Canada.

## SUMMARY

Sometime in the mid 1980's, Canadians will complete a process begun in 1871 and will be predominantly using the metric system of units. The reason for the recent emphasis on converting is to improve Canada's position in economic markets other than the U.S. The government has set up the required infrastructure to co-ordinate a voluntary system of conversion.

Hydraulic engineering is involved in many sectors of the economy thus there is no one conversion plan. In general, though, conversion has gone smoothly for hydraulic engineers with very few exceptions. Some of the reasons advanced for the smooth conversion include the nature of hydraulic engineering, the committment of the profession, the conversion planning and the large number of European trained engineers.

## ACKNOWLEDGEMENTS

The author would like to acknowledge and thank the following for their contributions to this paper: Mr. G. Seropian, Metric Commission Canada; Mr. Peter Wall, B.C. Metric and Metric Standards Information Office; Dr. W. Bell and Mr. Leviu Sachter, B.C. Hydro; Mr. Fred Parkinson, LaSalle Hydraulic Laboratories; Messrs R. Cote and M. Drouin, James Bay Corporation; Mr. D. Dobson, Water Survey of Canada; Mr. R. Carson, Crippen Acres Limited; Mr. S. Potter, MONENCO; and Prof. M. Quick, University of British Columbia.

## REFERENCES

1.      White Paper on Metric Conversion in Canada, Ministry of Industry, Trade and Commerce, January 1970.

2.      Metric Conversion Plan, Metric Commission Canada, November, 1975.

# COMPARISON OF THIRD-ORDER TRANSPORT SCHEMES

By Ross W. Hall[1] and Raymond S. Chapman,[2] A.M. ASCE

## ABSTRACT

A comparison of two explicit third-order accurate finite difference schemes is presented for the solution of the transient advective
transport equation in both one and two dimensions. The numerical
schemes examined are QUICKEST, which is based on a conservative, control volume integral formulation, and a Lagrangian technique referred
to as the 12-POINT scheme. Results presented show that both schemes
possess favorable amplitude and phase characteristics; however,
unlike QUICKEST, which is mass conservative, the 12-POINT scheme
exhibits mass conservation errors that are directly attributable to
the time step employed.

## INTRODUCTION

Numerous finite difference schemes have been applied to transient advective transport; however, until recently low-order spatial
and temporal discretization techniques were predominately used in
practical applications. Typically weighted combinations of second-
order central differencing and first-order upwind differencing were
commonly employed. Unfortunately, it is well known that behavioral
errors (10) such as the artificial diffusion associated with upwind
differencing, and the parasitic oscillations characteristic of central differencing often render these techniques unsuitable for applications in advective transport models. Consequently, the need to
progress to higher order schemes is apparent.

The relative merit of steady state applications of spatially
third-order accurate schemes are well illustrated in the literature
(1, 2, 5, 6, 7, 8, 9). The comparisons of the spatially third-order
accurate QUICK (Quadratic Upstream Interpolation for Convective Kinematics) technique (6) with central and upwind differencing show QUICK
to be far superior in steady advective transport calculations where
practical grid spacings are used. However, Leonard et al. (5) and
Leschziner (7) have shown that QUICK suffers from a boundedness problem when applied to transient advection. Leonard (6) however, has
presented the approximate temporally third-order accurate extension

---

1  Research Limnologist, USAE Waterways Experiment Station, Vicksburg, MS.
2  Research Associate, Virginia Polytechnic Institute and State
   University, Blacksburg, VA, on contract to USAE Waterways Experiment Station, Vicksburg, MS.

of the QUICK technique called QUICKEST (Quadratic Upstream Interpolation for Convective Kinematics with Estimated Streaming Terms), which is specifically designed to address uni-directional transient transport problems.

In this paper, the explicit QUICKEST finite difference technique is extended to two-dimensional advective transport, and its performance is compared with the existing Lagrangian 12-POINT scheme of Hinstrup et al. (4) for both one- and two-dimensional transient advective transport. Performance criteria examined include numerical diffusion/amplitude, phase, and mass conservation errors.

## MODEL EQUATION

The two-dimensional advective transport of an arbitrary conservative scalar, $\emptyset$, is described with appropriate initial and boundary conditions by:

$$\frac{\partial \emptyset}{\partial t} + \frac{\partial(u\emptyset)}{\partial x} + \frac{\partial(v\emptyset)}{\partial y} = 0 \qquad (1)$$

where

$$t = \text{time},$$

$$x, y = \text{two-dimensional Cartesian coordinate directions, and}$$

$$u, v = \text{velocity components in the } x \text{ and } y \text{ directions, respectively}$$

## COMPUTATIONAL METHODS

The third-order accurate computational methods examined in the present work are the two-dimensional extension of the explicit QUICKEST scheme of Leonard (6) and the Lagrangian 12-POINT scheme of Hinstrup et al. (4).

QUICKEST, which is based on a conservative control cell formulation, uses a spatial six-point upstream weighted interpolation surface with a temporal advective correction to obtain third-order approximations to cell and cell face averaged quantities. Although a detailed derivation is not presented herein, the two-dimensional version of QUICKEST is simply a direct extension of the work of Leonard (6).

In contrast, 12-POINT is a Lagrangian procedure which updates local values of the scalar function $\emptyset$ according to

$$\emptyset^{n+1}(x,y) = \emptyset^{n}(x - u\Delta t, y - v\Delta t)$$

where the function on the right hand side is approximated by a 12-point Everett interpolation formula (3).

It is interesting to note that for spatially constant Courant number, QUICKEST and 12-POINT are algebraically equivalent in one dimension. However, proof of equivalency in two dimensions is yet to be established.

## TEST SIMULATIONS AND RESULTS

A systematic program of numerical experimentation was performed in two phases. The first phase consisted of the pure advection of a Gaussian distribution in a uniform velocity field in one dimension. As previously noted for spatially constant Courant number, QUICKEST and 12-POINT are algebraically equivalent in one dimension. Therefore, results presented for the uni-directional advection tests are applicable to both QUICKEST and 12-POINT. The second phase consisted of the solid body rotation of a Gaussian distribution.

The uni-directional advection of a Gaussian distribution was conducted at Courant numbers ($u\Delta t/\Delta x$) of 0.25, 0.50, 0.75, and 1.0. Each test computation was run for 200 time steps. In addition, grid density effects were investigated by distributing 95 percent of the initial distribution over 5, 9, or 17 grid points, which corresponds to variances of $(\Delta x)^2$, $4(\Delta x)^2$, and $16(\Delta x)^2$, respectively.

Figure 1 presents the results for an initial Gaussian distribution of variance $4(\Delta x)^2$ and varying Courant number. Examination of the figure reveals negligible phase error; however, a noticeable but slight amplitude difference is seen between Courant numbers 0.5 and 0.25/0.75.

The large amplitude differences observed between Courant number 1.0, which reproduces the continuum solution through exact point-to-point transport, and Courant numbers 0.25, 0.50, and 0.75 are due to the grid density selected. Figure 2 is a comparison of model simulations performed at Courant number equal to 0.5 for varying grid density. This figure clearly demonstrates the importance of selecting an appropriate grid density.

The two-dimensional solid body rotation calculations were conducted using a symmetric bivariate Gaussian distribution of variance $16(\Delta x)^2$ in a velocity field of constant angular frequency equal to $\pi/1000$ rad sec$^{-1}$. Courant numbers defined at the center of mass of the distribution ranged from 0.34 to 0.01 which required 200 to 6400 time steps.

The results of the solid body rotation calculations are presented in Figures 3 and 4, a comparison of the simulation results of 12-POINT and QUICKEST with the exact solution, respectively. Examination of Figures 3 and 4 reveals that both 12-POINT and QUICKEST exhibit good phase and amplitude characteristics. However, unlike QUICKEST, whose control cell formulation insures mass conservation, 12-POINT exhibited a 5 percent mass loss. This mass conservation error was subsequently investigated by repeating the test simulation with varying time step. Figure 5, a plot of mass error, in percent, versus time step, $\Delta t$, clearly illustrates that the ability of

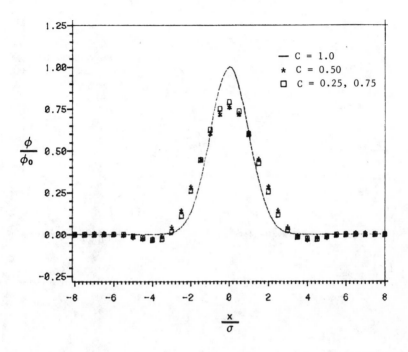

Figure 1. Effects of Courant Numbers on the Uni-Directional Advection of a Gaussian ($\sigma = 2\Delta x$) Distribution.

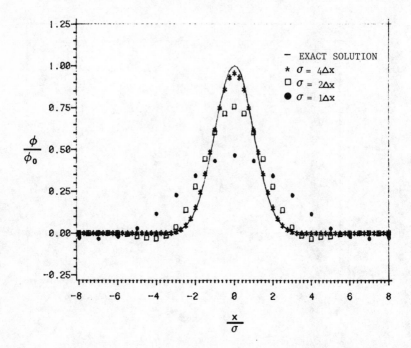

Figure 2.   Effects of Grid Density on the Uni-Directional
Advection (C = 0.5) of a Gaussian Distribution.

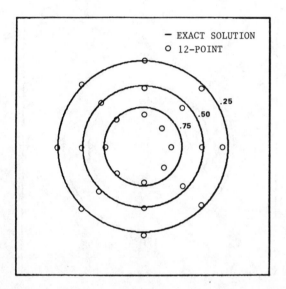

Figure 3.  Comparison of 12-POINT with the Exact Solution for
Solid Body Rotation of a Bivariate Gaussian Distribution.

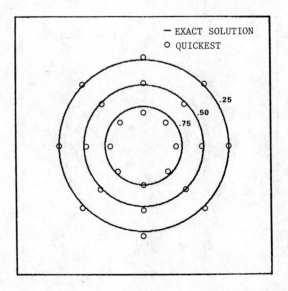

Figure 4. Comparison of QUICKEST With the Exact Solution for Solid Body Rotation of a Bivariate Gaussian Distribution.

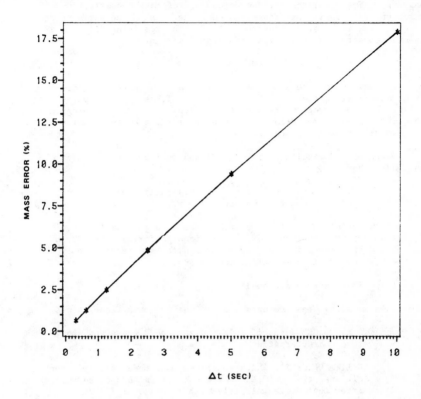

Figure 5.   Mass Conservation Error of the 12-POINT Scheme
as a Function of  $\Delta t$.

12-POINT to conserve mass is strongly dependent upon the time step employed. For completeness, the dependence of mass conservation error on grid density was investigated by decreasing the variance of initial Gaussian distribution to 4 $(\Delta x)^2$. The resulting difference in mass conservation error was less than 1 percent.

## CONCLUSIONS

The purpose of the study was to critically compare two third-order numerical schemes, QUICKEST and 12-POINT, as applied to the solution of the advective transport equation. A systematic program of numerical experimentation was conducted for both one-dimensional and two-dimensional test cases.

Based on the results of the one-dimensional test case, it may be concluded that:

    i)    QUICKEST and 12-POINT are algebraically equivalent for spatially constant Courant number.

    ii)   Variations in the Courant number result in small amplitude differences; however, simulations performed with varying grid densities show that inadequate grid resolution results in large amplitude errors.

   iii)   Phase errors are negligible.

The two-dimensional test results suggest the following conclusions:

    i)    Both QUICKEST and 12-POINT possess good phase and amplitude characteristics.

    ii)   Unlike QUICKEST, which conserved mass under all test conditions, the 12-POINT scheme exhibits a mass conservation error that varies directly with the size of the time step employed.

Finally, the work presented here is preliminary to the development of a general purpose, depth-integrated water quality model. Realizing that the mass conservation property is a fundamental requirement, it is clear that of the two third-order finite difference techniques examined, QUICKEST is far superior for engineering applications where practical grid spacing and time steps are essential.

1.  Chapman, R. S. and C. Y. Kuo, "Application of a High Accuracy Finite Difference Technique to Steady Free Surface Flow Problems," Presented at the Joint ASME/ASCE Mechanics Conference, Boulder, CO, June 1981.

2.  Han, T., J. A. C. Humphrey, and B. E. Launder, "A comparison of Hybrid and Quadratic-Upstream Differencing in High Reynolds Number Elliptic Flows," Computer Methods in Applied Mechanics and Engineering, Vol. 29, 1981.

3.  Hildebrand, F. B., Introduction to Numerical Analysis, McGraw-Hill Book Co., Inc., New York, 1956.

4.  Hinstrup, P., A. Kej, and U. Kroszynski, "A High Accuracy Two-Dimensional Transport-Dispersion Model for Environmental Applications," Paper B17, 17th IAHR Congress, Baden-Baden, Germany, 1977.

5.  Leonard, B. P., M. A. Leschziner, and J. McGuirk, "Third-Order Finite-Difference Method for Steady Two-Dimensional Convection," Proceedings of the International Conference on Numerical Methods in Laminar and Turbulent Flow, Swansea, July 1978.

6.  Leonard, B. P., "A Stable and Accurate Convective Modelling Procedure Based on Quadratic Upstream Interpolation," Computer Methods in Applied Mechanics and Engineering, Vol. 19, 1979.

7.  Leschziner, M. A., and W. Rodi, "Calculation of Strongly Curved Open Channel Flow," Journal of the Hydraulic Division, ASCE, Vol. 105, No. HY10, 1979.

8.  Leschziner, M. A., "Practical Evaluation of Three Finite Difference Schemes For the Computation of Steady-State Recirculating Flows," Computer Methods in Applied Mechanics and Engineering, Vol. 23, 1980.

9.  Leschziner, M. A., and W. Rodi, "Calculation of Annular and Twin Parallel Jets Using Various Discretization Schemes and Turbulence-Model Variations," Journal of Fluids Engineering, Transactions, ASME, Vol. 103, 1981.

10. Roache, P. J., Computational Fluid Dynamics, Hermosa Publishers, Albuquerque, NM, 1972.

# MODELING OF RAINFALL-RUNOFF PROCESSES

By

Farouk M.Abdel-Aal*,Member ASCE

## ABSTRACT

There are three conceptual approaches that have been used in developing models for rainfall-runoff processes. Those are stochastic, parametric, and deterministic modeling. If there are no information of cause-effect, then the process must be regarded as purely random and must be treated as stochastic. On the other hand, if almost complete information on cause-effect is available, then the process may be treated as deterministic. Seldom, however, the process is on those two extreems, and the parametric approach is used as a compromise solution.

A survey of the available knowledge which leads to the prediction of runoff hydrographs from rainfall information and basin characteristics is made. Several stochastic,parametric, and deterministic models are discussed. The importance of computer simulation in producing runoff data for ungaged areas and exptrapolating short records is pointed out.

## INTRODUCTION

It is becomming increasingly obvious that the social and economic planning in all parts of the world must seriously consider the management of water as a priority. An integral part of any water management program must be the ability to predict river flows. This is achieved by hydrologic modeling.

Hydrologic systems, in general,have both stochastic and deterministic components. The stochastic components are parameters defined by means of probability distributions, whereas the deterministic components are processes that can be modeled mathematically or graphically without probabilistic statements. Combined modeling could lead to advances in many areas in hydrology (Laurenson, 1974).

Two different types of models may be distinguished for rainfall-runoff processes; those used for long-range studies, and those used for short-range analysis. Longrange studies have generally been performed with stochastic models, whose function is to produce a set of traces

*Associate Prof.of Civil Engg.KAAU,P.O.Box 9027, Jeddah, Saudi Arabia

sufficiently long and, in the limit, statistically indistinguishable from the historical record. Short-range analysis covers the input-output type of modeling in which parametric and deterministic models have generally been used. For any selected rainfall-runoff model, the theory and computations are greatly simplified if the relationships are assumed to be linear and time invariant. Stationarity in stochastic models corresponds to time invariance in deterministic models.

This paper discusses the various types of stochastic, parametric, and deterministic models used for rainfall-runoff processes.

## STOCHASTIC AND PROBABILITY MODELS

Stochastic processes might be used to generate a synthetic record of rainfall, which could be transformed to runoff by an appropriate method. Such a procedure could be used if a runoff record were too short to provide a basis for stochastic generation or if more details were desired about the runoff than could be obtained from monthly flow. As far as the quality of the information is concerned, the new data are no better than the data from which they were generated. However, the synthetic record has the advantage that it is longer than the historical record and thus more amenable to analysis.

Two main approaches have been adopted in the generation of runoff data. First it may be assumed that the data are purely random, in which case Monte Carlo methods are used. More realistically it may be assumed that any runoff value is in part determined by the preceeding runoff values at that point, in which case the process known as the Markov chain is used.

Pattison (1965) demonstrated the feasibility of developing stochastic flow data for a small basin by generation of a single-station record of rainfall. Using a Markov model, he successfully reproduced the flood frequency curve of a stream. Franz (1970) employed a multivariate normal analysis to generate compatible data at several rainfall stations.

The runoff record from a basin may be analyzed without reference to the generating rainfall. Using a record of N years duration, the recurrence interval or return period, $T_p$, is defined as,

$$T_p = \frac{N+1}{M} \quad \dots\dots\dots\dots\dots\dots\dots\dots\dots\dots\dots\dots \quad (1)$$

where M is the rank of flood in order of magnitude in the record, the largest flood having M = 1.

The calculated values of $T_p$ may be plotted against the magnitude of their respective discharges, Q, to give a convenient method of relating the two variables. It is obvious that the result will not permit the prediction of runoff from a given rainfall, but it allows the design engineer to estimate the peak discharges likely to happen within the life span of a given structure.

The approach is somewhat improved if rainfall rather than runoff is the subject of the analysis, since rainfall has usually been recorded for much longer time than runoff. Recurrence intervals of given rainfalls may then be used in conjunction with one of the input-output models to predict runoff from a basin where runoff data are limited.

## PARAMETRIC MODELS

These models adopt a parameter to link input and output which is not necessarily a function of the basin. They are also known as input-output models or black-box models. Usually, rainfall and runoff characteristics are chosen and correlated by least square regression analysis to find the best mathematical function which describes the relationship. Such models can only be applied to basins where rainfall and runoff records are available, and parameters found will be applicable to those basins alone.

The predictive ability of such models will be limited since runoff will not vary with rainfall alone but will depend upon rainfall intensity, storm duration, and most important, antecedent basin moisture. The antecedent precipitation index I may be calculated on the basis of a logarithmic recession during periods of no precipitation thus,

$$I_t = I_o \, k^t \quad \dots\dots\dots\dots\dots\dots\dots\dots\dots\dots\dots\dots\dots \quad (2)$$

where $I_o$ is the initial value of antecedent precipitation index, $I_t$ is the reduced value t days later, and k is a recession factor ranging normally between 0.8 and 0.98.

It is possible therefore, to predict runoff from the combined effect of rainfall and antecedent precipitation index for a given basin by using multiple regression analysis. Weyman (1975) gave for a basin in England the following relationship,

$$Q = - 3.16 + 0.35 \, i_o + 0.34 \, I_{30} \quad \dots\dots\dots\dots\dots \quad (3)$$

where Q is storm runoff in mm, $i_o$ is storm rainfall in mm, and $I_{30}$ is a 30 day antecedent precipitation index.

One of the best known methods to estimate runoff is the rational formula. This method assumes that the maximum rate

of runoff $Q_p$ from a drainage area A will occur when the entire area is contributing to the runoff. It may be expressed
in the form (Chow, 1964),

$$Q_p = C \, i_o \, A \quad .................................. \quad (4)$$

where C is a runoff coefficient indicating the percentage of
rainfall which appears as surface runoff, and $i_o$ is the
rainfall intensity. The rational formula is used for small
areas but its accuracy diminishes very much when it is
applied to areas larger than 10 acres.

In the so called co-axial correlation, the gross rainfall is adjusted for the expected losses on entering the
basin until the net rainfall is directly related to runoff.
Although the solution is more or less a graphical version of
the multiple regression techniques discussed before, it is
essentially rational, Fig. 1. This approach often has high
degree of prediction, but the limitation of its use is due
to the fact that it does not contain parameters related to
basin characteristics.

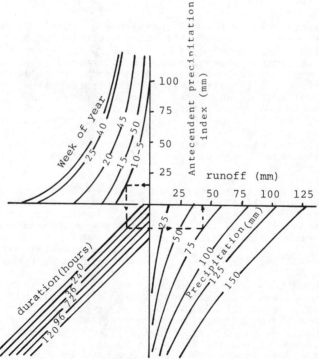

Fig.1. Coaxial correlation for estimating runoff

In order to produce general rainfall-runoff relationships, basin characteristics must be included. The procedure
is known as regional flood frequency analysis. Dury (1959)
in an analysis of a system, developed recurrence interval
graphs for each basin and then correlated the discharge of
the 2.33 year flood against basin area. The following
formula was found,

$$Q_{2.33} = 5.1 \ A^{0.98} \qquad \dots\dots\dots\dots\dots\dots\dots \quad (5)$$

where $Q_{2.33}$ is the peak discharge of the flood with a return
period of 2.33 years, and A is the basin area. For an areal
extension of Eq. 5, it is necessary to include an input
variable, since the rainfall of a given return period will
vary in space, and to include some measure of the internal
organization of the basin. Rodda (1967) gave the following
empirical formula in foot units for a drainage density D,

$$\log Q_{2.33} = 1.8 + 0.77 \ \log A + 2.92 \ \log i_{2.33} + 0.81 \ \log D \ \dots \quad (6)$$

where $i_{2.33}$ is the average annual madimum daily rainfall.

## UNIT HYDROGRAPH TECHNIQUES

Sherman (1932) proposed that since a stream hydrograph
reflects many of the physical characteristics of the drainage basin, similar hydrographs will be produced by similar
rainfalls occuring with comparable antecedent conditions.
Thus, once a typical or unit hydrograph has been determined
for certain conditions, it is possible to estimate runoff
from a rainfall of any duration or intensity. The unit hydrograph is the hydrograph with a volume of 1 inch or 25mm
of direct runoff resulting from a rainfall of specified
duration and areal pattern. Hydrographs from other rainfalls
of like duration and pattern are assumed to have the same
time base, but with ordinates of flow in proportion to the
runoff volumes.

When the unit hydrograph of a given effective-rainfall
duration is available, the unit hydrographs of other dura-
tions can be derived, Fig. 2. This is done by the principle
of superposition and is known as the S - hydrograph or S -
curve method, Fig. 3. The unit hydrograph for any rainfall
duration $t_0$ hours, may be obtained by subtracting two
S-curves with their initial points displaced by $t_0$ hours.

The unit hydrograph technique has been used with considerable success but, as a description of the basin system,
has some theoritical limitations. Large errors may result
from simplifying assumptions about variations of infiltration capacity, uniformity of rainfall over the basin in
time and space, and the invariant linear response of the
basin to these inputs. The basin however, is a non-linear,
time variant system.

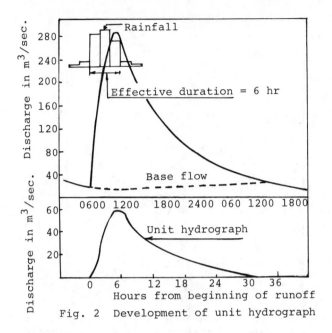

Fig. 2   Development of unit hydrograph

Sometimes, it may be necessary to determine unit hydrographs for basins with few if any runoff records. In such
cases, close correlation between the physical characteristics of the basin and the resulting hydrograph are necessary.
Snyder (1938) attempted to develop a synthetic unit hydrograph based on the basin size and topography, which could be
used for ungaged basins. The synthetic unit hydrograph was
drawn from some hydrograph parameters which were defined for
a given basin as,

$$t_p = c \ (L \ L_c)^{0.3} \quad \dots\dots\dots\dots\dots\dots\dots\dots\dots\dots\dots\dots \quad (7)$$

where $t_p$ is the time lag from rainfall centroid to peak discharge in hours, L is the basin length in miles, $L_c$ is the
distance from the center of the basin to its exit, and c is
a coefficient varying from 1.9 to 2.2 as the average basin
slope decreases.  For rainfalls of standard duration $t_d$, it
was found that,

$$t_d = \frac{t_p}{5.5} \quad \dots\dots\dots\dots\dots\dots\dots\dots\dots\dots\dots\dots\dots\dots \quad (8)$$

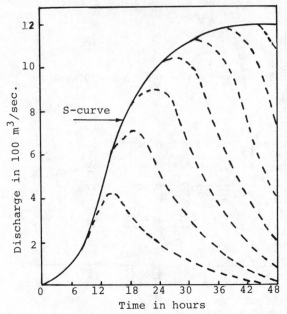

Fig.3 Graphical illustration of S-hydrograph

The peak discharge, $Q_p$, of the unit hydrograph is given by,

$$Q_p = \frac{645.6 \ c \ A}{t_p} \quad \dots\dots\dots\dots\dots\dots\dots\dots\dots\dots\dots\dots\dots \quad (9)$$

where A is the basin area in square miles and c is a co-efficient ranging from 0.56 to 0.69. Finally the duration of the unit hydrograph, T, in days is given as,

$$T = 3 + 3 \ (\frac{t_p}{24}) \quad \dots\dots\dots\dots\dots\dots\dots\dots\dots\dots\dots\dots \quad (10)$$

If the duration of the effective rainfall becomes in-finitesimally small, the resulting unit hydrograph is called an instantaneous unit hydrograph. By the principle of superposition in the linear unit hydrograph theory, when an effective rainfall of function $I(\tau)$ of duration $t_o$ is applied, each infinitesmal element of rainfall will produce a discharge equal to the produce of $I(\tau)$ and the instanta-neous hydrograph expressed by $u(t-\tau)$. Thus the outflow run-off $Q_t$ at any time t, is related to the inflow of rainfall excess $I(\tau)$ by the convolution equation,

$$Q_t = \int_{o}^{t \ \leqslant \ t_o} u \ (t-\tau) \ I(\tau) \ d\tau \quad \dots\dots\dots\dots\dots\dots \quad (11)$$

The equation defines the impulse - response function of the system $u(t)$, i.e. the instantaneous hydrograph(Chow, 1964).

The instantaneous hydrograph can be converted into a unit hydrograph for any duration, t, by averaging ordinates t units of time apart, and plotting the average at the end of the period, Fig. 4. The major advantage of the instantaneous hydrograph is that it is independent of the duration of effective rainfall, and thus eliminating one variable in hydrograph analysis. It is better suited for theoritical investigations on rainfall-runoff relationships in drainage basins. Although the determination of the instantaneous hydrograph is numerically more tedious than that of unit hydrograph, the computations can be simplified through the use of electronic computers.

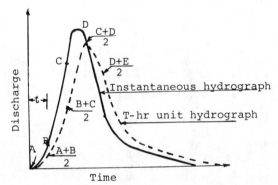

Fig.4 Converting instantaneous hydrograh to one of finite duration

## MECHANICS OF RAINFALL-RUNOFF RELATIONSHIPS

Henderson and Wooding (1964) studied the runoff over a sloping area with length L and subject to a rainfall of intensity $i_o$ and duration $t_o$. Neglecting infiltration and using the method of characteristics, they gave for the time of concentration $t_c$ at the end of slope,

$$t_c = \left(\frac{L}{\alpha\, i_o{}^{m-1}}\right)^{1/m} \quad \dots\dots\dots\dots\dots\dots\dots\dots\dots\dots\dots \quad (12)$$

where $\alpha$ is a coefficient incorporating slope and friction, and m is an exponent. The discharge at the end of slope for any time t is,

$$q = \alpha\,(i_o\, t)^m \quad \dots\dots\dots \quad t \leqslant t_c \leqslant t_o \quad \dots\dots\dots\dots \quad (13)$$

and the equilibrium discharge,

$$q_e = \propto (i_o \, t_c)^m \quad \ldots\ldots \quad t_c \leqslant t \leqslant t_o \quad \ldots\ldots\ldots \quad (14)$$

For subsidence of runoff following the cessation of rainfall, the discharge is determined from,

$$(t - t_o) = \frac{L - (q/i_0)}{\propto m (q/\propto)^{\frac{m-1}{m}}} \quad \ldots\ldots \quad t > t_o \quad \ldots\ldots\ldots \quad (15)$$

The runoff hydrograph for the complete rainfall period can
therefore, be computed and plotted. However, infiltration
rate may be subtracted from rainfall intensity to give
more meaningful results.

## COMPUTER SIMULATION

Another alternative to the traditional methods of modeling rainfall-runoff processes involves the construction of
drainage area models using digital computers. The parameters
of the model may be determined from field measurements of
hydrologic data or by means of successive optimization procedures. This is continued until the model behavior approximates most closely that of the drainage basin.

The most important work in this field is the Stanford
Watershed Model IV, whose principle input data are hourly
rainfall and daily potential evapotranspiration (Crawford
and Linsley, 1966). The model is calibrated to a basin by
trial until the observed flows are reproduced accurately.
In general, a good calibration can be obtained with 3 to 5
years of data.

The value of computer simulation lies in its ability to
produce runoff data for ungaged points, or to extrapolate
short runoff records. Possibly more important is the ability
to project flows for assumed future conditions in the basin.
However, transfer of any program to another basin may be
quite unsatisfactory unless the program realistically describes the hydrologic processes.

## CONCLUSIONS

This paper discusses the stochastic, parametric, and
deterministic models of rainfall-runoff processes. The
following conclusions are reached,

1.  Stochastic processes may be used to generate synthetic
    record of rainfall, which could be transformed to run-
    off by an appropriate method. The synthetic record has

the advantage that it is longer than the historical
record and thus more amenable to analysis.

2. Regression analysis can be applied to basins where
rainfall and runoff records are available. The parameters found will be applicable to those basins alone.

3. The rational method is used for small drainage areas
and its accuracy diminishes very much when applied to
areas larger than 10 acres.

4. The co-axial correlation has a high degree of prediction. However, the limitation of its use is due to
the fact that it does not contain parameters related
to basin characteristics.

5. Regional flood frequency analysis permits the inclusion
of basin characteristics. Therefore, general rainfall-
runoff relationships may be produced.

6. The unit hydrograph technique has some theoritical
limitations. The model assumes that the rainfall is
uniform in time and space, and that the basin operates
as a time invariant linear system.

7. The instantaneous hydrograph is independent of the duration of effective rainfall. It is therefore, better
suited for theoritical investigations of rainfall-
runoff relationships.

8. The mechanics of rainfall-runoff relationships permit
the determination of the flow hydrograph from a
sloping area.

9. Computer simulation enables the production of runoff
data for ungaged points, and extrapolating short runoff records. However, transfer of any program to another
basin may be quite unsatisfactory.

REFERENCES

1. Chow, V.T. 1964. Handbook of Applied Hydrology. McGraw-
Hill, New York.

2. Crawford, N.H. and Linsley, R.K.1966. Digital Simulation
in Hydrology: Stanford Watershed Model IV. Stanford
University, Dept. Civil Eng., Tech. Rept.39.

3. Dury, G.H. 1959. Analysis of Regional Flood Frequency in
the Nene and Great Ouse. Geographical Journal 125,
pp. 223-229.

4. Franz, D.D. 1970. Hourly Rainfall Synthesis for a Network of Stations. Stanford University, Dept. Civil Eng.,
Tech. Rept. 126.

5. Henderson, F.M. and Wooding, R.A. 1964. Overland Flow
and Groundwater Flow from a Steady Rainfall of Finite
Duration. Journal of Geophysical Research, Vol.69,No.8.

6.  Kohler, M.A., and Linsley, R.K. 1951. Predicting the
    Runoff from Storm Rainfall. U.S.Weather Bureau Research
    Paper 34.

7.  Laurenson, E.M. 1974. Modeling of Stochastic-Determinis-
    tic Hydrologic Systems. Water Resources Research, Vol.10,
    No.5, pp. 955-961.

8.  Pattison, A. 1965. Synthesis of Hourly Rainfall Data.
    Water Resources Research, Vol. 1, pp.489-498.

9.  Rodda, J.C. 1967. The Significance of Characteristics
    of Basin Rainfall and Morphometry in a Study of Floods
    in the United Kingdom. UNESCO Symposium on Floods, and
    their Computation, Leningrad, 11 pp.

10. Sherman, L.K. 1932. Streamflow from Rainfall by Unit-
    graph Method.Engineering News Record,108,pp.501-505.

11. Snyder, F.F.1938. Synthetic Unit-graphs. American
    Geophysical Union, Vol. 19, pp. 447-454.

12. Weyman, D.R. 1975. Runoff Processes and Streamflow
    Modeling. Oxford University Press, London.

# A FLOW MODEL FOR ASSESSING THE TIDAL POTOMAC RIVER

By Raymond W. Schaffranek [1], M. ASCE

## ABSTRACT

The tidal Potomac River from the head-of-tide in the northwest quadrant of Washington, D.C., to Indian Head, Maryland, including its major tributaries and tidal inlets is modeled using a generally applicable, one-dimensional, network-type, flow-simulation model. Water-surface elevations and discharges can be computed at any desired location throughout the network of channels using the model which is formulated on a weighted, four-point implicit finite-difference method. The flow model was calibrated using recorded water-surface elevations as well as measured discharges from throughout the network. The utility of the model in evaluating the flushing capability and the transport behavior of the tidal Potomac River is illustrated. The general model is a proven viable and economical flow assessment tool that is applicable to a wide range of hydrologic conditions and varying field situations. Through use of such techniques water managers and scientists involved in similar comprehensive flow assessments can certainly attain a better understanding of the interrelationship of predominant riverine and estuarine processes.

## INTRODUCTION

Man's pervasive activities are continually altering riverine and estuarine systems such as the tidal Potomac River and Estuary. Upland river flows are being controlled and reduced by increased removal of water for industrial processes and domestic use. On occasion channels are dredged and tidal marshlands reclaimed. All the while, the waters are being used for effluent transport and disposal, for recreation, and for commercial fishing. Unfortunately, oftentimes such varied activities are not complementary. Altering riverine inflow by water withdrawal or changing the estuarine volume by dredging will effect the mixing processes and circulation patterns and subsequently change the flushing characteristics of the system. Consequently, the capacity of the system to accommodate and dispose of effluent will be impacted. Such interrelationships must be identified and should be considered in order that the many diverse and interrelated needs of riverine and estuarine systems can be mutually satisfied or optimized through an effective and enlightened management process.

However, in order for water managers to understand and evaluate the impact of changes in the present ecology and hydrology of such systems, knowledge of their complex hydrodynamic behavior is a fundamental requirement. Understanding the flow dynamics is vital, for the waters of these systems serve both as the transporting vehicle and as the media in which most of the chemical

---

[1] Research Hydrologist, National Center, U.S. Geological Survey, Reston, Virginia

and biological processes occur. Data and operational methods for appraising the impact of various management decisions are therefore required in order to conduct such flow assessments.

The development of operational methods for conducting comprehensive flow assessments in riverine and estuarine environments has been actively pursued by the U.S. Geological Survey for more than two decades (l). This research has resulted in a set of powerful, efficient, problem-oriented, digital computer models for simulating unsteady open-channel flow. One such model, a one-dimensional, branched-network type, capable of simulating flow in singular or multiply connected channels (3), has been applied to the uppermost tidal-river segment of the Potomac River as first reported in the Proceedings of the Specialty Conference on Computer and Physical Modeling in Hydraulic Engineering (2). Subsequent to this initial effort the model implementation has been extended to include the entire 50-kilometer (3l-mile) tidal-river segment with its numerous side-channel tributaries and embayments. This paper briefly describes the branch-network flow model, its implemention to the tidal Potomac River, and its calibration, before illustrating its utility in providing basic information for conducting comprehensive flow assessments in such riverine network environments.

## BRANCH - NETWORK FLOW MODEL

The branch-network flow model solves the one-dimensional equations of unsteady flow consisting of the equations of continuity and of motion, i.e.,

$$B\frac{\partial Z}{\partial t} + \frac{\partial Q}{\partial x} - q = 0 \tag{1}$$

$$\frac{\partial Q}{\partial t} + \frac{\partial (\beta Q^2/A)}{\partial x} + gA\frac{\partial Z}{\partial x} + \frac{gk}{AR^{4/3}} Q|Q| - \xi BU_\alpha^2 \cos \alpha = 0. \tag{2}$$

In these equations, the water-surface elevation, $Z$, and the discharge, $Q$, are the dependent variables, and the longitudinal distance along the channel, $x$, and time, $t$, are the independent variables. The cross-sectional area, channel top width, gravitational acceleration, lateral inflow, hydraulic radius, and wind velocity occurring at an angle $\alpha$ with respect to the positive x-axis are given by $A$, $B$, $g$, $q$, $R$, and $U_\alpha$, respectively. The coefficient $\beta$, known as the momentum or Boussinesq coefficient, which can be expressed as

$$\beta = \frac{\int u^2 dA}{U^2 A} \tag{3}$$

is present to adjust for any nonuniform velocity distribution over the channel cross section. In this expression $u$ represents the velocity of water passing through some finite elemental area, $dA$, whereas $U$ represents the mean flow velocity in the entire cross-sectional area, $A$. The coefficient $k$ is a function of the flow-resistance coefficient, $\eta$ (similar to Manning's n), which can be expressed as $(\eta/1.49)^2$ in the inch-pound system of units or simply as $\eta^2$ in the metric system.

The coefficient $\xi$ is the dimensionless wind-resistance coefficient which can be expressed as a function of the water-surface drag coefficient, $C_d$, the water density, $\rho$, and atmospheric density, $\rho_a$, as

$$\xi = C_d \frac{\rho_a}{\rho} \tag{4}$$

Equations (1) and (2) constitute the basic equations governing one-dimensional unsteady flow in open channels sometimes referred to as the full dynamic equations. The equations contain terms that account for effects of wind forces and (or) continuous lateral inflow. In their derivation it is assumed that the flow is substantially homogeneous in density and that hydrostatic pressure prevails at any point in the channel. The channel is assumed to be sufficiently straight, its geometry sufficiently simple, and its gradient mild and uniform throughout. Futhermore, frictional resistance is assumed to be amenable to approximation by the Manning formula. The resultant set of nonlinear partial-differential equations defies analytical solution. However, approximate solutions can be obtained by finite-difference techniques such as the one presented herein for the branch-network flow model.

In the branch-network flow model a weighted, four-point, finite-difference approximation of the nonlinear, partial-differential flow equations is employed. The finite-difference technique is described in detail in Schaffranek et al. (3). A "weighted four-point" implicit scheme is used because it can readily be applied with unequal distance steps, it can easily be varied throughout the range of approximation depicted by a box-centered scheme on the one hand to a fully forward scheme on the other, and its stability-convergence properties can be controlled.

In order to effect a solution by implicit means the flow equations are first linearized. The equation set is rendered linear by assigning values, initially through quadratic extrapolation and subsequently through iteration, to one unknown quantity in those terms involving products of dependent variables. Iteration is available in the model and serves to diminish any difference between the values used to compute such products. The efficiency of the method is primarily dependent on the accuracy of the initial approximation of the unknown in the nonlinear terms. In general it has been found that the extrapolation procedure yields sufficiently accurate initial values to assure convergence of the solution within one to three iterations.

After the flow equations for all segments--a segment being the primary subdivision of a branch--within the network are developed and appropriate boundary conditions are defined, a system of equations is formed that is determinate. Solution of this linear equation set can be accomplished directly by matrix methods once appropriate coefficient matrices have been constructed. In the branch-network model formulation, however, transformation equations that relate the unknowns at the ends of the branches are developed from the segment flow equations. These transformation equations are used in place of the segment flow equations to develop the coefficient matrices. The reason for using the branch-transformation equations instead of segment flow equations is the resultant dramatic reduction in computer memory and execution time requirements. After initial solution of the branch-transformation equations produces the unknown water-surface elevations and discharges at the ends of the branches of the network, intermediate values of the unknowns at ends of the segments are derived by back substitution of the branch-transformation equations. Details on the equation transformation technique are presented in Schaffranek et al. (3).

The model uses values computed at the current time level as initial conditions for computation of the next time step quantities and proceeds step-by-step to the designated end of the simulation through successive solution of the coefficient matrices. Initial values of the unknowns are required to start the simulation; these can be obtained by measurement, computed from another source, derived from a previous unsteady flow simulation or simply estimated. A Gaussian elimination method based on maximum pivot strategy is employed to repetitively solve the system of simultaneous equations.

## THE TIDAL POTOMAC RIVER

The Potomac River is the second largest source of fresh water to Chesapeake Bay. Its average fresh water discharge of 325 $m^3$/s (11,500 $ft^3$/s) is second only to that of the Susquehanna. As it wends its southeasterly course to the Bay, the Potomac River traverses the fall line separating the Piedmont from the Atlantic Coastal Plain. A series of falls and rapids, the last of which is immediately upstream of Chain Bridge in the northwest quadrant of the District of Columbia, indicate where the River intersects the fall line and define the location of the head-of-tide.

The River downstream of Chain Bridge is confined for a short distance (approximately 5 km (3 mi)), to a narrow, deep, but gradually expanding channel bounded by steep, rocky banks and high bluffs. Further downstream the River consists of a broad, shallow, and rapidly expanding channel confined between banks of low-to-moderate relief. The segment of the River, hereinafter referred to as the tidal Potomac River, extending from the head-of-tide near Chain Bridge to Indian Head, Maryland (a distance of nearly 50 km (31 mi)) as shown in figure 1, is of primary concern in this report. Also included in this segment are the Anacostia River, Roosevelt Island channel, Washington Channel, the Tidal Basin, and several small inlets, downstream of Wilson Bridge, formed where small streams discharge into the river thus creating irregularities and indentations in the channel.

Flow in the narrow channel of the upstream portion of the tidal Potomac River is typically unidirectional and pulsating whereas bidirectional flow induced by semi-diurnal tidal fluctuations occurs in the broader downstream portion. However, the location of the transition from one flow pattern to the other varies primarily in response to changing inflow at the head-of-tide as well as to changing tidal and meteorological conditions. Although the channel geometry is quite irregular field investigations confirm that the flow consists of predominantly longitudinal ebb and flood currents. The channel cross-sectional area increases rapidly with downstream distance, thus contributing to the generally observed amplification of the tide wave as it propagates upstream. The flow is vertically well mixed throughout the segment.

## TIDAL POTOMAC RIVER FLOW MODEL

The initial phase of implementation of the branch-network flow model to the tidal Potomac River was application of the model to the upper 19.7 km

(12.3 mi) portion of the tidal River from Chain Bridge to Wilson Bridge and including the Anacostia River, Roosevelt Island channel, Washington Channel, and the Tidal Basin as reported previously (2). The second phase involved application of the model to the lower 30.1 km (18.7 mi) portion of the tidal River from Wilson Bridge to Indian Head, Maryland, and including the Broad Creek, Piscataway Creek, Dogue Creek, Gunstone Cove, Pohick Bay, and Accotink Bay tidal inlets. In the final phase of implementation these two subset models were combined to form a complete model of the 49.8 km (31.0 mi) tidal Potomac River including its major side-channel tributaries and tidal embayments.

## Network Schematization

The tidal Potomac River system shown in figure 1 is schematized for model purposes as shown in figure 2. The network system is composed of 25 branches (identified by Roman numerals) that either join and (or) terminate at 25 junction locations (identified by numbered boxes). Those junctions that do not constitute tributary or inlet locations in figure 2 were deemed necessary in the network schematization in order to accommodate potential nodal flows (point source inflows or outflows such as sewage treatment outfalls or pump withdrawals) or abrupt changes in channel characteristics. In addition to junction locations, cross-sectional profiles were developed to describe the channel geometry at selected locations between junctions in order to properly account for irregularities. A total of 64 cross-sectional profiles, located at unequal intervals, were used to depict the irregular geometry of the 25-branch network of channels with its side-channel tributaries and embayments. After the cross-sectional profile locations were established, branch lengths required by the model were determined by measuring distances along the thalweg centerline of the channels as interpolated from hydrographic charts.

In order to develop the cross-sectional profiles characterizing the channel geometry, hydrographic data consisting of over 40,000 soundings in the tidal-river segment were obtained from the National Ocean Survey, National Oceanic and Atmospheric Administration (NOS/NOAA). From this set of data, soundings within a narrow envelope of influence surrounding each cross-sectional transect were selected and used to derive, through polynomial surface interpolation, the desired cross-sectional profiles.

Out of the total 64 cross-sectional profiles, 25 were used to describe the highly variable geometry of the Potomac River itself, the remaining 39 were used to describe the geometry of the other channels of the network. The cross-sectional area of the Potomac River expands over 40-fold between Chain Bridge and Indian head, increasing from approximately 232 $m^2$ (2,500 $ft^2$) at Chain Bridge to 3,810 $m^2$ (41,000 $ft^2$) at Wilson Bridge to 9,960 $m^2$ (104,000 $ft^2$) at Indian Head. The corresponding channel width increases from a minimum width of 44.2 m (145 ft) at Chain Bridge to 1,160 m (3,800 ft) at Wilson Bridge to 1,950 m (6,400 ft) at Indian Head. These values serve to point out the drastically changing cross-sectional properties of the tidal Potomac River.

## Data Requirements

Like many unsteady flow models, the branch-network model requires that one of the time-dependent variables--either the water-surface elevation or flow discharge--at the physical extremities of the network being simulated be known

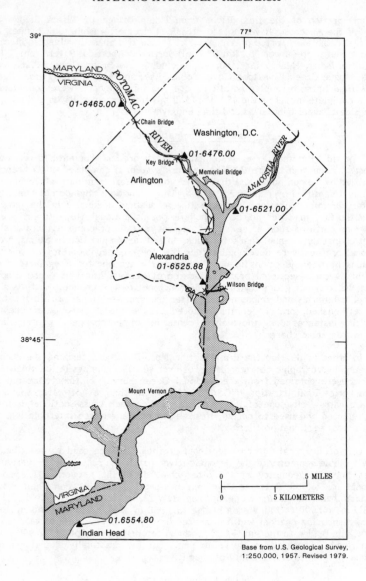

Base from U.S. Geological Survey,
1:250,000, 1957. Revised 1979.

Fig. 1.--Map of Potomac River Near Washington, D.C.

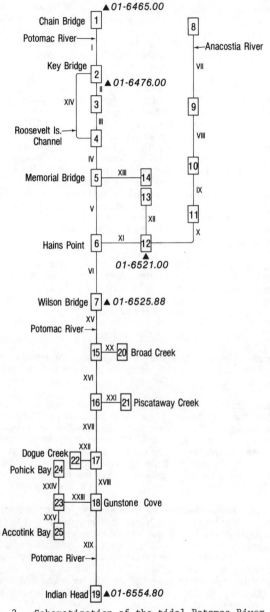

Fig. 2.--Schematization of the tidal Potomac River
System for the Branch-Network Flow Model.

throughout time and given. Typically, sequences of discrete, water-surface elevations are recorded at these locations and supplied as time-series of values to the model in order to satisfy these boundary condition data requirements. However, flux rates (discharges) can also be, and frequently are, used as boundary-value data, e.g., the null discharge at the closed end of a dead-end channel.

As can be seen from the network schematization of figure 2, boundary conditions must be specified for the tidal Potomac River model at Chain Bridge (junction 1) and Indian Head (junction 19) as well as for the Anacostia River (junction 8) and for junction locations 13, 14, 20, 21, 22, 24, and 25. The model uses discharges derived from a rating curve for a gaging station (station number 01-6465.00 in figure 2) located 1.9 km (1.2 mi) upstream of Chain Bridge as its upstream boundary values at junction 1. Water-surface elevations continuously recorded at a gaging station (station number 01-6554.80 in figure 2) located at Indian Head, Maryland, are utilized as the downstream boundary values at junction 19. All other boundary condition requirements, including the Anacostia River, are fulfilled by assuming zero discharge conditions prevail at the upstream tidal extent of the channel or embayment. It is also possible, of course, to use known inflow rates or water-surface elevations at any of these locations as boundary values for conducting the model simulations.

In addition to these external boundary conditions the model requires that boundary conditions be specified at the internal junctions of the network. To satisfy these requirements the model automatically generates equations, based on discharge continuity and stage compatibility conditions, that constitute consistently prevailing conditions at these locations.

## Simulation Results

Before model simulations could be conducted, values for the frictional-resistance coefficient, $n$, and other computational-control variables had to be determined and specified.

Frictional-resistance coefficients had to be designated for all branch segments within the network. These are typically initially estimated and subsequently adjusted in the model calibration process to produce agreement between simulated and measured data. The tidal Potomac River flow model calibration process has resulted in frictional-resistance coefficient values ranging from 0.0275 at Chain Bridge to 0.019 at Indian Head for the Potomac River. Coefficient values for all other segments within the network likewise fall within this range.

Additionally, flow-simulation controls such as the computation time step, finite-difference weighting factors, and convergence conditions had to be assigned. In a fashion similar to the frictional-resistance coefficients, but to a lesser extent, these parameters are also subject to modification during the model calibration process. The most recent flow simulations have been conducted using a 15-minute time step and a value of 0.75 for the discretization weighting factors. These simulations have satisfied convergence criteria set at 0.46 cm (0.015 ft) and 0.71 m$^3$/s (25.0 ft$^3$/s) for water-surface elevations and for discharges, respectively, in an average of 3 iterations.

In order to calibrate and subsequently verify the model, water-surface elevations continuously recorded near Key Bridge (station number 01-6476.00), near Wilson Bridge (station number 01-6525.88), and near Hains Point (station number 01-6521.00) are being utilized. (See figure 2.) Model-computed discharges are also being compared with discharges measured for complete tidal cycles near National Airport, Broad Creek, and at Indian Head. These tidal-cycle discharge measurements were also augmented, and in some instances extended in duration by correlation, with concurrently operated, bottom-moored, current meters that provided speed and direction-of-flow data at selected locations throughout the network.

In figure 3, simulated discharges are compared with measured discharges for a complete tidal cycle from 2015 hours on June 3, 1981 to 0830 hours on June 4, 1981 at Indian Head, Maryland. As this plot shows simulated discharges compare favorably with the measured discharges. Measured discharges range from a high of 3,960 m$^3$/s (140,000 ft$^3$/s) to a low of -4,390 m$^3$/s (-155,000 ft$^3$/s) whereas simulated discharges range from a high of 4,020 m$^3$/s (142,000 ft$^3$/s) to a low of -4,530 m$^3$/s (-160,000 ft$^3$/s). The mean simulated discharge is approximately 3.8

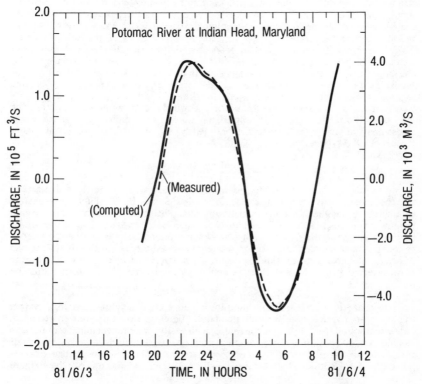

Fig. 3.--Model-Generated Plot of Simulated Versus Measured
Discharges for the Potomac River at Indian Head, Maryland.

percent more than the mean measured discharge; the phase of the simulated discharge is slightly (on the order of 15 minutes) ahead of that measured. A comparison of measured and model-computed mass transport through the measurement cross section reveals a 6.6% excess of mass simulated for the ebb volume and a 5.6% excess for the flood volume. Other tidal-cycle discharge measurements and recorded water-surface elevations were similarly used to calibrate and subsequently verify the model's performance.

Some additional work remains with respect to model calibration and verification particularly as pertains to assessing the impact of wind stress as a forcing function on the flow. Wind velocity and direction data collected at Indian Head are presently being processed for this purpose.

## USING THE MODEL FOR FLOW ASSESSMENT

Determination of the flow dynamics occurring throughout a network system, such as the tidal Potomac River, is, of course, fundamental to evaluating its mixing and flushing capabilities, both vital elements in assessing water quality changes as a result, for instance, of contaminants introduced into the system. Such knowledge is also important for predicting water treatment requirements and for evaluating the amounts of municipal and industrial wastewaters that can be tolerated by the system. However, the hydrodynamics of the tidal Potomac River are difficult both to measure and to analyze for they do not evolve to steady-state phenomena but rather undergo ceaseless change in response to the ebb and flood of the tides, variations in freshwater inflow, erratic effects of winds, movement of weather fronts, and irregularities in geometry. In this regard, information in the form of flow-interchange summaries and particle transport rates available from the branch-network model, as herein depicted for the tidal Potomac River, can provide useful operational information for assessment of the quantity, and subsequently the quality, of water available throughout the system.

### Flow-Interchange Summaries

Frequently, it is desirable, as in the computation of nutrient loads and suspended sediment concentrations, to evaluate the flow-volume interchange throughout a network system such as the tidal Potomac River in order to gain insight into the tidal-cycle variability in the concentration and dispersal of nutrients and sediments. Also such flow-volume information is frequently needed to locate major sources or sinks for nutrients and sediments. In the tidal Potomac River, as in most similar tidal systems, it is the net seaward flux due to the freshwater riverine inflow that, in large measure, provides the important flushing action.

In order to evaluate the volume interchange capability of a network system the branch-network model not only computes the basic flow information required, but also assimilates it in a convenient and easily comprehensible format. One sample output containing flow volumes computed for successive tidal cycles of the Potomac River during the month of January 1980 at Wilson Bridge (station number 01-6525.90) is illustrated in figure 4. Shown in the figure are flood (negative) and ebb (positive) volumes of flow in millions of cubic feet (multiply by 0.02832 to convert to cubic meters) past Wilson Bridge during the month of January. (Flood volume is the total quantity of water that flows in the upstream direction in the period between the slack waters of successive ebb and flood tides; similarly ebb volume is the reciprocal total quantity of flow in the downstream direction.) These accumulated flow volumes separated by the time, approximated to the nearest computational time step, of slack water preceding the flow reversal are tabulated on a daily basis for the month as shown in figure 4. The first volume shown for any given day in figure 4 is the volume of water accumulated from the beginning of the day to the first reversal; similarly the last volume for any given day is the volume of water accumulated from the last reversal to the end of that day.

January 1980 Flow in Millions of Cubic Feet at Station 01-6525.90

| Day No. | Flow Vol. | (Rev.) (Time) | Flow Vol. | (Rev.) (Time) | Flow Vol. | (Rev.) (Time) | Flow Vol. | (Rev.) (Time) | Flow Vol. |
|---|---|---|---|---|---|---|---|---|---|
| 1 | 254 | (0300) | -327 | (0630) | 915 | (1445) | -345 | (1845) | 642 |
| 2 | 379 | (0345) | -237 | (0715) | 967 | (1600) | -371 | (2000) | 411 |
| 3 | 486 | (0445) | -274 | (0815) | 801 | (1615) | -382 | (2030) | 369 |
| 4 | 553 | (0500) | -329 | (0900) | 800 | (1700) | -344 | (2115) | 248 |
| 5 | 626 | (0545) | -285 | (0915) | 809 | (1745) | -380 | (2145) | 166 |
| 6 | 765 | (0700) | -330 | (1115) | 537 | (1745) | -356 | (2215) | 89 |
| 7 | 674 | (0630) | -379 | (1045) | 750 | (1900) | -227 | (2245) | 48 |
| 8 | 787 | (0830) | -105 | (1115) | 720 | (2030) | -216 | | |
| 9 | -2 | (0015) | 654 | (0815) | -333 | (1245) | 566 | (2015) | -231 |
| 10 | -23 | (0030) | 673 | (0845) | -309 | (1315) | 640 | (2145) | -153 |
| 11 | -67 | (0130) | 623 | (0930) | -327 | (1400) | 517 | (2130) | -87 |
| 12 | -103 | (0230) | 752 | (1100) | -227 | (1415) | 884 | | |
| 13 | 22 | (0045) | -122 | (0345) | 558 | (1130) | -286 | (1545) | 685 |
| 14 | 41 | (0115) | -214 | (0445) | 579 | (1200) | -314 | (1615) | 680 |
| 15 | 58 | (0115) | -204 | (0445) | 946 | (1400) | -182 | (1730) | 768 |
| 16 | 268 | (0245) | -127 | (0515) | 1235 | (1445) | -160 | (1715) | 1105 |
| 17 | 583 | (0400) | -62 | (0545) | 1846 | (1600) | -84 | (1800) | 1129 |
| 18 | 803 | (0500) | -109 | (0700) | 1554 | (1630) | -181 | (1900) | 828 |
| 19 | 932 | (0530) | -80 | (0730) | 1949 | (1845) | -58 | (2030) | 463 |
| 20 | 1126 | (0645) | -54 | (0815) | 1692 | (1845) | -90 | (2045) | 424 |
| 21 | 1388 | (0745) | -60 | (0915) | 1783 | (2015) | -58 | (2145) | 202 |
| 22 | 1411 | (0815) | -86 | (1015) | 1491 | (2030) | -86 | (2245) | 81 |
| 23 | 1333 | (0845) | -149 | (1115) | 1371 | (2130) | -103 | (2400) | 3 |
| 24 | 1174 | (0945) | -122 | (1215) | 1082 | (2200) | -129 | (2400) | 2 |
| 25 | -44 | (0130) | 1097 | (1045) | -175 | (1330) | 1080 | (2315) | -23 |
| 26 | -127 | (0215) | 991 | (1130) | -249 | (1430) | 999 | | |
| 27 | 4 | (0015) | -170 | (0330) | 901 | (1215) | -202 | (1545) | 879 |
| 28 | 80 | (0130) | -197 | (0445) | 788 | (1245) | -256 | (1630) | 793 |
| 29 | 141 | (0200) | -233 | (0530) | 818 | (1345) | -270 | (1730) | 747 |
| 30 | 253 | (0315) | -163 | (0600) | 802 | (1500) | -207 | (1900) | 428 |
| 31 | 274 | (0315) | -253 | (0715) | 796 | (1545) | -205 | (1930) | 465 |

Fig. 4.--Model-Produced Table of Accumulated
Flow Volumes at Wilson Bridge for
January 1980.

During the month of January 1980, Potomac River inflows at Chain Bridge ranged from a low of 241 m $^3$/s (8,500 ft $^3$/s) on the 11th to a high of 1,190 m$^3$/s (42,100 ft $^3$/s) on the 17th.  The simulated flow volumes shown in figure 4 ranged from a minimum flood volume of -382 million cubic feet between 1615 and 2030 hours on the 3rd to a maximum ebb volume of 1,949 million cubic feet between 0730 and 1845 hours on the 19th.  As is readily apparent the times of low and high freshwater inflows at Chain Bridge are not reflected as the minimum and maximum flow volumes at Wilson Bridge.  This can be, and in all likelihood is, due to the variable tidal influence at the downstream end of the channel, although wind and meteorological conditions could also be contributing factors.

Flow-volume summaries as illustrated in figure 4 can be produced by the model for any location within the network.  Thus, it is possible to analyze simultaneously the flux of mass through many designated control sections.  With this capability it is possible to directly evaluate the flow-interchange capabilities and subsequently the flushing capacity of the entire network system.

## Particle Transport

The variable flushing capacity of a network system, as a function of freshwater inflow and tidal influences, can also be readily illustrated by using the branch-network model to track the movement of neutrally buoyant conservative-type substances (or injected dye).  Such monitoring of the movement of index particles can also answer questions pertaining to the travel time between specific locations within the network.

Figures 5 and 6 present the results of two simulations made to illustrate the particle tracking capability of the model as used to evaluate the flushing characteristics of the tidal Potomac River.  The central graph in both of these figures is the desired plot of constituent travel time versus longitudinal distance along the main Potomac River channel.  River mile 0.0 in the graph refers to the Potomac River at Chain Bridge near Washington, D.C., and river mile 30.8 corresponds to the Potomac River at Indian Head, Maryland.  The locations of Key Bridge and Wilson Bridge in Washington, D.C., and the location of Mount Vernon, Virginia, are also identified on the graph.  The inflow discharge hydrograph representing the upstream boundary condition in the model that was derived from the rated gaging station (01-6465.00)  near the Little Falls pumping station is plotted to the left of the time-of-travel graph.  The stage hydrograph representing the downstream boundary condition in  the model that was obtained from the gaging station (01-6554.80) at Indian Head, Maryland, is plotted to the right of the time-of-travel graph.  The time-of-travel graph depicts the movement of nine simultaneously injected index particles (labeled A through I) along the main Potomac River channel in response to the indicated boundary conditions.

Presented in figure 5 are the paths-of-travel of index particles for the 2½ day period from noon on February 23, 1981, to midnight of February 25, 1981. During this time period the inflow hydrograph at Chain Bridge ranged from 917 m³/s (32,400 ft³/s) to 1,350 m³/s (47,500 ft³/s) as depicted in the figure. All particles, except particle A injected at Chain Bridge, are completely displaced from the system within the 2½ days of travel time depicted in the figure. The diminished effect of tidal influence on particle transport can be readily seen in the figure. Only very minimal upstream particle displacement occurs upstream of Wilson Bridge as a result of tidal fluctuations. Travel times in the upstream section of the river are significantly less than those of the downstream section. Illustrating this fact in a converse manner particle A injected at Chain Bridge is transported 20.2 miles downstream to Mount Vernon in approximately the same time it takes particle G injected at Mount Vernon to traverse the 10.7 miles downstream to Indian Head.

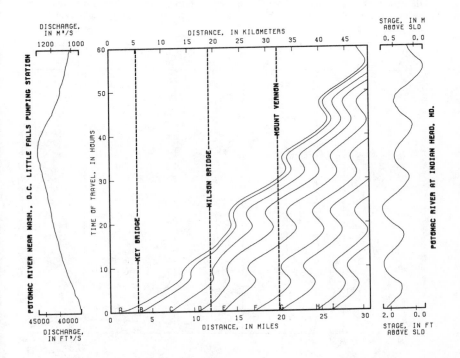

Fig. 5.--Time-of-Travel Plot of Injected Particle Movement
for the Potomac River from Noon of Feb. 23, 1981,
to Midnight of Feb. 25, 1981.

Depicted in figure 6 are the paths-of-travel of index particles injected at the same locations as those in figure 6 but for the 7½ day period from midnight of August 17, 1981, to noon on August 25, 1981. Inflow at Chain Bridge during this time period varied from a high of 49.3₃ m ³/s (1,740 ft ³/s) at the start of simulation to a low of 27.6 m ³/s (973 ft ³/s) at the end. For the 7½ day duration of this simulation none of the injected particles is displaced from the system. The increased impact of tidal influence on particle transport is vividly illustrated in the figure. In fact, particle D injected at river mile 10.6 which is about 1½ miles upstream of Wilson Bridge actually remains in the vicinity of Wilson Bridge for nearly 120 hours (5 days). The net displacement of particles E through I for the first 90 hours (3 3/4 days) of simulation is in the upstream direction. These particle movements are direct consequences of the changing tidal fluctuations at the downstream end of the channel. During the first 90 hours of this time period a gradual increase in the mean elevation of the semi-diurnal tides occurs at Indian Head. This increase in elevation results in the storage of water in the system. The resultant influx of water from the downstream end exceeds the magnitude of the freshwater inflow and thereby causes the upstream particle displacements.

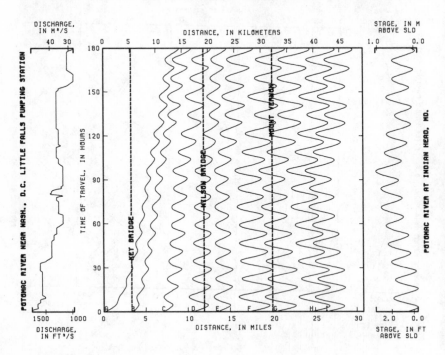

Fig. 6.--Time-of-Travel Plot of Injected Particle
Movement for the Potomac River from Midnight
of Aug. 17, 1981, to Noon of Aug. 25, 1981.

The flow-volume summary of figure 4 and the time-of-travel plots of figures 5 and 6 are presented as examples of the ways in which flow-simulation models, such as the branch-network model, can be used to compute and present flow information to describe the hydrodynamic behavior of riverine networks, such as the tidal Potomac River. Other flow assessment techniques are certainly tractable given the availablity of comprehensive flow data such as can be computed by the branch-network flow model. Presentation of such techniques is beyond the scope of this report. However, several have been reported elsewhere (1). Hopefully, the presentation of these examples has served to emphasize the role of model utilization as a viable and economical means of conducting such flow assessments.

## SUMMARY

A generally applicable one-dimensional model capable of simulating the unsteady flow in a singular channel, as well as, that occurring throughout a network of simply or multiply connected channels has been implemented to depict the flow dynamics of the tidal Potomac River. The 50-kilometer (31-mile) segment of the Potomac River between Indian Head, Maryland, and the head-of-tide immediately upstream of Chain Bridge in the District of Columbia, the Anacostia River, Roosevelt Island channel, Washington Channel, the Tidal Basin, and several small embayments downstream of Wilson Bridge constitute the system of channels that comprise the 25-branch network model. Boundary conditions for the model consist of discharges derived from a rated gaging station located upstream of Chain Bridge, water-surface elevations recorded at Indian Head, Maryland, and assumed null discharge conditions at all other extremities of the network. Comparison of simulated and measured tidal-cycle discharges clearly demonstrate the validity and accuracy of the model implementation. Flow-volume summaries and particle paths-of-travel computed and compiled by the model illustrate its utility in providing operational information immediately useful for water-resources assessment and subsequently available for water-resources management.

## REFERENCES

1.    Lai, C., Schaffranek, R. W., and Baltzer, R. A., An operational system for implementing simulation models, A case study: Seminar on Computational Hydraulics, Proceedings, ASCE Hydraulics Division, 26th Annual Hydraulics Speciality Conference, 1978, p. 415-454.

2.    Schaffranek, R. W., and Baltzer, R. A., A one-dimensional flow model of the Potomac River: Speciality Conference on Computer and Physical Modeling in Hydraulic Engineering, Proceedings, ASCE Hydraulics Division, 1980, p. 292-309.

3.    Schaffranek, R. W., Baltzer, R. A., and Goldberg, D. E., A model for simulation of flow in singular and interconnected channels: Chapter C3, Book 7, Automated Data Processing and Computations, TWRI, U.S. Geological Survey, 1981, 110 p.

PHOTODENSIMETRIC METHODS IN TIDAL FLUSHING STUDIES

by

Ronald E. Nece[1], F. ASCE and H. Norman Smith[2]

ABSTRACT

Photodensimetric methods for measuring tidal exchange in small-
scale physical models are described. Application of the technique is
illustrated by a study of a proposed small boat basin in Alaska.

INTRODUCTION

The purpose of models, physical or numerical, is to provide
answers to questions asked of and by designers. This paper addresses
one type of project - namely, small harbors - where simple, inexpensive
physical hydraulic models still provide the most efficient mechanism
for producing required information. Photographic methods for quantify-
ing tidal exchange in models of small harbors are described.

Tidal exchange (flushing) has been considered a significant factor
in the quality of water within small boat harbors relative to quality of
ambient waters. Interest in tidal flushing of small harbors in the
Pacific Northwest has been keyed to concerns of fishery and regulatory
agencies; poor water quality could lead to fish kills of juvenile salmon
which occupy these harbors temporarily on their seaward migration.
Questions are raised about both overall flushing performance and spatial
variability in local water exchange rates within the harbors. The
methods described here provide answers to these questions, and provide
an economical basis on which decisions that might be dictated by tidal
flushing questions can be made on whether or not a project can be
allowed to proceed.

Use of small-scale distorted models for this purpose still has a
place because the development of numerical models which provide adequate
details of the complex, three-dimensional, unsteady flows in such small
harbors is still an ongoing process. Tidal circulation patterns in
these harbors are dominated by angular momentum effects associated with
flows having pronounced separation due to tidal inflows through relatively small entrances where wave protection for the basin is the first
requirement. Numerical models for bays and harbors are usually two-
dimensional, and many do not now include transverse shear stress terms
needed at separation surfaces (e.g., tidal inflows through an opening
in a breakwater) and for the resultant circulation cells or gyres
established within the harbor. Such gyres are predicted by models without specifically including horizontal eddy viscosity terms (e.g.,(2)),

[1] Professor of Civil Engineering, University of Washington, Seattle,
Washington.
[2] Research Engineer, Department of Civil Engineering, University of
Washington, Seattle, Washington.

but typically having very large grid size: water depth ratios.  Actual
water exchange calculations would require that appropriate diffusion coefficients be incorporated in the transport models.  This has been done
recently with encouraging results for schematic horizontal-bottom harbors
having some planform features common to the small boat basins of concern
(3).

Distorted physical models accomodating complex bathymetries are
still commonly used, even though limitations of such models for determining diffusion and mixing in tidal flows are well acknowledged (4).
Vertical distortion and low Reynolds numbers in a single-fluid model
rule out equivalence of local diffusion characteristics in model and
prototype.  However, transport in the harbors studied has been advection dominated.  Earlier field studies at an existing harbor have shown
that in the absence of stratification the model reproduces the complex
tidal circulation pattern of the prototype (5).  It is felt that water
exchange is also adequately reproduced.  Where exchange performances of
alternative designs are to be compared, relative flushing values obtained from model tests of the different geometries should be sufficiently accurate to define trends and to be used in water quality models
which may be developed in the future.

The harbors studied have met these criteria.  Local ambient waters
are well mixed, and only harbors with no freshwater inflow have been investigated; stratification effects would be minimal in the prototypes.
Breakwaters or spits on the seaward side of the harbor minimize wind
driven currents in the harbor, and typical narrow entrances minimize
effects of exterior currents although major features of these currents
should be simulated when they may affect water exchange.

The use of photo-densimetric methods to measure tracer concentrations in water is not new. Photo-densimetric procedures have been used
with aerial photography to study outfall plumes in the ocean(1) and to
measure concentration dilutions of solutes dispersing under oscillatory
flow conditions in a laboratory channel of constant depth (9).  The
procedures described in this paper provide a simple and economical
method for tracking variations in space and time of tracer concentrations in small bodies of water with irregular depths and boundaries
under conditions of tide-driven circulation.

## TEST FACILITY AND MODEL SCALES

Planform areas of the small harbors tested have not exceeded 25-30
acres.  These sizes have allowed models with a typical 1:500 horizontal
scale to be accommodated in a small laboratory basin.  The testing basin
(Fig. 1) has an overall plan size of 8 ft. by 12 ft., with an 18-inch
working depth.  Repetitive constant amplitude, constant period tides are
produced by a tide generator which is a variable elevation waste weir,
driven by a small motor through a variable speed gear box and a Scotch
yoke mechanism to obtain harmonic motion, and fed by a constant rate
water supply entering the tank through a perforated manifold.  Tide
ranges, mean water levels, and tide periods are ajustable.

Models typically have been constructed with the commonly used 10:1
vertical distortion.  Prototype tides tested have been within the 3 to
15-ft. range, so that the 1:50 vertical scale produces model ranges of
0.06 - 0.30 feet.

As shown in Fig. 1, models are placed at the far end of the tank
from the tide generator to maximize ambient water area.  Bathymetries
can be reproduced to typical depths of 50-60 feet below mean lower low

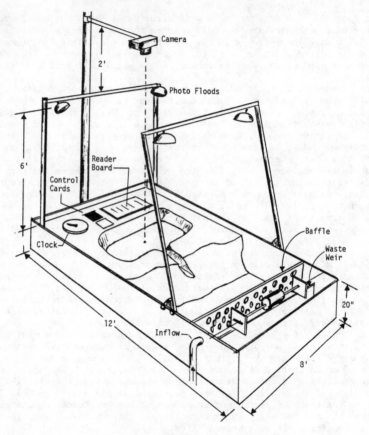

Figure 1. Experimental Facility

water, the tidal datum for the region.  Models are oriented in the tank
so that major effects of ambient currents can be simulated.  If the harbor is located on an "estuary" where ambient currents are significant,
extensions can be added to the tank "upstream" from the harbor entrance
to provide the proper tidal prism volume.

The models themselves are generally constructed of fiberglass
cloth and polyester resin laid up on fine mesh chicken wire, hardware
cloth, or wire screen molded over plywood contour sections providing the
structural frame; the resulting light-weight models can be removed
readily from the tank and stored for possible reuse.  The bottom,
beaches, and all boundaries in contact with the water are painted with a
white primer enamel to an elevation above highest tide levels to be
tested, in order to provide a flat white surface compatible to the
photo-densitometer background reference used in the measurement procedures described in the section.  Roughness strips are placed in the harbor entrance to generate turbulent flow in the flood tide currents
entering the basin.

EXPERIMENTAL PROCEDURES

The specific information to be quantified and for which the experimental procedures were developed is the relative exchange of water
within a harbor basin with ambient water due to tidal flushing.  Results
are presented in the form of an "exchange coefficient".

The average per-cycle exchange coefficient, which indicates that
fraction of water in a basin or a segment of the basin which is removed
(flushed out) and replaced with ambient water during each tidal cycle
(defined in this report as the time from low water to following low
water), is represented by the equations:

$$E = 1 - R$$
$$\text{and} \quad R = (C_n/C_o)^{1/n}$$

where  $E$ = average, per cycle, exchange coefficient
       $R$ = average, per cycle, retention coefficient
       $C_o$ = initial tracer concentration
       $C_n$ = tracer concentration after n cycles, where n
             is an integer

The above equations assume repetitive, identical tides and consequent
identical exchanges on each of the cycles.  The local value of exchange
coefficient for a particular segment of a basin is designated by $E$,
while the spatial average coefficient applicable to an entire basin is
designated as $\bar{E}$.

The low water definition was used in the representative study described later because at this position on the tidal cycle the residual
currents caused during the flood portion, very much in evidence at high
water, have diminished during the subsequent ebb flow.  This condition
presents two advantages: first, from an experimental standpoint, conditions are more likely to be repeated in more detail from one cycle to
the next; second, the spatial distribution at low tide of relative concentrations of water originally in the basin at the prior low tide
should be more indicative of the effective flushing within the basin
over an entire tidal cycle.

The photographic technique tracks the change of tracer dye
concentration with time at selected grid points in a basin at specified
times on sequent tidal cycles.  Dye density values are measured directly
from 35-mm black-and-white negatives, using a Tobias Associates Model
TBX photo-densitometer equipped with a digital readout, and a choice of
1, 2, or 3-mm diameter aperature, depending upon spatial resolution
desired.  The camera, mounted above the basin (Fig. 1), is automated to
take a photo at the specified times; it can also be triggered manually.
The film used is Kodak Plus-X pan (ASA 125).  The camera setting is f2
at an exposure time of 1/15 second; a red filter is used.  The B-2 photo
flood lamps with reflectors providing the necessary lighting are positioned so as to provide uniform lighting over the basin, are controlled
by the same circuit, and are illuminated only long enough for the picture
to be taken by the camera before they are extinguished; the objective is
to minimize the possibility of heating the water in the model by the
lamps and introduce stratification effects which could destroy the unstratified conditions desired.  The procedural steps utilized in the
tests are:

1.  With water in the tide tank at low water (low tide) elevation, in-
    sert a temporary barrier dam across the entrance, separating basin

and ambient waters.

2.  Photograph the model when filled with clear water (at low tide level)
    to establish a background light level, at C = zero percent. Gray and
    white control cards are placed in the camera field of vision for
    control purposes (Fig. 1).

3.  The basin is dosed with a suitable water-soluble dye (Mrs. Stewart's
    bluing was used). Insert dye in four equal increments until the
    final dosage ($C_o$=100%) is reached; a pre-selected dye concentration
    of 200-300 ppm is used, depending on depth. A photograph is taken
    at each dosage level after the dye has been thoroughly mixed into the
    basin and the water has then been allowed to become quiescent. This
    produces at each grid point in the basin a 5-point calibration curve
    of densitometer reading vs. dye concentration. Separate calibration
    curves must be determined for various points in the model because of
    differences in lighting from point to point, depth effects, etc. The
    relationship between the logarithm of the optical density and dye
    concentration is nearly linear so interpolation between points on
    the calibration curve results in very small errors.

4.  The model tank is completely flushed, and refilled to high tide level.
    This precautionary move minimizes possibilities that the water in the
    basin, which is shallower than the rest of the tank, might warm dif-
    ferentially from the ambient water and cause stratification problems.

5.  The temporary barrier dam is re-inserted, and the proper amount of
    dye (determined from measurements of basin volumes at low and high
    tide levels) is added to the basin to bring the dye concentration in
    the basin to the $C_o$ value. The dye is again mixed thoroughly and the
    basin waters are allowed to become quiescent.

6.  Remove the barrier dam and simultaneously start the tide generator.
    Tests are initiated at high tide so that the model will be in a dy-
    namic state when the low tide reference concentration ($C=C_o$) is
    photographed when the first low water stage is reached.

7.  The tide generator runs for 4 complete cycles from the time of the
    initial reference condition. At low water slack corresponding to
    n=4, take a photograph, stop the tide generator and simultaneously
    replace the barrier dam. Thoroughly mix the water in the basin,
    allow it to become quiescent, and take a final photograph giving the
    spatially averaged tracer dye concentration in the basin.

8.  Develop the negatives, set up a grid overlay, and determine dye den-
    sities $C_o$ and $C_n$ with the photodensitometer. This last operation
    could range from entirely manual to fully computerized, depending on
    equipment available. Also, the calibration data obtained in step 3
    can be plotted as regular calibration curves for graphical solution
    or inserted into an appropriate computer program for numerical
    solution. Equations given earlier are used to determine E.

    This sequence was repeated in the example study for three replicate
runs for each basin configuration-tide range combination. The results
for the three replicates did indeed agree well, verifying that the testing
procedure does give satisfactorily reproducible results. Results for
replicate runs usually do agree reasonably well unless stratification is
present. Stratification tends to increase tidal flushing; if replicate
runs do produce different $\bar{E}$ values, the lowest value, and not the three-
run average, is taken as the conservative value to be applied to predicted prototype performance. During one of the three runs photographs were
taken at each quarter-cycle to obtain a visual record of the circulation

dynamics throughout the tidal cycle and to provide a data source for more detailed analysis if this should be desired. The use of a 4 cycle test, and consequent use of n=4 in the equations, has been verified for the apparatus used (6).

The negatives constitute an easily-stored permanent data record. A reader board and an automated "clock" located within the camera field of view contain necessary particulars for each test: date, tidal range, time and tidal cycle for each photograph, harbor identification, and camera settings. One tidal range sequence of tests can be taken on a roll of 36-exposure film which contains the three replicate runs and one calibration series.

Complete details of the testing procedures described have been given elsewhere (8).

## EXAMPLE CASE

The experimental methods are illustrated by some results obtained in a recent study of tidal exchange in a proposed small boat harbor at Sitka, Alaska.

The proposed harbor would be located in an existing lagoon connecting to ambient waters via a natural opening. The lagoon would be deepened by dredging, the water area decreased by dredged material serving as landfill for service areas, and the entrance channel would be dredged. The harbor, with a 12-acre surface area at mean tide level, would be dredged to 10 feet (MLLW datum). Incorporation of another existing, shallow brackish lagoon of comparable surface area (deepest points about - 10 feet, approximately 40 percent of bottom above MLLW) into the boat harbor circulation system would be a possible mitigative alternative replacing wetlands lost in the dredging of the harbor. A schematic planform drawing, showing the lagoon-harbor linkage, is shown in Fig. 2.

Tests were run at three tide ranges: 3 feet ("neap"), 7.7 feet (local mean range), and 12.0 feet ("spring"). All three sinusoidal tides were operated about the local mean tide level of 5.3 feet (MLLW datum). The temporary barrier dam required in the following tests was located in the entrance, at the outer boundary of the basin, and the entire water volume behind the dam was dyed.

Tidal exchange data for the dredged basin for the case of the basin only are shown in Fig. 3, and for the lagoon-basin configuration (with dredged connecting channel) in Fig. 4. The 7.7-foot mean tide range was used in each case. The dashed lines show the water edge at low tide, and delineate the sampled areas.

Figures 3 and 4 show the distribution of local E values, based on exchange with ambient water. Each E value shown was calculated from a densitometer reading (1-mm aperature) from the negative of the photograph taken at low tide, n=4. A 1-mm x-y grid was used so that the negative areas covered were tangent. Contours of per-cycle E values are shown in incremental steps of 0.05. A histogram showing the frequency of readings for each local E value is included with each figure; the histogram provides a visual index of the uniformity of mixing, and the standard deviation gives a statistical measure. The standard deviation is defined as

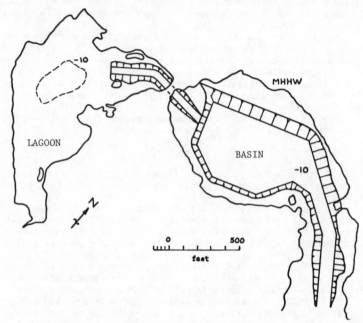

Figure 2. Planform Sechematic of Basin-Lagoon Configuration

$$S = \sqrt{\frac{\sum_{i=1}^{n}(E_i - \overline{E})^2}{n-1}}$$

where

n = number of local E values measured

Fig. 3 indicates that mixing is quite uniform in the basin. Lowest local E values occur in the center of the basin, with the larger E values around the edges; this situation is typical for the single-gyre circulation pattern produced for this configuration.

For the lagoon-basin configuration (Fig. 4) the $\overline{E}$ value for the lagoon is, as anticipated, lower than in the basin. The plot of local E values in the basin indicates that the gyre pattern obtained in the absence of the lagoon has been altered by the flow into and, particularly, out of the lagoon. Uniformity of tidal exchange within these embayments is greater than in most small harbors that have been tested previously. Actual exchange is low, however; the $\overline{E}$ values of 0.31 and 0.22 on Fig. 4 for the basin and lagoon, respectively, compare with corresponding tidal prism ratios of 0.47 and 0.57 for the 7.7-foot tide. Further details, including quantification of the tidal exchange between the basin and lagoon, are given in the project report (7).

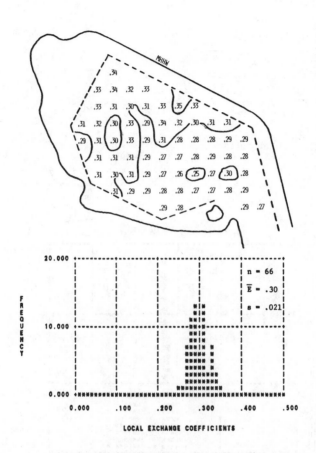

Figure 3. Exchange Contours and Frequency Histogram, Dredged Basin, Tide Range = 7.7 feet.

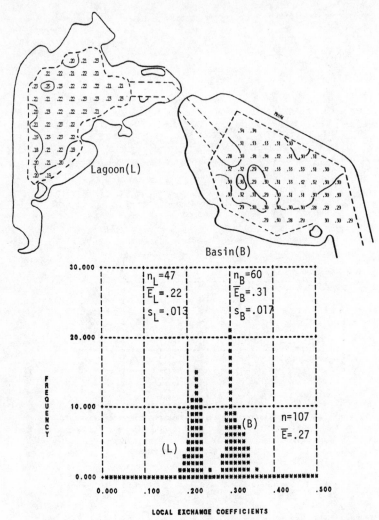

Figure 4.  Exchange Contours and Frequency Histogram, Dredged
Basin - Lagoon Configuration, Tide Range 7.7 feet.

REFERENCES

1. Burgess, F.J. and James, W.P., "Airphoto Analysis of Ocean Outfall Dispersion", U.S. Environmental Protection Agency, Program No.16070 ENS, Corvallis, Oregon, June, 1971.

2. Chiang, W.L. and Lee, J.J., "Simulation of Large-Scale Circulation in Harbors", Journal of the Waterway, Port, Coastal and Ocean Division, ASCE, Vol. 108, No. WW1, Proc. Paper 16841, Feb., 1982, pp. 17-31.

3. Falconer, R.A., "Modelling of Planform Influence on Circulation in Harbors", Proceedings of the Seventeenth Coastal Engineering Conference, Sydney, Australia, March, 1980, pp. 2726-2744.

4. Fischer, H.B. et al, "Mixing in Inland and Coastal Waters", Academic Press, New York, N.Y., 1979.

5. Nece, R.E. and Richey, E.P., "Flushing Characteristics of Small-Boat Marinas", Proceedings of the Thirteenth Coastal Engineering Conference, Vancouver, Canada, July, 1972, pp. 2499-2512.

6. Nece, R.E., Richey, E.P., Rhee, J. and Smith, H.N., "Effects of Planform Geometry on Tidal Flushing and Mixing in Marinas", Technical Report No. 62, Charles W. Harris Hydraulics Laboratory, Department of Civil Engineering, University of Washington, Seattle, WA., December, 1979.

7. Nece, R.E. and Smith, H.N., "Tidal Exchange in Proposed Sitka, Japonski Lagoon, Small Boat Harbor", Technical Report No. 71, Charles W. Harris Hydraulics Laboratory, Department of Civil Engineering, University of Washington, Seattle, WA, June, 1981.

8. Smith, H.N., "The Analysis of Circulation in Small Boat Harbors by Photodensimetric Methods", Thesis submitted to the University of Washington, Seattle, WA, in partial fulfillment of the requirements for the degree of Master of Science in Civil Engineering, 1982.

9. Ward, P.R.B., "Measurement of Dye Concentrations by Photography", Journal of the Environmental Engineering Division, ASCE, Vol. 99, No. EE3, Proc. Paper 9768, June, 1973, pp. 165-175.

# DISPERSIVE TRANSPORT IN INLET CHANNELS: CASE STUDY

By Elias Sanchez-Diaz,[1] M. ASCE and Ashish J. Mehta,[2] M. ASCE

## ABSTRACT

The role of a tidal inlet as a "gate" for the exchange of waters between the sea and the bay is important from the point of view of reducing pollutant levels in coastal waters. The nature of transport through the inlet channel, particularly a long one, is a controlling factor which influences the rate of constituent exchange, and therefore the renewal rate of bay waters. An investigation of dispersive transport in the 2.62 km long Blind Pass channel in Florida was carried out through dye injection experiments. The dispersion coefficient was determined by comparing the measured temporal and spatial distributions of the dye in the channel with predictions based on a numerical solution of the transport-diffusion equation for a conservative constituent. The value of the dispersion coefficient is found to agree, to a reasonable degree, with the same derived previously using analytic means. It is concluded that transport in the channel is primarily advective. In simple exchange models, channel transport, under conditions similar to those considered, may be assumed to be advective only.

## INTRODUCTION

Tidal inlets have been utilized traditionally as navigation routes. Their role as "gates" for the exchange of waters between the sea and the bay has become important in recent times (5, 7). In cases where a conservative constituent such as a pollutant occurs in the bay or is transported to the bay from an upstream source, tidal exchange enables the transport of the constituent to the sea through the inlet, which thereby results in the flushing of the bay waters. The exchange mechanism is influenced by the length of the inlet channel. Fig. 1 depicts two cases, that of an inlet with a "short" channel and one with a "long" channel. A definition of channel length may be stated in terms of the particle excursion length within the channel. For a progressive shallow water wave, the maximum excursion length, $\xi_m = 2a\sqrt{g}/\sigma\sqrt{h}$, where a = tidal amplitude, g = acceleration due to gravity, $\sigma$ = tidal frequency and h = depth of flow. A channel may be considered to be "short" if $\hat{\xi}_m = \xi_m/\ell \gg 1$, and "long" if $\hat{\xi}_m < 1$. While in both cases the influence of the tidal prism is likely to be localized near the bayward end of the channel as suggested by the dashed line in Fig. 1 (4), the case

---

[1]Assoc. Prof., Hydr. Engrg. Dept., Univ. of Carabobo, Valencia, Venezuela.

[2]Assoc. Prof., Coastal and Oceanographic Engrg. Dept., Univ. of Florida, Gainesville, FL.

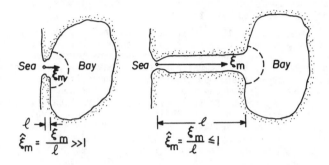

Fig. 1.   Definitions of "Short" and "Long" Inlet Channels.

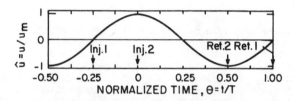

Fig. 2.   Dye Injection and Monitoring Methodology.

Fig. 3.   Blind Pass on the Gulf of Mexico Coast of Florida.

of a long channel is one in which the nature of transport in the channel itself becomes one of the controlling factors influencing constituent exchange between the bay and the sea.

In a tidal inlet with a relatively long channel, the mouth, i.e. the inlet segment seaward of the throat section (minimum flow area), is the region where mixing processes are enhanced by the action of waves. In the channel, i.e. the segment bayward of the throat, wave influence is generally much smaller, and transport is controlled by advection and dispersion under tide-induced flows. A question arises concerning the importance of dispersive transport relative to advection. If dispersion is small, transport in the channel itself can, in some simple cases, be considered to be advective. The rate of constituent exchange in this case will be controlled by the mixing processes adjacent to the channel, in the receiving waters. In simple exchange models, such an assumption concerning channel transport is usually implicit (7).

The main objectives of the present investigation were: a) determination of the turbulent dispersion coefficient in a relatively long tidal inlet channel, and b) estimation of the importance of dispersive transport relative to advection in the channel.

METHODOLOGY

The study was organized to inject a slug of dye near the inlet throat, and to monitor dye concentration in the channel during the subsequent transport of the dye. As shown in Fig. 2 where $\hat{u}$ is equal to the current velocity $u(t)$ in the channel divided by a characteristic maximum velocity $u_m$, and $\theta$ is equal to time $t$ divided by the tidal period $T$, instantaneous injection (across the flow-section) can be carried out at slack before flood (Inj. 1) or during the strength of flood (Inj. 2). In the first case the dye peak will return (Ret. 1) to the throat approximately at the time of the following slack before flood, whereas in the second case it will return (Ret. 2) approximately during the following strength of ebb. The latter procedure was utilized in this study.

The time and space history of the concentration $C(x,t)$ of a conservative constituent can be obtained from the solution of the transport-diffusion equation for a non-stratified, one-dimensional channel (representing cross-sectional average conditions):

$$\frac{\partial}{\partial \theta}(\hat{A}\hat{C}) + \left(\frac{u_m T}{\ell}\right)\frac{\partial}{\partial \hat{x}}(\hat{u}\hat{A}\hat{C}) = \frac{\partial}{\partial \hat{x}}\left(\frac{E_L T}{\ell^2}\hat{A}\frac{\partial \hat{C}}{\partial \hat{x}}\right) \tag{1}$$

where $\hat{C} = C/C_0$, $C_0$ = initial constituent concentration at the time of injection, $\hat{A} = A/A_c$, $A(x,t)$ = channel cross-sectional area, $A_c(t)$ = throat area, $\hat{x} = x/\ell$, and $E_L$ = turbulent dispersion coefficient. The importance of the dispersive term relative to the advective term in Eq. 1 can be evaluated from the ratio $E_L/\ell u_m$.

Expressions relating $E_L$ to easily measurable flow parameters have been proposed for the case of dispersion in a channel of finite width in which shear flow is the dominant mechanism for dispersion (2). Assuming a straight channel of constant dimensions, Taylor (6) proposed an expression which may be described in functional form as

$$\hat{E}_L = \frac{E_L}{u_m^2 T} = f(\hat{T}_z, \hat{T}_c) \qquad (2)$$

where $\hat{E}_L$ = a normalized dispersion coefficient, $\hat{T}_z = T_{cz}/T$, $\hat{T}_c = T_{cz}/T_{cy}$, $T_{cz}$ = vertical mixing time and $T_{cy}$ = lateral mixing time. The values $T_{cz}$ and $T_{cy}$ can be obtained from: $T_{cz} = \bar{h}^2/K_z$ and $T_{cy} = (\bar{w}/2)^2/K_y$, where $\bar{h}$ = mean depth of flow (msl) and $\bar{w}$ = mean channel width. The lateral and vertical eddy diffusivities, $K_y$ and $K_z$, respectively, are obtained from: $K_y = 0.6\langle u_*\rangle\bar{h}$ and $K_z = 0.068\langle u_*\rangle\bar{h}$, where $\langle u_*\rangle$ = flood or ebb-mean friction velocity (2).

The following procedure was adopted: 1) A "one-dimensional" hydro-dynamic model was developed and utilized for the generation of the temporal and spatial distributions of the velocity, u(x,t), in channel (1, 5). This model was calibrated against measurements of tides and currents in the channel by adjusting the Chézy friction coefficient, $C_z$. 2) Eq. 1 was solved for C(x,t) with the required initial conditions, C(x,0), and boundary conditions, C(0,t) and C(ℓ,t), using an implicit finite difference scheme (1, 5). 3) The predicted solution was "matched" with measurements by adjusting the value of $E_L$.

EXPERIMENTS

Blind Pass Channel (Fig. 3) which connects the Gulf of Mexico to Boca Ciega Bay, Florida, was selected for the study. This inlet has an "overlapping offset" configuration (3). Such a configuration results in a "diagonal" channel across the barrier island, so that the updrift shore extends seaward and downdrift to overlap the channel. The mouth of the inlet was a 240 m long channel (Figs. 3 and 4). The 25 m wide and 1.6 m (msl) deep throat was followed by a 2,620 m long channel. The channel width ranged from 40 m just beyond (60 m bayward of) the throat, to 460 m at the bayward end. The corresponding depths were 2.6 m and 2.7 m, respectively. The maximum depth of 3.7 m occurred at a distance of 1,460 m from the throat. Average channel dimensions are given in Fig. 4. The two "arms" of the channel (Fig. 3) at distances of 850 m and 2,300 m from the throat were considered to be parts of the channel in the hydrodynamic model. Given (see Table 1) a = 0.22 m, h = 2.62 m and σ = 1.36x10$^{-4}$ rad/sec, $\xi_m$ = 6,260 m is obtained so that $\hat{\xi}_m = \xi_m/\ell$ = 2.4. If, however, particle excursion beginning the strength of flood (as opposed to slack before flood) to the subsequent strength of ebb is considered, the maximum value would be $\xi_m/2$ = 3,130 m, and the ratio $\xi_m/2\ell$ = 1.2, which is close to unity. The channel was therefore consid-ered to be nearly sufficient in length for a study in which injection occurred at the strength of flood.

Hydrographic measurements were carried out during the period August 6-14, 1975 (5). The tidal prism for the spring range of tide was found to be 9.77x10$^5$ m$^3$. Two dye dispersion experiments were carried out, one on August 7 (expt. 1) and another on August 12 (expt. 2). Approximately 1.9 liters of Rhodamine-WT flourescent dye (20% by weight solution in water, 1.2 gm/cm$^3$ density) was injected almost instantaneously near the throat (Symbol TH in Figs. 3 and 4) at the strength of flood. The lateral position of the injection point was 4 m in expt. 1 and 12 m in expt. 2 from the eastern bank (Fig. 3). The injection was near the water surface in expt. 1 and 0.6 m below the surface in expt. 2. Dye

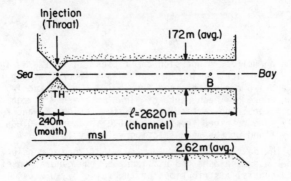

Fig. 4.   Schematic Description of Blind Pass Channel.

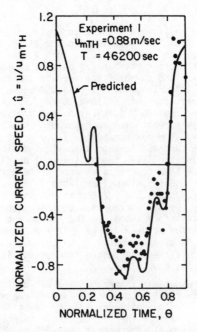

Fig. 5.   Temporal Variation of Cross-sectional Mean
Current at the Throat during Experiment 1.

was sampled between the throat and the bridge across the channel at
point B (Figs. 3 and 4).

RESULTS

Table 1 gives the salient hydraulic parameters.  During both experiments the tidal range, 2a, was moderate, and the period was greater
than that for a semi-diurnal tide, inasmuch as the tide is mixed in this
region.  The maximum velocity, $u_{mB}$, at B was an order of magnitude lower
than the value, $u_{mTH}$, at the throat.  In general, velocities in the
channel were considerably lower than that at the throat.  Fig. 5 shows
the temporal variation of the velocity at the throat.  The model predicted values compare reasonably with measurements, although the latter
had slightly lower magnitudes during ebb.  The model was calibrated
using $C_Z = 14.4$ m$^{1/2}$/sec.  Injection occurred at $\theta = 0$.

TABLE 1. - Flow Conditions during Experiments

| Expt. | Date | 2a | T | $u_{mTH}$(m/sec) | | $u_{mB}$(m/sec) | |
|---|---|---|---|---|---|---|---|
| (1) | (2) | (m) (3) | (sec) (4) | flood (5) | ebb (6) | flood (7) | ebb (8) |
| 1 | Aug. 7 | 0.44 | 46200 | 0.97 | 0.80 | 0.055 | 0.038 |
| 2 | Aug. 12 | 0.63 | 58800 | 0.58 | 1.01 | 0.031 | 0.050 |

Dye injection and transport parameters are given in Table 2.  The
initial concentration, $C_0$ (in parts per billion, by weight), was 35%
lower in expt. 1 than in expt. 2, inasmuch as injection during expt. 1
took a few seconds longer than during expt. 2.  As a result there was
greater than desired longitudinal spreading of the dye in expt 1, and
consequently lower $C_0$.

TABLE 2. - Dye Injection and Transport Parameters

| Expt. | Injection | $C_0$ | $\hat{\xi}_{cm}$ | | $E_L$ |
|---|---|---|---|---|---|
| (1) | (hr) (2) | (ppb) (3) | (measured) (4) | (predicted) (5) | (m$^2$/sec) (6) |
| 1 | 1200 | 17 | 0.57 | 0.40 | 0.093 |
| 2 | 0340 | 27 | 0.38 | 0.37 | 0.093 |

Given $\hat{\xi}_c (= \xi_c / \ell)$ as the normalized value of the dye excursion
distance from the throat, the maximum value, $\hat{\xi}_{cm}$, is given in Table 2.
In Fig. 6, the temporal variation of $\hat{\xi}_c$ is shown for expt. 2.  The
predicted maximum compares reasonably well with the measured value for
expt. 2, although the measured value was 3% larger.  In expt. 1, the
peak was transported a distance which was 45% greater than the predicted
value.  This discrepancy is attributed to the procedure for dye injection.  The inset of Fig. 7 shows the lengths, $\ell_1$ and $\ell_2$, required for
the dye (injected at a point) to mix completely over the depth and the
width of the channel, respectively.  Order of magnitude values of $\ell_1$
and $\ell_2$ can be estimated from: $\ell_1 = \langle u \rangle T_{cz}$ and $\ell_2 = \langle u \rangle T_{cy}$,
where $\langle u \rangle$ is the flood-averaged current velocity.  The values were found
to be $\ell_1 = 177$ m and $\ell_2 = 21,630$ m, which are applicable to both the

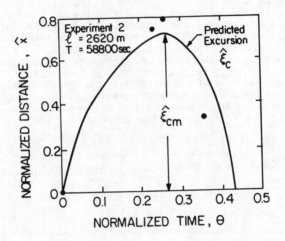

Fig. 6.  Excursion of Peak Dye Concentration during
         Experiment 2.

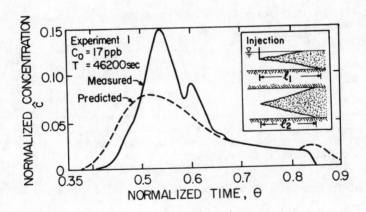

Fig. 7.  Temporal Distribution of Dye Concentration at the
         Throat during Experiment 1.

experiments. The corresponding ratios, $\ell_1/\xi_m$ and $\ell_2/\xi_m$ should be much less than unity. These conditions were not met in particular with respect to lateral mixing. This limitation possibly contributes to an explanation for the discrepancy between the predicted and the measured dye peak excursions during expt. 1. Initially, the point-injected dye moved faster and farther, without much spreading, with the local surface velocity which was greater than the cross-sectional mean velocity, at all times. In expt. 2 in which the dye was injected 0.6 m below the surface, the velocity was closer to the cross-sectional mean value. The discrepancy resulting from point injection (as opposed to cross-section injection through several vertical "line" sources) also resulted in a more concentrated measured dye peak than what was predicted. See for example Figs. 7 and 8.

Values of the dispersion coefficient, $E_L$, given in Table 2 are those that were required to calibrate the solution to Eq. 1. In Table 3, calculations for $E_L$ values based on the analytic results of Taylor (6) are given, making use of Fig. 9. The predicted values of $E_L$ and of $\hat{E}_L$ are two to three times as large as the measured values in both experiments, but may be considered to be of the same order of magnitude. Measured and predicted values of $\hat{E}_L$ are plotted in Fig. 9, with $\hat{T}_c = 8.07\times10^{-3}$ and $\hat{T}_z = 5.52\times10^{-2}$ in expt. 1, and $\hat{T}_c = 8.2\times10^{-3}$ and $\hat{T}_z = 4.41\times10^{-2}$ in expt. 2. Considering the range of possible values of $\hat{E}_L$ shown in Fig. 9 (i.e. from 0 to 330), and the limitations of the dye experiments, the agreement must be considered to be reasonable.

TABLE 3. - Summary of Calculations for $E_L$

| Expt. | $u_m$ | $K_y$ | $K_z$ | $E_L$ (predicted) | $\hat{E}_L \times 10^5$ | |
|---|---|---|---|---|---|---|
| (1) | (m/sec) (2) | (m$^2$/sec) (3) | (m$^2$/sec) (4) | (m$^2$/sec) (5) | (predicted) (6) | (measured) (7) |
| 1 | 0.108 | 0.0058 | 0.00063 | 0.273 | 50.7 | 17.2 |
| 2 | 0.091 | 0.0053 | 0.00062 | 0.212 | 43.6 | 19.1 |

Cross-sectional velocity distributions measured at the throat indicated that the flow velocity in the region where the dye was injected was approximately 17% higher than the cross-sectional mean value during both experiments (5). If this correction (i.e. decreasing the velocity by 17%) is applied to the predicted values, $E_L = 0.207$ m$^2$/sec and 0.166 m$^2$/sec result for expt. 1 and 2, respectively. These values are in better agreement with the measurements.

CONCLUSIONS

1) A value of the dispersion coefficient due to shear flow, $E_L = 0.093$ m$^2$/sec was measured in two dye injection experiments at Blind Pass, Florida, which has a 2.62 km long channel.

2) Considering a) the expected limitations inherent in the analytic approach (6) and b) limitations in the field experimental procedure, the agreement between the measured and the predicted (0.273 m$^2$/sec and 0.212 m$^2$/sec in expts. 1 and 2, respectively) values of $E_L$ must be considered to be reasonable.

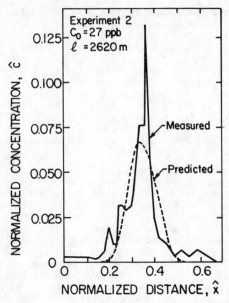

Fig. 8. Spatial Distribution of Dye Concentration
at $\theta = 0.24$ in the Channel during Experiment 2.

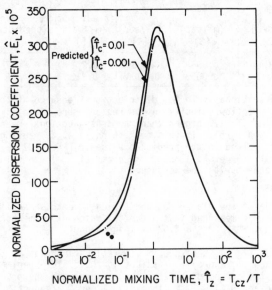

Fig. 9. Normalized Dispersion Coefficient, $\hat{E}_L$ as a
Function of Normalized Mixing Times, $\hat{T}_Z$ and $\hat{T}_C$.

3)   The ratio $E_L/\ell u_m$ can be computed to be $3.29 \times 10^{-4}$ in expt. 1 and $3.90 \times 10^{-4}$ in expt. 2. It is concluded that the transport is primarily advective. Therefore, in simple models for the exchange of sea water with the bay water, transport in the inlet channel itself may be assumed to be advective.

ACKNOWLEDGMENT

Dean M. P. O'Brien was the principal investigator for the project which was supported by the Office of Naval Research, Geography Programs, Contract N00014-68-A-0020, Project NR-388-106. Mr. A. Damsgaard assisted in the development of the numerical schemes.

REFERENCES

1.   Abbott, M. B., Computational Hydraulics, Pitman Publishing Company, London, 1979.

2.   Fischer, H. B., List, E. J. Koh, R. C. Y., Imberger, J., and Brooks, N. H., Mixing in Inland and Coastal Waters, Academic Press, New York, 1979.

3.   Galvin, C. J., "Wave Climate and Coastal Processes," Symposium on the Water Environment and Human Needs, Department of Civil Engineering, Massachusetts Institute of Technology, Cambridge, Massachusetts, October, 1971, pp. 44-78.

4.   Mehta, A. J., and Zeh, T. A., "Influence of a Small Inlet in a Large Bay," Coastal Engineering, Vol. 4, No. 1, 1980.

5.   Sanchez-Diaz, E., "A Dye Dispersion Study at Blind Pass, Florida," M. S. Thesis, University of Florida, Gainesville, Florida, 1975.

6.   Taylor, R. B., "Dispersive Transport in River and Tidal Flows," Proceedings of Fifteenth Coastal Engineering Conference, ASCE, Vol. IV, Ch. 192, Honolulu, Hawaii, July, 1976, pp. 3336-3357.

7.   Taylor, R. B., and Dean, R. G., "Exchange Characteristics of Tidal Inlets," Proceedings of Fourteenth Coastal Engineering Conference, ASCE, Vol. III, Ch. 132, Copenhagen, Denmark, June, 1974, pp. 2268-2289.

ENTRAINMENT OF BED MATERIAL

by

Peter J. Murphy[1], A.M. ASCE

ABSTRACT: A theoretical investigation of the process by which bed
material becomes suspended is presented. The particle dynamics of the
motion of bed particles is developed and applied to the entrainment
process. One result is the development of a new equation for the fluid force on a low Reynolds number particle in a turbulent flow. Another result is an equation for the time required for the suspension
of sand and gravel. This entrainment time is compared to the duration
of the saltation process and is used to quantify the distinction between suspended load and bed load.

INTRODUCTION

The goal of this paper is to explain the particle dynamics of the
process in which a sediment particle becomes suspended in a turbulent
open-channel flow. This entrainment process is important because it
can be combined with the suspension deposition process to provide an
improved bottom boundary condition for the analysis of suspended sediment transport problems in rivers.

An improvement in this bottom boundary condition is clearly needed. The common experimental procedure of measuring the sediment concentration $C(a)$ near the river bottom and calculating the bed load
with a bed-load formula avoids the problem of the connection between
the bed load and the suspended load. Einstein's (3) theoretical procedure neglects the differences between bed load and suspended load by
essentially equating them at a somewhat arbitrary level above the bed.
Garde and Ranga Raju (4) provide a recent review of the problem and
conclude that even with its faults the Einstein procedure can be considered the state-of-the-art.

The entrainment process will be studied by examining the dynamic
behavior of a single sediment particle in a turbulent flow. The forces acting on any particle are the force of gravity, the bed impact
force and the force of the fluid. These forces determine the motion
of each sediment particle and so are the basis for an understanding of
the physics of the entrainment process. The results of such particle
dynamics analysis have already been used by Hinze (5), Soo (12) and
others to explain the mechanics of turbulent suspensions. In this

---

[1]Assistant Professor, Department of Civil Engineering, University of
Massachusetts, Amherst, MA 01003.

paper this particle dynamics method will be applied to the problem of the entrainment of river bed material.

## FLUID FORCE EQUATIONS

Consider particle movement within a steady horizontal flow above a bed of loose material. The particle motion can be classified as "rolling" or "sliding" if the bed continuously supports the net weight of the particle, and as "bouncing" or "saltating" if the bed's support is intermittent. The particle is said to move in suspension when it hits the bed so infrequently that the particle's net weight is supported principally by the fluid rather than by the bed. In typical open channel flow the fluid motion is turbulent and this random fluid motion provides the fluid force required to suspend silt, sand, and even gravel. In contrast, particles that are only rolling, sliding, and bouncing are not substantially affected by the turbulence; rather the mean velocity of the fluid drags these particles downstream and supplies the energy required to keep them moving. Thus a major difference between the bed load and suspended load types of particle motion is which fluid force causes the particle's motion. Forces due to the mean fluid velocity typify bed load, and forces due to the fluctuating fluid velocity typify suspension.

The force that the fluid exerts on the moving particle is due to the pressure and shear stresses associated with a variety of elementary fluid forces. The buoyancy force is an upward pressure force equal to the weight of the displaced fluid. The added mass force is an unsteady pressure force that opposes the acceleration of an object if that object accelerates through the fluid. In addition there is a similar unsteady pressure force that accelerates a particle if the particle is surrounded by fluid that is accelerating past the particle. The lift force is a pressure force that acts in the direction perpendicular to the fluid flow far from the particle. The lift force may be due to the circulation around a spinning particle, to the vorticity of the fluid around a particle that is not spinning, or to the presence of a solid wall near the particle. The drag force is a frictional force due to both pressure and shear stress. For steady flow it acts in the direction of the velocity of the fluid relative to the particle, but in unsteady flow the history of the motion must also be considered. Since the lift and drag forces are complex, they are frequently described with lift and drag coefficients based on experiments. Since most experiments have used spherical particles, the force on a sphere is studied here. The resulting fluid force on a moving sphere of radius a is

$$F_i = \left(\frac{4\pi a^3}{3}\right) \rho g \frac{\partial z}{\partial x_i} - \frac{1}{2} \left(\frac{4\pi a^3}{3}\right) \rho \frac{dU_i}{dt} + \frac{3}{2} \left(\frac{4\pi a^3}{3}\right) \rho \left(\frac{\partial U_{ci}}{\partial t} + U_j \frac{\partial U_{ci}}{\partial x_j}\right) \tag{1}$$

$$+ C_L \frac{1}{2} \rho \left|U_{ci} - U_i\right|^2 \pi a^2 n_i + C_D \frac{1}{2} \rho \left|U_{ci} - U_i\right| (U_{ci} - U_i)\pi a^2$$

where $U_i$ is the velocity of the moving sphere, $U_{ci}$ is the velocity

that the fluid would have at the location of the center of the sphere
if the sphere were absent, $n_i$ is the direction of the lift force, and
$C_L$ and $C_D$ are the lift and drag coefficients. This equation for the
fluid force is similar to those used previously by Tsuchiya (14) and
Reizes (10) with the exception of the term due to the acceleration of
the fluid. Following the development of that acceleration presented
by Taylor (13), Batchelor (2), and Newman (9), the component due to
the gradient of the fluid velocity along the path of the particle was
added to the unsteady term of the fluid acceleration.

In order to separate the fluid force into the components due to
the mean fluid flow and the components due to the fluctuating flow,
the sphere's velocity $U_i$ and the fluid velocity $U_{ci}$ are divided into
two parts, their ensemble average components $\overline{U}_i$ and $\overline{U}_{ci}$ and their random, fluctuating components $U_i'$ and $U_{ci}'$. Introducing these mean and
fluctuating velocities into the equation for the fluid force and then
ensemble averaging produces a mean fluid force equation. Subtracting
this mean force equation from the total force equation gives an equation for the fluctuating fluid force. However, the Reynolds number,
$Re = |U_{ci} - U_i| d/\nu$, has an important effect on this separation process.
If the Reynolds number is low enough $(Re < 1.0)$, the lift force becomes negligible and the drag force becomes linear. The resulting
equation for the mean fluid force at low Reynolds numbers is

$$\overline{F}_i = (\frac{4\pi a^3}{3}) \rho g \frac{\partial z}{\partial x_i} - \frac{1}{2} (\frac{4\pi a^3}{3}) \rho \frac{d\overline{U}_i}{dt} + \frac{3}{2} (\frac{4\pi a^3}{3}) \rho \overline{U}_j \frac{\partial \overline{U}_{ci}}{\partial x_j}$$

$$+ 6\pi\mu a (\overline{U}_{ci} - \overline{U}_i) + \pi\mu a^3 \nabla^2 \overline{U}_{ci}$$

$$+ \frac{6\pi\mu a^2}{\sqrt{\pi\nu}} \int_{-\infty}^{t} (\overline{U}_j \frac{\partial \overline{U}_{ci}}{\partial x_j} - \frac{d\overline{U}_i}{dt'}) (t - t')^{-\frac{1}{2}} dt' \qquad (2)$$

while the corresponding random fluid force is

$$F_i' = - \frac{1}{2} (\frac{4\pi a^3}{3}) \rho \frac{dU_i'}{dt} + \frac{3}{2} (\frac{4\pi a^3}{3}) \rho (\frac{\partial U_{ci}'}{\partial t} + \overline{U}_j \frac{\partial U_{ci}'}{\partial x_j})$$

$$+ 6\pi\mu a (U_{ci}' - U_i') + \pi\mu a^3 \nabla^2 U_{ci}'$$

$$+ \frac{6\pi\mu a^2}{\sqrt{\pi\nu}} \int_{-\infty}^{t} (\frac{\partial U_{ci}'}{\partial t'} + \overline{U}_j \frac{\partial U_{ci}'}{\partial x_j} - \frac{\partial U_i'}{\partial t'}) (t - t')^{-\frac{1}{2}} dt' \qquad (3)$$

If the Reynolds number is high enough so that the advective fluid
acceleration is important, the lift and drag forces are non-linear
functions of the fluid velocity. In order to separate the mean and
fluctuating components of these non-linear forces the assumption will
be made that $|\overline{U}_{ci} - \overline{U}_i| \gg |U_{ci}' - U_i'|$. This "weak suspension"
assumption will be valid for saltation and near the threshold of sus-
pension, $V_*/w = 1$. In particular it will be valid when the fluid is
accelerating a slow particle that is entering the suspended state.
The result of this weak suspension assumption is that equation 1 is

applied only to the mean velocities, $\overline{U}_{ci}$ and $\overline{U}_i$, and is used in its drag and lift coefficient form.

## SALTATION AND SUSPENSION

The saltation process will be examined first to obtain an estimate of the time scale associated with bed load. That time scale is most easily calculated by assuming it to be the duration of a saltation which occurs when the lift and drag forces are weak in comparison with the forces of weight, buoyancy and added mass. When lift and drag are small, equation 1 becomes identical to the dynamic equation for the mean motion of a particle moving in a vacuum under the influence of an equivalent gravitational acceleration $g_e$. The duration of a saltation under this assumption is

$$T_s = \frac{2V_0}{g_e} \qquad (4)$$

with

$$g_e = \frac{2g\,(\rho_s - \rho)}{2\,\rho_s + \rho} \qquad (5)$$

where $\rho_s$ is the particle density, $\rho$ is the fluid density and $V_0$ is the initial vertical velocity of the particle.

The suspension entrainment process at high Reynolds numbers will now be examined to obtain an estimate of the time required to accelerate a sediment particle into the weak suspension state. The slow particle is affected by both the mean and fluctuating fluid forces, but during entrainment the mean fluid force dominates. If the particle starts from rest, then $|\overline{U}_{ci} - \overline{U}_i| \gg |U_{ci}' - U_i'|$ since $|\overline{U}_{ci}| \gg |U_{ci}'|$. If the particle starts with some initial velocity due to its transition from bed load to suspended load, the bed impact still causes a dominant mean velocity difference. Both of these entrainment processes are shown in Figure 1.

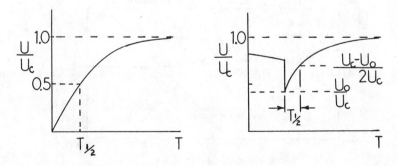

Figure 1 - Suspension Entrainment: a) from rest, b) from saltation.

The time required for the mean velocity difference $|\overline{U}_{ci} - \overline{U}_i|$ to be reduced by a factor of 2 is a characteristic half-life of the entrainment process, since suspended particles must move with roughly the mean velocity of the fluid. Figure 1 shows this half-life or entrainment time.

At high Reynolds numbers a simple calculation of this entrainment time can be made based only on the steady drag and added mass components of the downstream fluid force. The drag coefficient in equation 1 is approximated by $C_D = 0.40 + 24/Re$ and the resulting half-life is

$$T_{\frac{1}{2}} = \frac{(2\,\rho_s + \rho)a^2}{9\mu} \ln\left(\frac{Re_0 + 120}{Re_0 + 60}\right) \qquad (6)$$

where $Re_0 = 2a|U_c - U_0|/\nu$, a is the radius of the spherical particle, $\mu$ is the dynamic viscosity and $\nu$ is the kinematic viscosity.

In order to evaluate the validity of $T_S$ and $T_{\frac{1}{2}}$ the Reynolds numbers associated with moving sand and gravel are now studied. Figure 2, from Murphy and Amin (7), shows that coarse sand and gravel first begin to move when the streamflow is strong enough to cross the Shield's (11) threshold of motion. The sediment movement is then bed load only. If the streamflow becomes even stronger the high $Re_*$, weak suspension threshold is reached. On the other hand fine sand and silt first begin to move when the streamflow is strong enough to cross the Bagnold (1) threshold of suspension. Since this occurs at low $Re_*$ neither $T_S$ nor $T_{\frac{1}{2}}$ are expected to be valid for fine sand or silt. However Figure 2 indicates that equations 4 and 6 should be valid for coarse sand and gravel.

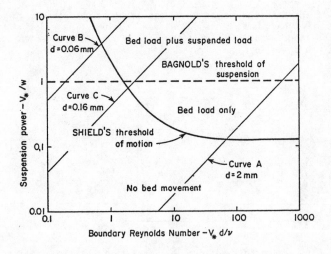

Figure 2 - Sediment Transport Thresholds.

The experimental results of Murphy and Hooshiari (8) indicate that the formula for the duration of saltation at high $Re_*$ is correct. They studied the movement of spheres the size of gravel ($D = 1.5$ cm) in a flume. Figure 3 shows a typical set of sphere trajectories for this saltation process. A sample of 41 trajectories gave an average value of $T_s \, g_e/2V_0 = 1.54$ with a standard deviation of 0.82. Since the accuracy of the measured $V_0$ was low, the rise H of the particle above the bed was used to provide a more accurate value than the direct $V_0$. This gave $T_s \, g_e/2\sqrt{2g_e H} = 0.94$ with a standard deviation of only 0.14.

The validity of the suspension entrainment time can be studied by considering the ratio of $T_{\frac{1}{2}}$ to $T_s$, given by

$$\frac{T_{\frac{1}{2}}}{T_s} = \frac{a^2 g \,(\rho_s - \rho)}{9\mu V_0} \, \ln \, (\frac{Re_0 + 120}{Re_0 + 60}) \tag{7}$$

By introducing the friction velocity $V_*$ and the fall velocity w this ratio can be expressed as

$$\frac{T_{\frac{1}{2}}}{T_s} = \frac{1}{48} \frac{V_*}{V_0} \, (\frac{w}{V_*}) (0.4 \, \frac{w}{V_*} \, Re_* + 24) \, \ln \, (\frac{Re_* + 120 \, V_*/\Delta V_0}{Re_* + 60 \, V_*/\Delta V_0}) \tag{8}$$

When reasonable assumptions about the magnitude of the initial particle velocity are made, the ratio $T_{\frac{1}{2}}/T_s$ can be plotted as a function of $Re_*$ and $V_*/w$ and compared with Figure 2. Figure 4 shows the ratio of the characteristic times of entrainment and saltation for values of the initial particle velocity that were typical of the experiments of Murphy and Hooshiari (8).

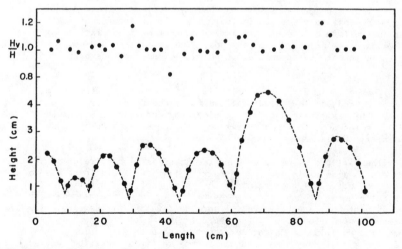

Figure 3 - Saltation Trajectories.

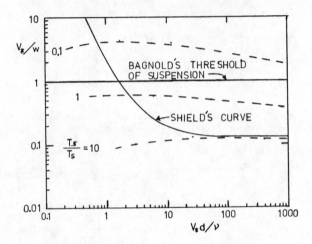

Figure 4 - Comparison of Suspension Entrainment Half-life and Duration of Saltation, $V_0/V_* = 3$, $\Delta V_0/V_* = 4.5$.

CONCLUSION

This study of the particle dynamics of suspension entrainment presents new information but it is still incomplete. The forces given by equations 2 and 3 are an improvement in the description of the force at low Reynolds numbers, but they omit entirely such high Reynolds number phenomena as lift and boundary layers. The major improvement that they provide is that they account for forces due to the non-uniform nature of turbulent flow that were not included in previous descriptions of suspensions. The derivation and application of these equations for general turbulent suspensions has been submitted for journal publication.

The particle dynamics of high Re saltation are accurately described by considering only the fluid force due to the mean fluid motion. This saltation analysis predicts the duration of saltation well, as shown by the good agreement with the data of Murphy and Hooshiari (8).

The particle dynamics of the high Re suspension entrainment process is also explained by considering just the fluid force due to the mean motion. The entrainment analysis predicts the half-life of the entrainment process and that half-life is in good agreement with the data of Lane and Kalinske (6) and the data of Shields (11). The threshold of motion occurs when the entrainment time is roughly ten times the duration of saltation, the threshold of suspension occurs when the entrainment time is equal to the saltation duration, and a strong suspension occurs when the entrainment time is one tenth of the duration of a saltation. Thus $T_{\frac{1}{2}}$ provides an improvement over Einstein's interaction time, $d/w$, and begins the process of explaining the particle dynamics of suspension entrainment.

REFERENCES

1) Bagnold, R. A., "The Nature of Saltation and of Bed-Load Transport in Water," Proceedings of the Royal Society, London, England, Series A, Vol. 332, 1973, pp. 473-504.

2) Batchelor, G. K., An Introduction to Fluid Dynamics, Cambridge University Press, Cambridge, 1967, pp. 398-409.

3) Einstein, H. A., "The Bed Load Function for Sediment Transportation in Open Channel Flows," Soil Conservation Service Bulletin No. 1026, United States Department of Agriculture, Washington, DC, September, 1950.

4) Garde, R. J., and Ranga Raju, K. G., Mechanics of Sediment Transportation, Wiley, New York, 1977.

5) Hinze, J. O., "Turbulent Fluid and Particle Interaction," Progress in Heat and Mass Transfer, Vol. 6, Pergamon, 1972, pp. 433-452.

6) Lane, E. W., and Kalinske, A. A., "The Relation of Suspended to Bed Material in Rivers," Transactions of the American Geophysical Union, Vol. 20, 1939, pp. 637-641.

7) Murphy, P. J., and Amin, M. I., "Compartmented Sediment Trap," Journal of the Hydraulics Division, ASCE, Vol. 105, No. HY5, May, 1979, pp. 489-500.

8) Murphy, P. J., and Hooshiari, H., "Saltation in Water: Dynamics," Journal of the Hydraulics Division, ASCE, Vol. 108, No. HY11, November, 1982.

9) Newman, J. N., Marine Hydrodynamics, The MIT Press, Cambridge, 1978, pp. 135-154.

10) Reizes, J. A., "Numerical Study of Continuous Saltation," Journal of the Hydraulics Division, ASCE, Vol. 104, No. HY9, 1978, pp. 1305-1321.

11) Shields, A., "Anwendung der Ähnlichkeitmechanik und Turbulenzforschung auf die Geschiebebewegung," Mitteil. Preuss. Versuchsanst. Wasser, Erd, Schiffsbau, Berlin, No. 26, 1936.

12) Soo, S. L., Fluid Dynamics of Multiphase Systems, Blaisdell, Waltham, Massachusetts, 1967.

13) Taylor, G. I., "Motion of Solids in Fluids when the Flow is not Irrotational," Proceedings of the Royal Society, London, England, Series A, Vol. 93, 1917, pp. 99-113.

14) Tsuchiya, Y., "On the Mechanics of Saltation of a Spherical Sand Particle in a Turbulent Stream," Proceedings 13th Congress of the IAHR, Vol. 2, 1969, pp. 191-198.

# ULTRASONIC MEASUREMENT OF SEDIMENT RESUSPENSION

by

Keith W. Bedford[1], AM ASCE; Robert E. Van Evra III[2];
and Hedayatollah Valizadeh-Alavi[2]

## ABSTRACT

Recognizing the need for improved measurement and parameterization of sediment resuspension, this paper presents a review of the major methods now in use for alleviating this need. Special attention is devoted to reviewing methods for obtaining sediment concentration profiles by acoustic scattering methods and the microcomputer based C-DART Tower developed by the authors to simultaneously measure wave, current, and sediment profile data. The presence of spatial data requires a new approach to its analysis. A generalized pattern recognition program for geophysical flows is discussed.

## INTRODUCTION

Resuspension of bottom sediment is a major factor in the coastal distribution of toxic substances. Recent experimentation in this field has shown that the possibility exists of a rational structure for bounded shear flows and were it to be confirmed would consist of intermittent but coherent events governed by distributions of vorticity. Thus turbulent shear for resuspension would come in bursts over short periods of time. Recent ultrasonic transducer technology advances now permit the direct high frequency profiling of near bottom sediment concentration to be taken and thereby provide the information to confirm the intermittency hypothesis. The primary objective of this paper is to review the methods of ultrasonic profiling of sediment concentration and its implementation in a microcomputer controlled C-DART (Coastal Data Acquisition and Retrieval Tower) data collection system for coupled benthic sediment velocity studies.

### A Brief Review of Methods for Performing Sediment Resuspension Research

The approaches to measuring and predicting resuspension fall into three categories. First is the mean flow critical shear approach in which bottom stress is found by assuming a known steady Karmen-Prandtl boundary layer profile and calculating various velocity and shear values from fixed point velocity data. Many investigators have used this method including, among others, Nece and Smith (1970), Channon and Hamilton

---

[1] Associate Professor, Civil Engineering Department, The Ohio State University, Columbus, Ohio 43210

[2] Research Associate, Civil Engineering Department, The Ohio State University, Columbus, Ohio 43210

(1971), and Wimbush and Munk (1970), with Sternberg's (1972) being the most comprehensive study available for shear and drag coefficients.

The eddy correlation method is the second approach wherein the shear is also measured from fixed point data as from above but evaluated as a Reynolds stress, i.e. the shear is set equal to the correlation of the fluctuating components of the velocity, measured at two fixed points. Bowden and Fairbairn (1956) published the first direct measurements of Reynolds stresses in the benthic boundary layer using electromagnetic current meters while Gordon and Dohne (1973) measured time traces of tidal velocities.

The third approach to the analysis of velocity influences on sediment entrainment has come from intermittency analysis. These studies assume that turbulence is generated periodically near the bottom and determining the statistical character of the burst becomes a major analytical goal. The works of Heathershaw (1974, 1976), Jackson (1976) and Gordon and Witting (1977) are fundamental. An important aspect of the work of Heathershaw (1976) in the Irish Sea was the confirmation that the turbulent energy production occurs intermittently in a bounded shear flow of geophysical proportions. Heathershaw calculated that as much as 70% of the Reynolds stress was accumulated in about 5% of the total time of record with an event duration in the bottom boundary layer of 5-20 seconds. In another somewhat controversial study in the Chesapeake Bay, Gordon and Witting (1977) also confirmed the bursting phenomena and performed a more detailed delineation of this activity.

With regard to combined high frequency velocity and sediment near bottom studies, Lesht (1976) has made point measurements of sediment concentration based on the light reflection principle and did experiments correlating near bottom velocities and turbidity over both mud and sand bottoms on the Long Island Inner Shelf.

A most thorough set of the point measurmenet of sediment resuspension was taken by Lavelle, Young, Swift and Clarke (1978) who investigated the relationship between near-bottom fluid velocity and suspended sediment matter on the inner Continental Shelf south of Long Island, New York. From different types of spectral analyses, Lavelle et al. confirmed the hypothesis that at some high frequencies, surface waves are driving the mean concentration field past a stationary measuring level.

Very little work has been done using acoustic systems to monitor those naturally occurring suspensates originating from or found in the vicinity of the benthic boundary layer. Orr and Hess (1978) used acoustic backscattering systems near Nantucket Island to obtain continuous quantitative measurements of sediment concentration and vertical and horizontal distributions of particles. Data were also presented which showed that acoustic systems can be used to remotely detect slope/shelf water frontal zone. Perhaps the largest and most comprehensive sequence of nearshore experiments being performed to date are tasks being performed under the sponsorship of NOAA (National Oceanic and Atmospheric Administration), ranging from model development to detailed measurements. Young et al. (1982) developed an acoustic (3 MHz) backscattering profilo-

meter for study of sediment transport in the bottom boundary layer.
Field experiments on the inner shelf using the profilometer combined
with velocity profiles showed individual profiles (taken at 1 Hz) during
erosion events seemed to occur in pulses characterized by highly variable concentrations and apparently rapid vertical mixing of particles.
Surface wave groups appeared to be correlated with groups of profiles
with high concentrations extending 25-100 cm above bottom.  A 5 MHz
system originated by Orr and Hess was also designed for moored applications, which can "look" up or down depending on the physics to be investigated.  This instrument is currently being used by these authors in
the Navy's research program entitled "The High Energy Benthic Boundary
Layer Experiment" (HEBBLE) at Woods Hole Oceanographic Institute.

## OSU C-DART DATA ACQUISITION TOWER

### Introduction

The C-DART system is designed primarily to record acoustic scattering profiles of suspended sediment and related boundary layer conditions.  Unlike the existing system, a variety of equipment limitations
have been removed which makes it usable for detailed study of lake and
nearshore lake turbulent structures:  current  meters  can be positioned at specific intervals; direct wave height and direction information is obtained; sampling intervals are easy to effect by software and
requires no rewiring of the electronics; a sufficient number of horizontal and vertical data points are obtained to permit direct signal
analysis, which is important particularly with the spatial data collected by the profilometer; and finally, the equipment can be moved easily
from site to site, to help isolate and identify effects of any local
influences.

### Tower Configuration

The C-DART system is based on a steel tower truss that sits atop
a 1.5 ton floatable steel cylinder triangular base (Figure 1).  The tower is placed by towing it in a floating position behind a work boat to
the study location and evacuating air from the cylinders that form the
base.  The rate of descent is diver-controlled by a single valve at the
top of the buoyancy collar.  Removal from the site is equally simple, by
evacuating water from the base using compressed air from pressure cylinders.  Thus the tower can be moved frequently with little more than a
day's work.

To support near bottom instruments away from the effects of water
moving around the base, two arms below the water surface hold a vertical post away from the base, over an undisturbed patch of the bottom.
The near bottom instruments are clamped in position to the vertical
post.

The instruments used with C-DART are:  A Weathertronics wind vane
and anemometer (models 2020 and 2030, respectively); four (4) Ocean
Data Equipment model WS-1000 6-meter long water level gauges; four (4)
Marsh McBirney model 511 dual-axis electromagnetic current meters; one
(1) Edo-Western Model 563 3 MHz acoustic concentration profilometer; and

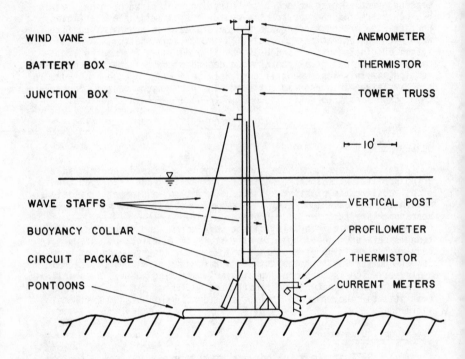

WIND VANE

BATTERY BOX

JUNCTION BOX

ANEMOMETER

THERMISTOR

TOWER TRUSS

⊢— 10' —⊣

WAVE STAFFS

BUOYANCY COLLAR

CIRCUIT PACKAGE

PONTOONS

VERTICAL POST

PROFILOMETER

THERMISTOR

CURRENT METERS

Figure 1.  Schematic of OSU-C-DART System Showing Location
and Position of Components

three (3) Yellow Springs Inc. model YSI 44018 thermistor composites
housed in protective sheaths.

## Microcomputer

The data collection system is controlled by a microcomputer built
by our Coastal Engineering Lab (CEL) that converts all instrument signals to 8-bit binary data.  The integrated circuitry is housed inside a
6 foot long, 10 1/2 inch inside diameter PVC pressure cylinder that has
been laboratory tested to a depth of nearly 400 feet.  One end of the
cylinder is a PVC cap that is attached to the inside chassis that holds
all of the integrated circuitry.  Through four (4) Burton Electrical
Engineering Inc. multi-conductor bulkhead receptables attached to the
cap, all lines to the instruments and external battery power supplies
run from the circuit package to a junction box situated above water level, to which all of the instrument cables are connected.

The microcomputer is controlled by an RCA CDP 1802-based CDP 18S691
protyping system consisting of model CDP18S621 VI memory boards, CDP
18S601 microboard computer and CDP 18S640 control/display board.  CEL-
fabricated circuit boards convert the individual instruments' output
signals to binary numbers stored in separate RCA CDP 1852 input/output
ports for recording.  The analog instrument signals are converted to
frequencies by using National Semiconductor LM 331 precision voltage/
frequency converters, then using microprocessor control, the frequencies are converted to binary numbers.  The profilometer signal is fast-
A/D-converted by two handshaking Analog Devices AD 7574 A/D converters
that are read directly by the microprocessor.  Data points are ready by
comparing channel codes outputted by the microprocessor to the code of
each instrument on the A/D circuit boards:  each instruments's code
is set on its A/D board by a set of DIP switches feeding one side of its
comparator.  The memory boards of the RCA system are used as a buffer
memory which temporarily stores all data before the microprocessor feeds
it onto tape.  A Sea Data model 633 microcprocessor-controlled high
speed digital tape unit  records the digital data onto Verbatim cassettes; the tape transport itself is located above the water surface so
tapes can be changed without disturbing the U/W circuit package, which
contains the transport control circuitry.

## Software

The software routine of the system is programmed into the CDP 18S691
by an RCA CDP 18S021 microterminal, a small keyboard unit, using 1802
op codes.  The operating program is based on a time unit of one second;
in one second, all of the instruments and a time-keeping counter are read
into buffer memory, the previous second's data are read onto tape and
the system is reset to take the next second's readings.  The microcomputer switches on and reads a current meter every 0.25 second, to avoid
"cross talk" caused when two current meters are operated simultaneously
near each other.  Beyond the one-second time unit, program control determines the length of the sampling period.  The program can be changed
by changing a few program statements (which can be done in the field) to
select the sampling routine.  Sampling can be:  continuous; periodic,
for several minutes separated by some "off" period; or conditional,

comparing readings at pre-set intervals to determine if the system
should start sampling in some desirable condition. The program can be
set for different cassette lengths so the system will automatically shut
down at the end of a cassette record to conserve battery power. A 450'
10-megabit cassette will hold slightly less than four hours of continuously collected data.

Parameters/Sampling Intervals

For data sets representing the relation between surface and bottom
events at the study location, C-DART records wind, wave, nearbottom velocity and suspended sediment data simultaneously at a frequency of 1 Hz.
The Weathertronics wind speed and direction sensors are mounted atop the
tower approximately 25 feet above the SWL and register winds out of any
direction to an accuracy of better than 2° ranging from still air to
100 MPH gales. The wave data is obtained from simultaneous readings of
four water level gauges arranged in a trihedral pattern with their mid-
points located at the SWL. The water level gauges are accurate to one
inch and extend 3 meters above and below the SWL.

Each Marsh McBirney current meter measures two perpendicular components of velocity ranging $\pm$ 62.5 cm/s with an accuracy of .5 cm/s.
The current meters can be mounted in nearly any vertical spacing along
the water column and measure several combinations of u, v and w velocity
components due to swivel mounting arms for each probe.

The acoustic profilometer is "aimed" at the point on the bottom directly below the current meters at a slant of 35° and samples 110 sediment concentrations at 1.14 cm intervals 125 centimeters along the slant
above the bottom. The signal strengths vary according to several different characteristics of the sediment, including size, density and concentration of the particles, but the output is in the 0-5 volt standard
A/D conversion range. More fine tuning of ultrasonic profilometer signals using more than one frequency is planned with Woods Hole soon as
an extension of the work done by Orr and Hess (1978).

Thermistors measure temperature of the air at the height of the
wind sensors and water temperature at the level of the profilometer.
The thermistors register temperatures ranging from +4°C to +29.6°C with
an accuracy of $\pm$ 0.1°C. A third thermistor is located inside the circuit package to monitor circuit temperature: in case of any rapid temperature fluctuations during sampling, this record will aid in any necessary circuit equation corrections.

Data are currently being sampled 1/2 mile offshore of the breaker
zone in 25 feet of water at the Old Woman Creek Estuarine Sanctuary, a
federal/state research program on Lake Erie. The bottom materials at
the site are mostly clays and silts, and the bottom slopes very gently
and smoothly for almost two miles offshore, with no bottom obstructions
nearby. This area is the southernmost reach of Lake Erie and winds from
the northwest, north to the northeast produce significant wave-induced
sediment movement episodes. Winds blow from the northwest more often,
but the northeast fetch is much longer, and winds from that direction
result in more fully-developed swells at the study site.

## DATA ANALYSIS PROCEDURES

The availability of exceptionally dense data as are being collected here require a great deal more care and  thought to be invested in the development of data analysis programs.  Traditional analysis procedures such as Gaussian statistics and correlation, auto, cross and rotary spectral analysis, and wave directional spectra will be applied to these data.  However, these programs do not,without great labor,extract fundamental structural relationships about the sediment resuspension and transport processes.  To elicit such relationships an extensive pattern recognition program for geophysical flows is developed.

Pattern recognition is a formal method of analyzing the seemingly random events occurring in turbulent flows and categorizing them into bursts of important activity and periods of no activity.  It has become clear that overall average values cannot provide the details necessary for understanding the mechanisms of turbulent shear flow and consequently, that of sediment resuspension.  A very extensive sequence of work in pattern recognition is that performed by Wallace, Brodkey and Eckelman (1977), Eckelman, Nychas, Brodkey and Wallace (1977), Eckelman, Wallace and Brodkey (1978) and Blackwelder (1978).  Wallace et al., developed a pattern recognition method for detecting ensemble averages and descriptions of the velocity signal signatures of coherent structures in the wall region of a bounded turbulent shear flow.  The patterns of streamwise fluctuation velocity signal "u" were characterized by a gradual deceleration from a local maximum followed by a strong sharp acceleration.  The acceleration within the u pattern was, on the average, over twice as strong as the deceleration in the region near the wall.  Occuring over 65% of the total sample in the region of high Reynolds-stress production, this pattern was believed to be the u signal signature of the turbulent bursting process in the wall region (Wallace et al., 1977).

The pattern recognition technique of Wallace et al. is modified for use in geophysical flow in order to detect, analyze, and describe the turbulent signal signatures of coherent structures and corresponding suspended sediment concentration profiles in the benthic boundary layer of Lake Erie.  Because of the availability of the suspended sediment profile data and simultaneous measurements of velocity at different locations, it is now possible to confirm and determine not only the frequency of these coherent events, but also to obtain additional information about the spatial extension of such bursting phenomena. Blocks of continuous or discontinuous data for different signals are fed into the computer program.  The statistical characteristics of all the overall data signals are computed.  The horizontal component of velocity (u) signal upon which the pattern recognition technique will be triggered is filtered using a Butterworth lowpass filter.  After removing the mean from all signals, the repeatedly occurring patterns of streamwise fluctuation velocity signal 'u', characterized by a gradual deceleration from a local maximum followed by a strong sharp acceleration, are detected using the concept of a short-time temporal average (TPAV).  The quadrant analysis technique is used to determine different pattern recognized coherent fluid mechanical events, i.e., ejections, sweeps, interaction wallward, interaction outward, as well as the total event. To interpret the mechanism of the flow and sediment resuspension, during the same interval that u signal is recognized, up to ten other recognized, signals (sediment concentration profiles, water level fluctuations, temperature, components of velocity fluctuations at different heights, etc.) are also normalized and ensemble averaged with respect to different quadrants for each fluid mechanical event. The r.m.s. and cross-correlations of all signals for each pattern recognized-quadrant analyzed event, as well as the total, are also normalized and ensembled. The results are tabulated and all pattern recognized events (all quadrants and total) for the ensembled averages, r.m.s. and cross-correlations are plotted. The pattern recognized ensembled averages of concentration profiles corresponding to the different events are plotted as density plots. The variable interval time averaging (VITA) technique of Blackwelder and Kaplan (1976) is also employed to determine some bursting statistics and to do a one-to-one comparison with results obtained from the above procedure. The computer program is now being applied to NOAA data as well as to the oil channel data of Wallace et al. (1977) and of course the OSU data.

## SUMMARY

The OSU C-DART system has been built to record the high-density multi-faceted data needed to couple wind wave conditions to lake bottom turbulent responses. The automated procedure developed here is necessary due to the enormous quantity of data generated in this study. Simultaneous patterns of different signals are being studied carefully to determine pulses, predominant factors and significant characteristics of these patterns contributing to the bottom sediment resuspension.

## REFERENCES

Blackwelder, R.F., "Pattern Recognition of Coherent Eddies," Proceedings of the Dynamic Flow Conference, Marseille, France and Baltimore, USA 1978, pp. 173-190.

Blackwelder, R. F., and Kaplan, R.F., "On the Wall Structure of the Turbulent Boundary Layer," Journal of Fluid Mechanics, Vol. 78, Part 1, 1976, pp. 89-112.

Bowden, K.F., and Fairbairn, L. A. "Measurements of Turbulent Fluctuations and Reynold Stresses in a Tidal Current," Proceedings of the Royal Society of London, Series A, 237, 1956, pp. 422-438.

Bowden, K. F., and Ferguson, S.R., "Variations with Height of the Turbulence in a Tidally-Induced Bottom Boundary Layer," in Marine Turbulence (Ed., J. C. J. Nihoul). New York: Elsevier Scientific Publishing Company, Series 28, 1980, pp. 259-286.

Channon, R.D., and Hamilton, D., "Sea Bottom Velocity Profiles on the Continental Shelf Southwest of England," Nature, Vol. 231, 1971, pp. 383-385.

Eckelmann, H., Nychas, S.G., Brodkey, R.S., and Wallace, J.M., "Vorticity and Turbulence Production in Pattern Recognized Turbulent Flow Structures," The Physics of Fluids, Vol. 20, No. 10, Part II, October 1977, pp. S225-S231.

Eckelman, H., Wallace, J. M., and Brodkey, R.S., "Pattern Recognition, a Means for Detection of Coherent Structures in Bounded Shear Flows," Proceedings of the Dynamic Flow Conference, Marseille, France, and Baltimore, USA, 1978, pp. 161-172.

Garden, C. M., and Dohne, C.F., "Some Observations of Turbulent Flow in a Tidal Estuary," Journal of Geophysical Research, Vol. 78, 1973, pp. 1971-1978.

Gordon, C.M., and Witting, J., "Turbulent Structure in a Benthic Boundary Layer," in Bottom Turbulence (Ed., J.C. J. Nihoul). New York: Elsevier Scientific Publishing Company, 1977, pp. 59-81.

Heathershaw, A.D., "Bursting Phenomena in the Sea," Nature, Vol. 248, March 1974, pp. 394-395.

Heathershaw, A.D., "Measurements of Turbulence in the Irish Sea Benthic Boundary Layer," in The Benthic Boundary Layer, (Ed., I.N. McCave), New York: Plenum Press, 1976, pp. 11-31.

Jackson, R.G., "Sedimentological and Fluid-Dynamic Implications of the Turbulent Bursting Phenomenon in Geophysical Flows," Journal of Fluid Mechanics, Vol. 77, Part 3, 1976, pp. 531-560.

Lavelle, J.W., Young, R.A., Swift, D.J.P., and Clarke, T.L., "Near-Bottom Sediment Concentration and Fluid Velocity Measurements on the Inner Continental Shelf, New York," Journal of Geophysical Research, Vol. 83, No. C12, December 1978, pp. 6052-6062.

Nece, R.E., and Smith, J.D., "Boundary Shear Stress in Rivers and Estuaries," Journal of Waterways and Harbors Division, ASCE, Vol. 96, No. WW2, May 1970, pp. 335-358.

Orr, M.H., and Hess, F.R., "Remote Acoustic Monitoring of Natural Suspensate Distributions, Active Suspensate Resuspension, and Slope/Shelf Water Intrusions," Journal of Geophysical Research, Vol. 83, No.C8, August 1978, pp. 4062-4068.

Sternberg, R.W., "Predicting Initial Motion and Bedload Transport of Sediment Particles in the Shallow Marine Environment," in Shelf Sediment Transport: Process and Pattern (Editors: Swift, D.J.P., Duane, D.B., and Pilkey, O.H.), Stroudsburg, Pennsylvania: Dowden, Hutchinson & Ross, Inc. 1972, pp. 61-82.

Wallace, J.M., Brodkey, R.S., and Eckelman, H., "Pattern-Recognized Structures in Bounded Turbulent Shear Flows," Journal of Fluid Mechanics, Vol. 83, Part 4, 1977, pp. 673-693.

Wimbush, M., and Munk, W., "The Benthic Boundary Layer," In the Sea, New York: John Wiley, Vol. 4, Part 1, 1970, pp. 731-758.

Young, R.B., Merrill, J.T., Clarke, T.L., and Proni, J. R., "Acoustic Profiling of Suspended Sediments in the Marine Bottom Boundary Layers," Geophysical Research Letters, Vol. 9, No. 3, March 1982, pp. 175-178.

# TESTING AND CALIBRATION OF TURBULENCE MODELS FOR TRANSPORT AND MIXING

By Wolfgang Rodi[1]

## ABSTRACT

The paper reports on the testing and calibration of one particular turbulence model, namely the depth-average version of the k-ε model. This model is described briefly: it relates the eddy-viscosity and diffusivity through two parameters characterizing the local depth-average state of the turbulence, the turbulent kinetic energy k and the dissipation rate ε, and determines these two parameters from transport equations. These equations contain terms that account for the production of turbulence by both the horizontal velocity gradients and by the bed shear, which is an important feature of the model that makes it applicable to near-field phenomena. The determination of the empirical constants from simple laboratory flows and from dye-spreading experiments is discussed briefly. Several test calculations and input-sensitivity studies are reported: one for the case of a side discharge into rectangular channel flow, one for coaxial buoyant discharges into channel flow and two applications of the model to real-life discharges of sewage and cooling water into natural rivers.

## INTRODUCTION

The mathematical simulation of transport and mixing processes in hydraulic flow situations requires the introduction of assumptions about the effects of turbulence, that is the introduction of a turbulence model. No matter whether the model employed is entirely empirical or only partly so, it always involves empirical coefficients that need to be determined from experiments. In many of the simple models that are often used, the empirical input is site-specific so that these models have to be calibrated for each application. Considerable research has been undertaken in recent years to develop models that can be applied to different flow situations without changing the empirical coefficients involved which would be determined once from a standard data base and would not be site-specific. However, turbulent phenomena are so manifold that an entirely general method is unlikely to evolve and the models will only be suitable for a certain range of flows. It is important therefore to always determine the range of applicability of each model by applying it to as many different flows as is possible and comparing the results with experiments.

The present paper discusses the calibration and testing of one particular model that is of the more universal kind. This model is a version of the k-ε turbulence model developed for use in depth-average calculations. The parent k-ε model has been applied in many different areas of fluid mechanics to vastly different flow problems with reasonable success, and the depth-average

---

[1]Professor, University of Karlsruhe, F.R. Germany, Member ASCE

version has also been used successfully in a number of calculations of horizontal transport and mixing processes in rivers, bays and similar water bodies. The paper shows how empirical information enters in this model and where the information is extracted from. It also demonstrates how sensitive the results are to certain of the empirical coefficients and gives an indication of the quality of the predictions.

## The Depth-Average Model

The aim of a depth-average calculation is to determine the distribution of the depth-average velocity components $\bar{U}$ and $\bar{V}$ and the depth-average quantity $\bar{\Phi}$ (which may stand for either temperature or concentration) by solving the following depth-average mean-flow equations:

$$\frac{\partial}{\partial x}(h\bar{U}) + \frac{\partial}{\partial y}(h\bar{V}) = 0 \tag{1}$$

$$\bar{U}\frac{\partial \bar{U}}{\partial x} + \bar{V}\frac{\partial \bar{U}}{\partial y} = -g\frac{\partial H}{\partial x} + \frac{1}{\rho h}\frac{\partial(h\bar{\tau}_{xx})}{\partial x} + \frac{1}{\rho h}\frac{\partial(h\bar{\tau}_{xy})}{\partial y} + \frac{\tau_{sx}-\tau_{bx}}{\rho h} \tag{2}$$

$$\bar{U}\frac{\partial \bar{V}}{\partial x} + \bar{V}\frac{\partial \bar{V}}{\partial y} = -g\frac{\partial H}{\partial y} + \frac{1}{\rho h}\frac{\partial(h\bar{\tau}_{xy})}{\partial x} + \frac{1}{\rho h}\frac{\partial(h\bar{\tau}_{yy})}{\partial y} + \frac{\tau_{sy}-\tau_{by}}{\rho h} \tag{3}$$

$$\bar{U}\frac{\partial \bar{\Phi}}{\partial x} + \bar{V}\frac{\partial \bar{\Phi}}{\partial y} = \frac{1}{\rho h}\frac{\partial(h\bar{J}_x)}{\partial x} + \frac{1}{\rho h}\frac{\partial(h\bar{J}_y)}{\partial y} + \frac{q_s}{\rho h} \tag{4}$$

where H is the water level above datum. These equations have been obtained by formally integrating the original three-dimensional equations over the water depth h. In general, this procedure introduces dispersion terms which express the vertical non-uniformity of the flow quantities. These terms have been neglected in Eqs. 2 to 4 by assuming that the water body is vertically well mixed so that the quantities are indeed nearly uniform in depth and that secondary motions are not essential. For the case of a buoyant discharge, the validity of this assumption will be discussed later in the paper.

The depth-averaging procedure introduces the bottom and surface shear stresses $\tau_b$ and $\tau_s$ and the surface heat flux $q_s$. Model assumptions need to be introduced for determining these quantities. Here only the bottom shear stress will be considered which is related to the depth-average velocity components by the usual quadratic friction law.

$$\tau_{bx} = \frac{1}{\cos\theta}c_f\rho\bar{U}(\bar{U}^2 + \bar{V}^2)^{1/2} \quad , \quad \tau_{by} = \frac{1}{\cos\theta}c_f\rho\bar{V}(\bar{U}^2 + \bar{V}^2)^{1/2} \tag{5}$$

$\tau_{bx}$ and $\tau_{by}$ are forces per projected horizontal area, $\theta$ is the inclination of the bed in the lateral direction and $c_f$ is a friction coefficient. How this can be determined will be discussed later. $\bar{\tau}_{xx}$, $\bar{\tau}_{xy}$ and $\bar{\tau}_{yy}$ appearing in the momentum equations 2 and 3 are depth-average turbulent shear stresses expressing the horizontal momentum transport by the turbulent motion. Similarly, $\bar{J}_x$ and $\bar{J}_y$ in the scalar-transport equation 4 are the depth-average heat or mass fluxes due to the turbulent motion. These quantities must be specified with the aid of a turbulence model. For this purpose, Rastogi and Rodi (6) have adapted the popular k-ε model for use in depth-average open-channel flow calculations. The k-ε model uses the eddy-viscosity/diffusivity concept which yields the following relations for the turbulent stresses and heat or mass fluxes:

$$\frac{\bar{\tau}_{ij}}{\rho} = \vartheta_t \left( \frac{\partial \bar{U}_i}{\partial x_j} + \frac{\partial \bar{U}_j}{\partial x_i} \right) - \frac{2}{3} \bar{K} \delta_{ij} \quad , \quad \frac{\bar{J}_i}{\rho} = \tilde{\Gamma}_t \frac{\partial \bar{\Phi}}{\partial x_i} \tag{6}$$

In analogy to the original k-ε model, Rastogi and Rodi (6) assume that the local depth-average state of turbulence can be characterized by two parameters, namely the turbulent energy and dissipation parameters $\bar{K}$ and $\tilde{\varepsilon}$. The eddy-viscosity $\vartheta_t$ and diffusivity $\tilde{\Gamma}_t$ can then be related to these parameters by dimensional analysis:

$$\vartheta_t = c_\mu \frac{\bar{K}^2}{\tilde{\varepsilon}} \quad , \quad \tilde{\Gamma}_t = \frac{\vartheta_t}{\sigma_t} \tag{7}$$

where $c_\mu$ and $\sigma_t$ are empirical constants, the later one being called turbulent Prandtl or Schmidt number. The variation of the turbulence parameters $\bar{K}$ and $\tilde{\varepsilon}$ is determined from the following transport equations:

$$\bar{U}\frac{\partial \bar{K}}{\partial x} + \bar{V}\frac{\partial \bar{K}}{\partial y} = \frac{\partial}{\partial x}\left( \frac{\vartheta_t}{\sigma_k} \frac{\partial \bar{K}}{\partial x} \right) + \frac{\partial}{\partial y}\left( \frac{\vartheta_t}{\sigma_k} \frac{\partial k}{\partial y} \right) + P_h + P_{kv} - \tilde{\varepsilon} \tag{8}$$

$$\bar{U}\frac{\partial \tilde{\varepsilon}}{\partial x} + \bar{V}\frac{\partial \tilde{\varepsilon}}{\partial y} = \frac{\partial}{\partial x}\left( \frac{\vartheta_t}{\sigma_\varepsilon} \frac{\partial \tilde{\varepsilon}}{\partial x} \right) + \frac{\partial}{\partial y}\left( \frac{\vartheta_t}{\sigma_\varepsilon} \frac{\partial \tilde{\varepsilon}}{\partial y} \right) + c_{\varepsilon 1}\frac{\tilde{\varepsilon}}{\bar{K}} P_h + P_{\varepsilon v} - c_{\varepsilon 2}\frac{\tilde{\varepsilon}^2}{\bar{K}} \tag{9}$$

where $\quad P_h = \vartheta_t \left[ 2\left(\frac{\partial \bar{U}}{\partial x}\right)^2 + 2\left(\frac{\partial \bar{V}}{\partial y}\right)^2 + \left(\frac{\partial \bar{U}}{\partial y} + \frac{\partial \bar{V}}{\partial x}\right)^2 \right] \tag{10}$

is the production of turbulent kinetic energy due to interaction of turbulent stresses with horizontal mean velocity gradients, and $c_{\varepsilon 1}$, $c_{\varepsilon 2}$, $\sigma_k$ and $\sigma_\varepsilon$ are further empirical constants. The source terms $P_{kv}$ and $P_{\varepsilon v}$ have been added to equations 8 and 9 in order to account for the significant turbulence production by the interaction of vertical velocity gradients near the bottom and the turbulent shear stress. This production is in addition to the production $P_h$ due to horizontal velocity gradients and depends strongly on the bottom friction. Because they are governed by near-bottom processes, the additional vertical-production terms are related to the resultant bottom shear stress through the friction velocity in the following way

$$P_{kv} = c_k \frac{U_*^3}{h} \quad , \quad P_{\varepsilon v} = c_\varepsilon \frac{U_*^4}{h^2} \tag{11}$$

where the friction velocity $U_* = [c_f(\bar{U}^2 + \bar{V}^2)/\cos\theta]^{1/2}$. The determination of the empirical constants $c_k$ and $c_\varepsilon$ will be discussed below.

## Determination of Coefficients

The coefficient $c_f$ in the friction law (5) depends of course on the roughness of the bed and, in the case of a smooth bed, also somewhat on the Reynolds number. For smooth beds, $c_f$ can be determined from the channel formula of Schlichting (9)

$$c_f = 0.027 \left( \frac{\nu}{\bar{U} R} \right)^{\frac{1}{4}} \tag{12}$$

where R is the hydraulic radius of the flow cross-section. For the high Reynolds number situations of practical importance it is usually sufficient to employ a constant $c_f$, and a typical value is 0.003. For rough beds, the friction coefficient can be calculated from Manning's law

$$c_f = \frac{n^2 g}{h^{1/3}} \qquad (13)$$

where n is the Manning roughness factor. Manning's relation (13) only provides some global estimate for the friction coefficient. For non-rectangular channels, $c_f$ may vary considerably along the wetted perimeter. Pavlovic (5) has developed a method for determining this variation and has made successful calculations for elliptic, trapezoidal and triangular channels. He further determined the integral value over the wetted perimeter at each cross-section in such a way that it corresponded to the total shear stress found from a one-dimensional backwater profile calculation. The details of this procedure are described in Ref. 5, and this has been used in the calculations of natural rivers to be presented later.

The constants $c_\mu$, $c_{\varepsilon 1}$, $c_{\varepsilon 2}$, $\sigma_k$, $\sigma_\varepsilon$ and $\sigma_t$ appearing in the depth average k-ε model have been taken from the parent model. The determination of these constants will now be discussed briefly. In grid turbulence, lateral gradients including the mean-velocity gradients are zero so that the lateral diffusion and also the production term P is zero. Further, the streamwise development of the turbulence is so slow that the streamwise diffusion can also be neglected so that only the first term on the left hand side and the last term on the right hand side in the k- and ε-equations 8 and 9 are left over so that $c_{\varepsilon 2}$ is the only constant appearing in these equations. The equations describe the decay of the turbulent kinetic energy k with distance from the grid, and $c_{\varepsilon 2}$ can therefore be determined directly from the measured decay rate and was found to lie in the range 1.8 to 2.0. For local-equilibrium shear layer situations, the dissipation rate ε can be replaced by the production $P = \nu_t (\partial U/\partial y)^2$ so that together with Eq. 6, which reads for shear layers $\tau_{xy}/\rho = \nu_t \partial U/\partial y$, there follows $c_\mu = (\tau_{xy}/\rho k)^2$. Measurements in these flows yielded $\tau_{xy}/\rho k \approx 0.3$ so that $c_\mu = 0.09$. In near-wall regions of shear layers, a logarithmic velocity profile prevails, P is approximately equal to ε and the left-hand side of the ε equation is negligible; with these conditions inserted, the ε equation reduces to the following relation between various constants

$$c_{\varepsilon 1} = c_{\varepsilon 2} - \frac{\kappa^2}{\sigma_\varepsilon \sqrt{c_\mu}} . \qquad (14)$$

where $\kappa$ is the von Karman constant appearing in the logarithmic velocity law. Relation 14 fixes the value of the constant $c_{\varepsilon 1}$ once the values of the other constants have been chosen. The diffusion constants $\sigma_k$ and $\sigma_\varepsilon$ were assumed to be close to unity and they as well as $c_{\varepsilon 2}$ were attained by computer optimization in such a way that well documented free shear layers (mixing layers, jets) were simulated well by the k-ε model. The values determined this way are listed in Table 1 below. The turbulent Prandtl-Schmidt number $\sigma_t$ was found from heat and mass transfer experiments in free and wall layers to vary from 0.5 in free shear flows to 0.9 in regions near walls. Since it is the immediate vicinity of jet-like discharges where this number has the greatest effect in a depth-average calculation (further downstream the diffusivity is governed mainly by the value of the dimensionless diffusivity $e^*$ to be introduced shortly) a value of $\sigma_t = 0.5$ was chosen.

## Table 1:   Constants in k-ε model

| $c_\mu$ | $c_{\varepsilon 1}$ | $c_{\varepsilon 2}$ | $\sigma_k$ | $\sigma_\varepsilon$ | $\sigma_t$ | $\kappa$ |
|---|---|---|---|---|---|---|
| 0.09 | 1.44 | 1.92 | 1.0 | 1.3 | 0.5 | 0.435 |

A sensitivity study has shown that the calculations are most sensitive to the values of $c_{\varepsilon 1}$ and $c_{\varepsilon 2}$; for example a 5 percent change of either $c_{\varepsilon 1}$ or $c_{\varepsilon 2}$ resulted in a 20 percent change of the spreading rate of a jet. On the other hand a change in the values of $\sigma_k$ and $\sigma_\varepsilon$ did not have a significant effect on the jet spreading and only altered the lateral profiles somewhat.

The coefficients $c_k$ and $c_\varepsilon$ appearing only in the depth-average version of the model (in Eq. 11) were determined by considering uniform flow in which all gradients are absent. In this situation, the k and ε equations 8 and 9 reduce respectively to

$$c_k \frac{U_*^3}{h} - \tilde{\varepsilon} = 0 \quad , \quad c_\varepsilon \frac{U_*^4}{h^2} - c_{\varepsilon 2}\frac{\tilde{\varepsilon}^2}{2\bar{k}} = 0 \quad . \tag{15}$$

In this case, the energy dissipation $\tilde{\varepsilon}$ is related to the energy slope, S, via $\tilde{\varepsilon}=Sg\bar{U}$, and the friction velocity $U_*$ to S via $U_*=\sqrt{Sgh}$. With this and the relation for $U_*$ given above, the elimination of k via Eq. 7 and the introduction of the dimensionless diffusivity $e*=\Gamma_t U_* h$, there follows from Eq. 15:

$$c_k = \frac{1}{\sqrt{c_\mu}} \quad , \quad c_\varepsilon = \frac{c_{\varepsilon 2}c_\mu^{1/2}}{(e^* \sigma_t)^{1/2}c_f^{3/4}} \tag{16}$$

The only new empirical input required is therefore the dimensionless diffusivity e*. Dye-spreading measurements in wide laboratory flumes have yielded a value of approximately 0.15 (recommended by Fischer et al (1) as average value) while similar experiments in real rivers yielded typically e*=0.6 which is the value recommended by Fischer et al (1) for practical purposes. It should be mentioned here that a diffusivity determined from such dye-spreading measurements accounts not only for the turbulent transport but also for the dispersion transport due to vertical non-uniformities of scalar quantities and velocity components that was neglected in the depth-average scalar transport equation 4. When significant secondary motions in cross-sectional planes are present, a relatively small non-uniformity of temperature or concentration may cause relatively large dispersion contributions to this equation. In natural rivers, such secondary motions may arise from large-scale irregularities in the river bed and, of course, also from river bends. As will be shown below, the value of 0.6 appears suitable for applications of the model to natural rivers.

### Test Calculations

A few calculations will now be presented that show the performance of the model and its sensitivity to the input parameters. The first two examples concern idealized flow situations in laboratory flumes, and here a value e*=0.15 has been used whereas the latter two examples are applications to natural rivers. In these cases, e*=0.6 has been used, but in one case the sensitivity to the value of e* was investigated. The first example is the side discharge of cooling water into rectangular channel flow as sketched in Fig. 1, which also shows calculated streamlines for one particular case. Owing to the restriction of entrainment of the discharge jet by the near bank, a recirculation

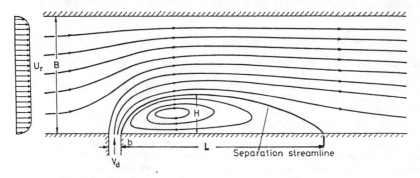

**Fig. 1:** Side-discharge situation and calculated streamline pattern

region with decreased surface elevation if formed behind the jet. Ref. 3 shows that dependence of the size of this region on the ratio of jet to channel momentum is predicted correctly and also the trajectory of the jet and the isotherms. Here, attention is focused on the dependence of the recirculation-eddy-size parameters on various input quantities. For one particular situation ($V_d^2 b/U_r^2 B=1$, $V_d/U_r=3.16$, $b/B=0.105$), Table 2 shows the response of the results to changes in the friction coefficient $c_f$ and in the

**Table 2:** Model input sensitivity test for side-discharge calculations

| Run | Model details | H/B | L/B |
|---|---|---|---|
| 1 | k-ε model, $c_f$ = 0.003 | 0.33 | 1.8 |
| 2 | as 1 but $c_f$ = 1/2 $c_f$ | 0.34 | 1.9 |
| 3 | as 1 but $c_f$ = 2 $c_f$ | 0.324 | 1.71 |
| 4 | as 1 but $\tilde{v}_t$ = 0 | 0.43 | 3.0 |
| 5 | as 1 but $\tilde{v}_t$ = ν | 0.43 | 3.0 |

eddy-viscosity $\tilde{v}_t$. Halfing or doubling the original value of $c_f$=0.003 can be seen to cause less than 5 percent change in the recirculation eddy parameters. This insensitivity is due to the fact that the channel bed was considered smooth, in which case bottom friction does not play a major role. In run 4 physical diffusion was put to $O(\tilde{v}_t=0)$ and in run 5 it was kept small by replacing $\tilde{v}_t$ by the laminar viscosity ν. The results of these two runs are identical. This indicates that there was (unintended) numerical diffusion in the calculations that was considerably larger than the laminar diffusion so that it made no difference whether the physical diffusion was equal to zero or was given a small number. Ref. 3 discusses at length that the numerical diffusion is however significantly smaller than the turbulent diffusion in those regions where diffusion is important. A comparison of runs 4 and 5 with run 1 shows a significant sensitivity to the model used for the physical viscosity. It was found that $\tilde{v}_t$ determined from the k-ε model was 20 to 1000 times larger than the laminar viscosity $v$, depending on the position in the flow. Some calculations were carried out with a constant value $\tilde{v}_t=200v$, and for the flow conditions given above, this produced eddy parameters very similar to those obtained with the k-ε model. For other momentum flux ratios however this value of $\tilde{v}_t$ gave very poor agreement with the experimental eddy parameters. This indicates that a constant turbulent viscosity is too crude an assumption to use for the calculation of the near field of discharges. A site-specific coefficient would have to be used and the superiority of the k-ε model in such situations becomes apparent as this model can cope with a much wider range of applications with the same empirical constants.

In the second example, the influence of secondary motions on the warm water spreading in channel flows is examined under varying bed roughness conditions. To this end, calculations were carried out for a full-depth co-axial slot discharge of heated water at the center of a rectangular channel. This situation was chosen because it could be calculated also without excessive effort with a three-dimensional method (see Ref. 6) in which secondary motions can be recovered. Such motions develop because the discharged warm water rises to the surface where it is deflected and and moved outward. By continuity, cold ambient water has to move inward near the bottom. This buoyancy-induced motion in the cross-sectional plane may be important for the lateral warm water spreading if the densimetric Froude number characterizing the bouyancy effects is low. That this is indeed the case is shown on the left hand side of Fig. 2 which presents depth-averaged isotherms obtained with both the 3D and the 2D depth-average model for the case of the densimetric Froude number $F=U_d/(gh(\rho_r - \rho_d)/\rho_r)^{1/2}=5$. The 3D model can be seen to produce much more spreading than the depth-average model. In the later the effect of secondary motion would enter through the dispersion terms, but since these have been neglected in the present calculations, there is no effect of the secondary motion at all. The left hand side of Fig. 2 is for a smooth channel bed. When the calculations were repeated for a rough bed with a Manning factor of n=0.025 as shown on the right hand side of Fig. 2, the 2D and 3D methods produced fairly similar results because in this case strong vertical mixing reduces the secondary motion and the vertical temperature differences, and thus the effect

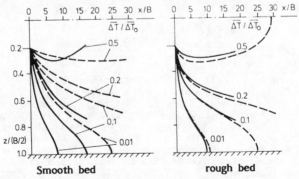

Fig. 2:   Comparison of 2D depth-average (- - -) and 3D(——) predictions for co -axial center discharge

of buoyancy. Since most real rivers have fairly rough beds, it appears that the depth-average procedure is sufficient for most problems of discharges into relatively fast moving streams. It is also clear from Fig. 2 that, when the 2D calculations for the smooth and the rough bed are compared, the rough bed causes faster spreading of the $_\sim$warm water. This is because, with the same value for e*, the diffusivity $\tilde{\Gamma}_t$ is larger in the rough channel because the friction velocity $U_*$ is larger.

The next 2 examples concern discharges into natural rivers, the first being a simulation of the discharge from the Stuttgart sewage plant into the river Neckar. The details of this study are described in Refs. 5 and 7. The discharge is located between two locks and the river is relatively slow flowing (0.15 m/s) with a discharge of 24 m³/s. The discharge from the sewage plant was 2.4 m³/s perpendicular to the river bank. In a measurement campaign, the sewage discharge was seeded with Rhodamin B, and the concentration profiles as well as velocity profiles were measured at certain cross-sections downstream of the discharge. The actual measurement points are shown in Fig. 3 but they had to be corrected because they are in conflict with the laws of conservation of mass and species, which is mainly due to the fact that Rhodamin B attaches itself to suspended particles and settles with these to the river bed. A 1D backwater profile calculation was carried out first which yielded the Manning factor n=0.035 and the total shear stress distribution as input for the depth-average calculation. The velocity and concentration profiles obtained from these calculations are shown in Fig. 3. Calculations with both e*=0.15 and e*=0.6 as empirical inputs are included. When a value of 0.15 is used, the spreading of the warm water plume is predicted too small, but this is significant only in the far field. In the near field, the turbulence generated by the discharge jet seems to dominate over that generated by bed shear so that the value of e* has little influence. In the far field the value e*=0.6 produces satisfactory agreement with the corrected measurements. It should be mentioned that the value of e* had very little influence on the velocity profiles which appear well predicted, including the near field where a recirculation region develops.

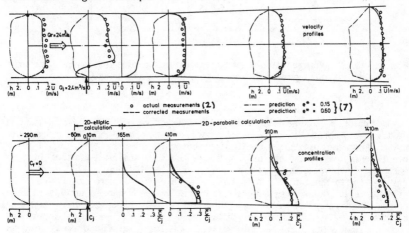

Fig. 3: Depth-average velocity and concentration profiles in Neckar near discharge of Stuttgart sewage plant

The last example is presented here in order to demonstrate that the same model with the same constants works well also for an entirely different discharge situation. Here, the cooling-water discharge from the power station Piacenza into the river Po is considered where the river velocity is of the order 1.5 m/s, the river discharge is 440 m³/s, and 25 m³/s of cooling water are rejected at an excess temperature of 8°C. Detailed velocity and temperature measurements were taken in the near field as reported in Ref. 8. The details of the calculations are described in Ref. 5. The 1D backwater profile calculations yield a Manning's factor n=0.035. The velocity and temperature profiles obtained with the depth-average method are compared with the field measurements in Fig. 4. The agreement can be seen to be very satisfactory, and it should be noted that in this case no recirculation region develops because the discharge is at a relatively small angle to the main stream. Ref. 5 reports on further applications of exactly the same model to cooling water discharges into the rivers Rhine and Danube, and the results were equally satisfactory in these cases also.

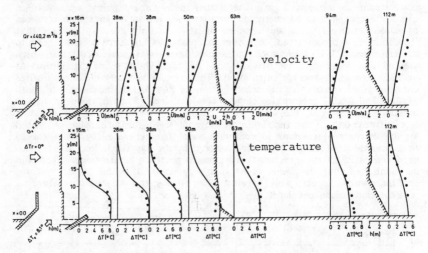

Fig. 4: Depth-average velocity and temperature profiles in Po near discharge of Piacenza power station, — calculations (5), data (8)

## CONCLUSIONS

For the particular case of the depth-average version of the k-ε turbulence model, the calibration and the input sensitivity was discussed. This model does not use constant exchange coefficients but calculates the variation of these coefficients over the flow field with the aid of transport equations for two parameters characterizing the local state of turbulence. The model was developed with the aim that it should not require case-to-case calibration. The evidence collected so far suggests that the model is indeed applicable to a much wider range of situations than are models using constant exchange coefficients. The dimensionless diffusivity e* enters as empirical input, and for laboratory situations a value of e*=0.15 was found suitable while for natural rivers e*=0.6 has been found to work well for a number of entirely different sites. Of course, no absolute universality can be claimed and there may indeed be rivers with strong geometrical irregularities and bends where a different value for e* is necessary, but the value of 0.6 appears to work well for many rivers and is perhaps the best starting point in the absence of any further information.

## ACKNOWLEDGEMENTS

This paper was written while the author was a visitor at the University of California at Davis. The support of the University of California is greatfully acknowledged. The author should also like to thank Nancy Nelson for typing the manuscript.

## REFERENCES

1. Fischer, H.B., List, E.J., Koh, R.C.Y., Imberger, J., and Brooks, N.H., Mixing in Inland and Coastal Waters, Academic Press, Inc., New York, 1979.
2. Kaser, F., Todten, H., Hahn, H.H., und Kupper, L., "Naturmessungen uber die Konzentrationsverteilung von konservativen und suspendierten Wasserinhaltstoffen unterhalb einer Einleitung in einen Fluss", University of Karlsruhe, Report SFB 80/ME/157, 1980.
3. McGuirk, J.J. and Rodi, W., "A Depth-Averaged Mathematical Model for the Near Field of Side Discharges into Open Channel Flow", J. Fluid Mech., 86, 1978 pp. 761-781.
4. Mikhail, R., Chu, V.H. and Savage, S.B., "The Reattachement of a Two-Dimensional Turbulent Jet in a Confined Cross Flows", Proc. 16th IAHR Congress, Sao Paulo, Brazil, Vol. 3, 1975, pp. 414-415.
5. Pavlovic, R.N., "Numerische Berechnung der Warme-und Stoffausbreitung in Flussen mit einem tiefengemittelten Modell", Ph.D. thesis, University of Karlsruhe, 1981.
6. Rastogi, A.K. and Rodi, W., "Prediction of Heat and Mass Transfer in Open Channels", Journal of the Hydraulics Division, ASCE, HY3, 1978.
7. Rodi, W., Pavlovic, R., and Srivatsa, S.K., "Prediction of Flow and Pollutant Spreading in Rivers", in Transport Models for Inland and Coastal Waters, ed. H.B. Fischer, Academic Press, 1981.
8. Rota, A. and Zavatorelli, R., "Misure sistematiche delle temperature e delle velocita del fiume Po in prossimita dello scarico dell' acqua dei condensatori della centrale termoelecttrica ENEL di Piacenza-Periodo agosto-settembre 1972", MISTER No. 5A/B, CISE-Centro Informazioni Studi Esperienze, Milano, 1973.
9. Schlichting, H., Boundary Layer Theory, McGraw Hill, New York, 1960.

# PROPERTIES OF DEPOSITED KAOLINITE IN A LONG FLUME

By Ashish J. Mehta,[1] M. ASCE, Emmanuel Partheniades,[2] M. ASCE,
Jayawantkumar G. Dixit,[3] and William H. McAnally,[4] M. ASCE

## ABSTRACT

Recent results from three laboratory tests on the deposition of kaolinite in water in a 100 m long flume are reported. In each test, suspended sediment was allowed to deposit in the flume at selected values of discharge and depth of flow. The undeposited portion of the sediment passed out of the flume at the downstream end. Sediment accumulated in each 12.2 m segment of the 100 m long deposited bed was next resuspended in a 1.5 m diameter and 0.21 m wide annular rotating flume, and was allowed to deposit under a constant applied bed shear stress. The rates of deposition are found to follow a law similar to that derived from basic laboratory investigations of the depositional behavior of cohesive sediments reported previously. Parameters characterizing this law are found to exhibit significant variation along the length of the 100 m long flume, implying a corresponding variation in the rates of deposition of the suspended aggregates. Differences in aggregate properties resulting from corresponding differences in the particulate composition of the aggregates appear to be a causative factor. The results elucidate a mechanism for sediment sorting which commonly occurs in muddy estuaries.

## INTRODUCTION

Where a suspended sediment-laden tributary stream meets a relatively wide estuarial water body, some of the kinetic energy of the stream is dissipated, and deposition of the suspended sediment occurs over some reach of the estuary. A similar phenomenon also occurs on the estuarial shelf in the sea, where the estuary itself deposits its sediment load. The result is sorting in the longitudinal direction, in terms of physical and physico-chemical properties characterizing the sediment. Particle sorting (and associated sediment mineral sorting) in

---

[1]Assoc. Prof., Dept. of Coastal and Oceanographic Engrg., Univ. of Fla., Gainesville, FL.

[2]Prof. of Hydr. Structures, Aristoteles Univ. of Thessaloniki, Greece; and Prof., Dept. of Engrg. Sci., Univ. of Fla., Gainesville, FL.

[3]Grad. Assist., Dept. of Coastal and Oceanographic Engrg., Univ. of Fla., Gainesville, FL.

[4]Res. Hydr. Eng., Estuaries Div., Waterways Expt. Sta., Vicksburg, MS.

the estuarial depositional environment is a well-documented phenomenon (1).

When the suspended load is composed of cohesionless sediment, the degree of sorting in the deposit can be determined directly from standard granulometric and mineralogical analyses of bottom samples. In the case of cohesive sediments, flocculation of the primary particles due to salts in the fluid, and aggregation, i.e. build up of flocs in the turbulent environment, enhance the complexity of the depositional process. In this case, longitudinal variation in the bed properties can be quantified by determining parameters that characterize: 1) the critical shear stress and the rate of resuspension of the bed sediment, or 2) depositional properties of the bed sediment which, for this purpose, must be entirely resuspended, initially. Either of these two procedures may be utilized with relatively small segments into which the total deposited bed is sub-divided. The variation of each parameter characterizing bed sediment resuspension or redeposition with distance would characterize the longitudinal variation of the erosive and depositional properties of the deposited sediment.

A basic laboratory investigation utilizing the second approach for demonstrating longitudinal sorting of the bed sediment is described here. Kaolinite clay was deposited in an approximately 100 m long flume. The bed was sub-divided into 12.2 m long segments. The redepositional properties of the sediment from each segment were investigated in an annular rotating flume. A brief description of the investigation follows. Details are reported elsewhere (1, 2).

APPROACH

In an extensive series of tests conducted on kaolinite and on naturally occurring estuarial muds, Mehta and Partheniades (3) showed that the variation of suspended sediment concentration, C, with time, t, during deposition under a steady, uniform, turbulent flow follows the log-normal law:

$$C* = \frac{1}{\sqrt{2\pi}} \int_{-\infty}^{T} e^{-\frac{w^2}{2}} dw \qquad (1)$$

where $C*$ = fraction of the total depositable sediment which is deposited at time t, $T = \log(t/t_{50})^{1/\sigma_2}$, $t_{50}$ = time at which $C* = 0.50$, $\sigma_2$ = standard deviation and w = a dummy variable. Given $C_0$ = initial suspended sediment concentration, $C* = (C_0-C)/(C_0-C_{eq})$, where $C_{eq}$ = steady state concentration corresponding to the non-depositable portion of the initially suspended sediment. The fraction $C_{eq}^* = C_{eq}/C_0$ also exhibited a log-normal dependence on the bed shear stress, $\tau_b$. This dependence may be expressed functionally as

$$C_{eq}^* = f[\sigma_1, (\tau_b*-1)/(\tau_b*-1)_{50}] \qquad (2)$$

where $\sigma_1$ = standard deviation of the log-normal relationship, $\tau_b* = \tau_b/\tau_{bmin}$, $\tau_{bmin}$ = value of $\tau_b$ below which $C_{eq} = 0$, i.e. the bed shear stress below which the entire amount of the initially suspended sediment deposits eventually, and $(\tau_b*-1)_{50}$ = value of $\tau_b*-1$ when $C_{eq}^* = 0.50$.

For a given fluid-sediment mixture, the parameters $\tau_{bmin}$, $(\tau_b*-1)_{50}$, $t_{50}$, $\sigma_1$, and $\sigma_2$ characterize the rate of sediment deposition through Eqs. 1 and 2. The manner in which these five parameters vary with distance along the flume was of primary interest in the investigation.

APPARATUS, MATERIAL AND PROCEDURE

The 99.7 m long, 0.46 m deep and 0.23 m wide flume used for the preparation of the deposited bed is located at the U.S. Army Corps of Engineers Waterways Experiment Station, Vicksburg, Mississippi. The sediment-water slurry was prepared in a separate system and was injected into the flume headbay at the desired rate. Fig. 1 shows a portion of the flume near station 40, which was 13.1 m from the upstream end of the flume.

A description of the annular flume located at the University of Florida, in which redeposition tests for sediment characterization were conducted is found elsewhere (3). The two main components of this flume are: an annular fibreglass channel (0.21 m wide, 0.46 m deep, and 1.5 m in mean diameter) containing the fluid-sediment mixture, and an annular ring of slightly smaller width positioned within the channel and in contact with the fluid surface. A simultaneous rotation of the ring and the channel in opposite directions generates a uniform turbulent shear field free from the influence of secondary currents. In the system there are no aggregate-disrupting elements such as pumps and diffusors in which very high shearing rates usually prevail.

A commercial kaolinite with a cation exchange capacity of approximately 13 milliequivalents per hundred grams was used in the tests. The fluids were: a) water with a chloride concentration of 18 ppm, pH = 7.8 and Sodium Adsorption Ratio = 2.07 in the 100 m flume, and b) deionized water plus a small amount of water from the 100 m flume (which occurred as pore fluid in each bed sample) in the annular flume.

The procedure in the 100 m flume was as follows: 1) a sediment-water slurry of approximately 100,000 ppm concentration was prepared by mixing kaolinite with water over a period of 3 hours. 2) the slurry was injected into clear water which flowed from the headbay to the flume. The injection rate was controlled so as to obtain the desired sediment concentration in the flume in which the depth was maintained at approximately 15 cm. 3) slurry injection was continued at a near-constant rate for a period ranging from 1 to 3 hours. 4) clear water flow was then maintained until the suspended sediment "cloud" had passed out of the flume. 5) the flume was drained slowly over a period ranging from 15 to 36 hours to near the bed surface, without allowing the bed to become dry. 6) approximately 400 to 1,000 gm of bed samples were removed by suction at 12.2 m intervals and stored.

The procedure in the annular flume using each stored bed sample was as follows: 1) after introducing a pre-determined amount of sediment from the sample into the annular flume, deionized water was added in order to attain a flow depth of 23 cm. 2) the mixture was brought into suspension by rotating the flume at a high speed for a period of 1 hour. At this speed the bed shear stress was approximately 1.5 N/m$^2$. 3) the flume speed was lowered to a particular "depositing" bed shear stress, $\tau_b$. 4) the sediment was allowed to deposit for a period ranging

Fig. 1. A View of 100 m Long Flume near Station 40, 13.1 m Downstream.

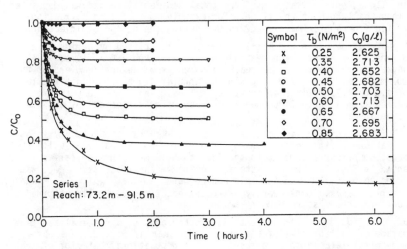

| Symbol | $\tau_b$ (N/m²) | $C_0$(g/$\ell$) |
|--------|-------------|------------|
| x | 0.25 | 2.625 |
| ▲ | 0.35 | 2.713 |
| □ | 0.40 | 2.652 |
| ○ | 0.45 | 2.682 |
| ■ | 0.50 | 2.703 |
| ▽ | 0.60 | 2.713 |
| ● | 0.65 | 2.667 |
| ◇ | 0.70 | 2.695 |
| ◆ | 0.85 | 2.683 |

Series 1
Reach: 73.2 m – 91.5 m

Fig. 2. Normalized Concentration, $C/C_0$, against Time t. Data from Annular Flume for Nine Values of Bed Shear Stress, $\tau_b$, Reach 73.2-91.5 m, Series 1.

from 6 to 23 hours while maintaining a constant $\tau_b$. 5) for each sediment sample, several deposition tests were carried out by varying $\tau_b$ from 0.15 to 0.85 N/m². 6) the grain size distribution of the sediment from each sample was measured using a standard hydrometer test.

RESULTS

A summary of conditions for the three reported tests in the 100 m flume is given in Table 1. In terms of the nominal initial concentration and water and slurry temperatures, tests 1 and 2 had similar conditions. However, the nominal flow rate and the slurry injection times were different. Conditions in test 3 were different from those in tests 1 and 2. The total volume of fluid through the flume during the injection period, obtained as the product of the flow rate and injection time, was maintained within a relatively narrow range of 28.1 to 36.7 m³, in the three tests.

TABLE 1. - Summary of Test Conditions in 100 m Flume

| Test No. (1) | Nominal Flow Rate (m³/sec) (2) | Nominal Initial Conc. (ppm) (3) | Injection Time (sec) (4) | Temperature | |
|---|---|---|---|---|---|
| | | | | Water (°C) (5) | Slurry (°C) (6) |
| 1 | 0.0070 | 10,000 | 3600 | 24 | 27 |
| 2 | 0.0052 | 10,000 | 5400 | 26 | 27 |
| 3 | 0.0034 | 5,000 | 10800 | 23 | 23 |

The shear stress, $\tau_0$, during deposition in tests 1, 2 and 3 was approximately 0.3, 0.2 and 0.1 N/m², respectively. These values are based on the wetted perimeter. As noted, due to the limited length of the flume, only a fraction of the total sediment injected was deposited along the length of the flume. For example, in test 1 this fraction varied from approximately 0.2 to 0.4 during the period of injection.

Experiments in the annular flume are identified by series 1, 2 and 3, which correspond with tests 1, 2, and 3 in the 100 m flume. Bed segments are identified by reach measured from the upstream end of the 100 m flume. The value of $C_0$ in the annular flume was within the range of 1.8 to 4.7 gm/liter in the three series. The fluid temperature varied from 12°C to 27°C, due to corresponding variations of the ambient temperature.

As a typical example of the obtained results, time-concentration variations during deposition for nine different values of $\tau_b$ using the bed segment corresponding to the 73.2 to 91.5 m reach from series 1 are shown in Fig. 2. The characteristic minimum shear stress, $\tau_{bmin}$, was found to be 0.180 N/m² for this bed segment, so that $\tau_b > \tau_{bmin}$ in each of the nine cases. Consequently, as observed, C(t) eventually attained a non-zero steady state concentration, $C_{eq}$.

In Fig. 3, C*, in percent, is plotted against the normalized time parameter, T, for five of the nine tests shown in Fig. 2. The observed log-normal relationship representing the depositional behavior of the sediment is expressed by Eq. 1. All the obtained results were found to conform to this relationship.

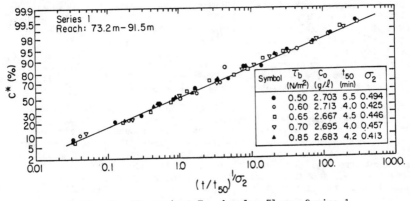

Fig. 3. C*, against T. Annular Flume, Series 1.

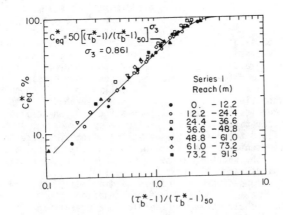

Fig. 4. $C^*_{eq}$ against $(\tau_b*-1)/(\tau_b*-1)_{50}$, Series 1.

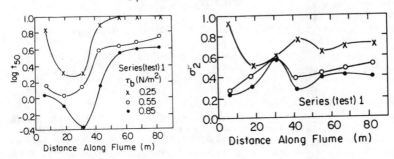

Fig.5a) $\log t_{50}$ against Distance, Series 1.

Fig. 5b) $\sigma_2$ against Distance, Series 1.

The dependence of $C_{eq}^*$, in percent, on the normalized bed shear stress parameter, $(\tau_b^*-1)/(\tau_b^*-1)_{50}$, is shown in Fig. 4 for all reaches in series 1. In the functional form this relationship is given by Eq. 2. The exponent $\sigma_3 = 0.861$ of the log-log relationship (Fig. 4), valid for all $C_{eq}^* < 90\%$, was found to be constant for all the reaches in series 1. The corresponding values in series 2 and 3 were 0.578 and 0.509, respectively. In the range, $90\% < C_{eq}^* < 100\%$, the curve is observed to become asymptotic to the line respresented by $C_{eq}^* = 100\%$.

In Figs. 5a, 5b, 6a and 6b, the variations of $t_{50}$ (in minutes), $\sigma_2$, $\tau_{bmin}$ and $(\tau_b^*-1)_{50}$ with distance along the 100 m flume are plotted. These results are based on series 1 measurements utilizing bed segments with reaches given in Fig. 4. Considering $t_{50}$, i.e. the "half time" for deposition, the variation of $\log t_{50}$ (and $\sigma_2$) with $\tau_b$ can be examined from Fig. 5a. For a given distance, $\log t_{50}$ decreases with increasing $\tau_b$, implying a corresponding increase in the rate of deposition. Values of $\tau_b^*=\tau_b/\tau_{bmin}$ were all greater than unity. The observed trend of decreasing $\log t_{50}$ with increasing $\tau_b^*$ is consistent with the earlier results of Mehta and Partheniades (3). Since the procedure for the initial mixing of sediment in the annular flume was the same for all the tests, the trend between $\log t_{50}$ and $\tau_b^*$ must be attributed to the process of aggregation of the sediment during the depositional phase of the experiments. In the range of suspended sediment concentrations in the annular flume, internal shearing due to local velocity gradients, and to some extent differences in the settling velocities of the different-sized aggregates, were likely to be the dominant mechanisms for inter-particle collision (1). These mechanisms therefore controlled the sizes and the shear strengths of the aggregates. The results indicate that increasing $\tau_b^*$, which also implies increasing inter-particle collision frequency throughout the fluid column, increased the "effective" settling velocity of the "depositable" portion of the sediment due to one or both of the following causes: a) an increase in the settling diameter and therefore the settling velocity of the aggregates, and b) decrease in the reflection or re-entrainment of the depositing sediment in the near-bed boundary layer region with high shearing rates, as a result of the enhanced rate of formation of aggregates with shear strength exceeding the prevailing shear stress. The amount of depositable sediment itself, as represented by the fraction $1-C_{eq}^*$, decreased considerably with increasing $\tau_b$, as indicated by the relationship of Fig. 4. The non-depositable portion, represented by the fraction $C_{eq}^*$, is that portion of the initially suspended sediment which was unable to form aggregates of sufficient strength for them to withstand the near-bed shearing (3).

For all three values of $\tau_b$, the variation of $\log t_{50}$ with distance along the flume appears to be similar in Fig. 5a. The initial trend is one of decreasing $\log t_{50}$ followed by a "leveling off" beyond 50 m. A minimum $\log t_{50}$ occurred between 18 m and 32 m. It can be surmised that the sediment that deposited in this reach under test 1 had a particularly high effective settling velocity, w. Given: u = mean flow velocity in the flume, h = depth of flow and $\ell$ = distance to the point of deposition, $w = uh/\ell$. Selecting u = 0.187 m/sec based on the nominal discharge rate, h = 0.177 m and $\ell$ = 25 m ( mean of 18 m and 32 m), w = 0.0013 m/sec is obtained.

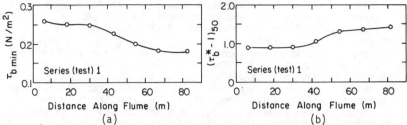

Fig. 6a) Variation of $\tau_{bmin}$ with Distance along 100 m Flume, Series 1.
Fig. 6b) Variation of $(\tau_b^*-1)_{50}$ with Distance along 100 m Flume, Series 1.

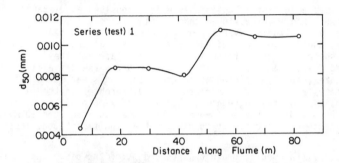

Fig. 7. Variation of Median Diameter, $d_{50}$, of Dispersed Kaolinite Particles with Distance along 100 m Flume, Series 1.

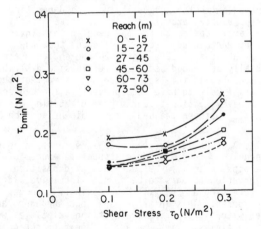

Fig. 8. Variation of $\tau_{bmin}$ with Shear Stress, $\tau_0$, during Deposition in 100 m Flume.

The significance of the variation of logt $_{50}$ with distance in Fig. 5a is with respect to a demonstration of the longitudinal variation of the deposited sediment properties in quantitative terms. It can be shown that the trends exhibited by $\sigma_2$, $\tau_{bmin}$ and $(\tau_b*-1)_{50}$ in Figs. 5b, 6a and 6b, respectively, corroborate the conclusions derived from Fig. 5a. In general, the shearing rates in the fluid column were likely to have been different in the two flumes due to two reasons: a) the bed shear stress ranges were different, and b) the flow structure, although turbulent, was different (3). As a result the sizes and the shear strengths of the aggregates could have been expected to be different. It is however evident from the annular flume results, that the observed longitudinal variation of the depositional properties is an indication of corresponding variation which occurred during deposition in the 100 m flume itself.

For example in test 1, aggregates which deposited at distances in excess of 25 m had w values lower than 0.0013 m/sec. This variation in w, implying a longitudinal variation of the deposited sediment properties, could have occurred due to two causes: a) changes in the sizes and shear strengths of the suspended aggregates as they were transported downstream, due to disaggregation under the prevailing shearing rates, and b) differences in the sizes and shear strengths of the aggregates resulting from corresponding differences in the particulate composition of the aggregates. Since, for example, in test 1 the mean residence time of the sediment in the flume was less than 10 min, the first cause does not appear to have been a significant contributing factor.

In Fig. 7, the median grain diameter, d $_{50}$, of the dispersed sediment is plotted against distance along the 100 m flume for test 1. These diameters may be compared with the much smaller diameter, 0.0011 mm, of the injected sediment. It is concluded that the 60 to 80% of the injected sediment which passed out of the flume consisted predominantly of the clayey portion, with the more silty portion depositing on the bed. In previously reported experiments on kaolinite deposition in an annular flume similar in design to the one used in this investigation but smaller in size, Partheniades et al. (4) noted that the suspended sediment aggregates at steady state contained the entire grain size range of the initially suspended sediment. Grain size distributions of the bed sediment measured in this study also suggested that the entire grain size range was likely to have been present in the bed sediment (1). There was, however, a significant degree of sorting (in terms of the median grain size) between the deposited sediment and the injected sediment.

Comparing Fig. 7 with Fig. 5a, there appears to be an approximate correlation between logt $_{50}$ and d $_{50}$ for distances in excess of 18 m. Larger grain size appears to imply slower rate of deposition. Although at first this observation appears to be counter-intuitive, it should be recognized that decreasing d $_{50}$ implies increasingly stronger interparticle cohesive forces. This in turn means that stronger, and perhaps larger, aggregates with higher effective settling velocities would be composed of finer particles. Figs. 5a and 7 appear to corroborate this hypothesis.

In Fig. 8, values of $\tau_{bmin}$ for all reaches in series 1, 2 and 3 are plotted against the shear stress, $\tau_0$, during deposition in tests 1, 2 and 3 in the 100 m flume. This plot is a demonstration of the influence of the depositional conditions in the 100 m flume, defined by $\tau_0$, on the properties of the deposited sediment. Overall, there is a trend of increasing $\tau_{bmin}$ with $\tau_0$. It has been shown previously (3), that increasing $\tau_{bmin}$ implies increasing rate of deposition of the sediment. According to Fig. 8, as $\tau_0$ increased, the redepositional rate of the deposited sediment increased, overall.

CONCLUSIONS

1) sediment deposited in the 100 m flume under a constant shear stress, $\tau_0$, showed a marked variation of the redepositional properties (as evaluated from the annular flume experiments) with downstream distance. This implies that longitudinal sorting occurred during the depositional process in the 100 m flume. 2) relatively strong and large aggregates with higher "effective" settling velocities (leading to deposition) were deposited in approximately the first 30 m reach. These aggregates were composed of sediment particles comprising the grain size range of the initially injected sediment, but were significantly finer, and therefore, more cohesive. 3) relatively weaker and smaller aggregates with lower effective settling velocities were deposited further downstream. The particles in these aggregates were coarser, and therefore less cohesive. 4) increasing the shear stress, $\tau_0$, during deposition in the 100 m flume caused the redepositional rates of the deposited sediment to increase.

ACKNOWLEDGEMENT

The work was funded by the U.S. Army Corps of Engineers research program: Improvement of Operation and Maintenance Techniques. The Chief of Engineers has granted permission to publish this paper.

REFERENCES

1.  Dixit, J. G., "Redepositional Properties of Kaolinite Deposited in a Long Flume," Report UFL/COEL-82/002, Coastal and Oceanographic Engineering Department, University of Florida, Gainesville, FL, 1982.

2.  McAnally, W. H., "Flume Experiments with Cohesive Sediments," Technical Report, U.S. Army Corps of Engineers Waterways Experiment Station, Vicksburg, MS, 1982.

3.  Mehta, A. J., and Partheniades, E., "An Investigation of the Depositional Properties of Flocculated Fine Sediments," Journal of Hydraulic Research, Vol. 13, No. 4, 1975, pp. 361-381.

4.  Partheniades, E., Kennedy, J. F., Etter, R. J., and Hoyer, R. P., "Investigations of the Depositional Behavior of Fine Cohesive Sediments in an Annular Rotating Channel," Report No. 96, Hydrodynamics Laboratory, Department of Civil Engineering, Massachusetts Institute of Technology, Cambridge, MA, June, 1966.

# CONFIRMATION OF WATER QUALITY SIMULATION MODELS

## BY

### Kenneth H. Reckhow[1]

ABSTRACT

Water quality simulation models, whether descriptive or predictive, must undergo confirmatory analyses if inferences drawn from the models are to be meaningful. Current practices in the confirmation of simulation models are examined and criticized from this perspective. In particular, labeling this process verification or validation (truth) probably contributes to the often inadequate efforts since these states are unattainable. The evaluation of scientific hypothesis, or water quality simulation models, may proceed according to inductive logic, the hypothetico-deductive approach, or perhaps according to a falsification criterion. The result of successful testing is at best confirmation or corroboration, which is not truth but rather measured consistency with empirical evidence. On this basis a number of statistical tests are suggested for model confirmation. The major difficulty to overcome, before confirmation becomes meaningful, is the generally inadequate data for establishing rigorous statistical tests.

INTRODUCTION

Mathematical model development begins with the conceptualization of the functions and relationships of the characteristics of the issue or system under study and proceeds with specification of the mathematical relationships, estimation of the model parameters, and validation of the model as a reliable representation. The process may be iterative. All steps are important; the validation step, however, may be most important because it provides confirmation (or lack thereof) that the conduct of the other steps resulted in a reliable model. Ironically, validation is also the step that most often is conducted inadequately in water resource model development.

In fact, validation, or the ascertainment of truth, is inconsistent with the logic of scientific research. As Anscombe (1967) notes, "The word valid would be better dropped from the statistical vocabulary. The only real validation of a statistical analysis, or of any scientific enquiry, is confirmation by independent observations." The testing of scientific models may be considered an inductive process, which means that, even with true premises, at best we can assign high probability to the correctness of the model. Philosophers of science have long debated the appropriate criteria for the effectiveness of arguments of this nature, considering characteristics

---

[1] School of Forestry and Environmental Studies, Duke University, Durham, North Carolina 27706.

such as the severity of tests and the goodness-of-fit.

How can this be translated into statistical terms for practical applications? Generalization of verifying criteria is possible only to a limited extent; beyond that, issue-specific criteria must be determined. Still, guidelines may be proposed for the composition of tests that are rigorous and for the selection of goodness-of-fit tests and acceptance levels. Both model developers and model users should benefit from careful consideration and application of criteria for model confirmation.

These issues are not strictly of academic interest. In the past 15 years, a number of water quality simulation models have been developed and then promoted as predictive methods to aid in the management of environmental quality. In most cases, however, these models have not been subjected to a rigorous validation or confirmation. Therefore, the model user often has no assurance that the model will yield reliable and informative predictions. This has potentially serious consequences since the inadequately confirmed planning model may lead to the implementation of economically inefficient or socially unacceptable water quality management plans. It is the purpose of this paper to outline some philosophical and statistical issues relevant to the problem of confirmation, and then to recommend appropriate applications of statistical confirmatory criteria.

PHILOSOPHY AND SCIENTIFIC CONFIRMATION

Until the 1950s, virtually all scientists and philosophers of science viewed the advancement of science and the scientific method as endeavors dominated by empiricism and logic. The empiricism of Hume and the deductive logic of Russell and Whitehead formed the basis for the approaches of the logical positivist and, later, the logical empiricist. In particular, the logical empiricists have enjoyed widespread support during the twentieth century.

Logical empiricism (see Hempel 1965) is based on the presupposition that observations and logic advance scientific knowledge. For example, when a scientific hypothesis (model) is proposed under logical empiricism, observations of relevant phenomena are acquired, and inductive or deductive logic is used to determine the degree of confirmational support. Inductive arguments, we may recall, cannot strictly be proven as true, but at best can be assigned a high likelihood of being correct. In statistical inference, inductive arguments often are associated with reasoning from the specific to the general. Deductive logic (reasoning from the general to the specific), on the other hand, must yield true conclusions if the premises are true and the arguments are valid.

Logical empiricists are divided on the importance and appropriate applications of deduction and induction in science. For example, under the hypothetico-deductive approach (Kyburg 1970), a scientific hypothesis is proposed and criteria that can be tested are deduced logically. The scientist then must be concerned with constructing rigorous tests which, depending on rigor and the results of testing, confer a degree of confirmation upon the hypothesis. When competing hypotheses are offered, philosophers have recommended acceptance of the simplest one that is consistent with the empirical evidence, possibly because it is most probable (Kyburg 1970).

Inductive logic, on the other hand, is important in a class of problems concerned with statistical explanation (Salmon 1971). Scientists and philosophers who subscribe to this approach argue that there are many scientific analyses in which the information content of the conclusion exceeds that of the premises. In those circumstances, inductive logic is appropriate, and we, at best, can assign high probability to the conclusion based on the misses.

Popper (1968) has proposed a variation on the hypothetico–deductive approach that has undergone a variety of interpretations since its introduction (Brown 1977). Popper rejects the notion that induction should be called logic, since the nature of induction is to support a conclusion that contains more information than the premises. Therefore, if we accept the logical empiricist view that science is based on logic, then deductive logic is necessary. Consistent with the hypothetico–deductivists, Popper requires the deduction of observational consequences of a scientific hypothesis; but in a break from previous thought, he bases scientific knowledge on a criterion of falsification rather than confirmation. This means, according to Popper, that scientific statements are distinguished, not by the fact that they can be confirmed by observation, but rather by the fact that they can be falsified by observation. Popper believes that candidate hypotheses should be subjected to severe tests, and from among the successful hypotheses, the one that is deemed most falsifiable is the one that tentatively should be accepted. Although this at first may seem counter–intuitive, it is reasonable since, following the application of severe tests, the hypothesis that was most likely to be falsified yet survived is the hypothesis receiving the greatest empirical support. Popper then would say that this highly falsifiable hypothesis had been corroborated through the application of rigorous tests. Like confirmation, corroboration has a vague quantitative meaning associated with the severity of the applied tests and the degree of success.

Independent of the preference we may have for a particular logical empiricist approach for evaluating scientific hypotheses, there are consistencies among the approaches that the scientist should note well. Without doubt, tests must be rigorous. This means that the hypothesis should be subjected to conditions that are most likely to identify its weaknesses or falsity. Mathematical simulation models must be tested with data that reflect conditions noticeably different from the calibration conditions; without this, there is no assurance that the model is anything more than a descriptor of a unique set of conditions (i.e., those representing the caliberation state). To assess the degree of confirmation or corroboration that a hypothesis or model should enjoy, a statistical goodness–of–fit criterion is necessary. Finally, the modeler should prepare a set of candidate formulations and then base the model choice in part on the relative performance of the models on statistical tests and on consistency with theoretical system behavior. Comparison of rival models/hypotheses is an important step in the testing of scientific hypothesis.

SOME PRACTICAL ISSUES

The selection of a statistical test for the confirmation of a mathematical model may be facilitated through consideration of the following issues:

1.  What characteristics of the prediction are of interest to the

modeler? The answer may be one or more of: mean values, variability, extreme values, all predicted values, and so forth. If one of the limited, specific responses is given, then the test statistical criterion should focus on that specific feature.

2. Is it intended that the model be primarily descriptive (identifying hypothesized cause-effect relationships) or primarily predictive? Different statistical tests are appropriate in each case.

3. What is the criterion for successful confirmation? In statistical inference, mean square error often is adopted, although many statistical tests (e.g., nonparametric methods) do employ other error criteria. In some situations, a decision theoretic approach such as regret minimization is warranted (see Chernoff and Moses 1959).

4. Are there any peculiar features to the model application of concern? This is a "catch-all" question intended to alert the model user to the fact that each application is unique, and therefore the confirmation process must be designed on that basis. For example:

   a. When prediction and observation uncertainty are considered, are all error terms quantified? It should be noted that model specification error rarely is estimated for water quality simulation models. This means that the corresponding prediction error is underestimated. Omission or mis-specification of any prediction/observation error terms will influence model confirmation.

   b. Are the assumptions behind any of the statistical tests violated? In particular, since time series or spatial series of data are often examined in model confirmation, autocorrelation may be a problem. When the validity of the statistical procedures is sensitive to a violated assumption (as is generally true for the assumption of independence), then some modifications or alternatives must be considered.

CONFIRMATION OF SIMULATION MODELS: LITERATURE REVIEW

Although there have been few, if any, rigorous attempts at confirmation of a water quality simulation model, this is not because of a complete lack of attention to this issue in the recent literature. General discussions on the importance of model confirmation or on confirmation as a step in simulation model development are noteworthy in this regard (Van Horn 1969, Naylor and Finger 1971, Mihram 1973, Davis et al. 1976, Caswell 1977). In addition, the Environmental Protection Agency recently sponsored a workshop on this topic (Hydroscience 1980). Unfortunately, the discussion groups convened as part of this workshop generally offered mixed or mild endorsement of rigorous statistical confirmatory criteria. While some of the papers presented at this workshop contained strongly-worded statements on the importance of confirmation (e.g., Velz) or presented statistical criteria for confirmation (e.g., Thomann, or Chi and Thomas), the workshop missed an opportunity to produce and to promulgate a set of confirmatory guidelines necessary for proper model development.

Despite the claims by some that statistical confirmatory criteria generally not be recommended because of the unique demands of each model application, there are in fact statistical tests that may be adapted to virtually any situation. A number of researchers (Mihram 1973, Shaeffer 1980, Thomann 1980, Thomann and Segna 1980, Thomann and Winfield 1976) have suggested various statistics useful for model confirmation. Thomann's work in particular stands out as one of the few statistical statements on confirmation specific to water quality simulation models. In the field of simulation modeling in hydrology, James and Burges (1981) have prepared a useful practical guide to model selection and calibration. Many of the statistical tests that they propose and apply to calibration are equally applicable for confirmation.

STATISTICAL METHODS FOR CONFIRMATION

Several statistical methods may be found useful for assessing the degree of confirmation of a mathematical model. Some of the more common techniques are listed in Table 1. Before selecting a technique (or, for that matter, before acquiring data for model development),

TABLE 1:   STATISTICAL METHODS FOR CONFIRMATION

1. Deterministic Modeling
    a. measures of error
    b. t-test
    c. Mann-Whitney-Wilcoxon test
    d. regression
    e. cross-correlation
    f. graphical comparisons - box plots
2. Stochastic Modeling
    a. deterministic methods for particular percentile
    b. probability density function "slices"
        i. Chi-square test
        ii. Kolmogorov-Smirnov test
        iii. comparison of moments
        iv. graphical comparisons - box plots

the investigator should consider the issues presented in previous sections of this chapter. For example, it is likely that model applications primarily are concerned with only certain features of the model. It is appropriate, then, for confirmation to focus on those features of concern. In addition, statistical assumptions must be considered. Common assumptions include normality, homogeneity of variance, and independence. Many procedures are robust to mild violations of the first two assumptions, but not to lack of independence. Transformations often may be applied to achieve approximate normality or to stabilize variance, while some robust and nonparametric procedures mentioned below may be useful under non-normality.

Violation of the independence assumption poses more difficult problems. Predictions and observations in water quality simulation often are time series, and autocorrelation may be present in one or both of these series. In a dependent (autocorrelated) series, the information content is less than in an equivalent-length, independent series because each data point to some degree is redundant with respect to the preceding point. This means that confidence intervals and significance tests that falsely assume independence will be overly optimistic, that is, the intervals will be too small.

Fortunately, when autocorrelation is found and quantified, there are some steps that can then be taken to permit application of many of the standard statistical tests. Yevjevich (1972) presents several relationships for

calculating the effective size, which is the size of an independent series
that contains the same amount of information as contained in the
autocorrelated series.  A less efficient but computationally easy alternative
to effective sample size is to use or to aggregate data covering intervals
greater than the period of autocorrelation influence.  For example, weekly
data may exhibit autocorrelation but monthly data may not; therefore, confine
the analysis to monthly data.

Following consideration of these application-specific and statistical
issues that help to determine the model terms and statistical methods to be
involved in confirmation, the investigator likely will employ one or more of
the techniques listed in Table 1.  Graphical examination of data sets or
series usually is a necessary part of any statistical analysis, and it
certainly can be helpful in model confirmation.  However, it is not listed in
Table 1 because of concern that confirmation will begin and end with a
graphical study and thus not advance beyond a visual comparison of predictions
and observations.

One view of model output and, hence, confirmation approaches leads to the
separate groupings of deterministic and stochastic modeling.  Most water
quality simulation models are deterministic, and this limits the set of
available statistical methods.  The trend toward error analysis in modeling is
important for model confirmation, although difficulties in the estimation of
model error may restrict the confirmation study.  Some of the methods listed
in Table 1 under deterministic modeling use aggregated data (prediction and
observation samples), and some of the methods are appropriate for data series.
Under "measures of error," we may include various weighting functions for the
difference between the prediction and the observations, such as the absolute
value or the square of the difference.

However, to assess the degree of confirmation (beyond a relative
comparison of models), we need to use a test of statistical significance such
as the t-test (parametric) or the Mann–Whitney–Wilcoxon test (nonparametric).
Both tests require assumptions of independent identically-distributed (i.i.d.)
observations, but the t-test adds a normality assumption.  While the t-test is
fairly robust to violations of the normality assumption (Box et al. 1978),
neither test is robust to violation of independence.

The nonparametric alternative to the t-test is the Mann–Whitney or
Wilcoxon test.  This procedure, which may be preferred under certain
conditions of non-normality, although perhaps not strongly (see Box et al.
1978), is based on the relative ranks achieved when the data are ordered.
Both the t-test and the Mann–Whitney test are described clearly in Snedecor
and Cochran (1976).

A second set of related statistical methods useful for model confirmation
are regression and correlation.  Here the method chosen would be used to
relate to one data series to another, and thus autocorrelation again is a
problem.  Following analysis and adjustment (if necessary) for
autocorrelation, the investigator may regress the predictions on the
observations.  The fit may be assessed through the standard error statistic,
or perhaps using the reliability index proposed by Leggett and Williams
(1981).  This index reflects on a plot of predictions versus observations, the
angle between the 1:1 line (line of best fit), and a line through each data
point from the origin.  Cross-correlation (see Davis 1973) is calculated in a

manner similar to that for the Pearson product moment correlation. Davis
(1973) provides a test statistic for assessing the significance of the
cross-correlation coefficient between two series. Remember that it is
important to adhere to the assumptions behind the statistical methods; failure
to do this under certain conditions can lead to faulty inferences concerning
confirmation.

The final method presented here for deterministic model confirmation
(from by no means an exhaustive list of options in Table 1) is the box plot
(McGill et al. 1978, Reckhow 1980). The box plot is an informative method for
graphing one or more sets of data for the purpose of comparing order
statistics. For each data set, the plot displays the median, the relative
statistical significance of the median, the interquartile range, and the
minimum and maximum points. With aggregated (non-series) data, the box plot
yields perhaps the best visual comparison of two or more data sets. A
detailed example illustrating box plot construction and several applications
is presented in Reckhow and Chapra (1982). McGill et al. (1978) and Reckhow
(1980) also describe the construction and interpretation of box plots.

Before examining some statistical options for confirming stochastic
models, consider one of the important practical issues that the model
developer must face with a multivariate deterministic model. Specifically,
the model developer may have calculated a confirmation statistic (e.g., a
cross-correlation coefficient) for each variable in the model and/or for
certain features of the model such as for extreme values. How might these
confirmatory statistics be aggregated into a single confirmation measure?

First, the modeler must realize that to aggregate statistics and to make
the confirmation decision on the basis of a single measure, means the loss of
potentially valuable model evaluation information. If this loss is
acceptable, then the modeler must decide on an aggregation scheme for the
individual confirmation statistics, for example, for the cross-correlation
coefficients. This decision should be based on the relative importance of the
model characteristics (e.g., model variables) for which confirmation
statistics are available. The confirmation statistics then are aggregated
using weights reflecting this importance. For example, if chlorophyll and
dissolved oxygen are deemed most important in an aquatic ecosystem model, then
the confirmatory statistics for these two variables should receive the highest
weights. The final confirmation measure is a weighted sum of, for example,
cross-correlation coefficients. This might take the form of $\sum w(x)\, r(x)$,
where, x is the model characteristic (variable), w is the weight reflecting
the importance of x, and r is the cross-correlation confirmation statistic for
x. The particular scheme presented above merely is meant to suggest one
option for aggregating statistics; others certainly are possible.

Less common at present than the deterministic simulation model, the
stochastic model nonetheless is important and quite amenable to a number of
statistical confirmatory approaches. In fact, all of the methods discussed
above for the deterministic model are appropriate for a number of features
(e.g., the time stream of mean values) of the three-dimensional prediction and
observation surfaces. In addition, we can take a two-dimensional slice of
these three-dimensional distributions. Several statistical goodness-of-fit
tests listed in Table 1 then can be employed. Of these tests, the
Kolmogorov-Smirnov test perhaps is preferred to the chi-square test because it
is based on the cumulative distribution function (cdf). This removes the

arbitrariness and investigator influence because cells are not required;
rather, the data are ordered and the deviations of the order statistics are
examined.

Either of these methods may be used when the data are arranged in
histogram or cumulative distributive form. Test statistics are available to
assess the degree of confirmation. Benjamin and Cornell (1970) present a
superb discussion on these goodness-of-fit methods for probability models.

Other statistics or procedures certainly may be considered to support
model confirmation. The comparison of moments, particularly higher moments,
can be useful in some situations (see Benjamin and Cornell 1970). In
addition, the box plot is quite effective for examining and displaying
differences between distributions.

In closing this discussion of statistical methods of confirmation, one
additional model type deserve mention: the cross-sectional regression model.
Confirmation of cross-sectional regression models often does not pose the
problem experienced with simulation models. This is particularly true if the
cases studied are truly representative (e.g., a random sample) of the
population of concern. Nonetheless, "shrinkage of the coefficient of multiple
correlation" (Stone 1974) between model development and application data sets
is to be expected. The methods of cross-validation[2] and the jackknife are
useful for both regression model confirmation and estimation of a
non-shrinking standard error or correlation. Mosteller and Tukey (1977) and
Stone (1974) discuss these methods in detail.

CONCLUDING COMMENTS

To end this discussion of the philosophy and statistics of simulation
model confirmation, a few points deserve restating.

1. Inadequate model confirmation increases the risks associated with
   the application of the model. Admittedly, there is a data cost and
   an analysis cost associated with model confirmation. This cost is
   to be compared with the risk resulting from the use of an
   unconfirmed model.
2. If confirmation is to be meaningful:
   a. rigorous statistical tests must be applied; and
   b. calibration-independent data are needed.
3. A number of plausible candidate models (or model sub-routines)
   should undergo confirmation. Comparison of the performance of the
   candidates aids the modeler in the determination of the degree of
   confirmation.

Finally, it must be recognized that the proposed confirmation criteria
rarely can be applied in practice to the extent outlined in this paper. This
realization does not reduce the importance of these criteria. Rather, a
confirmation goal has been proposed, and the modeler may assess the extent of
achievement of this goal. This degree of confirmation, estimated in terms of

---

[2] Calibrate on half of the data and confirm on the other half; if no
significant difference occurs, recalibrate using all the data.

test rigor, test success, and data set independence, represents a measure of confidence to be assigned to the model as a predictive tool.

REFERENCES

Anscombe, P.J. 1967. "Topics in the Investigation of Linear Relations Fitted by the Method of Least Squares," J.R. Statist. Soc. B. 29:1-52.

Benjamin, J.R. and C.A. Cornell. 1970. Probability, Statistics and Decision for Civil Engineers. (New York: McGraw-Hill Book Co.), 684 pp.

Box, G.E.P., W.G. Hunter and J.S. Hunter. 1978. Statistics for Experimenters: An Introduction to Design, Data Analysis, and Model Building. (New York: John Wiley and Sons, Inc.), 653 pp.

Brown, H.I. 1977. Perception, Theory, and Commitment. (Chicago: University of Chicago Press), 202 pp.

Caswell, H. 1977. "The Validation Problem." In Systems Analysis and Simulation in Ecology, Vol. IV. Ed. B.C. Patten. (New York: Academic Press), pp. 313-325.

Chernoff, H. and L.E. Moses. 1959. Elementary Decision Theory. (New York: John Wiley and Sons, Inc.), 364 pp.

David, D.R., L. Duckstein and C.C. Kisiel. 1973. "Model Choice and Evaluation from the Decision Viewpoint." In Theories of Decisions in Practice. (New York: Crane, Russak and Co., Inc.), pp. 341-351.

Davis, J.C. 1973. Statistics and Data Analysis in Geology. (New York: John Wiley and Sons, Inc.), 550 pp.

Hempel, C.G. 1965. Aspects of Scientific Explanation. (New York: The Free Press), 504 pp.

Hydroscience, Inc. 1980. "Workshop on Verification of Water Quality Models." U.S. Environmental Protection Agency. EPA-600/9-80-016.

James, L.D. and S.J. Burges. 1981. "Selection, Calibration, and Testing of Hydrologic Models." Amer. Soc. of Agr. Engrs. Monograph.

Kyburg, Jr., H.E. 1970. Probability and Inductive Logic. (London: The Macmillian Co.), 272 pp.

Leggett, R.W. and L.R. Williams. 1981. "A Reliability Index for Models." Ecol. Modelling. 13:303-312.

McGill, R., J.W. Tukey and W.A. Larsen. 1978. "Variations of Box Plots." Am. Stat. 32:12-16.

Mihram, G.A. 1973. "Some Practical Aspects of the Verification and

Validation of Simulation Models." Op. Res. Q. 23:17–29.

Mosteller, F. and J.W. Tukey. 1977. Data Analysis and Regression: A Second Course in Statistics. (Reading, Mass: Addison-Wesley), 588 pp.

Naylor, T. H. and J.M. Finger. 1971. "Validation." In Computer Simulation Experiments with Models of Economic Systems. Ed. T.H. Naylor. (New York: John Wiley and Sons, Inc.), pp. 153–164.

Popper, K.R. 1968. The Logic of Scientific Discovery. (New York: Harper & Row), 480 pp.

Reckhow, K.H. 1980. "Techniques for Exploring and Presenting Data Applied to Lake Phosphorus Concentration." Can. J. Fis. Sci. 37:290–294.

Reckhow, K.H. and S.C. Chapra. 1982. Engineering Approaches for Lake Management: Vol. 1, Data Analysis and Empirical Modeling. (Ann Arbor, MI: Ann Arbor Science Publishers).

Salmon, W.C. 1971. Statistical Explanation and Statistical Relevance. (Pittsburgh: University of Pittsburgh Press), 117 pp.

Schaeffer, D.L. 1980. "A Model Evaluation Methodology Applicable to Environmental Assessment Models." Ecol. Modelling 8:275–295.

Snedecor, G.W. and W.G. Cochran. 1967. Statistical Methods. (Ames, Iowa: The Iowa State University Press), 593 pp.

Stone, M. 1974. "Cross-validatory Choice and Assessment of Statistical Predictions." J.R. Statis. Soc. B. 36:111–147.

Thomann, R.V. 1980. "Measures of Verification." In Workshop on Verification of Water Quality Models. Hydroscience, Inc. U.S. Environmental Protection Agency. EPA-600/9-80-016.

Thomann, R.V. and J.S. Segna. 1980. "Dynamic Phytoplankton – Phosphorus Model of Lake Ontario: Ten Year Verification and Simulations." In Phosphorus Management Strategies for Lakes. Eds. R.C. Loehr, C.S. Martin and W. Rast. (Ann Arbor, MI: Ann Arbor Science), pp. 153–190.

Van Horn, R. 1969. "Validation." In The Design of Computer Simulation Experiments. Ed. T.H. Naylor. (Durham, NC: Duke University Press), pp. 232–251.

Yevjevich, V. 1972. Probability and Statistics in Hydrology. (Fort Collins, Colorado: Water Resources Publications), 302 pp.

TVA Procedures for Hydrodynamic Model Verification

by William R. Waldrop, Member, ASCE[1]

## ABSTRACT

Various techniques used to verify hydrodynamic computer models developed by the TVA Water Systems Development Branch are discussed. Several methods for acquiring field data for model verification are described and examples of comparisons are presented. These include:

- Flow measurements during pulsed dam releases for verifying a model of a wave progressing down a dry streambed;
- Dye studies to determine dispersion coefficients for a passive pollutant dispersion model;
- Temperature patterns recorded with stationary automated monitors at various locations throughout a reservoir for comparison with a two-dimensional model of the annual stratification cycle of a reservoir;
- Temperature patterns recorded from a boat near the intake/discharge region of a power generating plant for comparison with a three-dimensional model.

## INTRODUCTION

The Water Systems Development Branch (WSDB) of the Tennessee Valley Authority (TVA) performs a diversity of hydrodynamic analyses supporting water resource development and power generation in the Tennessee Valley. With the advent of new computers and numerical methods, computer models have proven to be an attractive and economical alternative to extensive field test programs for providing the basis for important and often very expensive decisions. In some instances, the nature of the problem precludes the use of any other method for analysis. The assessment of the potential impact of a hypothetical accident affecting ground water, streams, or reservoirs provides a case in point.

Because the computer models of WSDB are developed and utilized for the analysis of specific cases, the results must be defendable both within TVA and to regulatory agencies. The credibility of model predictions are closely linked to a demonstrated ability to compare favorably with observed data. Although it is sometimes essential to do so, managers are rightly skeptical about basing important decisions upon unverified computer generated results. In this regard, TVA is fortunate in that it is often possible to arrange for specific power plant operation and dam release patterns to provide

---

[1]Assistant Chief, Water Systems Development Branch, Tennessee Valley Authority, Norris, Tennessee

the appropriate conditions at a specific time and river reach for obtaining data for model verification.  These field investigations are designed to collect data to support computer model applications as opposed to provide data as the basis for analyzing the given problem; this distinction may be quite important.  The verified model with the appropriate well-defined boundary conditions can then be used to systematically vary parameters to simulate a wide range of possible events in an economical fashion.

Various field gathering techniques are used to provide data for model verification.  Four such techniques are described along with examples of how the data were used in model applications.

## FLOW INTO A DRY STREAMBED

Streambeds downstream from large hydroelectric dams operated for peak power generation can become essentially dry during extended periods of no releases.  Such operational patterns are not conducive to viable tailwater fish populations.  Short, pulsed releases during extended offpeak periods (e.g., weekends) are one option for providing water depths and quality needed to improve the tailwater fishery.

For other cases, flow is normally diverted from the streambed downstream from a dam by diverting the water through a tunnel or flume to a powerhouse several kilometers downstream.  The only flow in the streambed between the dam and powerhouse comes from natural inflow (usually small) or releases from flood gates.  Opening the flood gates often produces a relatively steep wave as the flow progresses down the dry bed.

Stages recorded at several locations downstream from TVA's Norris and Apalachia Dams during prearranged release patterns were used to calibrate a flow routing model described by Ferrick (1).  Velocities were recorded and integrated at selected cross-sections during steady flows (Figure 1) to equate flows with stage under those conditions.  Ideally, computed flow rates should be compared with recorded flows, but measuring such flows during highly transit conditions is almost impossible.  If the model rigorously conserves mass, then comparing flow rates during steady conditions and stages during both steady and transit conditions is sufficient.  A comparison of stages downstream from pulsed releases from Apalachia Dam is presented in Figure 2.  From this verification, it became evident that including local inflows in the model was important.

## DYE DISPERSION IN A RIVER

Dye concentrations are recorded as a means of calibrating and verifying models of the transport and dispersion of contaminants which may be accidentally released into rivers.  Steady flows are arranged for such field tests.  Rhodamine WT dye is injected from one or more boats at the upstream end of the study reach (Figure 3).  If the objective is to simulate an instantaneous release, then the injection process must be accomplished quickly and evenly.  The typical dye sampling procedure is to position boats downstream and record dye concentrations at various times as the dye cloud passes.  Samples are

Figure I : Velocity and Stage Measurements
in the Clinch River

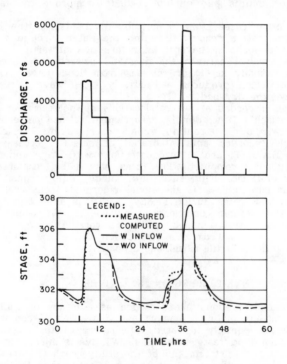

Figure 2 : Transient Flows in the Hiwassee River

**Figure 3 : Dye Injection for Dispersion Study**

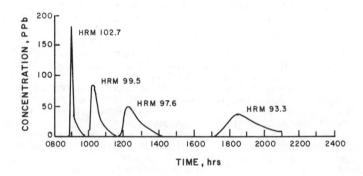

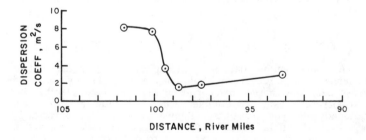

**Figure 4 : Dye Concentration and Dispersion Coefficients in Holston River**

usually recorded at several depths and lateral positions. Fluorometers are used to measure dye concentrations.

Results of such an experiment conducted by McIntosh and Ungate (2) on the Holston River are presented in Figure 4. The variation in the magnitude of the dispersion coefficients along the stream reflects the many natural changes of the physical and hydrodynamic characteristics of the stream. These coefficients can be used with a transport model to predict the fate of hazardous materials that may be accidentally spilled into the Holston River.

## PERMANENT WATER TEMPERATURE MONITORS

Permanent monitors have been installed in the reservoir in the vicinity of several power plants to provide a semicontinuous record of water temperatures. These monitors may be placed on either fixed or floating platforms (Figure 5) or attached to some type of permanent structure such as a skimmer wall of a plant intake. Each monitor has a vertical string of thermistors positioned at a specified depth or elevation which provide a vertical profile of the temperatures. These temperatures are transmitted at 1-hour intervals by telemetry to a meteorological station near the plant for establishing a permanent record on magnetic tape. Acquisition and storage of all data are controlled automatically by a minicomputer in the meteorological station.

Water temperatures from these monitors have proven useful in verifying a two-dimensional model of Wheeler Reservoir. This model, described by Harper and Waldrop (3, 4), was designed to isolate the far-field subtle effects of the thermal effluent of the Browns Ferry Nuclear Plant from the natural diurnal and seasonal temperature fluctuations. Hydrodynamics and water temperatures were simulated throughout the annual stratification cycle, beginning with unstratified conditions in early spring. Computed hourly temperatures were compared with recorded values near Wheeler Dam which is 31 kilometers downstream from the nuclear plant. A comparison of temperatures at two depths for the month of July is presented in Figure 6. In addition to tracking the diurnal and seasonal trend, the model adequately reflected a period of destratification resulting from a storm during late July.

## BOAT-ORIENTED DATA ACQUISITION SYSTEM

A boat-oriented data acquisition system provides detailed flow patterns and three-dimensional temperature distributions over large areas near the thermal discharges from TVA power plants. Most of the temperature data are obtained with a 7-meter aluminum work boat equipped with a minicomputer (Figure 7) which is programmed to perform all of the data acquisition. Temperature measurements are made using either a vertically traversing bathythermograph while the boat is stationary or by individual thermistors mounted at fixed depths on a strut which is towed through the water at speeds up to eight kilometers per hour. The position of the boat is established by monitoring continuous radio signals from four transmitters placed at locations around the area to be surveyed. An onboard receiver senses the four signals, then translates them into a set of special coordinates

Figure 5 : Permanent Water Temperature Monitors

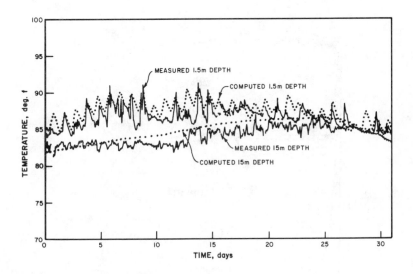

Figure 6 : Water Temperatures of Lower Wheeler Reservoir During June 1977

Figure 7 : Survey Boat with Computerized Data Acquisition System

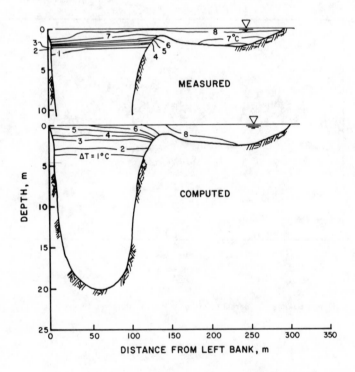

Figure 8 : Water Temperatures Near the TVA Gallatin Steam Plant

which are supplied in digital form directly to the minicomputer. The computer thus obtains an accurate position, ±1 meter, for each data point and stores the information on a cassette which is taken to a larger computer in the laboratory at Norris for plotting and further reduction.

Velocities are often recorded with an electromagnetic water velocity probe. The electromagnetic instrument operates without moving parts utilizing the electric field generated by the water moving through a magnetic field to determine two components of velocity which are vectorally summed and recorded with an onboard microprocessor.

Various portions of this versatile boat system have been used to obtain field data for verification of several models. The three-dimensional model data presented in Figure 8 as a sample were extracted from an analysis by Waldrop and Tatom (5) prepared for an environmental assessment of a TVA power plant. The philosophy of how limited data from a boat study can be used to complement a modeling effort was previously discussed by Waldrop (6).

## CONCLUSION

Computer models can provide an economical alternative to extensive field test programs for defining hydrodynamic processes in streams and reservoirs. However, a favorable comparison of model results with field data is essential to lending credibility to the model. Field tests can be planned to complement a modeling effort and thus provide the necessary verification.

## REFERENCES

1. Ferrick, M. G., "Flow Routing in Tailwater Streams," Proceedings, ASCE Hydraulics Division Specialty Conference on Computer and Physical Modeling, Chicago, Illinois, August 1980.

2. McIntosh, D. A., and C. D. Ungate, "Measurement of Longitudinal Dispersion in the Holston River," TVA Water Systems Development Branch, Report No. WR28-1-530-103, Norris, Tennessee, December 1980.

3. Harper, W. L., and W. R. Waldrop, "A Two-Dimensional, Laterally Averaged Hydrodynamic Model with Application to Cherokee Reservoir," Proceedings of Symposium on Surface-Water Impoundments, University of Minnesota, Minneapolis, Minnesota, June 1980.

4. Harper, W. L., and W. R. Waldrop (1980), "Numerical Hydrodynamics of Reservoir Stratification and Density Currents," IAHR, Trondheim, Norway, Proceedings of the Second International Symposium on Stratified Flows, July 1980.

5. Waldrop, W. R., and F. B. Tatom, "Analysis of the Thermal Effluent from the Gallatin Steam Plant During Low River Flows," TVA Water Systems Development Branch Technical Report 33-30, June 1976.

6.  Waldrop, W. R., "Hydrothermal Analyses Using Computer Modeling and Field Studies," 26th Annual ASCE Hydraulics Division Specialty Conference, College Park, Maryland, August 1978.

PERFORMANCE TESTING OF THE SEDIMENT-CONTAMINANT
TRANSPORT MODEL, SERATRA

Y. Onishi[1]            M. ASCE
S. B. Yabusaki[1]       A.M. ASCE
C. T. Kincaid[1]

## ABSTRACT

Mathematical models of sediment-contaminaint migration in surface
water must account for transport, intermedia transfer, decay and
degradation, and transformation processes. The unsteady, two-
dimensional, sediment-contaminant transport code, SERATRA (Onishi,
Schreiber and Codell 1980) includes these mechanisms. To assess the
accuracy of SERATRA to simulate the sediment-contaminant transport and
fate processes, we tested the code against one-dimensional analytical
solutions, checked for its mass balance and applied the code to field
sites. The field application cases ranged from relatively simple,
steady conditions to unsteady, nonuniform conditions for large, intermediate, and small rivers. We found that SERATRA is capable of simulating sediment-contaminant transport under a wide range of conditions.

## INTRODUCTION

The migration and fate of toxic contaminants in surface waters
are controlled by four complex mechanisms. These mechanisms are contaminant transport due to water and sediment movements; intermedia
transfer due to adsorption/desorption, precipitation/dissolution, and
volatilization; decay and degradation due to radionuclide decay or
chemical and biological degradation; and transformation due to the
yield of daughter, degradation, or chemical-reaction products.

Hence, mathematical models must include adequate mechanisms to
simulate these dynamic processes. The models must be tested to confirm that they are valid under a wide range of conditions before they
are actually applied. Because mathematical models tend to be complex,
and it is difficult to obtain enough field data, model testing is expensive and time consuming. This paper describes tests undertaken to
examine the integrity and applicability of the sediment-contaminant
transport model, SERATRA.

---

(1) Pacific Northwest Laboratory, Richland, Washington. The labora-
    tory is operated for the U.S. Department of Energy by Battelle
    Memorial Institute.

SERATRA DESCRIPTION

The sediment-contaminant transport model, SERATRA, is a finite element model that predicts time-varying longitudinal and vertical distributions of sediments and toxic contaminants (e.g., radionuclides, pesticides, and heavy metals) in rivers and some impoundments. The model consists of the following three coupled submodels, which describe sediment-contaminant interactions and migration:

- a sediment transport submodel
- a dissolved contaminant transport submodel
- a particulate contaminant transport submodel.

The sediment transport submodel simulates transport, deposition, scouring and armoring for three size fractions of cohesive and noncohesive sediments. The transport of particulate contaminants (i.e., contaminants absorbed by sediment) is also simulated for each sediment size. Dissolved contaminants are linked by the adsorption/desorption process to the sediment and particulate contaminants. The contaminant submodels account for 1) advection and dispersion of dissolved and particulate contaminants; 2) chemical and biological degradation resulting from hydrolysis, oxidation, photolysis, biological activities, and radionuclide decay where applicable; 3) volatilization; 4) adsorption/desorption; and 5) deposition and scouring of particulate contaminants. SERATRA also computes changes in riverbed conditions for sediment and contaminant distributions.

MODEL TESTING

Before a model is applied, it should be examined for its basic computational scheme, mass balance, and applicability to actual field conditions.

Testing of the Code Integrity

Although SERATRA is a two-dimensional code, it uses the marching solution technique to solve the longitudinal distribution, while it solves the entire vertical distribution simultaneously at each time step. The integrity of SERATRA has been confirmed through a two-phase test. The first phase examined how accurately the code algorithm portrays the vertical sediment and contaminant distributions. This was accomplished by applying the code to well-posed convection-diffusion boundary value problems having analytical solutions. One of the cases dealt with the following unsteady one-dimensional convection-diffusion equation:

$$\frac{\partial C}{\partial t} + U\frac{\partial C}{\partial y} = \epsilon_y \frac{\partial^2 C}{\partial y^2}$$

with an initial condition of

$$C(y,o) = \exp\left(-\frac{Uy}{2\epsilon_y}\right) \sin\left(\frac{\pi y}{\ell}\right)$$

and boundary conditions of

$$C(o,t) = 0 \text{ and } C(\ell,t) = 0 .$$

Parameter values employed in the test cases were length $(\ell)$ = 1, velocity $(U)$ = -2, and vertical dispersivity $(\epsilon_y)$ = 1 . The computed results and the analytical solution compared favorably (see Figure 1). Such agreement verifies that the finite-element computational scheme of the SERATRA code accurately solves the convection-diffusion equation.

In the second phase of the SERATRA integrity test, SERATRA's mass balance was examined over a series of model runs. A different mechanism was changed for each run, and the mass conservation of sediment and contaminants from one river reach to the next for each case was checked. For both instantaneous and continuous releases, changes were made in 1) velocity, 2) the dispersion coefficient, 3) sediment erosion and deposition parameters which control erosion and deposition of clean and contaminated sediments, 4) adsorption/desorption parameters, 5) degradation and decay rates, 6) volatilization rate, and 7) channel geometry. For all the runs covering these conditions, SERATRA maintained the proper mass balance of sediments and contaminants.

## Field Application Under Steady Flow Conditions

SERATRA was then applied to field sites on a large river--the Columbia River in Washington (Onishi and Wise 1979) and an intermediate river--the Clinch River in Tennessee (Onishi, Schreiber, and

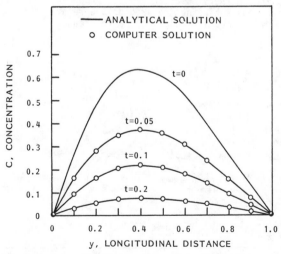

FIGURE 1. Comparison of SERATRA results with the Analytical Solution to the Unsteady Convection-Diffusion Equation

Codell 1980) under steady flow conditions. For the Columbia River case, the model was applied to a 170 km reach between Priest Rapids (River Kilometer 640) and McNary Dams (River Kilometer 470) to simulate the transport of sediments, radioactive $^{65}Zn$, and a heavy metal. The study reach contains both free moving and back-water regions. For the Clinch River case, the 37.2 km back-water reach between Melton Hill Dam (River Kilometer 37.2) and the river month was studied to simulate instantaneous and continuous releases of radioactive $^{137}Cs$ and $^{90}Sr$, which are discharged from Oak Ridge National Laboratory to the river through White Oak Creek (Figure 2). The reasonably good agreements between predicted and measured radionuclide concentrations in the Clinch River (see Figure 3) demonstrate that SERATRA is capable of handling sediment-contaminant transport under steady flow conditions in large and intermediate rivers.

## Field Application Under Unsteady Flow Conditions

The next step in the model testing was to apply SERATRA to small streams whose flows change rapidly. As such, the model was applied to Fourmile and Wolf Creek in Iowa to simulate the migration and fate of alachlor (a pesticide) and river sediments (Onishi et al. 1979). The model was applied to a 67.2-km reach between River Kilometer 19.3 in Fourmile Creek and the mouth of Wolf Creek (Figure 4). Fourmile Creek joins Wolf Creek at River Kilometer 48.3 of Wolf Creek. The simulation of sediment and alachlor migration in these streams was performed for a continuous 3-year period between June 1971 and May 1974. Alachlor was assumed to be released to Fourmile Creek only from a small portion of the Fourmile Creek watershed (see the hatched area in Figure 4). Since several runoff events introduced significant amounts of water, sediments, and alachlor into Fourmile Creek, the transport processes for these constituents were very dynamic, as shown in Figure 5. Although the simulation results under this unsteady flow condition seem reasonable, no measured data were available to examine the accuracy of the model prediction for this case.

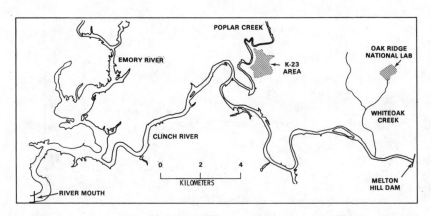

FIGURE 2. Map of the Clinch River, Tennessee

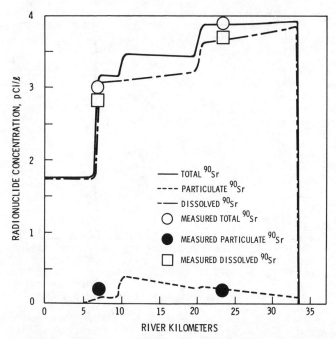

FIGURE 3.  Longitudinal Distributions of Dissolved, Particulate and
Total Strontium–90 Concentrations in the Clinch River,
Tennessee

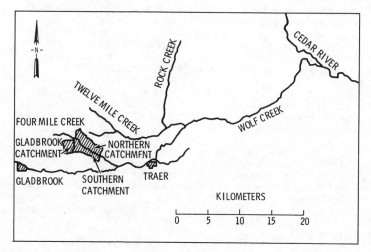

FIGURE 4.  Fourmile and Wolf Creeks in Iowa

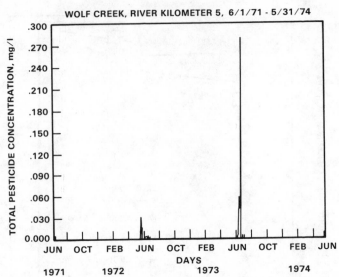

FIGURE 5. Predicted Time Variation of Total Alachlor Concentration at Wolf Creek River Kilometer 5 During the 3-Year Simulation Period

## Field Application With Calibration and Verification Data

A numerical model is critically dependent upon the quality of the measured data used in model input, calibration, and verification. Up to this point, too few measured data were available to extensively test SERATRA through comparisons of predicted results with measured data. Thus, from 1977 to 1979, an extensive data collection program was carried out in the Cattaraugus Creek watershed in New York (Ecker and Onishi 1979; Onishi, Walter and Ecker 1980). The 67-km study reach begins in Buttermilk Creek, just upstream of its confluence with Franks Creek, continues 4 km downstream to Cattaraugus Creek, and ends at the mouth of Cattaraugus Creek at Lake Erie (see Figure 6). Three flow events (November 1977, September 1978 and April 1979), each with distinctly different hydrologic characteristics, were monitored for discharge, sediment and radionuclide distributions at 12 sampling locations (see Figure 6).

SERATRA simulated the transport of sediment and four radionuclides, $^{137}Cs$, $^{90}Sr$, $^{239,240}Pu$ and $^{3}H$ in these streams. In this case there are only four model parameters and coefficients, that were adjusted to fit the simulation results to measured data for the model calibration. They were the vertical dispersion coefficient, and the erodibility coefficient and critical shear stresses used to calculate deposition and erosion rates of cohesive sediments. Thus only the sediment transport modeling required the calibration. The model calibration was performed for the April 1979 condition. After the calibration, without changing calibrated model parameters SERATRA was then applied to the other flow events (November 1977 and September

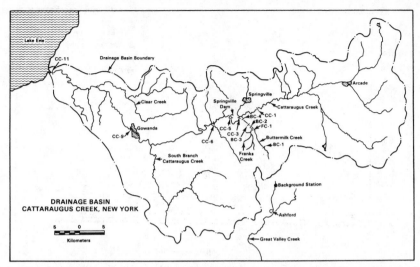

FIGURE 6. Map of Cattaraugus, Buttermilk, and Franks Creeks, New York

1978). Time-varying predicted sediment and measured data at 8 locations in Buttermilk Creek (BC-2 through BC-4) and Cattaraugus Creek (CC-3 through CC-11) were then compared for these two flow events. Although there were some discrepancies between predicted and measured values for all three flow events, considering the complexity of the modeling system and field data accuracy, agreement between predicted and measured sediments were judged to be reasonable. Examples of the calibration and testing runs are shown in Figures 7 and 8.

Since the contaminant transport submodels do not have adjustable model parameters for the model calibration, the radionuclide modeling for all three conditions were regarded as the model testing. Predicted time-varying concentrations of dissolved and particulate $^{137}$Cs, $^{90}$Sr, $^{239,240}$Pu and $^3$H at these 8 locations for three flow conditions were compared with measured data. Again, considering that the modeling system is very complex, that there are no adjustable model calibration parameters, and that some of the field data are not very accurate, agreement of the predicted and measured radionuclide concentrations were judged fairly well. Comparison of the predicted $^{137}$Cs with measured data at the mouth of Buttermilk Creek for the November 1977 and April 1979 cases are shown in Figures 9 and 10. Figure 11 shows the comparison of predicted and measured $^3$H at River Kilometer 61.6 in Cattaraugus Creek (3.2 km upstream of Springville Dam) for November 1977.

CONCLUSIONS

Mathematical models must be validated before they are used. Studies of the SERATRA code demonstrated that it is reasonably well tested. Although SERATRA may be now applied to a field site, we

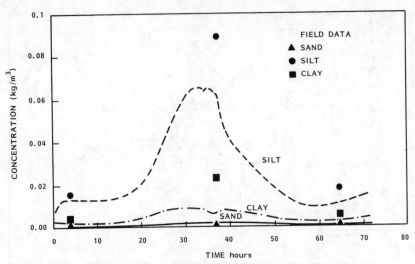

FIGURE 7. Comparison of Predicted and Measured Sediment Concentrations at the Mouth of Buttermilk Creek in April 1979

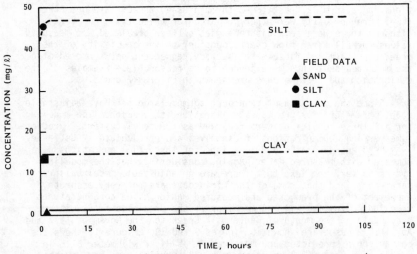

FIGURE 8. Comparison of Predicted and Measured Sediment Concentrations at the Mouth of Buttermilk Creek in September 1978

recommend that SERATRA be further validated through a comparison of predicted results with field data from a carefully controlled tracer test at a field site.

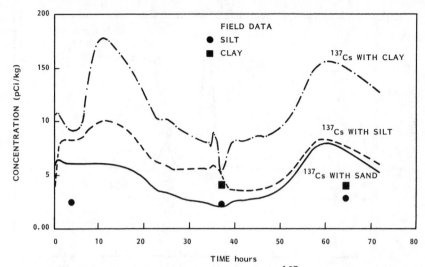

FIGURE 9.  Comparison of Predicted and Measured $^{137}$Cs Concentrations at the Mouth of Buttermilk Creek in November 1978

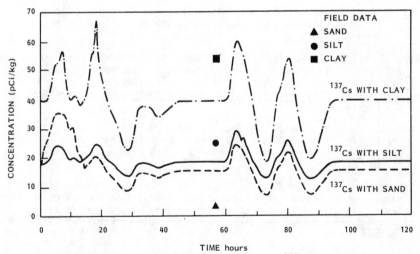

FIGURE 10.  Comparison of Predicted and Measured $^{137}$Cs Concentrations at the Mouth of Buttermilk Creek in April 1979

## REFERENCES

Ecker, R. M., and Y. Onishi. 1979. Sediment and Radionuclide Transport in Rivers, Phase 1: Field Sampling Program During Mean Flow Cattaraugus and Buttermilk Creeks, New York. NUREG/CR-0576, PNL-2551, Pacific Northwest Laboratory, Richland, Washington.

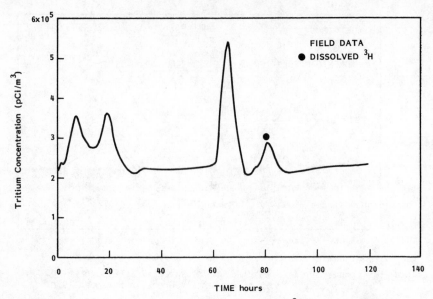

FIGURE 11.   Comparison of Predicted and Measured ³H Concentrations in Cattaraugus Creek (River Kilometer 58.5) in November 1977

Onishi, Y., and S. E. Wise. 1979. "Finite Element Model for Sediment and Toxic Contaminant Transport in Streams." Proceedings of ASCE Hydraulics and Energy Divisions Speciality Conference on Conservation and Utilization of Water and Energy Resources. San Francisco, California, pp. 144-150.

Onishi, Y., A. R. Olsen, R. B. Ambrose and J. W. Falco. 1979. "Pesticide Transport Modeling in Streams." Proceedings of the ASAE Hydrologic Transport Modeling Symposium. New Orleans, Louisiana, pp. 264-274.

Onishi, Y., D. L. Schreiber, and R. B. Codell. 1980. "Mathematical Simulation of Sediment and Radionuclide Transport in the Clinch River, Tennessee," Chapter 18 of Process Involving Contaminants and Sediment, edited by R. A. Baker, Ann Arbor Science Publisher, Inc., pp. 393-406.

Onishi, Y., W. H. Walters, and R. M. Ecker. 1980. Annual Progress Report October 1978 - September 1979--Sediment and Radionuclide Transport in Rivers; Field Sampling Program, Cattaraugus and Buttermilk Creeks, New York. NUREG/CR-1387, PNL-3329, Pacific Northwest Laboratory, Richland, Washington.

# NUMERICAL SOLUTION OF TRANSIENT-FLOW EQUATIONS

BY

M. Hanif Chaudhry*, M.ASCE

## ABSTRACT

This paper discusses the terminology and various methods used
for the numerical integration of partial differential equations
describing unsteady flows in closed conduits. The details of an
explicit, an implicit finite difference method and of the method of
characteristics are presented. The convergence, consistency and
stability of numerical schemes are then discussed. The paper concludes by comparing the results computed by using various schemes and
discussing the ability to simulate a steep wave front.

## INTRODUCTION

The continuity and dynamic equations describe one-dimensional
transient flows in closed conduits. These equations are a set of
partial differential equations if the system is considered a
distributed system while they are ordinary differential equations if
the system is considered as a lumped system. In the distributed-
system approach, the variation of dependent variables both in time
and space (in a one-dimensional case it is distance) are considered.
In the lumped-system approach, however, the disturbances are assumed
to travel instantaneously throughout the conduit and therefore only
variation with respect to time has to be taken into consideration.

The governing equations in both distributed- and lumped-system
approach are nonlinear and because of the presence of non-linearities
a closed-form solution is not usually available. Therefore, these
equations are integrated numerically.

---

Assoc. Professor, Department of Civil Engineering, Old Dominion University, Norfolk, VA, 23508, U.S.A.

Numerical methods for the analysis of lumped systems are well-established (8). In addition, it has been shown that a surge-tank system--normally analyzed using a lumped-system approach--may be analyzed using a distributed-system approach without significantly increasing the digital computer time. Therefore, only numerical methods suitable for the analysis of distributed systems are discussed herein.

In this paper, terminology is introduced first; numerical methods available for the solution of governing equations are then presented. This is followed by a discussion of the stability and convergence of numerical schemes. The paper concludes by a comparison of the ability of various schemes to simulate a steep wave front.

## GOVERNING EQUATIONS

The dynamic and continuity equations (1,2,12) describing one-dimensional, transient-state flows in closed conduits are:

$$\frac{\partial Q}{\partial t} + gA \frac{\partial H}{\partial x} + R\, Q|Q| = 0 \tag{1}$$

$$\frac{\partial H}{\partial t} + \frac{a^2}{gA} \frac{\partial Q}{\partial x} = 0 \tag{2}$$

in which Q = flow; H = piezometric head; a = waterhammer wave velocity; A = conduit cross-sectional area; g = acceleration due to gravity; x = distance along the conduit measured positive in the downstream direction; t = time; R = f/(2DA); D = conduit diameter, and f = Darcy-Weisbach friction factor. In these equations, Q and H are dependent variables while x and t are independent variables. The values of other parameters A, a, D, and f depend upon a particular system.

By neglecting friction losses, i.e. R = 0, differentiating Eq. 1 with respect to x and Eq. 2 with respect to t, and eliminating Q from the resulting equation, one obtains

$$\frac{\partial^2 H}{\partial t^2} = a^2 \frac{\partial H}{\partial x^2} \tag{3}$$

This is our well-known wave equation (9) which describes the time variation of piezometric head at all points in a pipeline.

## TERMINOLOGY

### Closed-form and Numerical Solution

In a closed-form solution, the dependent variables $Q$ and $H$ are expressed in terms of the independent variables $x$ and $t$ so that the values of $Q$ and $H$ can be computed for any specified value of $x$ and $t$. In a numerical solution, the solution of the governing equations is obtained only at discrete values of $x$ and $t$ known as grid or mesh points.

### Initial and Boundary Conditions

To solve Eqs. 1 and 2, it is necessary that $Q$ and $H$ be specified at some instant of time usually time $t = 0$. These distributions of $Q$ and $H$ with respect to $x$ at $t = 0$ are called initial conditions. Expressions specifying the dependent variables or their derivatives at both ends of a system for all times are referred to as boundary or end conditions. For example, if there is a constant-level reservoir at the upstream end, i.e., at $x = 0$, then this boundary condition may be written as

$$H(0,t) = H_r$$

in which $H_r$ = height of the water surface in the reservoir above the datum. Similarly, $H$ at the downstream end, i.e., at $x = L$, may be specified. In some systems, a boundary may be present at an interior location; thus requiring interior boundary conditions.

### Finite-difference Approximations

To simplify discussion let us first consider a function $f(x)$ which is a function of only one variable $x$. Let this function, and its derivatives with respect to $x$ be finite, single valued and continuous. Then by Taylor series

$$f(x_0 + \Delta x) = f(x_0) + \Delta x\, f'(x_0) + \frac{(\Delta x)^2}{2!} f''(x_0)$$

$$+ \frac{(\Delta x)^3}{3!} f'''(x_0) + O[(\Delta x)^4] \qquad (4)$$

Similarly

$$f(x_0 - \Delta x) = f(x_0) - \Delta x\, f'(x_0) + \frac{(\Delta x)^2}{2!} f''(x_0)$$

$$- \frac{(\Delta x)^3}{3!} f'''(x_0) + O[(\Delta x)^4] \qquad (5)$$

Adding Eqs. 4 and 5

$$f(x_0 + \Delta x) + f(x_0 - \Delta x) = 2 f(x_0) + (\Delta x)^2 f''(x_0)$$

$$+ 0[(\Delta x)^4] \qquad (6)$$

in which $0 [(\Delta x)^4]$ denotes terms containing fourth and higher powers of x. Assuming these are negligible when compared to lower powers of $\Delta x$, it follows from Eq. 6 that

$$f''(x_0) = \frac{d^2 f}{dx^2}\bigg|_{x = x_0} = \frac{f(x_0 + \Delta x) - 2 f(x_0) + f(x_0 - \Delta x)}{(\Delta x)^2}$$

$$+ 0[(\Delta x)^2] \qquad (7)$$

with a leading error of order $(\Delta x)^2$. Subtracting Eq. 5 from Eq. 4

$$f(x_0 + \Delta x) - f(x_0 - \Delta x) = 2 \Delta x f'(x_0) + 0[(\Delta x)^3]$$

or

$$f'(x_0) = \frac{f(x_0 + \Delta x) - f(x_0 - \Delta x)}{2\Delta x} + 0[(\Delta x)^2] \qquad (8)$$

with an error of the order of $(\Delta x)^2$. Thus we could write Eqs. 7 and 8 as

$$\frac{d^2 f}{dx^2}\bigg|_{x=x_0} = \frac{f(x_0 + \Delta x) - 2 f(x_0) + f(x_0 - \Delta x)}{(\Delta x)^2} \qquad (9)$$

and

$$\frac{df}{dx}\bigg|_{x=x_0} = \frac{f(x_0 + \Delta x) - f(x_0 - \Delta x)}{2 \Delta x} \qquad (10)$$

Eqs. 9 and 10 represent approximation of derivatives by finite differences known as central finite difference approximation. Note that these approximations are second order accurate. Also, referring to Fig. 1, we may say that these approximate the slope of the tangent at P by the slope of chord AB.

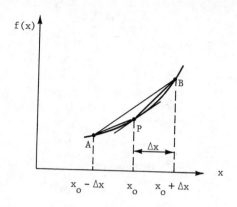

$$f(x)$$

$$x_0 - \Delta x \qquad x_0 \qquad x_0 + \Delta x$$

Fig. 1

From Eq. 4 it follows that

$$f'(x_0) = \frac{f(x_0 + \Delta x) - f(x_0)}{\Delta x} + O(\Delta x)$$

or

$$f'(x_0) \simeq \frac{f(x_0 + \Delta x) - f(x_0)}{\Delta x} \qquad (11)$$

and from Eq. 5 it follows that

$$f'(x_0) = \frac{f(x_0) - f(x_0 - \Delta x)}{\Delta x} + O(\Delta x)$$

$$\simeq \frac{f(x_0) - f(x_0 - \Delta x)}{\Delta x} \qquad (12)$$

Geometrically, Eq. 11 approximates the tangent at P by the slope of the chord PB and Eq. 12 approximates it by the slope of the chord AP. The former is called <u>forward difference</u> and the latter is called the <u>backward-difference</u> formula. Note that both of these are accurate to the first order.

Let us now consider a function, f, which is a function of independent variables x and t. Let us say we have divided the x-t plane into a grid having spatial spacing of $\Delta x$ and time-wise spacing of $\Delta t$ as shown in Figure 2. We shall designate the function f and

approximate its derivatives with respect to x at a point P having
coordinates $x_0$ = i $\Delta x$ and $t_0$ = j $\Delta t$ as follows:

$$f_p = f(x_0, t_0) = f(i\Delta x, j\Delta t) = f_i^j \tag{13}$$

Central finite difference:

$$\left.\frac{\partial^2 f}{\partial x^2}\right|_p = \left.\frac{\partial^2 f}{\partial x^2}\right|_i^j = \frac{f_{i+1}^j - 2f_i^j + f_{i-1}^j}{(\Delta x)^2} \tag{14}$$

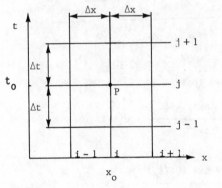

Fig. 2

Forward finite difference:

$$\left.\frac{\partial f}{\partial x}\right|_p = \left.\frac{\partial f}{\partial x}\right|_i^j = \frac{f_{i+1}^j - j_i}{\Delta x} \tag{15}$$

Backward finite difference:

$$\left.\frac{\partial f}{\partial x}\right|_p = \left.\frac{\partial f}{\partial x}\right|_i^j = \frac{f_i^j - f_{i-1}^j}{\Delta x} \tag{16}$$

Again, central finite-difference has a leading error of the order of
$(\Delta x)^2$ and both forward and backward finite differences have a leading
error of the order of $\Delta x$.

Similar equations can be written for $\frac{\partial^2 f}{\partial t^2}$ , and $\frac{\partial f}{\partial t}$ .

## AVAILABLE NUMERICAL METHODS

The following numerical methods may be used to integrate Eqs. 1
and 2 numerically since a closed-form solution is not possible because
of the presence of nonlinear term:

1. Method of characteristics
2. Finite difference methods
The finite-difference methods (1,9) may be further classified into:
1. Explicit finite difference methods
2. Implicit finite difference methods

In the method of characteristics (1-5,7,12), the partial differential equations are first converted into ordinary differential equations which are then solved by a finite-difference method. Since the characteristics represent the path of travelling waves or disturbances, this method is the most appropriate method for analyzing hyperbolic systems (9). In the finite difference methods, the partial derivatives of the governing equations (Eqs. 1 and 2) are replaced directly by finite-difference approximations thus resulting in a set of algebraic equations. These equations are then solved to obtain the solution.

### METHOD OF CHARACTERISTIC

Let us consider a linear combination of Eqs. 1 and 2, i.e.

$$L = L_1 + \lambda L_2 \tag{17}$$

or

$$\left( \frac{\partial Q}{\partial t} + \lambda a^2 \frac{\partial Q}{\partial x} \right) + \lambda gA \left( \frac{\partial H}{\partial t} + \frac{1}{\lambda} \frac{\partial H}{\partial x} \right) + R\,Q\,|Q| = 0 \tag{18}$$

If $H = H(x,t)$ and $Q = Q(x,t)$ are solutions of Eqs. 1 and 2, then the total derivatives may be written as

$$\frac{dQ}{dt} = \frac{\partial Q}{\partial t} + \frac{\partial Q}{\partial x} \frac{dx}{dt} \tag{19}$$

and

$$\frac{dH}{dt} = \frac{\partial H}{\partial t} + \frac{\partial H}{\partial x} \frac{dx}{dt} \tag{20}$$

By defining the unknown multiplier $\lambda$ as

$$\frac{1}{\lambda^2} = \frac{dx}{dt} = a^2 \tag{21}$$

or

$$\lambda = \pm \frac{1}{a} \tag{22}$$

and, on the basis of Eqs. 19 and 20, Eq. 18 may be written as

$$\frac{dQ}{dt} + c \frac{dH}{dt} + R \, Q|Q| = 0 \qquad (23)$$

if

$$\frac{dx}{dt} = a \qquad (24)$$

and

$$\frac{dQ}{dt} - c \frac{dH}{dt} + R \, Q|Q| = 0 \qquad (25)$$

if

$$\frac{dx}{dt} = -a \qquad (26)$$

Equation 23 is valid if Eq. 24 is satisfied and Eq. 25 is valid if Eq. 26 is satisfied. Thus by imposing the relations given by Eqs. 24 and 26 we have transformed the partial differential equations (Eqs. 1 and 2) into ordinary differential equations (Eqs. 23 and 25) in the independent variable in t.

In the above transformation we have not made any approximation at all and Eqs. 23 and 25 are as valid as Eqs. 1 and 2. The only difference is that Eqs. 1 and 2 are valid throughout the x-t plane while Eqs. 23 and 25 are valid only along the lines defined by Eqs. 24 and 26. These lines are called characteristic lines. Mathematically they divide the x-t plane into regions such that each region may have different solutions from that in the adjacent region, i.e. the solution may be discontinuous along these lines. Physically, they trace the path of a disturbance in the x-t plane.

Eqs. 23-26 may be solved using a first- or higher-order finite-difference approximations. Herein we shall use the following first-order (or linear) approximation which is simple to use and has given good results:

$$\int_{x_0}^{x_1} f(x) \, dx \simeq f(x_0) \, (x_1 - x_0) \qquad (27)$$

Referring to Fig. 3 and based on Eq. 27, Eqs. 23 - 26 may be written as

$$a(t_P - t_A) - (x_P - x_A) = 0 \tag{28}$$

$$Q_P - Q_A + c_A (H_P - H_A) + R_A (t_P - t_A) Q_A |Q_A| = 0 \tag{29}$$

$$a(t_P - t_B) + (x_P - x_B) = 0 \tag{30}$$

$$Q_P - Q_B - c_B(H_P - H_B) + R_B (t_P - t_B) Q_B|Q_B| = 0 \tag{31}$$

in which subscripts A, B, and P refer to various quantities at these points in the x-t plane (Fig. 3).

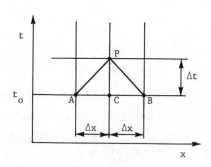

Fig. 3

Eqs. 28-31 may be used to obtain a numerical solution of Eqs. 1 and 2. For this purpose, grid of characteristics or specified intervals may be employed (4). In the grid of characteristics, the location of the discrete points is not fixed in the x-t plane, and their position is determined as the solution progresses. In the method of specified intervals, however, the position of the grid points is specified by the analyst .

Since the method of specified intervals requires fewer computations than the grid of characteristics, gives the pressure and flow histories at specified locations directly rather than interpolating or extrapolating it from the computed results, the former has become more popular and is discussed herein.

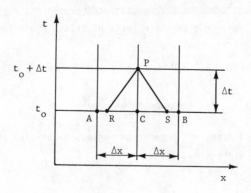

Fig. 4

## METHOD OF SPECIFIED INTERVALS

Let the values of pressure head and flow be known at time $t_o$ (i.e., j level) at all grid points and we have to compute their values at time $t_o$ + $\Delta t$ (i.e., j+1 level). The known values at time $t_o$ may be either the initial steady state values or they may have been computed during the previous time step. Referring to Fig. 4, flows and pressure heads at A, B, and C are known and their values are to be determined at P. Let us pass characteristics lines through P and let us assume that they intersect line AC at R and line BC at S.

Since $x_p$ and $t_p$ are specified by the analyst, the coordinates of R and S can be determined from

$$x_R = x_p - a(t_p - t_R) = x_p - a\Delta t \qquad (32)$$

$$x_s = x_p + a(t_p - t_s) = x_p + a\Delta t \qquad (33)$$

To compute conditions at P, we have to know their values at R and S. However, their values are known only at grid points A, B, and C. Their values at R and S may be computed using a linear or higher-order interpolation from the known conditions at A, B, and C. These interpolations have been reported to cause numerical dispersion and attenuation (12) and a number of computational procedures have been reported (3,10,11) to minimize the effects of these interpolations. By using linear interpolation

$$Q_R = Q_C - \frac{a\Delta t}{\Delta x} (Q_C - Q_A) \tag{34}$$

$$Q_S = Q_C - \frac{a\Delta t}{\Delta x} (Q_C - Q_B) \tag{35}$$

$$H_R = H_C - \frac{a\Delta t}{\Delta x} (H_C - H_A) \tag{36}$$

$$H_S = H_C - \frac{a\Delta t}{\Delta x} (H_C - H_B) \tag{37}$$

The values of $Q_P$ and $H_P$ may now be determined from the following equations (which are obtained by replacing A by R in Eq. 29 and replacing B by S in Eq. 31 since the characteristics through P pass through R and S instead of A and B):

$$Q_P - Q_R + c(H_P - H_R) + R \Delta t \, Q_R|Q_R| = 0 \tag{38}$$

and

$$Q_P - Q_S - c(H_P - H_S) + R \Delta t \, Q_S|Q_S| = 0 \tag{39}$$

Eqs. 38 and 39 may be written as

$$Q_P = c_p - c \, H_P \tag{40}$$

$$Q_P = c_n + cH_P \tag{41}$$

in which

$$c_p = Q_R + cH_R - R \Delta t \, Q_R|Q_R| \tag{42}$$

$$c_n = Q_S - cH_S - R \Delta t \, Q_S|Q_S| \tag{43}$$

Note that Eq. 40 is valid along the positive characteristic line RP and Eq. 41 is valid along the negative characteristic line SP; these equations are referred to as positive and negative characteristic equations respectively. The value of constant c depends upon conduit properties while constants $c_p$ and $c_n$ may be determined from known conditions at each time step.

## Interior Nodes

At each interior node or grid point there are two unknowns $Q_p$ and $H_p$. The value of these unknowns may be determined by solving simultaneously Eqs. 40 and 41, i.e.,

$$Q_p = 0.5 (c_p + c_n) \tag{44}$$

The value of $H_p$ may now be computed from either Eq. 40 or 41.

## Boundary Conditions

For the upstream boundary, we have Eq. 40 while at the downstream boundary we have Eq. 41. However, we have two unknowns at each boundary. To have a unique solution, these equations are solved simultaneously with the conditions imposed by the boundary.

### EXPLICIT FINITE DIFFERENCE METHOD

In the explicit finite difference method, the partial derivatives are replaced by finite difference approximations such that the unknown conditions at a point at the end of a time step are expressed in terms of the known conditions at the beginning of the time step. Several explicit finite difference schemes have been reported in the literature; however, details of one of these schemes known as diffusive scheme (1) are presented in the following paragraphs.

## Interior Nodes

Referring to Fig. 5, let us assume that the conditions at time $t_o$ (i.e., j level) are known and we have to compute their values at time $t_o + \Delta t$ (i.e., j+1 level). Let us approximate the partial derivatives as follows:

$$\frac{\partial H}{\partial t} = \frac{H_i^{j+1} - \overline{H}_i}{\Delta t} \tag{45}$$

$$\frac{\partial Q}{\partial t} = \frac{Q_i^{j+1} - \overline{Q}_i}{\Delta t} \tag{46}$$

$$\frac{\partial Q}{\partial x} = \frac{Q_{i+1}^j - Q_{i-1}^j}{2\Delta x} \tag{47}$$

$$\frac{\partial H}{\partial x} = \frac{H_{i+1}^j - H_{i-1}^j}{2\Delta x} \tag{48}$$

in which

$$\overline{H}_i = 0.5(H_{i+1}^j + H_{i-1}^j) \tag{49}$$

and

$$\overline{Q}_i = 0.5\ (Q_{i+1}^j + Q_{i-1}^j) \tag{50}$$

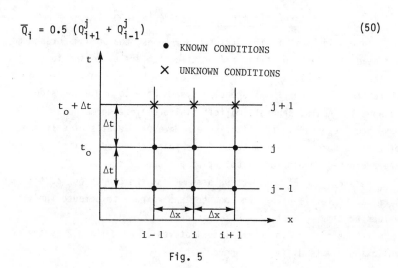

Fig. 5

Substituting Eqs. 45 to 48 into Eqs. 1 and 2 and writing the friction-loss term in terms of $\overline{Q}_i$, one obtains

$$Q_i^{j+1} = \frac{1}{2}(Q_{i-1}^j + Q_{i+1}^j) - \frac{1}{2}\ gA\ \frac{\Delta t}{\Delta x}\ (H_{i+1}^j - H_{i-1}^j) - R\ \Delta t\ \overline{Q}_i\ \left|\overline{Q}_i\right| \tag{51}$$

and

$$H_i^{j+1} = \frac{1}{2}(H_{i-1}^j + H_{i+1}^j) - \frac{1}{2}\ \frac{\Delta t}{\Delta x}\ \frac{a^2}{gA}\ (Q_{i+1}^j - Q_{i-1}^j) \tag{52}$$

Thus we have expressed explicitly the unknown variables $H_i^{j+1}$ and $Q_i^{j+1}$ at section i in terms of the known values of Q and H at sections i-1 and i+1. Hence $Q_i^{j+1}$ and $H_i^{j+1}$ can be directly computed from Eqs. 51 and 52.

Boundary Conditions

Eqs. 51 and 52 are valid for the interior grid points. At the boundaries, however, we cannot write Eqs. 49 or 50 since there are no grid points on one side of the boundary. Therefore, to determine conditions at the boundaries, several computational procedures have been reported. Of these procedures, solving the characteristic equations (Eqs. 40 or 41) simultaneously with the conditions imposed by the bondary appears to be most suitable.

## IMPLICIT FINITE DIFFERENCE METHOD

In the implicit finite difference methods, the unknown discharge and head at a section at the end of time step are expressed in terms of the unknown values of these variables at the neighboring sections. Therefore, equations for the entire system have to be solved simultaneously. Again, several implicit finite difference methods (1,4,9) have been reported in the literature. However, details of only one of these schemes, referred to as "four point centered implicit scheme" are presented herein for illustration purposes.

Referring to Fig. 5, let us assume that the conditions at time $t_0$ (i.e., j level) have been computed and we have to compute their values at time $t_0 + \Delta t$, (i. e., j+1 level).

Let us replace the partial derivatives of Eqs. 1 and 2 by the following finite-difference approximations:

$$\frac{\partial H}{\partial x} = \frac{(H_{i+1}^{j+1} + H_{i+1}^{j}) - (H_{i}^{j+1} + H_{i}^{j})}{2\Delta x} \tag{53}$$

$$\frac{\partial H}{\partial t} = \frac{(H_{i+1}^{j+1} + H_{i}^{j+1}) - (H_{i+1}^{j} + H_{i}^{j})}{2 \Delta t} \tag{54}$$

$$\frac{\partial Q}{\partial x} = \frac{(Q_{i+1}^{j+1} + Q_{i+1}^{j}) - (Q_{i}^{j+1} + Q_{i}^{j})}{2 \Delta x} \tag{55}$$

$$\frac{\partial Q}{\partial t} = \frac{(Q_{i+1}^{j+1} + Q_{i}^{j+1}) - (Q_{i+1}^{j} + Q_{i}^{j})}{2 \Delta t} \tag{56}$$

$$Q = 0.5 (Q_{i+1}^{j} + Q_{i}^{j}) \tag{57}$$

Substituting these equations into Eqs. 1 and 2 and simplifying, we obtain

$$Q_i^{j+1} + Q_{i+1}^{j+1} - C_1 H_i^{j+1} + C_1 H_{i+1}^{j+1} + C_2 = 0 \tag{58}$$

and

$$Q_i^{j+1} - Q_{i+1}^{j+1} - C_3 H_i^{j+1} - C_3 H_{i+1}^{j+1} + C_4 = 0 \tag{59}$$

in which

$$C_1 = \frac{gA\Delta t}{\Delta x} \tag{60}$$

$$C_2 = C_1(H_{i+1}^j - H_i^j) - (Q_i^j + Q_{i+1}^j) + \frac{1}{2}R \, \Delta t \, (Q_i^j + Q_{i+1}^j)$$
$$(Q_i^j + Q_{i+1}^j) \tag{61}$$

$$C_3 = \frac{gA \, \Delta x}{a^2 \Delta t} \tag{62}$$

$$C_4 = - Q_{i+1}^j + Q_i^j + C_3(H_i^j + H_{i+1}^j) \tag{63}$$

In Eqs. 58 and 59, there are four unknowns, $Q_i^{j+1}$, $Q_{i+1}^{j+1}$, $H_i^{j+1}$, and

$H_{i+1}^{j+1}$, and for a unique solution we need four equations. The other two equations in addition to Eqs. 58 and 59 are provided by the adjacent sections i-1 and i+1. If a system is divided into n reaches, then there will be n+1 sections. Since there are two unknowns, $Q_i^{j+1}$ and $H_i^{j+1}$, for each section i (i = 1, 2, ..., n+1), there will be 2(n+1) unknowns and for a unique solution we will need as many equations. Of these, 2(n-1) equations are provided by writing Eqs. 58 and 59 for each interior section and of the remaining equations two are provided by the specified end conditions and two by writing Eqs. 58 and 59 for the upstream end.

The above algebraic equations may be written in the matrix form as

$$B \, \underline{y} = \underline{b} \tag{64}$$

in which $\underline{y}$ is a column vector comprised of $Q_i^{j+1}$ and $H_i^{j+1}$, i = 1,2...
n+1; $\underline{b}$ is a column vector comprised of coefficients $C_2$ and $C_4$ for
each interior node and the conditions imposed by the upstream and
downstream boundaries; and B is a matrix comprised of coefficients $C_1$
and $C_3$. By properly arranging the governing equations, B may be made
a banded matrix. With a banded matrix, the solution of Eq. 64
requires lesser computer storage and lesser computational time.

For illustration purposes, let us consider the piping system
shown in Fig. 6 in which flow at the valve is instantaneously stopped
at t = 0. Let us divide the pipeline into five reaches; thus there
are six nodes and twelve unknowns. Therefore, we need twelve equations to have a unique solution. Ten of these equations will be
obtained by writing Eqs. 58 and 59 for the node numbers 1 to 5; one
equation will be written by specifying that the head always remains
constant at section 1, i.e. $H_1^{j+1} = H_r$ in which $H_r$ = height of the
reservoir water surface above datum; and one equation will be
obtained by specifying that $Q_{n+1}^{j+1} = 0$. (If the valve had been gradually closed, then a relationship between the head and discharge
through the valve would have been specified). The resulting algebraic
equations may be arranged as shown in Fig. 7. Note that by writing
the condition imposed by the upstream boundary as the first equation,
B is a banded matrix. If this had not been done, then B would not be
banded.

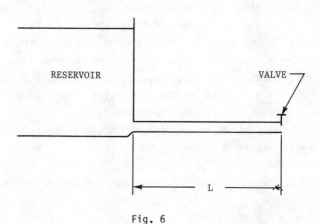

Fig. 6

$$
\begin{bmatrix}
0 & 1.0 & 0 & 0 & 0 & 0 & 0 & 0 & 0 & 0 & 0 & 0 \\
1.0 & -c_1 & 1.0 & c_1 & 0 & 0 & 0 & 0 & 0 & 0 & 0 & 0 \\
1.0 & -c_3 & -1.0 & -c_3 & 0 & 0 & 0 & 0 & 0 & 0 & 0 & 0 \\
0 & 0 & 1.0 & -c_1 & 1.0 & c_1 & 0 & 0 & 0 & 0 & 0 & 0 \\
0 & 0 & 1.0 & -c_3 & -1.0 & -c_3 & 0 & 0 & 0 & 0 & 0 & 0 \\
0 & 0 & 0 & 0 & 1.0 & -c_1 & 1.0 & c_1 & 0 & 0 & 0 & 0 \\
0 & 0 & 0 & 0 & 1.0 & -c_3 & -1.0 & -c_3 & 0 & 0 & 0 & 0 \\
0 & 0 & 0 & 0 & 0 & 0 & 1.0 & -c_1 & 1.0 & c_1 & 0 & 0 \\
0 & 0 & 0 & 0 & 0 & 0 & 1.0 & -c_3 & -1.0 & -c_3 & 0 & 0 \\
0 & 0 & 0 & 0 & 0 & 0 & 0 & 0 & 1.0 & -c_1 & 1.0 & c_1 \\
0 & 0 & 0 & 0 & 0 & 0 & 0 & 0 & 1.0 & -c_3 & -1.0 & -c_3 \\
0 & 0 & 0 & 0 & 0 & 0 & 0 & 0 & 0 & 0 & 1.0 & 0
\end{bmatrix}
\begin{Bmatrix}
Q_1 \\ H_1 \\ Q_2 \\ H_2 \\ Q_3 \\ H_3 \\ Q_4 \\ H_4 \\ Q_5 \\ H_5 \\ Q_6 \\ H_6
\end{Bmatrix}
=
\begin{Bmatrix}
H_r \\ -c_2 \\ -c_4 \\ -c_2 \\ -c_4 \\ -c_2 \\ -c_4 \\ -c_2 \\ -c_4 \\ -c_2 \\ -c_4 \\ Q_{valve}
\end{Bmatrix}
$$

Fig. 7

## CONVERGENCE, STABILITY, CONSISTENCY

In order to have a reasonable accurate numerical solution of a partial differential equation, the finite difference approximations have to satisfy convergence and stability conditions.

### Discretization Error

Let us assume that $U(x,t)$ is an exact solution of a partial differential equation having x and t as independent variables and that $u(x,t)$ is an exact solution of the finite difference equation approximating the partial differential equation. Then the difference $(U-u)$ is referred to as the discretization error (9).

### Convergence

A finite-difference scheme is said to be convergent if u tends to U as $\Delta t$ and $\Delta x$ both tend to zero.

It is difficult to directly develop convergence conditions. However, procedures have been developed to investigate convergence of linear, hyperbolic partial differential equations through stability and consistency conditons. A finite-difference scheme is said to be

convergent if it is consistent with the differential equation and
satisfies the stability conditions (9).

## Truncation Error

Let $F_i^j(u) = 0$ represent the finite difference equation at grid
point (i, j). Now if we substitute the exact solution of the partial
differential equation, U, into the finite difference equation, then
$F_i^j(U)$ is called the local truncation error at grid point (i, j).

## Consistency

If the truncation error tends to zero as both $\Delta t$ and $\Delta x$ tend to
zero, then the difference equation is said to be consistent with the
partial differential equation (9).

## Stability

While solving the finite-difference equations, an exact solution
$u(x,t)$ will be obtained only if computations were performed to an
infinite number of significant figures or decimal places. However,
since the calculations are carried out to a finite number of decimal
places even with the use of the modern computers, round-off errors
are introduced at each time step. Therefore, the numerical solution
we obtain is different from the exact solution.

A numerical scheme is said to be stable (9) if the amplification
of the round-off error remains bounded for all sections i as j tends
to infinity.

As shown in Appendix II, the diffusive scheme is stable if

$$\frac{\Delta x}{\Delta t} \geqslant a$$

This is referred to as the Courant-Friedrich-Lewy condition. Similar
analysis shows that the above stability condition also applies to the
method of characteristics with specified intervals; and that the
implicit finite difference method is unconditionally stable i.e.
there is no restriction on the ratio of $\Delta x$ and $\Delta t$ for stability.

Note that the above stability condition requires that the characteristics through P must intersect the AB line between AC and CB
(see Fig. 4)

## COMPARISON OF NUMERICAL METHODS
General

   In the implicit finite-difference method, the algebraic equations
for the whole system are solved simultaneously. Since the friction
term in this equation is nonlinear, this may require solution of a
larger number of nonlinear algebraic equations. In addition, the
analysis of complex boundary conditions by iterative procedures may
require large computing time since the entire system would have to be
analyzed for each iteration. The main advantage of the method is
that the time interval does not have to be restricted for the scheme
to be stable. However, the time interval cannot be arbitrarily
increased to avoid violation of the validity of the replacement of
the partial derivatives by finite-difference approximations. Because
of these limitations, the implicit finite-difference method has not
become very popular for the analysis of transients in closed conduits.

   In the method of characteristics, each boundary and each conduit
section are analyzed separately during a time step. Therefore, the
method is particularly suitable for the analysis of systems having
complex boundary conditions. The main disadvantage of the method is
that the stability conditions restrict the size of the time step. In
addition, interpolations may be necessary to use this method to
analyze systems having more than one pipe. These interpolations
cause numerical dispersion and attenuation.

### Simulation of Steep Waves or Shocks

   To investigate the ability of the above finite-difference
schemes to simulate a steep wave or a shock, transients in a simple
piping system shown in Fig. 6 were analyzed. In this system, the
pipeline is 16,100 ft long, and has a cross-sectional area of 1.0 $ft^2$.
The waterhammer wave velocity is 3,220 ft/sec. To produce a steep
wave, a flow of 1 cfs was instantaneously stopped at the downstream
end. This produced a 100-ft high positive wave which travelled in
the upstream direction.

   Figs. 8 to 10 show the distribution of computed piezometric
heads in the pipeline at various times by using different finite-
difference schemes and for different values of Courant number, $C_r$,
which is defined as the ratio of the actual wave speed, a, to the

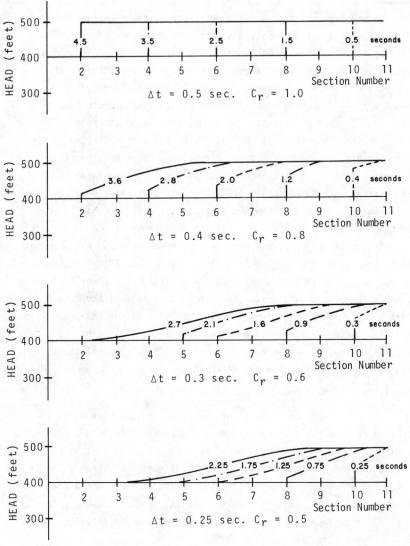

Fig. 8   Method of Characteristics
[$\Delta x$ = 1610 feet]

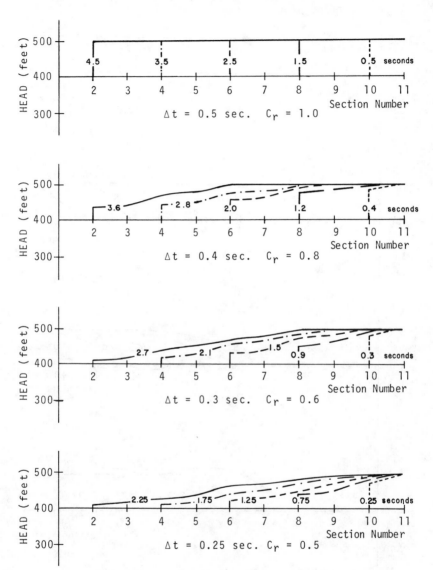

Fig. 9  Explicit Finite Difference Method
[Δx = 1610 feet]

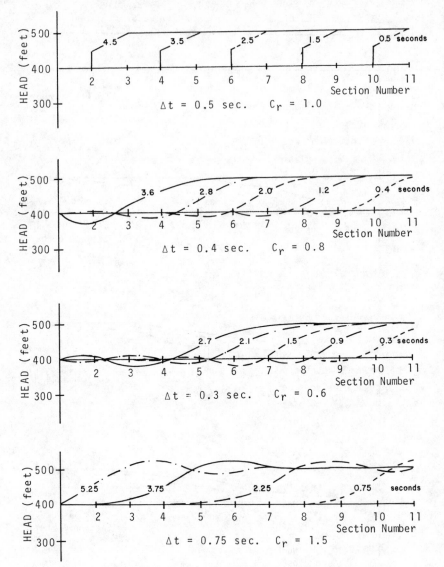

Fig. 10    Implicit Finite Difference Method
[Δx = 1610 feet]

computational wave speed, $\Delta x/\Delta t$. To eliminate the effect of friction on the dispersion of the positive wave front, friction losses were assumed to be zero. (Later investigations indicated that there is insignificant difference between the wave shapes computed by including or neglecting friction losses).

Fig. 8 is for the method of characteristics using specified intervals. In these runs, $C_r$ is a measure of the amount of interpolation. For $C_r = 1.$, characteristics pass through the grid points and no interpolations are involved. Lower the value of $C_r$, higher is the amount of interpolation. It is clear from this figure that with $C_r = 1.$, the speed and shape of the wave front are correctly reproduced. However, with $C_r < 1.$, the wave front is dispersed, and the amount of dispersion depends upon the difference $(1. - C_r)$.

Fig. 9 shows the computed results by using the diffusive scheme. The boundary conditions were developed by solving the positive and negative characteristic equations with the conditions imposed by the boundary. With $C_r = 1.$, the wave shape remains unchanged during propagation while this is not the case with $C_r < 1$.

The results computed by using implicit method are presented in Fig. 10. Since this method is unconditionally stable, runs were made with $C_r = 0.8, 1.0$ and $1.5$. With $C_r = 1.$, the wave shape is slightly changed during propagation; with other $C_r$ values, however, not only the wave does not travel at the correct speed but the wave is dispersed and the piezometric heads at other locations in the system are changed prior to the arrival of the wave front.

## ACKNOWLEDGEMENTS

Figs. 8 - 10 are based on the computer programs written by Michael Holloway, John Fowler, and Robert Fennema for an assignment in a graduate course "Unsteady Closed-Conduit Flow".

## APPENDIX I - REFERENCES

1.  Chaudhry, M.H. "Numerical Methods for Solution of Unsteady Flow Equations," in Closed-Conduit Flow, Chaudhry, M.H. and Yevjevich, V. (eds.), Water Resources Publications, Fort Collins, Colorado, pp. 167-191.

2.  Chaudhry, M.H., Applied Hydraulic Transients, Van Nostrand Reinhold Co., New York, 1979.

3.  Kaplan, M., Belonogoff, G., and Wentworth, R.C., "Economic Methods for Modelling Hydraulic Transient Simulation," Proc., First International Conference on Pressure Surges, published by British Hydromechanics Research Assoc., Bedford, England, 1972, pp. A4-33 to A4-38.

4.  Lister, M., "The Numerical Solution of Hyperbolic Partial Differential Equations by the Method of Characteristics," Chapt. 15 in Mathematical Methods for Digital Computers, edited by Ralston, A. and Wilf, H.S., John Wiley & Sons, New York, 1960.

5.  Martin, C.S., "Method of Characteristics Applied to Calculation of Surge-Tank Oscillations," Proc., First International Conference on Pressure Surges, published by British Hydromechanic Research Assoc., Sept. 1972, pp. E1-1 to E1-12.

6.  Mesinger, F. and Arakawa, A., "Numerical Methods Used in Atmospheric Models," vol. 1, Global Atmospheric Research Programme (GARP), August 1976.

7.  Perkins, F.E., Tedrow, A.C., Eagleson, P.W., and Ippen, A.T., "Hydro-Power Plant Transients, Part II, Response to Load Rejection," Report No. 71, Hydrodynamics Lab, Massachusetts Institute of Technology, Sept. 1964.

8.  Bullough, J.B.B., and Robbie, J.F., "The Accuracy of Certain Numerical Procedures When Applied to the Solution of Ordinary Differential Equations of the Type Used in the Digital Computer Prediction of Mass Oscillations in Closed Conduits," Proc., First, International Conference on Pressure Surges, published by British Hydromechanic Research Assoc., Sept. 1972, pp. A6-53 to A6-75.

9.  Smith, G.D., Numerical Solution of Partial Differential Equations, Second Edition, Clarendon Press, Oxford, England, 1978.

10. Vardy, A.E., "On the Use of the Method of Characteristics for the Solution of Unsteady Flows in Networks," Proc., Second International Conference on Pressure Surges, Published by the British Hydromechanic Research Association, Bedford, England, 1977.

11. Wiggert, D.C. and Sundquist, M.J., "On the Use of Fixed Grid Characteristics for Pipeline Transients," Jour., Hyd. Div. Amer. Soc. of Civil Engrs., vol. 103, Dec. 1977, pp. 1403-1416.

12. Wylie, E.B. and Streeter, V.L., Fluid Transients, McGraw-Hill Book Co., New York, 1978.

## APPENDIX II - STABILITY OF FINITE-DIFFERENCE SCHEMES

The stability of a finite-difference scheme may be investigated by using von Neuman or Fourier Series Method (7,9). In this method, which may be used only for linear equations, errors in the numerical solution at an instant of time are expressed in a Fourier series. Then it is determined whether these errors decay or grow as time increases. A scheme is said to be stable if the errors decay with time while it is said to be unstable if the errors grow as time increases.

The procedure is illustrated by discussing the analysis of the stability of the diffusive scheme. Let $Q_k^j$ and $H_k^j$ (at time $t = t_0$) contain small errors $q_k^j$ and $h_k^j$, i.e.,

$$Q_{k_{comp}}^j = Q_{k_{true}}^j + q_k^j \tag{65}$$

$$H_{k_{comp}}^j = H_{k_{true}}^j + h_k^j \tag{66}$$

in which the subscript "comp" indicates actual numerical solution while "true" indicates exact solution. Note that subscript k is used for the grid points in the x-direction instead of subscript i to avoid confusion since i will be used later for $\sqrt{-1}$ .

Neglecting friction (i.e., R = 0), substituting the above equations into Eqs. 51 and 52 and noting that true solution will also satisfy these equatons, one obtains

$$q_k^{j+1} = \frac{1}{2} (q_{k-1}^j + q_{i+1}^j) - \frac{1}{2} gAr (h_{i+1}^j - h_{i-1}^j) \tag{67}$$

$$h_k^{j+1} = \frac{1}{2} (h_{k-1} + h_{k+1}^j) - \frac{1}{2} r \frac{a^2}{gA} (q_{k+1}^j - q_{k-1}^j) \tag{68}$$

in which $r = \Delta t/\Delta x$. Let us assume that we can express the errors q and h in a Fourier Series, then

$$q_k^j = \sum_{n=0}^{N} A_n' e^{\frac{n\pi i x}{L}}$$

or

$$q_k^j = \sum_{n=0}^{N} A_n' e^{i\theta_n x} \tag{69}$$

and similarly

$$h_k^j = \sum_{n=0}^{N} B_n e^{i\theta_n x} \tag{70}$$

in which $\theta_n = n\pi/L$; $L = N\Delta x$; and $L$ = length of the pipe. The values of N+1 unknowns, $A_n$, n=0,1,...N, may be determined from N+1 equations given by Eq. 69. Since the Fourier-Series method is valid only for linear finite-difference equations, propagation of error in a single term only, say $A_n e^{i\theta_n x}$ may be considered. To simplify let us drop subscript n and let us assume that

$$q_k^{j+1} = A' e^{\alpha\Delta t} e^{i\theta x} = \xi A' e^{i\theta x} \tag{71}$$

and

$$h_k^{j+1} = B e^{\alpha\Delta t} e^{i\theta x} = \xi B e^{i\theta x} \tag{72}$$

in which $\xi$ = amplification factor. In addition, noting that $x_{k+1} = x_k + \Delta x$ and $x_{k-1} = x_k - \Delta x$, we may write

$$q_{k-1}^j = A' e^{i\theta x} e^{-i\theta\Delta x} \tag{73}$$

$$q_{k+1}^j = A' e^{i\theta x} e^{i\theta\Delta x} \tag{74}$$

$$h_{k-1}^j = B e^{i\theta x} e^{-i\theta\Delta x} \tag{75}$$

$$h_{k+1}^j = B e^{i\theta x} e^{i\theta\Delta x} \tag{76}$$

Substituting Eqs. 71-76 into Eq. 67 and simplifying, we obtain

$$[\xi - \frac{1}{2}(e^{i\theta\Delta x} + e^{-i\theta\Delta x})] A' + \frac{1}{2} gAr (e^{i\theta\Delta x} - e^{-i\theta\Delta x}) B = 0 \tag{77}$$

or

$$(\xi - \cos \delta)\ A' + i(gAr\ \sin \delta)\ B = 0 \tag{78}$$

in which $\delta = \theta\Delta x$. Similarly, substituting Eqs. 71 - 76 into Eq. 68 and simplifying, we obtain

$$i(\frac{a^2}{gA}\ r\ \sin \delta)\ A' + (\xi - \cos \delta)\ B = 0 \tag{79}$$

For a non-trivial solution, A and B are not both equal to zero. Hence

$$\begin{vmatrix} \xi - \cos \delta & i\ gAr\ \sin \delta \\ i\ \dfrac{a^2}{gA}r\ \sin \delta & \xi - \cos \delta \end{vmatrix} = 0 \tag{80}$$

which upon expansion becomes

$$(\xi - \cos \delta)^2 + a^2 r^2 \sin^2\delta = 0 \tag{81}$$

or

$$(\xi - \cos\delta)^2 = -(a\ r\ \sin\delta)^2 \tag{82}$$

or

$$\xi - \cos\delta = \pm\ i\ a\ r\ \sin \delta \tag{83}$$

or

$$\xi = \cos\delta\ \pm\ i\ a\ r\ \sin \delta \tag{84}$$

The errors will not increase as time increases if the amplification factor

$$|\xi| \leqslant 1 \tag{85}$$

i.e.

$$\sqrt{\cos^2\delta + a^2\ r^2\ \sin^2\delta}\quad \leqslant 1 \tag{86}$$

or

$$a^2\ r^2\ \sin^2\delta\quad \leqslant 1 - \cos^2\delta \tag{87}$$

or

$$a^2 r^2 \sin^2\delta \leqslant \sin^2\delta \qquad (88)$$

or

$$\left( a \frac{\Delta t}{\Delta x} \right)^2 \leqslant 1 \qquad (89)$$

or

$$\Delta x \geqslant a \Delta t \qquad (90)$$

This is the necessary condition for the diffusive scheme to be stable. This is known as Courant-Friedrich-Lewy condition.

## APPENDIX III-NOTATIONS

The following symbols are used in this chapter:

| | |
|---|---|
| A | cross-sectional area of pipe; |
| a | velocity of pressure wave; |
| C | $gA/a$; |
| $C_1$ to $C_4$ | constants defined by Eqs. 31 to 34; |
| $C_p$, $C_n$ | constant defined by Eqs. 60 and 61; |
| f | Darcy-Weisbach friction factor; |
| g | acceleration due to gravity; |
| H | transient-state piezometric head above datum at the beginning of time interval; |
| $H_o$ | steady state piezometric head above datum; |
| $H_p$ | transient-state piezometric head above datum at the end of time interval; |
| L | length of pipe; |
| Q | transient-state discharge at the beginning of time interval |
| $Q_p$ | initial steady-state discharge; |
| R | $f/(2DA)$; |
| t | time; |
| $\Delta x$ | distance along the pipeline; |
| $\Delta t$ | time interval |

# MODEL INTERCOMPARISON OF LAKE ERIE STORM SURGE RESONANCE PREDICTIONS

by

Keith W. Bedford[1], AM ASCE: and

J. Steven Dingman[2]

## ABSTRACT

Stress band resonant coupling in Lake Erie is tested using three numerical storm surge models. The results of the models are compared to the theoretical results obtained by Rao (1967) in his method of characteristics solution of the shallow water equations. Resonant effects are observed in the non linear models for southwest to northeast stress bands travelling at speeds equal to the average speed of a gravity wave in Lake Erie. All models failed to predict resonant effects in the water levels for northeast to southwest travelling stress bands.

## INTRODUCTION

Lake Erie storm surges are a relatively common occurrence and can be excessive at times. Statistical studies (e.g., Pore et al., 1975) have shown that a ten foot set-up between Buffalo and Toledo can be expected to occur every two years, and thus forecasting of such events becomes increasingly important due to residential and commercial development along the lake. A model intercomparison is now underway whose objective is to test the accuracy of various storm surge forecasting techniques against existing Lake Erie field data. The following paper reports on the preliminary results of one aspect of the intercomparison, the ability to forecast stress band resonance phenomena.

Sudden increases of the water level in Lake Erie, as well as other enclosed bodies of water and along coasts can occur as a result of resonant coupling between a moving atmospheric disturbance and the body of water. In general, when the speed of propagation of the atmospheric disturbance is nearly equal to the speed of free or gravity waves on the lake, an exchange of energy between the atmospheric disturbance and the lake will result in higher than expected water levels at one end of the lake or at other localized areas. The energy supplied to the lake by a moving atmospheric disturbance is a combined result of the pressure gradient and wind stress force. For small scale disturbances, sudden atmospheric pressure changes, such as pressure jumps associated with squall lines, can produce higher than expected waves which can temporarily cause sudden water level rises in shoaling areas of lakes. For

---

[1] Associate Professor, Civil Engineering Department, The Ohio State University, Columbus, Ohio 43210

[2] Research Associate, Civil Engineering Department, The Ohio State University, Columbus, Ohio 43210

large scale distrubances such as storm surges, wind stress forces predominate (Rao, 1967).

Early accounts of small scale resonance can be found in Whittlessey
(1874), as well as Gilbert (1897) who found that waves or surges could
be produced by pressure jumps occurring at another part of the lake during thunderstorms or the passage of squall lines and surface fronts.  An
early mathematical treatment on pressure resonance in a body of water
was discussed by Proudman (Douglas, 1929) in an attempt to explain the
cause of large waves being formed in the English Channel during a squall
on 20 July, 1929.  Proudman (1953) gives a more detailed mathematical
discussion on this subject.  The most frequently cited example of resonant coupling is the discussion given by Ewing et al. (1954).  A sudden and unexpected water rise of over six feet along the Chicago shoreline of Lake Michigan, resulting in the drowning of seven persons,
occurred on 26 June 1954 and was associated with the passage of a pressure jump line across the Lake.  Harris (1957), and Platzman (1958)
presented numerical solutions to the 1954 surge.  Donn (1959) reported
that higher than normal waves, which occurred during the northeast
storm of 5 May 1952, were produced by a moving pressure jump travelling
across the Western Basin of Lake Erie at approximately the same speed
as a gravity wave in the area.  Chaston (1979) commented on a 1.5-3.0
foot rise in the water level of Lake Ontario near Rochester, New York
within ten minutes, which may have been produced by the passage of a
tornado across Lake Ontario between Toronto and Buffalo.

Reports of large scale resonance are not well documented.  The primary discussion is by Rao (1967), who studied the response of a lake of
uniform depth and finite length to a time dependent wind stress.  Rao's
results are based on the solution of the one dimensional linearized
shallow water wave equations by the method of characteristics.  A brief
mathematical treatment of the method of characteristics and the results
of Rao's work will be discussed below.  The second part of the paper
will concern itself with a summary of the preliminary numerical experiments performed to simulate large scale resonance using storm surge computer models.  A discussion of these results will also be presented.

## RESONANCE PHENOMENA ANALYSIS

Large scale resonance phenomena is an important ingredient in the
creation of storm surges in Lake Erie.  Unusually high storm surges have
been produced by wind velocities whose magnitudes were less than gale
force;  in other cases, whole gale to hurricane force winds have produced surges with similar magnitudes as above.  A simple analytical
treatment of resonant coupling can be obtained by the solution of the
shallow water equations by the method of characteristics, and the results appear below.

### General Characteristic Equations for Shallow Water Equations

The phenomena of resonance between a travelling wind stress band
and a lake can be analyzed by the solution of the one dimensional shallow water equations by the method of characteristics.  The general form
of the equations are (Liggett and Cunge, 1975)

$$B \frac{\partial h}{\partial t} + A \frac{\partial u}{\partial x} + uB \frac{\partial h}{\partial x} = q \tag{1}$$

$$\frac{\partial u}{\partial t} + u \frac{\partial u}{\partial x} + g \frac{\partial h}{\partial x} = g(S_0 - S_f) - \frac{qu}{A} \tag{2}$$

where h is the free survace elevation; u is the fluid velocity; A is the cross sectional area of channel; B is the top width of channel; $S_0$ and $S_f$ are the water and friction slopes; g is the acceleration of gravity; and q is the lateral inflow. Using known solution procedures, the resulting characteristics are

$$\{(u+c) \frac{\partial}{\partial x} + \frac{\partial}{\partial t} \}(u+w) = \frac{q}{A} (c-u) + g(S_0 - S_f) \tag{3}$$

along $$\frac{dx}{dt} \bigg|_+ = u+c \tag{4}$$

for the forward characteristic and

$$\{(u-c) \frac{\partial}{\partial x} + \frac{\partial}{\partial t} \}(u-w) = \frac{-q}{A} (c-u) + g(S_0 - S_f) \tag{5}$$

along $$\frac{dx}{dt} \bigg|_- = u-c \tag{6}$$

for the backward characteristic where,

$$w = \int_0^h \sqrt{\frac{gB}{A}} \, dz \tag{7}$$

If a rectangular channel is assumed, $w = 2c$, c is the celerity of a gravity wave in the channel, $c = \sqrt{gH}$. Equations 3-6 can be integrated by numerical methods (Liggett and Cunge, 1975) in order to find the free surface fluctuations. Equations 1 and 2 can be adapted easily for resonance studies in lakes.

## Linear Analysis

In Rao's work, a lake of uniform width and depth is assumed. The lake is of finite length, L, and is bounded by boundaries at $x = 0$ and $x = L$. An external force, the wind stress, is included in the equations and is prescribed as a function of distance and time. Ignoring the non linear terms, non-hydrostatic pressure forces, and the bottom stress, the shallow water equations in a vertically integrated form become

$$\frac{\partial M}{\partial t} = -c^2 \frac{\partial \xi}{\partial x} + R \tag{8}$$

$$\frac{\partial \xi}{\partial t} = - \frac{\partial M}{\partial x} , \qquad\qquad M \equiv \int_{-h}^0 u \, dz \tag{9}$$

where M is the volume transport across a vertical section, u is the horizontal velocity in the x direction. R is defined as $\rho^{-1}\tau$, $\tau$ being the wind stress force and $\rho$ the water density. $\xi$ is the fluctuation of the free surface elevation from the mean level and $c^2 = gH$, H being the depth of the lake. The boundary conditions are that M = 0 at x = 0, L, indicating total reflection at the boundaries. The initial conditions considered are that the lake is at rest, M = $\xi$ = 0 at t = 0. For a prescribed R, equations 8 and 9 are solved for the set-up that is defined as the difference of water levels, $\xi$, at x = 0 and x = L. Equations 8 and 9 are non-dimensionalized and when these equations are added and subtracted, the resulting characteristics are

$$\frac{d}{dt} (\overline{M} \pm \overline{\xi}) = R \qquad \text{for} \quad \frac{dx}{dt} = \pm 1 \qquad (10)$$

where,   $\overline{M} = \dfrac{CM}{R_0 L}$ ,     $\overline{\xi} = \dfrac{c^2 \xi}{R_0 L}$ ,     $\overline{R} = \dfrac{R}{R_0}$ ,   $R_0$ = scale value of the wind stress.

Equation 10 is integrated for two special cases, a semi-infinite stress band and finite stress band. A summary of these results are summarized below.

The semi-infinite stress band is represented as a step function moving at a constant translation speed, V, nondimensionalized by the speed of a gravity wave, v/c = $\overline{V}$. When this relationship is applied in 10 and numerically integrated, Rao found that the maximum set up was a function of the propagation of the speed of the stress band. For speeds $\geq$ C, the maximum set-ups occur with the maximum value occurring as $\overline{V} \to \infty$. For speeds less than C, the set ups were smaller. Rao also showed in his lake that a lag time existed between the high and low water levels at each end of the lake. For a stress band travelling left to right at twice the speed of a gravity wave, the maximum water level occurred at the time when the first gravity wave reached the right side of the lake. The minimum water level at the left side occurred at a time equal to 1.5 times his period, this resulting in a lag time of 1/2 the time for the gravity wave to cross the lake. The finite stress band is represented by the superposition of two semi-infinite stress bands of the type mentioned above: a positive stress band and a negative stress band of the same intensity, both moving with the same speed and direction, but with the jumps separated by a finite distance. When this case is numerically solved, Rao found that the set-up magnitudes were slightly smaller than the infinite stress band and depended on the width of the band. For all band widths less than the length of the lake, the maximum set-up occurs when the propagation speed of the stress band equals C. The magnitudes of the set-up decrease as the width of the band decrease; note, however, that even for a band width of zero, which corresponds to a situation similar to a pressure jump line, moving at a velocity V = C, the magnitude of the set-up is still significant compared to larger band widths propagating at velocities greater than or less than C. When the band length was greater than the length of the lake, the maximum set-up occurs when the time required for the stress band to pass a fixed point is equal to the time required for a gravity wave to cross the lake. Thus, resonance can occur when these band widths travel at speeds which are greater than the celerity of a gravity wave. One of Rao's assumptions in his derivation is the omission of the non-linear term in the momentum equation. Due to the shallowness of the lake in the Western Basin, non-linear effects can be important, especially during times of storm events.  Thus, the inclusion of the non linear term in the method of characteristics solution may be beneficial.

## Non Linear Analysis

The inclusion of the non linear term in the shallow water wave equation, u $\partial u/\partial x$, and its solution using the method of characteristics was performed.  The solution method is the same as in Liggett and Cunge (1975).  Although the results are not reproduced here, it was found that with the same initial and boundary conditions used by Rao, the water surface elevation and set-up profiles for various widths and propagation speeds of stress bands did not show any significant differences than the results obtained by Rao.  The inclusion of a variable bottom topography will be reported in a later paper.

## LAKE ERIE RESONANCE SIMULATION

To simulate resonance in Lake Erie, a wind stress band of a specified magnitude and width is numerically propagated at a resonant speed across Lake Erie in three different storm surge models:

   a) One Dimensional Non Linear Model similar to Sykes and Bedford (1981)
   b) Leendertse Type Alternating Direction Implicit (ADJ) Model
   c) Linear Unit Convolution Model (Schwab, 1978)

A description of each model appears below.

## One Dimensional Model

The one-dimensional hydrodynamical model is based on the solution of the one dimensional nonlinear time varying St. Venant type equations for river channels.  Wind shear stress sets the lake in motion while atmospheric pressure gradient effects are neglected, as well as in the other models.  A full Newton-Raphson solution of these equations is employed with the equation created from four point implicit discretization.  Based on navigational chart information with a datum set at 571.0 ft above sea level, Lake Erie is divided into eighty (80) cross sections where the cross-sectional area, top width, and average depth for each cross section is determined.  Eighty nodal points are chosen at the thalweg of each cross section and all numerical calculations are performed at these nodes.  The distance between each node is roughly 3 miles (5 km).  A map of Lake Erie showing land based water level and meteorological stations as well as the depth contours in each basin is presented in Figure 1.  These same nodal point locations are used for reducing two dimensional results to centerline profiles for intercomparisons.

## Leendertse Type Model

The two dimensional ADI model is derived from the vertical integration of the momentum and continutiy equations (Leendertse, 1970; Leendertse and Gritten, 1971).  The finite difference approximations of the fully nonlinear equations are detailed on the usual space staggered grid with a 5 km x 5 km size.

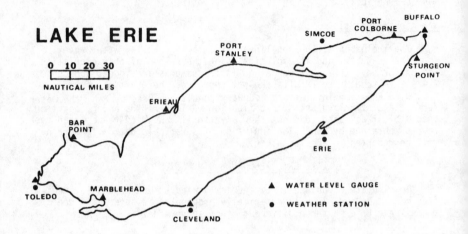

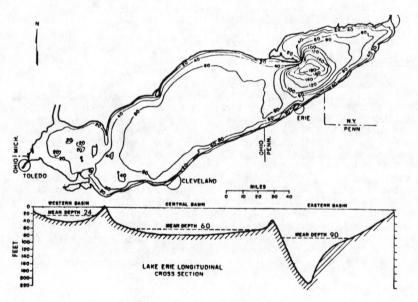

Figure 1.   Map of Lake Erie

## Unit Convolution Model

The hydrodynamic model of Schwab (1978) is based on the same set of governing equations as in the Leendertse model with the non linear terms in the x and y momentum equations removed. The bottom stress parameter corresponds to the shallow water limit of the full Ekman layer treatment of bottom friction as discussed by Platzman (1963) and the variables are discretized on a 10 km x 10 km grid using a Richardson lattice scheme. In order to simplify the solution procedure Schwab linearized the transport equation by eliminating the inertial acceleration. A unit impulse response function convolution approach is used to create a much faster but linear prediction for water levels. Since the system of equations is linear, the free surface solution for arbitrary uniform forcing, $\tau(t)$, is given by the convolution integral,

$$h(x,y,t) = \int_{-\infty}^{t} g(x,y,t-t')\tau(t')\lambda t \qquad (11)$$

where g is the impulse response function. The function g is the Green's function for water levels. These functions are calculated by letting the individual stress components be 1 dyne/cm$^2$ for the first time step only, integrating forward in time with no forcing, and recording the g functions at points of interest. Since the forcing function $\tau$ is usually given at intervals longer than the time resolution of the impulse response function, the calculation can be simplified by presuming g over that period, which in this case is equivalent to one hour.

To include spatial dependence in the wind field, an interpolation scheme is used that assigns a weighting factior, $w_i$, to the discrete forcing terms such as

$$\tau(x,y,t) = \sum_{i=1}^{M} w_i(x,y)\ \tau_i(t); \text{ where } \sum_{i=1}^{M} w_i(x,y) = 1 \qquad (12)$$

and M is the number of wind stations. This is the identical scheme used to calculate center lake water level values in the two dimensional models. The impulse response functions are calculated for given wind stations, i, surrounding Lake Erie and are calculated once and stored on tape or disk. These same functions can then be used for a particular wind field that produces a storm surge.

## Input Conditions

Resonant coupling between wind shear bands and the lake is simulated by the application of a step function for infinite stress bands and box functions for finite band sizes (Figure 2). The magnitude of the stress band is taken as 10 dyne/cm$^2$ which corresponds to a wind of roughly 35 MPH and is capable of producing a sizable set-up in the lake.

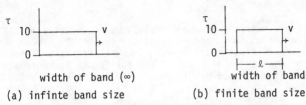

width of band (∞)                    width of band

(a) infinite band size              (b) finite band size

Figure 2.  Stress Band Sizes

These wind stress bands were propagated in a SW-NE or NE-SW direction
across the lake and at varying speeds. These directions correspond to
a wind blowing from 240° (for SW-NE Band) and 30° (for NE-SW Band); and
are the directions from which the maximum setups in Lake Erie occur.
This direction roughly corresponds to the orientation of Lake Erie from
the horizontal and thus the longest overwater fetch occurs with winds
from these directions. In Schwab's model, a wind field for each of the
wind stations are read in every hour; the wind will either be a velocity
equivalent to a 10 dyne/cm$^2$ wind stress or zero, and this will depend
on whether the wind stress front has passed the particular wind station.
x and y component wind stress values are computed and are used in the
calculation of the storm surge. For the Leendertse model the x and y
wind stress components for a 10 dyne/cm$^2$ wind stress band moving in a
SW-NE or NE-SW direction are calculated. As the band propagates across
the lake, the wind stress values corresponding to its location on the
numerical grid will be activated or shutoff, and the calculation of the
storm surge proceeds.

Output Preparation and Portrayal

For both two-dimensional models, water surface fluctuations at the
corresponding land data collection stations are calculated for each time
step in the model (see Figure 1). These stations are then transformed
into center line water elevations at the corresponding nodal points in
the one-dimensional model by using an interpolation procedure based on
the inverse square of the distance between the nodal points and land
stations (Platzman, 1963)

Table 1 is a partial listing of the resonant simulations that were
performed. In general, wind stress bands of magnitude 10 dynes/cm$^2$ for
various widths were propagated in a SW-NE or NE-SW direction at a speed
of 28 MPH which is the speed of a gravity wave in Lake Erie, based on
an average depth of 60 feet. Figures 3-6 are representative 3-D profiles of the center line water levels for a particular simulation. Each
line represents hourly water level profiles. A complete simulation
lasted 120 hours; this length was chosen in order to analyze the oscillation nature of each model and to fix the bottom coefficients so that
die away times of the surges for each model are similar. For the infinite stress band simulations, the wind stress is shut off after 72
hours and the lake levels oscillate for 48 hours. Thus in Figures 3-5
the first 14 hours of the simulations are plotted, the portion of the
simulation where steady state conditions are achieved are eliminated,
and the dieoff of the surge after the 72 th hour is plotted. A discussion of the results follows.

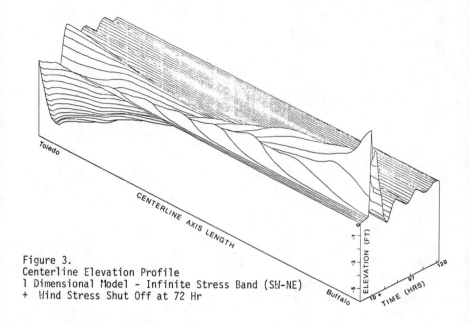

Figure 3.
Centerline Elevation Profile
1 Dimensional Model - Infinite Stress Band (SW-NE)
+ Wind Stress Shut Off at 72 Hr

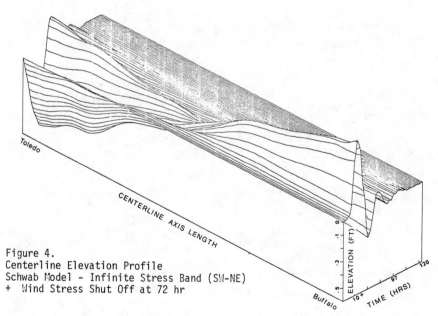

Figure 4.
Centerline Elevation Profile
Schwab Model - Infinite Stress Band (SW-NE)
+ Wind Stress Shut Off at 72 hr

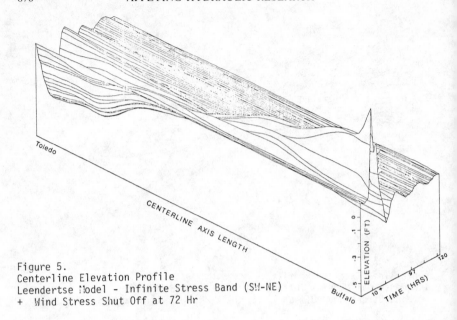

Figure 5.
Centerline Elevation Profile
Leendertse Model - Infinite Stress Band (SW-NE)
+  Wind Stress Shut Off at 72 Hr

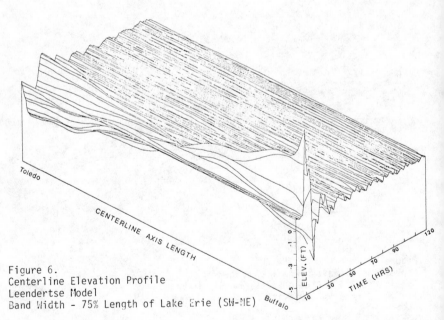

Figure 6.
Centerline Elevation Profile
Leendertse Model
Band Width - 75% Length of Lake Erie (SW-NE)

Table 1.  Resonant Simulation[1]

| Stress Propag. Direction | Stress Band Width | Model | Max. Elev. | Time | Min. Elev. | Time |
|---|---|---|---|---|---|---|
| SW-NE | ∞ | Schwab | 5.6 ft | 10 hr | -5.9 ft | 15 hr |
| SW-NE | ∞ | Leendertse | 7.8 ft | 9.5 hr | -3.9 ft | 14.5 hr |
| SW-NE | ∞ | 1-D | 6.6 ft | 9 hr | -6.0 ft | 15 hr |
| SW-NE | 187 mi[2] | Schwab | 5.6 ft | 10 hr | -4.2 ft | 7 hr |
| SW-NE | 187 mi | Leendertse | 8.2 ft | 9.5 hr | -2.3 ft | 7.5 hr |
| SW-NE | 187 mi | 1-D | 6.8 ft | 9 hr | -3.2 ft | 7 hr |
| SW-NE | 17 mi | Leendertse | 3.3 ft | 9.5 hr | -- | -- |
| NE-SW | ∞ | Schwab | 6.7 ft | 13 hr | -4.0 ft | 6 hr |
| NE-SW | ∞ | Leendertse | 5.5 ft | 12.5 hr | -3.5 ft | 13.5 hr |
| NE-SW | ∞ | 1-D | 5.06 ft | 13 hr | -4.8 ft | 12 hr |

[1] Basin gravity wave speed = 28 MPH.
[2] Band width equals 75% the length of Lake Erie.

## DISCUSSION

Several sets of simulations for various band widths were tested in the three models and are summarized in Table 1.  One object of the tests was to intercompare the results of each model for similar test cases. A second object is to see if any similarities existed between the computed water level fluctuations and the theoretical results obtained by Rao for his 'universal' lake.  Several striking differences occur between each model.  For the southwest to northeast infinite stress band moving at an average Lake Erie resonance speed, the Leendertse model shows a large 'flip' in water level at the eastern end of the basin that is 1-2 feet larger than the other two models.  A smaller flip can be observed in the one dimensional model (Figures 3 - 4), but no flip is evident in Schwab's model.  This so called flip is the resonance contribution to the actual storm surge produced by the wind stress band.  Since the whole lake is eventually subjected to a constant wind stress, each model predicts a set-up magnitude similar to each other if the flip contribution is eliminated.  When resonance occurs, the exchange of energy between the lake and the stress band will produce a set up that is noticeably larger.  When the wind stress band was propagated at a velocity slower than c ($\sqrt{gH}$), no eastern basin flip in water levels occurred in any of the models.  The exclusion of the non linear terms in Schwab's model is one explanation for the absence of the resonance contribution to the storm surge, however, when the non linear term was added to the method of characteristics solution, no flip at the eastern end of the lake was evident in the solution.  A second reason is that the calculation of the wind stress band is determined from land based winds and the fact that the center line elevations are determined from calculations based solely on the spatial dependence of an impulse function for these land based stations and not on calculations over the whole lake grid.

The time to maximum and minimum water elevations are compared for the infinite stress band runs.  Peak water level rises for each model occur between the 9th and 10th hour in the simulation. Assuming an average longitudinal length for Lake Erie of around 250 miles, a gravity wave travelling at 28 MPH will take about 9 hours to travel across the lake. Thus, the peak water level fluctuations occur at the approximate time for a gravity wave to travel across the lake, a similar result obtained by Rao. In actuality, because of variable topography in Lake Erie, the gravity wave speed will very from basin to basin. In comparing the time to minimum water level at the west end of the lake, this time occurs between the 14.5 and 15th hour of simulation, thus indicating a lag of around five hours for each model. This compares favorably to Rao's theoretical results in his constant depth lake. Discrepancies in the lag times can be attrivuted to changes in lake topography and irregular lake boundaries.

Wind stress bands of finite width were tested in the models. For a band width of 75% the length of Lake Erie, a large flip greater than for the infinite band case occurred in the Leendertse model, and no flip is evident in the Schwab model. The one dimensional profile is similar to the infinite band case. The times for maximum water level elevations are the same as for the infinite case. This occurrence is contrary to Rao's findings since his results indicated that maximum elevation will be produced by infinite stress bands. For finite band sizes greater than the length of the lake, resonant flips occurred in the Leendertse and one dimensional model at resonant band speeds and speeds greater than the resonant speed. For speeds less than C, no resonance flips are evident in the models. To test the sensitivity of the Leendertse model to band widths of a very small size, a seventeen (17) mile band width, which is equal to the smallest widths that can be calculated due to the time step of the model, was propagated across the lake. The results of these calculations show that even small band widths can produce water level changes of greater than two feet.

Northeast to southwest propagating stress bands were also calculated in the model. These are important in Lake Erie because they often induce severe flooding and erosional problems along the shores of the Western Basin of the lake and are often misforecasted by models due to the improper handling of the non linear effects in the Western Basin. The standout feature in these simulations is that a resonant flip at the Western end of the basin is not evident in any of the models for resonant, subresonant, and superresonant propagation speeds. The times to high and low water elevations are similar in the non linear models but the time to minimum water level in the Schwab model occurs after only six hours with another minimum water level of the same magnitude occurring 8 hours later. Apparently the models fail to detect resonance phenomena moving in a east to west direction. Although the actual occurrences of east to west moving disturbances across Lake Erie are rare, the consequences of such events can be damaging due to the small depths of the Western Basin and very flat terrain surrounding the lake.

## CONCLUSIONS

In general, the non linear models detected resonance phenomena for southwest to northeast stress bands of various widths travelling at speeds greater than or equal to the average celerity of a gravity wave. The magnitudes of the set ups for finite band widths were as large as the infinite stress bands. All northeast to southwest computer runs failed to detect resonance. To improve the results of the models new terms will be added, such as a non-hydrostatic pressure assumption in the models, and the inclusion of the non linear terms in Schwab Unit Convolution Model.

## REFERENCES

Chaston, Peter R., 1979: The Rochester Seiche. Weatherwise, 32, 211.

Donn, William L., 1959: The Great Lakes Storm Surge of May 5, 1952. J. Geophys. Res., 64, 191-198.

Douglas, C. K. M., 1929: The Line Squall and Channel Wave of July 20, 1929. Meterol. Mag, 64, 187-189.

Ewing, M., F. Press, and W. L. Donn, 1954: An Explanation of the Lake Michigan Waves of 26 June 1954. Science, 120, 654-686.

Gilbert, G. K., 1897: Modification of the Great Lakes by Earth Movements, Nat. Geog. Mag., 8, 233-247.

Harris, D. Lee, 1957: The Effects of a Moving Pressure Disturbance on the Water Level in a Lake. Meterol. Monographs, 2(10), 46-57.

Leendertse, J. J., 1970: A Water Quality Simulation Model for Well Mixed Estuaries and Coastal Seas. RM-6230-RC, Principles of Computation, Vol. 1, The Rand Corporation, 70 pp.

Leendertse, J. J., and E. C. Gritton, 1971: A Water Quality Simulation Model for Well Mixed Estuaries and Coastal Seas, R-708-NYC, Computation Procedures, Vol. 2, The Rand Corporation, 53 pp.

Liggett, James A., and Jean A. Cunge, 1975: Numerical Methods of Solution of the Unsteady Flow Equations, Chapter 4 in Unsteady Flow in Open Channels, (K. Mahmood and V. Yevjevitch, eds.) Water Res. Publications, 89-182.

Platzman, George W., 1958: A Numerical Computation of the Surge of 26 June, 1954 on Lake Michigan. Geophysica, 6, 407-438.

Platzman, George W., 1963: The Dynamical Prediction of Wind Tides on Lake Erie, Meterol. Monographs, 4(26), 1-41.

Pore, N. Arthur, H. P. Perrotti, and W. S. Richardson, 1975: Climatology of Lake Erie Storm Surges at Buffalo and Toledo, NOAA Tech. Memo, NWS TDL-54, 26 pp.

Proudman, J., 1953: Dynamic Oceanography, London, Methuen, 409 pp.

Rao, Desiraju, 1967: The Response of a Lake to a Time Dependent Wind Stress, J. Geophys. Res., 72, 1697-1708.

Schwab, Davis J., 1978: Simulation and Forecasting of Lake Erie Storm Surges, Mon. Wea. Rev., 96, 1476-1487.

Sykes, Robert M., Keith W. Bedford, and Charles Libicki, 1981: A Dynamic Advective Transport Model for Storm Water Quality Assessment. Inter. Symp. Urban Hydrology Hydraulics and Sediment Control, 219-227.

Whittlessey, Charles, 1874: Sudden Fluctuations of Levels in Quiet Waters - Records of Observations. Amer. Assoc. Advan. Science, 23, 139-143.

# MODEL VERIFICATION USING TIDAL CONSTITUENTS

Douglas G. Outlaw[1] and H. Lee Butler[2]

ABSTRACT

Tidal model verification using tidal constituents provides an improved technique for accurate and reliable calibration and verification of numerical tidal circulation models. The amplitude and phase of the constituent tidal surface elevation and velocity at selected stations in the estuarine area and in the open-ocean remote from the influence of the estuary can readily be compared in both the model and prototype. Using verification by constituents, the model is initially calibrated for the predominant tidal constituent in the study area. The relative phase lag between stations in the model data quickly shows areas where further model calibration is required.

The verification technique includes acquisition of time series tidal data (normally a minimum of 29 days), data editing, band-pass filtering to remove nontidal trends, harmonic analysis, and analysis of the data residual after completion of the harmonic analysis to ensure that all significant constituents have been included in the analysis. The technique, previously used in a limited verification of a physical model of Murrells Inlet, South Carolina, has been applied during a tidal circulation study of Lake Pontchartrain and vicinity.

Tidal analysis results showed that the diurnal O1 and K1 constituents have the largest amplitude in Lakes Pontchartrain and Borgne and indicate that the diurnal tides in Lake Pontchartrain are co-oscillating with little change in constituent amplitude over the lake. Prototype data and model results from an implicit finite-difference model of the study area indicated excellent agreement for both the O1 constituent and over mean, spring, and 14-day tidal cycles.

## INTRODUCTION

Prior to use of a tidal model, either physical or numerical, for evaluation of proposed changes to a tidal estuary or harbor, the model must be verified using known data for existing conditions. After demonstrating the capability to reproduce tidal data for existing conditions, a reasonable degree of confidence can be placed in the model results. Usually, limited tidal data for existing conditions will be available and a prototype data acquisition and analysis program must be conducted as a part of a model study.

---

[1]Research Hydraulic Engineer, Wave Dynamics Division, Hydraulics Laboratory, U. S. Army Engineer Waterways Experiment Station, CE, Vicksburg, Miss.

[2]Research Physicist, Wave Dynamics Division, Hydraulics Laboratory, U. S. Army Engineer Waterways Experiment Station, CE, Vicksburg, Miss.

Verification techniques using tidal constituents for elevations and currents provide an improved method for accurate and reliable model verification. The technique was used during a study of tidal circulation in Lake Pontchartrain and vicinity. Previously, the constituent verification technique had been used during verification of a physical model (3) of Murrells Inlet, South Carolina, using a single constituent only.

PROTOTYPE OBSERVATIONS

Lake Pontchartrain, Fig. 1, is located north of and adjacent to the city of New Orleans, Louisiana. The two major natural outlets of the lake are The Rigolets and Chef Menteur Pass between Lakes Pontchartrain and Borgne. Lake Pontchartrain also is connected with the Gulf through the Inner Harbor Navigation Canal (IHNC), Intracoastal Waterway, and the Mississippi River-Gulf Outlet Channel (MR-GO). Lake Maurepas, west of Lake Pontchartrain, and Lake Pontchartrain are connected by Pass Manchac. Lake Pontchartrain is about 25 miles (40.23 km) wide at its widest point and is about 40 miles (64.37 km) long. Tides are primarily diurnal in Lake Pontchartrain and the surrounding study area. Mean tide range (2) at Long Point in Lake Borgne is 1.0 ft (0.31 m) and is 0.4 ft (0.12 m) in Lake Pontchartrain.

Prototype data acquisition included an intensive tidal elevation and current data program for a duration of approximately 30 days and a long-term water quality program lasting for approximately one year immediately following the intensive current data program. Data stations for the intensive program included:

a.  Tidal elevation measurements at 18 stations.

b.  Tidal current measurements at 21 stations.

c.  Conductivity and temperature measurements at 9 stations.

d.  25-hour current survey over 6 ranges.

e.  Wind speed and direction measurements at 2 stations.

f.  Conductivity, temperature, dissolved oxygen, and pH measurements at three depths at periodic distances along a set of predetermined transects in Lake Pontchartrain and Lake Borgne.

Water quality data stations included:

a.  Tidal elevation measurements at 18 stations.

b.  Tidal current, temperature, and conductivity measurements at 14 stations.

c.  Wind speed and direction measurements at 2 stations.

d.  Salinity measurements at 19 stations.

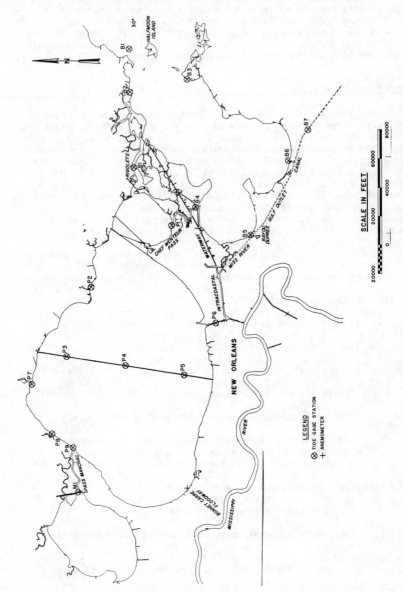

FIG. 1.—Vicinity map and location of tide gage and weather stations

MEASUREMENT LOCATIONS

WES tide gage locations, designated by prefixes B, M, P, and R for
Lake Borgne, Lake Maurepas, Lake Pontchartrain and The Rigolets, are
shown in Fig. 1 and were identical for both the intensive program and
the water quality program. For the intensive data acquisition program,
16 current meters (ENDECO Models 105 and 174) were installed at 5 stations along The Rigolets, 8 meters at 3 stations in Chef Menteur Pass,
5 meters at 5 stations along the MR-GO and connecting canals to Lake
Borgne, 1 meter in the Intracoastal Waterway, 1 meter near the mouth of
the Pearl River, and 5 meters in Lake Pontchartrain, near Tchefuncta and
Tangipahoa Rivers, and in Pass Manchac.

ENDECO Model 174 current meters were installed at middepth in the
water quality data acquisition program at 14 stations. Station locations were selected to record the variation in temperature and conductivity within Lake Pontchartrain and the principal inlets to and discharges from the lake. Weather stations were located near the southwest
shore of Lake Pontchartrain and near the east end of the lake and are
shown in Fig. 2. Six ranges in the 25-hour survey were selected to provide current, temperature, and conductivity data in the IHNC, Chef Menteur Pass, and The Rigolets.

Tide gages were installed in the study area between 10 August 1978
and 10 October 1978 and remained in place for approximately one year.
The intensive current data acquisition program was conducted during the
fall of 1978. The current data acquisition for the water quality program commenced following the intensive program.

DATA ACQUISITION EQUIPMENT

Fisher and Porter Type 150 surface elevation level recorders were
used at all tide stations and recorded at 6-min intervals. ENDECO Model
105 and 174 current meters used for current data acquisition in the
study area are axial flow, ducted impeller, tethered meters designed
specifically for use on the continental shelf and in estuarine areas.
The Model 105 meter monitored current only and recorded data on film.
The Model 174 meter monitored temperature and conductivity as well and
recorded data on magnetic tape. The recording rate for the Model 105
and 174 meters was 30 min and 2 min, respectively. Normally, the current meters were techered by a 5-ft (1.52 m) line from the current meter
to a mooring cable. The mooring cable was anchored at the bottom and
suspended vertically by a submerged float attached to the top of the
cable.

Wind speed (run), wind direction, and air temperature were recorded
using a Meteorology Research, Incorporated, Mechanical Weather Station
with data recorded on strip chart.

HARMONIC ANALYSIS

Similar analysis techniques were used for both the surface elevation and current data. The following steps were included in the analysis procedure:

a. Edit to remove data spikes and mean of the data record.

b. Filter to remove high- and low-frequency trends from the data.

   c. Harmonic analysis for tidal constituents.

   d. Analysis of data residual.

Editing of surface elevation data was necessary to correct for spikes in the data record and to fill in short sequences of missing data. A digital band-pass filter was applied to attenuate high and low frequencies. The period range considered in the harmonic analysis was approximately 3 to 28 hr. The tidal range in the study area is relatively low and is sensitive to meteorological effects. An eight-pole Butterworth filter, characterized by a smooth power gain with maximum flatness in the passband and the stop band along with a reasonably sharp cutoff, was applied to the surface elevation and current data. Half-power frequencies are $1.85(10^{-4})H_z$ and $0.0842(10^{-4})H_z$ or 1.5 hr and 33 hr, respectively.

Elevation of the prototype tide at a station can be represented (4) by

$$h(t) = H_o + \sum_{i=1}^{J} f_i H_i \cos\left[\bar{a}_i t + (V_o + u)_i - K_i\right] \quad (1)$$

where

      $h$ = elevation at time $t$

      $t$ = time reckoned from some initial epoch

      $H_o$ = mean height above reference datum

      $J$ = total number of constituents

      $f_i$ = factor to reduce mean amplitude to year of prediction

      $H_i$ = mean amplitude of $i^{th}$ constituent

      $\bar{a}_i$ = angular speed of $i^{th}$ constituent

  $(V_o + u)_i$ = equilibrium argument of the $i^{th}$ constituent for $t = 0$

      $K_i$ = local epoch of $i^{th}$ constituent

The coefficients $f_i$, $\bar{a}_i$, and the equilibrium argument can be calculated or obtained from tables (4). Equation 1 may be rewritten as

$$h(t) = H_o + \sum_{i=1}^{J} A_i \cos(\omega_i t + \emptyset_i) \quad (2)$$

where

    $A_i = f_i H_i$ = amplitude of the $i^{th}$ constituent

    $\omega_i$ = angular frequency of the $i^{th}$ constituent

    $\emptyset_i$ = phase of the $i^{th}$ constituent

Observed prototype surface elevation data $h_p$ can be represented (2) as

$$h_p(t) = \bar{h}_p(t) + \varepsilon(t) = a_o + \sum_{i=1}^{J} a_i \cos(w_i t)$$

$$+ b_i \sin(w_i t) + \varepsilon(t) \qquad (3)$$

where

$\bar{h}_p$ = the calculated tidal elevation represented by a harmonic series of known frequencies

$\varepsilon(t)$ = noise in observed data

$a_o, a_i$, and $b_i$ = coefficients

The noise level is not known and the unknown coefficients (amplitudes and phases) are solved for by minimizing the variance of the sum of the squared difference between the observed prototype tidal elevation data and the form represented by Equation 2 using a least squares procedure. The least squares procedure minimizes the variance E such that

$$E = \sum_{n=1}^{N} \varepsilon^2 (n\Delta t) = \sum_{n=1}^{N} \left[ \bar{h}_p(n\Delta t) - h_p(n\Delta t) \right]^2 \rightarrow \text{minimal} \qquad (4)$$

where N is the total number of data samples and $\Delta t$ is the time interval between consecutive samples. To minimize the variance, set

$$\frac{\partial E}{\partial a_i} = 0 \ , \quad i = 1, \ldots, J \qquad (5)$$

and

$$\frac{\partial E}{\partial b_i} = 0 \ , \quad i = 1, \ldots, J \qquad (6)$$

For each tidal constituent, the amplitude $A_i$ and the phase $\emptyset_i$ can be determined from

$$A_i = \left( a_i^2 + b_i^2 \right)^{1/2} \qquad (7)$$

and

$$\emptyset = \arctan \left( \frac{b_i}{a_i} \right) \qquad (8)$$

Results from the least squares analysis may then be expressed as the tidal amplitude and local epoch. Current data were analyzed similarly for the predominant direction of flood and ebb flow.

TIDAL CONSTITUENTS

Tidal constituents in the data analysis included diurnal, semidiurnal, and shallow-water overtide components. Previous harmonic analyses by the Coast and Geodetic Survey (5) of tidal elevation data from several stations in and near Lake Pontchartrain indicated that the principal tidal components were the diurnal constituents K1, O1, P1, and the semidiurnal constituents S2 and M2.

ANALYSIS RESULTS

Typical observed tidal elevation data for Julian days 320 through 350 in 1978 are shown in Fig. 2 gage P-4 near the center of Lake Pontchartrain. The elevation data show the effect of both tidal fluctuations and low-frequency trends in the data. The tidal range has decreased in Lake Pontchartrain and the low-frequency trends are more apparent.
The principal elevation tidal constituents were O1, K1, P1, M2, and S2. Variation of the O1 constituent mean amplitude and ranged from 0.48 ft (0.15 m) at gage B-1 (Mississippi Sound) to 0.07 ft (0.02 m) at gage M-1 between Lake Pontchartrain and Lake Borgne. The K1 constituent mean amplitude followed a similar trend and ranges from 0.58 ft (0.18 m) at gage B-1 to 0.07 ft (0.02 m) at gage M-1.
Analysis results for the O1, K1, P1 constituents indicate the formation of a co-oscillating tide or a forced oscillation in Lake Pontchartrain with a general rise and fall of the water level over the lake. The co-oscillating tide in Lake Pontchartrain results in a similar phase lag being observed for the O1, K1, and P1 constituents along the causeway bridge and at the western end of the lake.
The larger semidiurnal constituents, M2 and S2, decrease from approximately 0.05 to 0.10 ft (0.02 to 0.03 m) in Lake Borgne to about 0.01 ft (0.003 m) in Lake Pontchartrain. Amplitude of the Q1 diurnal constituent varies from about 0.10 to 0.05 ft (0.03 to 0.02 m) in Lake Borgne to 0.02 ft (0.01 m) in Lake Pontchartrain. Amplitudes of the less significant constituents are negligible in Lake Pontchartrain (less than 0.02 ft (0.01 m)). Amplitude data for the M4, M6, and M8 overtides indicate they are not significant (less than 0.02 ft (0.01 m)) in the study area.
Mean amplitudes of the O1 constituent current in the flood-ebb direction are variable from station to station and range from 1.35 fps (0.41 mps) in the Rigolets to 0.17 fps (0.05 mps) in the Intracoastal Waterway.

NUMERICAL MODEL APPLICATION

The numerical model used in this study is described by Butler (1). The theoretical background is summarized in the following paragraphs. Hydrodynamic equations used in the WES implicit flooding model (WIFM) are derived from the classical Navier-Stokes equations in a Cartesian coordinate system. By assuming vertical accelerations are small and the fluid is homogeneous, and intergrating the flow from sea bottom to water surface, the usual two-dimensional form of the equations of momentum and continuity is obtained.
A major advantage of WIFM is the capability of applying a smoothly varying grid to the study region, permitting simulation of complex

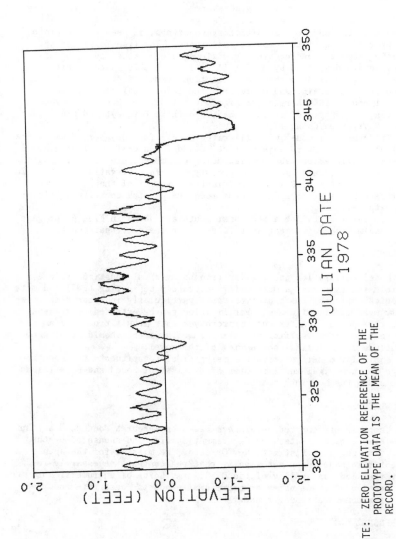

NOTE: ZERO ELEVATION REFERENCE OF THE PROTOTYPE DATA IS THE MEAN OF THE RECORD.

FIG. 2.--Observed tide elevation data at gage P-4 for day 320 to day 350 1978

landscapes by locally increasing grid resolution and/or aligning coordinates along physical boundaries. For each direction, a piecewise reversible transformation which takes the form

$$x = a + b\alpha^c \qquad (9)$$

where $a$, $b$, and $c$ are arbitrary constants, is used to map prototype or real space into computational space. The transformation and resulting momentum and continuity difference equations and solution technique are described by Butler (1). The variable spaced finite-difference grid used in the numerical tidal simulation included Lakes Borgne, Pontchartrain, and Maurepas. The major tidal boundary was at the east end of Lake Borgne northward across St. Joe Pass. The model tide was also specified at gages B-5, B-6 (Lake Borgne), and P-6 (Inner Harbor Navigation Channel).

The numerical model was calibrated over a four-day period using the O1 tidal constituent to adjust friction and flood cell coefficients. The model calibration was verified using a combined nine constituent tide. A comparison of model and prototype surface elevation is shown in Fig. 3 for gages M-1, R-1, and P-1 in the interior of the study area. After several days for spin-up, the model data match constituent elevation data well.

Similar results for a semilunar month are shown in Fig. 4 for gage P-1. Again, good agreement is obtained over the semilunar month.

CONCLUSIONS

Model calibration and verification using tidal constituents are feasible and represent a method for accurately separating tidal and meteorological effects from prototype data, particularly in areas with relatively weak tidal influence. Verification can be extended over any desired time period. An extensive prototype data acquisition program is necessary to obtain sufficient data for analysis and should be planned to include for all tidal boundaries of the model.

A similar model verification using tidal constituents has been completed for numerical model studies of Mississippi Sound and Fire Island Inlet along the New York coast.

ACKNOWLEDGMENT

The results reported herein are based on research conducted at the U. S. Army Engineer Waterways E..periment Station and sponsored by the U. S. Army Engineer District, New Orleans. Permission for the author to present this paper does not signify that the contents necessarily reflect the views and policy of the U. S. Army Corps of Engineers, nor does mention of trade names or commercial products constitute endorsement or recommendations for use.

REFERENCES

1.  Butler, H. L., "WIFM-WES Implicit Flooding Model, Theory and Program Documentation" (in publication), U. S. Army Engineer Waterways Experiment Station, CE, Vicksburg, Miss.

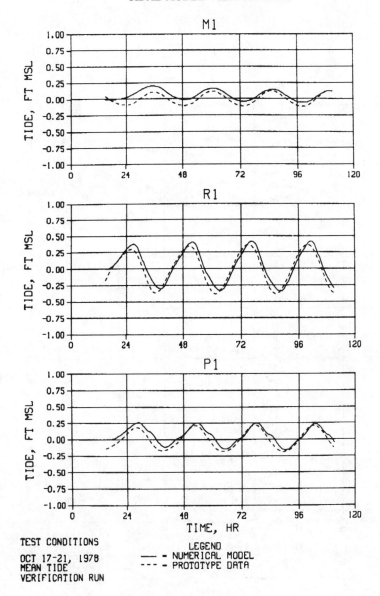

FIG. 3.—Model verification comparison for gages M1, R1, P1

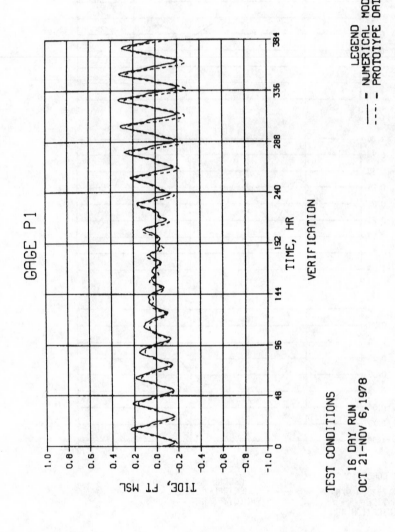

FIG. 4.--Comparison of model and calculated constituent tidal data for a semilunar month

2. Outlaw, D. G., "Lake Pontchartrain and Vicinity Hurricane Protection Plan; Prototype Data Acquisition and Analysis," Technical Report HL-82-2, Report 1, U. S. Army Engineer Waterways Experiment Station, CE, Vicksburg, Miss., Jan., 1982.

3. Perry, F. C., Jr., Seabergh, W. C., and Lane, E. F., "Improvements for Murrells Inlet, South Carolina; Hydraulic Model Investigation," Technical Report H-78-4, U. S. Army Engineer Waterways Experiment Station, CE, Vicksburg, Miss., Apr., 1978.

4. Shureman, P., "Manual of Harmonic Analysis and Prediction of Tides," Special Publication No. 98, U. S. Department of Commerce, U. S. Coast and Geodetic Survey, 1958.

5. U. S. Department of Commerce, Coast and Geodetic Survey, "Tidal Harmonic Constants, Atlantic Ocean Including Arctic and Antarctic," TH-1, Washington, D.C., Jan., 1942.

SYSTEMATIC EVALUATION OF THE OPTIMAL LOCATION
OF MULTILEVEL INTAKES FOR RESERVOIR RELEASE QUALITY CONTROL

by Jeffery P. Holland[1]

INTRODUCTION

As a result of increasing public awareness and State and Federal
legislation, water resources projects are being operated with a greater
priority on water quality considerations. Many proposed projects are
being designed to operate for given water quality objectives. Furthermore, many existing projects are being retrofitted in order to meet
water quality requirements. The use of a reservoir outlet works
incorporating multilevel selective withdrawal structures is a primary
method for the control of reservoir release quality. These structures
allow release of water from various vertical strata in the lake, thereby
allowing, through blending or direct release, greater water quality
control. Although reservoirs may be operated for a variety of water
quality objectives, the most common (and the objective of this discussion) is maintenance of the temperature of the release from the
impoundment by means of selective withdrawal in order to meet a prescribed downstream temperature. It is imperative, therefore, that the
selective withdrawal intakes be placed in such quantity and location as
to maximize the control of reservoir release temperature over a range of
hydrological, meteorological, and operational conditions. However,
these intakes should be sited in a manner which is also cost-effective.
An approach is therefore required that systematically determines the
optimal number and locations of these intakes required to meet downstream temperature objectives for a range of conditions.

The purpose of this report is to detail a procedure whereby, for a
given set of conditions (inflow, outflow, meteorology, etc), an optimum
selective withdrawal intake configuration may be obtained. This procedure is accomplished through the coupling of a reservoir thermal model
and a mathematical optimizer. The thermal model requires as inputs
hydrological and meteorological data to simulate the pattern of thermal
stratification, or "state", of the reservoir over a given simulation
period. Algorithms simulating operation of the selective-withdrawal
structure for a specified intake configuration are used in the model to
predict the reservoir outflow distributions and, subsequently, the
downstream release temperatures. A scalar index of performance (objective function) is used to measure the effectiveness of a specific intake
configuration at meeting downstream temperature objectives. The objective function is computed based on the deviation of the predicted downstream release temperatures from the given downstream objectives. Using

_____

[1]Research Hydraulic Engineer, US Army Engineer Waterways Experiment
Station, Vicksburg, MS  39180

the specified intake configuration and its corresponding objective
function value as inputs, the mathematical optimization routine considers alternative selective withdrawal intake configurations which,
when simulated, produce smaller objective function values. Minimization
of the objective function produces the optimal selective withdrawal
intake configuration whose operation results in the minimum deviation of
downstream release temperature from the given downstream temperature
objective. Comparison of results from repetition of this procedure for
varying numbers of intakes allows determination of the optimum number of
intakes required for the given set of conditions. An overview of the
components of this procedure appears below.

## Reservoir Thermal Model Description

Downstream release temperatures and in-lake temperature characteristics were predicted using a reservoir thermal model. The model,
WESTEX (1), used in conjunction with this investigation was developed at
the US Army Engineer Waterways Experiment Station (WES). The WESTEX
model, which is based on the solution of the one-dimensional (vertical)
thermal energy equation, provides a procedure for examining the balance
of thermal energy imposed on an impoundment. This energy balance and
lake hydrodynamic phenomena are used to map vertical profiles of temperature in the time domain. The model includes computational methods for
simulating heat transfer at the air-water interface, heat advection due
to inflow and outflow, and the internal dispersion of thermal energy.
In addition, a subroutine in the model, DECIDE, is used to simulate the
operation of the selective withdrawal structure. For each simulation
step (generally 1 day) the DECIDE subroutine evaluates the thermal
structure of the reservoir, the desired temperature objective, and the
total flow to be released downstream. Based upon the operational constraints of the selective withdrawal structure, the DECIDE subroutine
determines the combination of selective withdrawal intakes and the flows
to be released through those intakes such that the release temperature
is as close as possible to the downstream temperature objective. The
operational constraints of the selective withdrawal structure considered
by the DECIDE subroutine include hydraulic constraints on intake operations, such as minimum and maximum allowable flows, intake geometry,
number of wetwells, and floodgate capacity. A thorough discussion of
the WESTEX model and its input data appears in Holland (2).

## Mathematical Optimization Description

Discussion of the optimization routine used in the intake location
procedure must be prefaced by an explanation of the role of an optimizer.
Explanation of this role begins by first examining the following sequential manner in which the intakes were located:

a. Provide an initial estimate for the location(s) of the intake(s)
   needed for maintenance of downstream temperature objectives.

b. Use this intake configuration as an input into the WESTEX thermal model in order to simulate the in-lake and release characteristics resulting from operation of this intake scheme.

c.  Compute an objective function (described in the next section)
based on the deviation of downstream temperature objectives and
release temperatures computed in step b.

d.  Determine a new estimate for the location of the selective withdrawal intake(s) based on steps a, b, and c which satisfies the
downstream temperature objective more effectively than the
previous intake configuration.

Steps b, c, and d are continued until a minimum value for the objective
function of step c is reached.  At the minimum objective function value,
the estimated configuration of the selective withdrawal intakes is the
"optimum" location for maintenance of downstream temperature objectives
for the conditions simulated.  Although this procedure could be accomplished manually, techniques of mathematical optimization systematically
examine the possible estimates for intake location and, in general,
converge to the best decision with the least expense of computer
resources.  Thus, an optimization routine, requiring the output of
steps a, b, and c as input, is used to search for (locate) the "optimum"
selective withdrawal configuration.

Three non-linear optimization routines have been coupled with the
WESTEX thermal model to search for optimum selective withdrawal intake
elevations. Two of the routines are univariate searches that find the
minimum of a function of one variable.  The first, a parabolic interpolation routine obtained from Boeing Computer Services (BCS), has been
used to locate an additional level of intakes (two intakes at the same
elevation) for an existing system due to project reformulation.  The
second univariate routine, a Golden Section search, has been used as part
of a cyclic coordinate search in order to evaluate the efficiency of an
existing multilevel intake configuration.  In the cyclic search, an
optimum elevation is obtained for a given intake using the Golden
Section search while holding all other intake elevations constant.  This
action is continued for each of the intakes to be located until a given
convergence criterion has been met.  The third optimization routine, a
conjugate search which minimizes a function of multiple variables (i.e.,
multiple intake locations) by Powell's Method, was also used to evaluate
the existing multi-level intake configuration.  Results of the use of
three of these methods are presented in a subsequent section.  The
parabolic interpolation routine obtained from BCS is described in Holland
(2); this routine was not used to locate multiple intakes due to search
inefficiencies.  The numerical methodology used as a basis for the
latter two optimization routines is described in detail by Box, Davies,
and Swann (3).  The routines were obtained from Colorado State University.

Objective Function Description

An objective function is a scalar index which relates the performance
of one possible decision (i.e., a particular intake configuration) in
meeting a specified system goal.  The goal of this work is to minimize
deviation of project downstream release temperatures from specified
downstream temperature objectives through the optimal location of
selective withdrawal intakes.  Minimization of this objective function
produces the optimal intake configuration for the given downstream

objectives.  Although many formulations can be envisioned which relate
to the severity of these deviations (and thus the performance of a given
intake system at meeting downstream objectives) in a numerical sense,
the objective function used in this procedure is based on the sum of the
squared deviation of release and objective (target) temperature over the
simulated period.  This formulation was chosen because minimization of
the sum of squared deviations smooths deviations from the downstream
objective temperature over the simulation period.  While this formulation does allow a greater number of less severe deviations than other
formulations, the severity of these individual deviations is reduced
on the downstream system.

    Both the univariate routine from BCS and the multivariate Powell
routine are unconstrained optimizers and therefore may attempt to locate
intakes within a non-feasible region (such as above the water surface).
In order to incorporate physical constraints in the intake location
process while using an unconstrained optimizer, additional terms referred
to as penalty functions are added to the objective function at non-
feasible points so that the unconstrained optimizer, in an effort to
minimize the objective function, will find these points unattractive as
solutions.  The Golden Section optimizer is a constrained routine requiring definition of the feasible search region as input.  Formulation
of the objective function for use with this optimizer therefore requires
no incorporation of penalty terms.

## Coupling of Optimization Components

    The three components described above have been incorporated in a
systematic numerical procedure.  A simplified schematic of this procedure
appears in Figure 1.  The following data are required as input for the
procedure:   (a) those hydrological, meteorological, and physical data
which are required as input into the thermal model;  (b) specification of
the initial estimate of the optimum selective withdrawal configuration;
and (c) assignment of the feasible search region (permissible region
within the reservoir for location of intakes) to be considered by the
optimizers.  This region is defined through the use of penalty functions
in the unconstrained routines (BCS, Powell) and by direct numerical
input into the constrained routine (Golden Section).  The objective
function is considered minimized, and the optimal selective withdrawal
intake configuration subsequently defined, when successive simulations
of two given intake configurations produce objective function values
which are different by less than a pre-defined tolerance.  This tolerance,
or convergence criterion, is also input into the procedures.  The cyclic
coordinate Golden Section routine requires specification of a second
convergence criterion which is used to establish the optimum location
for a given intake while all other intakes are held constant.  Additional
parameters which limit the number of function evaluations allowed for
an both individual search steps and the entire procedure are also incorporated into the procedure to limit the exhaustion of computer
resources.

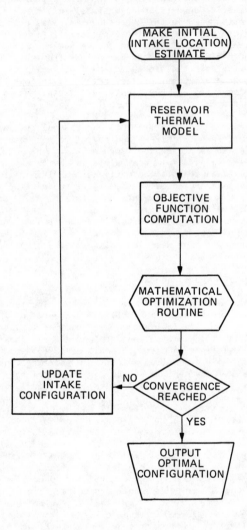

Figure 1:   Schematic Representation of Intake Location Procedure

APPLICATION OF PROCEDURE

The procedure described herein has been used in two different cases to optimally locate selective withdrawal intakes. The first use was for the retrofitting of an existing project (Project 1). The second use of the procedure was evaluation of an existing selective withdrawal structure (Project 2) in-house to show the inherent utility of this procedure. Results from each of these two applications are addressed.

Project 1

Reallocation of some flood control storage of Project 1 to water supply storage will result in increases of 30-40 ft in the normal pool elevation. These increases were deemed large enough to render the existing selective withdrawal system inadequate. The present system was designed for a normal pool depth of 45 ft (elevation 1045*). WES was requested to locate up to 6 intakes (as restricted by structural constraints) in addition to those present (2 intakes at centerline** el 1037.0; 2 intakes at el 1014.5) which would minimize deviation of downstream release temperatures from a downstream target temperature. This target was specified to be a $\pm 2.78^{\circ}$C ($5^{\circ}$F) band around the pre-project natural stream temperature for each day of simulation. The intake location procedure was run for four study years (1958, 1964, 1966, 1967) and three increased normal pool elevations (el 1075, el 1080, el 1085). The unconstrained optimization routine obtained from BCS was coupled with the WESTEX model for this effort.

Shown in Table 1 are the results from location of one additional level of intakes (two intakes at the same elevation, one in each of two wetwells in addition to those intakes already existing). As shown, the existing system was quite inadequate at meeting the downstream temperature objective. However, addition of one level of intakes at the elevations prescribed for each pool and year showed excellent improvement. Holland (2) was further able to show that the addition of a second intermediate set of intakes provided little improvement in maintenance of the temperature objective for the conditions simulated. This second additional set however, did provide system flexibility which could prove advantageous during filling and for conditions not specifically simulated. Based on these results, Holland (2) determined an optimum intake configuration for each of the three normal pool elevations which was effective for each of the study years at satisfying the downstream temperature objective. In addition, a single intake configuration was determined which was equally effective at satisfying the downstream objective for all pools and study years with the exception of the 1958/1085 pool condition. Thus, considerable resource savings can be realized at Project 1 through use of this procedure. The systematic evaluation of intake location provided by this procedure reduced the number of pool-specific intake configurations under consideration to a maximum of

---

* All elevations (el) cited herein are in feet referred to the National Vertical Geodetic Datum (NGVD).

** All intake elevations are centerline elevations in this paper.

three.  Further, with the exception of a single condition, one intake
configuration could be considered optimum for all the conditions
evaluated.  Finally, a maximum of four additional intakes was deemed
necessary, resulting in the savings of costs for two additional intakes
which could have been sited.

## Project 2

Project 2 has a gated intake tower which houses eight selective
withdrawal intakes in two wetwells at elevations ranging from el 547.5
to el 617.0.  In order to show further utility of the intake location
procedure described herein, operation of the existing project was
simulated to meet a specified downstream temperature objective.  Project
operations were then simulated for intake configurations consisting of
from 1 to 6 intakes whose optimum elevations had been determined by the
procedure.  A flood gate, at el 507.1, was used in all simulations.
Objective function values, consisting of the sum of squared deviation of
release and target temperatures, were computed for each intake configuration.  The downstream target temperatures were computed by first simulating operation of the existing project for a given set of hydrological and
observed meteorological conditions to produce predicted downstream
release temperatures.  A sine function was then fitted to these
temperatures.  This function was perturbated (target minimum and maximum
were made more extreme and the point of maximum temperature shifted to
earlier in the year) in order to make the target more difficult to meet
with operation of intake configurations exhibiting lesser flexibility
(fewer intakes) than the existing system.

Figure 2 shows the results of simulation of the existing system
when compared to optimum intake configurations for 1, 2, 3, 4, and 6
intakes obtained with the intake location procedure.  These results were
obtained with the Golden Section-cyclic coordinate search.  Use of
Powell's multivariate search produced similar results.  The elevations
of the intakes for each optimal configuration are given in Table 2.

Figure 2 further shows the utility of this intake location procedure.  Intake configurations consisting of from 2 to 6 optimally located
intakes met the specified target as or more effectively than the existing
design of 8 intakes.  While the effects simulated represent only one
condition, and therefore do not require the system flexibility necessary
for the design life of a project, the results do point out the possible
resource savings (fewer intakes required for objective maintenance)
associated with optimal location of selective withdrawal intakes.
Further, systematic evaluation of a variety of conditions with this
procedure should yield a more direct and cost-effective method of
intake location than presently exists.

A final utility of this procedure, and potentially the most important, is common to both retrofitted and new construction projects.
Siting of intakes in such a configuration which is optimal for a range
of conditions results, by definition, in the location of a number of
intakes which produce the smallest deviation of release and objective
temperatures over that range of conditions.  If this range is broad
enough to be representative of conditions which will actually occur at a
project, the intake configuration obtained with this procedure will be

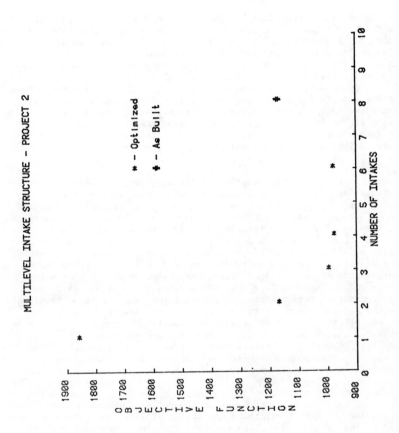

Figure 2: Comparison of Objective Function Values for Existing 8 Intake Configuration and Optimum Configurations of 1 to 6 Intakes, Project 2

quite effective for the majority of operating conditions.  This will
greatly lessen the likelihood of future retrofitting unless unanticipated
hydrological, meteorological, or operational conditions develop.

CONCLUSIONS

     A mathematical procedure has been developed which optimally locates
selective withdrawal elevations for a specified water quality objective,
(in this paper a downstream temperature objective).  This procedure
couples a reservoir thermal model with a mathematical optimization
routine to achieve this result.  The optimizer systematically evaluates
intake configurations whose operation minimizes the objective function
value (an index based on the sum of squared deviations of the downstream
objective and the predicted release temperature obtained by simulation
of the given intake configuration).  Minimization of this function
produces the optimal configuration of a given number of intakes for the
conditions simulated.  Further, repetition of this procedure for varying
numbers of intakes allows determination of the optimum number of intakes
required for the given conditions.

     The procedure has been applied to two projects, the first applica-
tion requiring the addition of selective withdrawal intakes beyond those
existing due to project reformulation.  Although six additional intakes
could be accommodated, this procedure found that only four were required.
The second application was formulated to show the ability of this procedure to optimally locate a specified number of intakes.  For the
specific conditions simulated, results from this procedure indicated
that operation of optimum configurations of from 2 to 6 intakes was as
or more efficient at meeting a downstream objective than the existing
8 intake configurations.  The possible savings in construction costs
point toward the attractiveness of locating the fewest intakes possible
while maintaining the required downstream water quality control.  Further,
the location of selective withdrawal intakes in a configuration which is
optimal for a representative range of conditions will significantly
lessen the probability of future retrofitting. The savings associated
with each of these points demonstrates the obvious utility of this
procedure.

     The procedure for the optimal location of selective withdrawal
intakes has recently been set up as one coordinated computer code.  The
code couples the WESTEX reservoir thermal model with the cyclic coordinate
search (using the Golden Section routine) in order to accomplish its
objective.  The Cyclic coordinate search with the Golden Section routine
was chosen over the Powell routine because it was easier to set up (no
penalty functions with the Golden Section routine, search limits read in
as input), had similar run times to those with the Powell routine, and
handled functional insensitivities more efficiently than the Powell
routine.  Further refinement of the procedure is now being completed.
Optimization of a multiple water quality objective remains a future
consideration.

TABLE 1

RESULTS OF SIMULATION OF EXISTING SELECTIVE WITHDRAWAL SYSTEM AND EXISTING SYSTEM WITH ONE ADDITIONAL SET OF INTAKES, PROJECT 1

| Year | Pool Elevation Above NGVD | Existing System | | Existing System Plus Additional 2 Intakes at Elevation Shown | | |
|---|---|---|---|---|---|---|
| | | Number of Days Release Fails to Meet Temperature Objective | Average Difference of Release Temperature From Objective (°C) | Optimum Elevation of Additional Intakes Above NGVD | Number of Days Release Fails to Meet Temperature Objective | Average Difference of Release Temperature From Objective (°C) |
| 1958 | 1085 | 148 | -4.18 | 1081.0 | 26 | -1.06 |
| 1964 | 1085 | 139 | -3.93 | 1067.0 | 5 | -0.50 |
| 1966 | 1085 | 161 | -5.74 | 1075.9 | 12 | -0.58 |
| 1967 | 1085 | 141 | -4.39 | 1075.4 | 9 | -0.54 |
| 1958 | 1080 | 142 | -4.34 | 1075.0 | 28 | -1.13 |
| 1964 | 1080 | 119 | -3.24 | 1063.0 | 5 | -0.39 |
| 1966 | 1080 | 151 | -4.95 | 1071.4 | 8 | -0.59 |
| 1967 | 1080 | 135 | -3.53 | 1076.4 | 8 | -0.40 |
| 1958 | 1075 | 147 | -3.93 | 1069.0 | 28 | -1.12 |
| 1964 | 1075 | 100 | -2.53 | 1062.0 | 2 | -0.21 |
| 1966 | 1075 | 142 | -4.15 | 1066.7 | 9 | -0.60 |
| 1967 | 1075 | 115 | -2.88 | 1070.0 | 8 | -0.39 |

APPLYING HYDRAULIC RESEARCH

ACKNOWLEDGMENTS

    The tests described and the resulting data presented herein, unless
otherwise noted, were obtained from research conducted under the Flood
Control Hydraulics Program and the Environmental and Water Quality
Operational Studies of the United States Army Corps of Engineers by the
US Army Engineer Waterways Experiment Station. Permission was granted by
the Chief of Engineers to publish this information.

REFERENCES

1.  Loftis, Bruce, "WESTEX - A Reservoir Heat Budget Model," In
    Preparation, U. S. Army Engineer Waterways Experiment Station,
    Vicksburg, MS.

2.  Holland, J. P., "Effects of Storage Reallocation on Thermal
    Characteristics of Cowanesque Lake, PA," Technical Report HL-82-9,
    U. S. Army Engineer Waterways Experiment Station, Vicksburg, MS.,
    May 1982.

3.  Box, M. J., Davies, D., and Swamm, W. H., "Non-Linear Optimization
    Techniques," Published for Imperial Chemical Industries, Limited by
    Oliver and Boyd, Ltd., Edinburg, 1969.

TABLE 2

OPTIMUM ELEVATION OF INDIVIDUAL INTAKES FOR CONFIGURATIONS OF
1, 2, 3, 4, AND 6 INTAKES, PROJECT 2

| Number of Intakes In Optimum Configuration | Elevation Above NGVD Of Intake Number as Shown | | | | | |
|---|---|---|---|---|---|---|
| | 1 | 2 | 3 | 4 | 5 | 6 |
| 1 | 617.1 | | | | | |
| 2 | 625.0 | 614.3 | | | | |
| 3 | 626.7 | 621.4 | 606.8 | | | |
| 4 | 627.1 | 623.2 | 605.6 | 563.3 | | |
| 6 | 627.0 | 622.7 | 604.7 | 575.3 | 565.5 | 516.5 |

# RESERVOIR DESTRATIFICATION: TECHNIQUES AND MODELS

Brian Henderson-Sellers[1]

## ABSTRACT

Inflows to bodies of fresh water may originate from any one of four basic sources: 1) natural streams, 2) heated effluents, e.g. into cooling ponds, 3) water input from a neighbouring catchment, usually by pipeline and 4) inflows as part of an oscillating two-lake pumped storage hydro-electric scheme.

The hydraulic forcing of natural streams is usually low, except in times of flood. This case is described by concatenating a catchment model with a reservoir model. Mathematical description of the remaining three categories often utilise theories of jets and plumes. The mixing of the influent jet may be described both in terms of the dilution of the incoming water and in terms of its effect upon the lake thermal structure.

As an alternative destratification technique the use of various air bubble systems is discussed. The use of air lines, bubble guns and Helixors are compared and two new modelling approaches outlined.

It is concluded that the method used for destratification depends upon many parameters including desired usage of the stored water, morphometric characteristics and climatic regime.

## INTRODUCTION

During Summer stratification, the physical and chemical characteristics of the hypolimnion of a lake or reservoir may change sufficiently for anoxic conditions to prevail, for treatment costs to be increased and release-water quality to adversely affect the downstream ecosystem. Stratification results from the non-linear interaction between buoyancy and wind induced mixing and may be prevented in cases of high wind speed or when there is a high degree of hydraulic forcing. Periods of non-stratification during Summer resulting from largescale wind mixing events (more frequent in certain climatically defined regions) are well described in the literature and these can be simulated by one-dimensional models. A second natural destratification can result from enhanced flows in tributaries which may cause an underflow or interflow current. Large volumes of influent water during flood peaks can be modelled simply in a one-dimensional stratification model by addition of slices at the appropriate level. The numerical simulation described here concatenates a catchment model with a 1-D thermocline model. In contrast artificial inflows may often be modelled using jet/plume theory. These include heated effluents, e.g. into cooling ponds; water pumped from a neighbouring catchment and inflows as part of a two-lake hydropower scheme.

---

[1]Lect. in Environmental Sci.Civ., Dept. of Div.Engrg., Univ. of Salford, Salford, England.

INFLOW STREAMS

The phenomena of underflow and interflow result when an inflowing stream is of a density greater than the lake surface water density. The incoming water downstream of the "plunge point" can be considered to have attained an equilibrium level, $z_e$. Assuming that it will disperse throughout the reservoir at this level, then the inflow volume, $V_i$, can be represented in a one-dimensional stratification model (e.g. Stefan and Ford, 1975, Henderson-Sellers, 1978b) by the addition of a volume $V_i$ to the model "slice" at the level $z_e$. This can be accomplished in terms of fixed slice models where the upper slice thickness is allowed to vary (e.g. Henderson-Sellers, 1978b) or by allowing the slice receiving the water to vary in thickness as in the WES model CE-QUAL-R1 (Robey, p.c., 1982). (The former approach is used here). A time series of the inflows is simulated using a catchment model. The results shown here concatenate the Pitman (1976) model with the one-dimensional University of Salford eddy diffusion (USED) model. To demonstrate the capabilities of this joint model, a time series for the Juskei River in the Transvaal is used as input. This gives a marked seasonality in the time series with peak flows (of the order of 5 $m^3$ $s^{-1}$) in late Spring and early Autumn. Although such a time series is not representative of <u>mean</u> U.K. conditions it does provide a first experiment in trying to understand and evaluate the destratifying effects of different flow regimes.

Figures 1 and 2 show the different effects on the stratification of an inflow at heights of 15 m and 40 m (above the lake floor) respectively for a 50 m deep lake. (In both cases a volumetric withdrawal of 0.75 $m^3$ s $^{-1}$ is incorporated from the second (finite difference) slice from the surface). It is evident that denser inflows affect the stratification little, whilst for the less dense inflow, the thermocline may suffer a downwards erosion of 10-20 m. Further numerical experiments are to be carried out with other flow regimes and for different climatic situations.

JET/PLUME MODELS

The three remaining inflow categories are related to different reservoir design aims. Water bodies acting as cooling ponds receive inflows in the form of heated discharges often below the surface (Figure 3). The hot water rises due to excess buoyancy, finally reaching the surface and then spreading laterally. Indeed many stratification models (e.g. Sundaram and Rehm, 1972) were designed for such an analysis, either with or without the sophistication of incorporating a plume entrainment scheme.

In the case of a "pumped storage" scheme (in which water is pumped into a reservoir from a nearby river or neighbouring catchment, viz not an impounding scheme), the inflow may be passed through a Venturi constriction and "jetted" into the reservoir. The angle of the jets (see Figure 1a of Burns, 1981) has been determined, largely empirically, by Steel (1975). In Wraysbury Reservoir (U.K.) angles of 0, $\pi/8$ and $\pi/4$ (to the horizontal) are available. The technique of jetting is also used successfully in e.g. Farmoor (Youngman, 1975), although this technique is usually regarded as being necessarily included at the

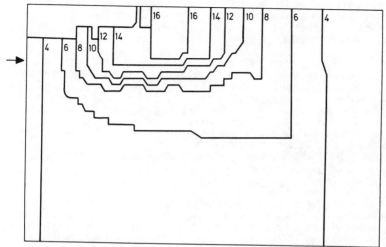

Computer simulation: Year 2 Inflow at 40m

Fig. 1. Stratification pattern for a "high density" influent stream impacting on a reservoir

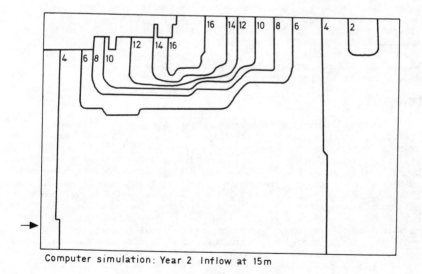

Computer simulation: Year 2 Inflow at 15m

Fig. 2. Stratification pattern for a "low density" influent stream impacting on a reservoir

design stage, rather than at a later date. Again, the entrainment
characteristics can be well modelled by applying jet theories to determine both entrainment (and hence dilution) rates and the effect on the
thermal structure (e.g. Henderson-Sellers, 1978a).

In a pumped storage hydropower scheme, water flows from upper to
lower reservoir by gravity to generate power. The inertia of the water
results in a jet type structure to the water discharged into the lower
lake. When electrical power is reconverted to potential energy, at
times of low demand, again it may be possible to utilise similar models
for the water inflow to the upper lake. Roberts (1981) has undertaken
a literature review of models available for this case. He identifies
16 possible combinations for inflow characteristics (Table 1) based on
the interrelationship between immersion (vs. boundary), jet/buoyancy,
ambient stratification and discharge port shape.

For all these three cases (above) it is evident that either a jet
or plume theory could usefully describe the flow field and mixing
characteristics. It is however not necessary to identify cases of plume
vs. jet flow since models are available (e.g. List and Imberger, 1973;
Henderson-Sellers, 1978a) which describe a "forced plume" (Morton, 1959)
possessing both momentum (jetlike) and buoyancy (plumelike) characteristics. The only caveat to such a description where these models have not
been validated is when the plume* impinges on the upper boundary. (The
description of buoyant surface plumes has in general been pursued
separately (e.g. Stolzenbach and Harleman, 1973; Harleman, 1975;
Muraoka and Nakatsuji, 1980). However, based on atmospheric studies of
plume rise suppression by the presence of elevated inversions (e.g.
Henderson-Sellers and Henderson-Sellers, 1981) together with the direct
effect of the inversion on buoyancy and plume shape (Foster, 1977;
Henderson-Sellers, 1979), it is reasonable to assume that an impinging
plume (at the free water surface) would be successfully modelled by
supposing an inversion of infinite strength to be present at the interface - work on this approach is currently in progress.

DESTRATIFICATION

The main aim of the inflows described above, with the exception of
"jetting", have not been to destratify. The advantages and disadvantages
of destratification are summarised by e.g. Dortch (1979); Henderson-
Sellers (1981); Roberts (1981); Burns (1981) and the effects on the biota
by Pastorok et al. (1980). Intentional destratification techniques may
be hydraulic (pumping) or pneumatic (air inflow). Hydraulic systems are
discussed by Steel (1975) (jets) and Dortch (1979). The latter author
describes experimental work on a pumping system whereby water is taken
out of the reservoir and recirculated to a different depth using 4
different configurations. He recommends withdrawing water from the
epilimnion and discharging it vertically upwards into the hypolimnion.
However, neither energy nor economic costs are discussed in detail, a
concept stressed by Tolland (1977) as vital for evaluating the usefulness of any destratification system.

---

*For simplicity the word plume will be used to describe a forced plume,
pure plume or jet.

TABLE 1.  Roberts (1981) 16 jet types showing the interrelationships possible.

| PRIMARY FLOW CLASSIFICATION | | | | | |
|---|---|---|---|---|---|
| Immersed jet | | Boundary jet | | SECONDARY FLOW CLASSIFICATION | |
| buoyant | non-buoyant | buoyant | non-buoyant | | |
| | | | | Round | Unstratified ambient |
| | | | | Slot | |
| | | | | Round | Stratified ambient |
| | | | | Slot | |

TABLE 2.  Velocity of rising water/air plume, $V_H$, in $ms^{-1}$ as a function of compression ratio, r, and entrainment constant, l (after Henderson-Sellers, 1981)

| | | Compression ratio r | | |
|---|---|---|---|---|
| | | 1.5 | 1.8 | 2.0 |
| Constant of proportionality for entrainment of water by air bubbles, $\ell$ | 4 | 3.5 | 4.1 | 4.5 |
| | 10 | 1.7 | 2.8 | 3.0 |
| | 30 | 1.1 | 1.6 | 1.9 |
| | 100 | 0.6 | 0.8 | 1.0 |

For air systems Tolland (1977) describes the use of oxygenation capacity (OC) and destratification efficiency (DE) which are defined by

$$OC = \frac{\text{net change in oxygen balance over time period } \Delta t}{\text{total energy input over time period } \Delta t} \tag{1}$$

$$DE = 100 \times \left(\frac{\text{net change in stability over time period } \Delta t}{\text{total energy input over time period } \Delta t}\right) \% \tag{2}$$

where the stability is measured as the difference between the actual PE and the PE of the equivalent isothermal reservoir. Tolland (1977) comments that although OC is preferable as an indicator of water quality improvement, data to calculate this parameter are more difficult to obtain than for DE - indeed neither is found to be entirely satisfactory and indeed some reservoir aeration/destratification systems have neither design parameter delineated as appropriate (e.g. Helixors - see also discussion below).

The three major aeration systems are

1) bubble guns
2) air lines, diffusers or bubble screens
3) Helixors

"Bubble guns", typified by the Aero-Hydraulics Gun (Bryan, 1964), consist of a vertical tube from which single bubbles (of pipe radius) are released intermittently. A similar device was evaluated in the laboratory by the Hydraulics Research Station (1978). Although this induces a constant water flow through the system, the bubble has a small surface volume ratio restricting the efficiency of both oxygen transfer and reservoir mixing. Bernhardt (1967) has calculated the OC for the system as 0.63 kg $O_2$/kWh. This value is in poor comparison with values of up to 1.95 kg $O_2$/kWh quoted by Symons et al (1970) for diffused air systems. Perhaps the simplest diffused air system is an air line at or just above the reservoir bottom. In the U.K. this system has been installed in 16 reservoirs (Davis, p.c., 1982) using a perforated pipe with holes 0.8 mm in diameter and 0.3 m apart. This system provides an air bubble screen which can be described in terms of release rates, bubble sizes etc. (e.g. Tekeli and Maxwell, 1978). The effect of bubble releases is analysed by Kranenburg (1979) in terms of a three layer reservoir system for near field, far field and transition zones. Other modelling approaches of air bubble release have been attempted using jet/plume theories. Based on work of Kobus (1968), Ashton (1978, 1979) calculated velocities, spread rates and induced flow of water at the surface for line and point sources respectively. The width b of the flow at the surface is given as

$$b = (H + x_o) \, C_c \, Q_a^{0.15} \tag{3}$$

where H is the depth of submergence, $x_o$ an empirical correction length = 0.8 m, $C_c = 0.182 \text{ m}^{-0.3} \text{ s}^{0.15}$ and $Q_a$ the air discharge rate in $\text{m}^2 \text{ s}^{-1}$ (line source), $\text{m}^3 \text{ s}^{-1}$ (point source). However it must be remembered that the rising air/water plume is in fact two superimposed plumes in which the bubbles rise faster than the water flow they have themselves

entrained. This has been recognised in the recent formulation of
McDougall (1978). This model describes a 'double-plume' structure
consisting of an inner bubble plume which rises to the surface and an
outer, largely water plume which may reach an equilibrium level (below
the surface) independently and spread laterally at that level. One
disadvantage in some reservoirs with unconfined bottom air releases is
that the locally induced flow field will stir up bottom muds which could
then be carried vertically upwards.

A device which, as does the Aero-Hydraulics Gun, contains the air/
water flow for the first part of the rise is the Helixor (Figure 4) which
is a polyethylene tube divided by a moulded helical component. The
length of travel between top and bottom is greater than in bubble guns
or perforated pipes and the emergent plume has an added swirl, which
complicates the mathematics of plume/jet entrainment (see e.g. Narain,
1974), whilst increasing oxygen transfer and reservoir mixing ability.
(In all destratification devices using air input, it is widely acknowledged that the re-oxygenation of hypolimnetic water is accomplished
primarily by air entrainment at the air-water interface and not from
direct solution of the air bubbles themselves. The rate and degree of
oxygen solution is also determined by climatic and topographic characteristics.) As yet no comprehensive theory of the Helixor is available
to provide a useful model. The manufacturers' claim, that the total
volume delivered to the surface, $Q_T$, can be calculated from the total
delivery through the pump, $Q_0$ (diameter d, distance below surface s)
using the simple relationship

$$Q_T = \frac{Q_0 \, s}{2.95d} \qquad\qquad (4)$$

is suspect if the distance from the lake surface to the top of the
Helixor is large (Henderson-Sellers, 1981). The manufacturers use
"guidelines" as design criteria, which have been, to a first order of
magnitude, substantiated by the work of Henderson-Sellers (1981). Two
simple models were proposed to calculate the velocity at the top of the
Helixor, $V_H$, which can be used as input to the (swirling) plume model
to calculate entrainment rates. The results of the first model are
given in Table 2 of the second model, which includes the different air
and water plume velocities, in Figure 5. This latter figure is in
excellent agreement with the empirical curve produced by the manufacturers (dotted curve in Figure 5) and gives induced flow rates of the same
order of magnitude as those derived by Ashton (1979) for point bubblers.

Hence, the jet/plume theories and models described above are found
to be useful in this case; but once again it should be stressed that it
is necessary to recognise that the system is an air and water plume.

CONCLUSIONS

For the majority of Robert's (1981) 16 jet types it would appear
that a single (or at most two) models would be sufficient. Such a
model would be capable of describing both the trajectory and dilution
characteristics of an inflow of water into a lake or reservoir and, if
coupled with a stratification model, of simulating the effected

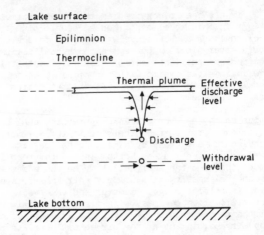

Fig. 3.   Buoyant plume of heated effluent
          discharged into reservoir (after
          Sundaram and Rehm, 1972)

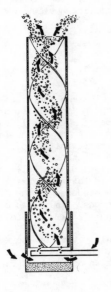

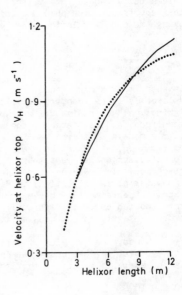

Fig. 4.   Schematic diagram
          of Helixor

Fig. 5.   Pumping capacity curve
          (observed and calculated)
          for a single Helixor as a
          function of Helixor
          length

destratification. The approach may be used for pumped storage schemes, natural inflows or cooling ponds.

When an aeration device is used for destratification a similar plume/jet formulation requires consideration of the fact that the air bubbles and entrained water form a 'double plume' (McDougall, 1978).

The type of destratification unit employed depends upon strength of stratification and extent of hypolimnetic deoxygenation (both controlled partly by prevailing climatic conditions) and by the lake morphometry and design operation (e.g. impounding or pumped storage).

REFERENCES

Ashton, G.D., 1978, Numerical simulation of air bubbler systems, Can.J. Civ.Eng., 5, 231-238.
Ashton, G.D., 1979, Point source bubbler systems to suppress ice, Cold Regions Science and Technol., 1, 93-100.
Bernhardt, H., 1967, Aeration of Wahnbach Reservoir, J.Am.Wat.Wks.Ass., 59, 943-964.
Bryan, J.G., 1964, Physical control of water quality, Br.Wat.Wks.Ass.J., 46, 546-564.
Burns, F.L., 1981, Hydraulic modelling of destratification, in Destratification of Lakes and Reservoirs to Improve Water Quality, ed F.L. Burns and I.J. Powling, Australian Govt. Publishing Service, Canberra, pp121-129.
Dortch, M.S., 1979, Artificial destratification of reservoir, Tech.Rept E-79-1., US Army Engineer Waterways Experiment Station, Vicksburg, Miss, 59pp.
Foster, P.M., 1977, Central Electricity Research Laboratories, Rept. RD/L/N 122/76.
Harleman, D.R.F., 1975, Heat disposal in water environment, Procs. ASCE, J.Hyd.Div., 101(HY9), 1120-1138.
Henderson-Sellers, A. and Henderson-Sellers, B., 1981, The effects of the urban environment on the trajectories of stack plumes, Appl.Math. Model., 5, 151-157.
Henderson-Sellers, B., 1978a, Forced plumes in a stratified reservoir, Procs.ASCE., J.Hyd.Div., 104(HY4), 487-501.
Henderson-Sellers, B., 1978b, The longterm thermal behaviour of a freshwater lake, Procs. Institution of Civil Engineers (Part 2), 65, 921-927.
Henderson-Sellers, B., 1979, The cross-sectional asymmetry of chimney plumes in inversion conditions, Appl.Math.Model., 3, 327-331.
Henderson-Sellers, B., 1981, Destratification and reaeration as tools for in-lake management, Water SA, 7, 185-189.
Hydraulics Research Station, 1978, Air bubbles for water quality improvement, Report OD/12, HRS, Wallingford, England.
Kranenburg, C., 1979, Destratification of lakes using bubble columns, Procs.ASCE, J.Hyd.Div., 101(HY1), 97-114.
List, E.J. and Imberger, J., 1973, Turbulent entrainment in buoyant jets and plumes, Procs.ASCE., J.Hyd.Div., 99(HY9), 1461-1474.
McDougall, T.J., 1978, Bubble plumes in stratified environments, J.Fluid Mech., 85, 655-672.
Morton, B.R., 1959, Forced plumes, J.Fluid Mech., 5, 151-163.

Muraoka, K. and Nakatsuji, K., 1980, Recent advances in hydrodynamic treatment for a buoyant surface discharge, in Advances in Environmental Science and Engineering, ed J.R. Pfafflin and E.N. Ziegler, Vol 3, 30-55, Gordon and Breach, N.Y.

Narain, J.P., 1974, Swirling shallow submerged turbulent plumes, Procs. ASCE., J.Hyd.Div., 100(HY9), 1229-1243.

Pastorok, R.A., Ginn, T.C., and Lorenzen, M.W., 1980, Review of aeration/ circulation for lake management, in Restoration of Lakes and Inland Waters; International Symposium on Inland Waters and Lake Restoration, Portland, Maine, Sept 8-12, pp124-133, USEPA 440/5-81-010.

Pitman, W.V., 1976, A mathematical model for generating daily flows from meteorological data in South Africa, Rept. No. 2/76, Hydrological Research Unit, University of the Witwatersrand, Johannesburg.

Roberts, P.J.W., 1981, Jet entrainment in pumped-storage reservoirs, Tech Rept E-81-3, Prepared by Georgia Institute of Technology, Ga, for the US Army Engineer Waterways Experiment Station, CE, Vicksburg, Miss., 63pp.

Steel, J.A., 1975, The management of Thames Valley reservoirs, paper 14, Water Research Centre Symposium on the Effects of Storage on Water Quality, March 24-26.

Stefan, H., and Ford, D.E., 1975, Temperature dynamics in dimictic lake, Procs.ASCE., J.Hyd.Div., 101(HY1), 97-114.

Stolzenbach, K.D. and Harleman, D.R.F., 1973, Three-dimensional heated surface jets, Water Resour.Res., 9, 129-137.

Sundaram, T.R. and Rehm, R.G., 1972, Effects of thermal discharges on the stratification cycle of lakes, Am.Inst.Aeronaut.Astronaut.J., 10, 204-210.

Symons, J.M., Carswell, J.K. and Robeck, G.G., 1970, Mixing of water supply reservoirs for quality control, J.Am.Wat.Wks,Ass., 62, 322-334.

Tekeli, S. and Maxwell, W.H.C., 1978, Behaviour of air bubble screens, Civil Eng. Studies, Univ.Illinois at Urbana-Champaign, Hydr.Eng.Res. Series, No. 33, UILU-ENG-78-2019.

Tolland, H.G., 1977, Destratification/aeration in reservoirs, Water Research Centre Tech Rept TR50, 37 pp.

Youngman, R.E., 1975, Observations on Farmoor; a eutrophic reservoir in the Upper Thames Valley, during 1965-1973, Water Research Symposium on the Effects of Storage on Water Quality, March 24-26.

BARNEGAT INLET HYDROGRAPHIC SURVEY COMPARISON

Samuel B. Heltzel,[1] M. ASCE, Victor E. LaGarde[1]
and John H. G. Shingler[1]

## Introduction

Tidal inlets are very dynamic with active sediment processes constantly remolding them. When a navigation project is required, it is
very important for the design engineer to be cognizant of both short-
and long-term modifications of the bathymetric nature in and surrounding
the inlet. An example of such a study was a hydrographic survey comparison of Barnegat Inlet conducted for the Philadelphia District by
the U. S. Army Engineer Waterways Experiment Station (WES).

The objectives of this paper are to briefly describe the computer
based data management system (DMS) used in performing the analysis for
this study, and to illustrate the application of the DMS by discussing
the results of the study and the problems encountered.

## Barnegat Inlet historical development

Barnegat Inlet is located on the Atlantic Coast of New Jersey
approximately 32 miles northeast of Atlantic City. The inlet serves as
a major passageway between Barnegat Bay and the Atlantic Ocean for many
recreational, commercial fishing, and U. S. Coast Guard vessels.

The existing Federal Navigation project constructed between 1939
and 1944, has failed to maintain a suitable channel for navigation
through the inlet. Considerable shoaling exists within the inlet and
in areas immediately outside the jetties. The authorized plan would
modify this project by providing a new jetty on the south side of the
inlet, aligned generally parallel to the existing north jetty and a
navigation channel 300 ft wide at 10 ft below mean low water through
the inlet.

## Purpose of project

The purpose of the project entitled "Barnegat Inlet Hydrographic
Survey Comparisons" was to compare nine hydrographic surveys that were
obtained for the inlet between the years 1932 and 1979. The comparisons
were made to determine volumetric changes inside and outside of the inlet
and to trace the development of scour and fill patterns of the inlet.

---

[1]U. S. Army Engineer Waterways Experiment Station, Vicksburg, MS 39180.

The project began with 9 hydrographic surveys and concluded with 9 contour maps; 8 difference maps; 72 area difference calculations, graphs, and plots; and 72 volumetric difference calculations and graphs. Sample results are discussed in later sections of this paper.

## Data management system

The DMS used by WES is described by LaGarde and Heltzel (1980). The process involved is to digitize and grid the elevation data, contour each set of data, compute the area and volume change between each hydrographic survey, and contour the differences. The process is straightforward but time consuming, especially in the initial phase of digitizing and editing the data.

## Metholology used and problems encountered

The basic areal unit for storing, retrieving, analyzing, and displaying the data is the grid cell. Figures 1 and 2 show the eight subregions and grid structure, respectively, for this study. A rectangular frame is placed over the region containing the study area. The frame is aligned with a rectilinear coordinate system such as a state planar grid system. The frame normally contains one or more irregularly shaped regions defined as area of interest for which analysis will be performed, and also other regions for which no subsequent analysis will be performed.

The region within the total frame is gridded, i.e., the information content of the input data is captured in the form of values for each grid. The size of the grid cell is dependent on the density and structure of the map data. A proper choice of grid size has a crucial bearing on the validity of the surface-fitting technique. If very coarse grid squares are chosen, many data points reside in each grid element. This causes data points to be ignored, resulting in loss of detail and accuracy. Likewise if the grid squares are chosen too small the grid points will not be well surrounded by data points. For best results, an attempt should be made to maintain approximately one data point per grid element. The grided data base thus obtained is used for all subsequent analysis. A grid spacing of 100 ft was used for this study.

Figure 3 shows the basic operations performed and the various software packages used in this study. The procedures used can be grouped into three broad classes: data recovery, data processing, and data retrieval and display. The DMS is configured so that the user progresses through the system with minimal manual handling of the data. Almost all user intervention takes place in the data recovery operation. A brief description of each computer program is given in Table 1.

The contour program fits a surface to the gridded data and produces a contour map of the data at all desired contour intervals. Difference contours are obtained by computing the changes between the similarly gridded data bases, then fitting a surface to this difference grid and producing a contour map. Other programs provide area and volume information necessary for water resource studies. Area and volume changes are calculated above and between bathymetric contours specified by the user and results reported in several forms. Calculations can be performed

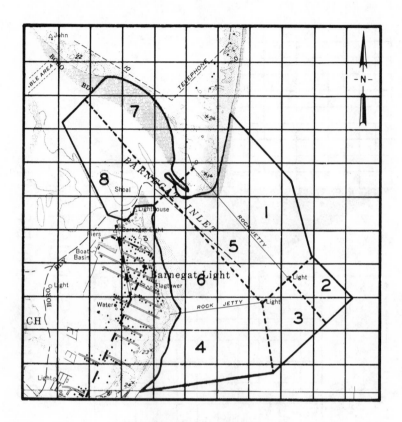

## SECTOR MAP
## Barnegat Inlet, NJ

Figure 1

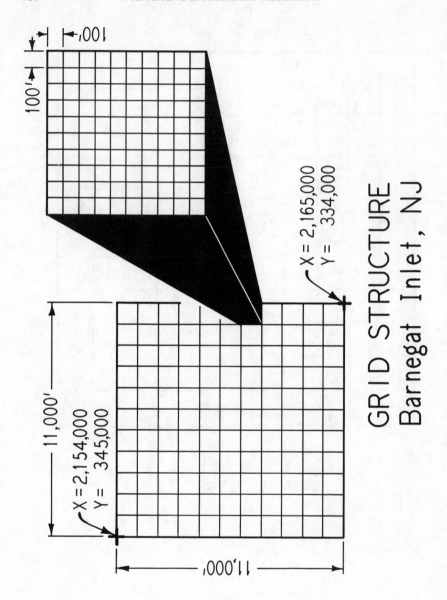

Figure 2

# PROCEDURES FOLLOWED IN USING THE DMS

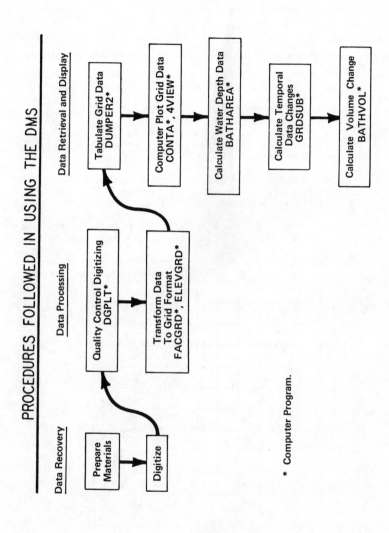

\* Computer Program.

Figure 3

Table 1

Description of Computer Programs

| Computer Program | Description |
| --- | --- |
| DGPLT | This program is used to perform quality control of the collected raw data. It checks the data form and format and consistency of the identification codes and prepares a computer plotted map of the data for visual inspection. |
| FACGRD | This program transforms any factor map type data from a linear string of XY coordinates defining the boundaries of factor patches to a grid array format. |
| ELEVGRD | This program transforms any variable surface data from a string of values at simple points into a three-dimensional description of the surface in a grid format. |
| 4VIEW | Used primarily for quality control, the program produces a series of computer plots, showing the three-dimensional structure of any gridded data. |
| CONTA | This program is used to contour map any gridded variable surface data. |
| SUMPER2 | Provides a high-speed printer map of any gridded data. |
| GRDSUB | This program aligns two grid maps using the master data base coordinate system and provides the difference between maps where there are data for both maps. |
| BATHAREA | Calculates the area of the underwater ground surface between any series of selected bathymetric contours. |
| BATHVOL | Calculates volume above and between specified bathymetric contour levels for the volume change in material between any series of selected bathymetric surveys. |

for the total area of coverage of the data base or for subregion patches the user specifies within the region covered by the data base. The area and volume computations were performed on the entire survey and for the eight individual subregions. The program presently handles up to 100 patches simultaneously in a single operation.

## Surface fitting procedure

Producing contour maps and computing volumes by hand is typically very tedious, making it quite advantageous to use a computer for this task. This involves, however, the difficult task of fitting a surface to a set of (X, Y, Z) data points. Where X and Y are horizontal (surface) coordinates, and Z is an elevation value. The task is much easier to do if the data are spaced on a rectangular grid. Unfortunately, few applications allow data to be supplied in this form. If the available (X, Y, Z) data points are to be artificially spaced on a rectangular grid, a surface fitting technique must be employed to obtain a set of Z values at the grid points consecutively and independently of each other. The calculation procedure followed at each grid point is as follows. A cartesian coordinate system is then centered on the grid location, oriented with the slope vector, and the nearest neighbor input data point in each quadrant is tagged. These nearest neighbors are used to calculate the grid-location value using the equation

$$Zg = \frac{\sum\limits_{i=1}^{4} \dfrac{Zi}{di^2}}{\sum\limits_{i=1}^{4} \dfrac{1}{di^2}}$$

where

    Zg = data value at the grid position

    Zi = data value of the nearest neighbor input data point in the ith quadrant

    di = distance between the grid and the ith nearest neighbor locations

## Results

The data recovery and processing phases were accomplished in a four-step procedure used with each of the nine surveys prior to the data retrieval and display phase. The survey sheets were first acquired and checked for distortion. This was accomplished by preparing these with the located points on the survey sheets. A rectangular grid was placed on these sheets and the state planar coordinates were placed on the grid. A sector map was also prepared in this phase. This was accomplished as step 1, prepare materials. Next, the data were digitized (step 2). Then an overlay of the digitized spatial locations was produced for quality control of digitizing (step 3). Step 4 was to transform the data to a grid format.  The data were then ready to be retrieved and displayed.

The study results isolated quite nicely the areas of active shoaling and showed the sequence of changes through the years.  As expected, the results confirmed the active shoaling occuring in this area.  It was also determined that due to slight digitizing differences between the patch overlays and the survey area of interest, a two percent error in totals was observed.  Therefore, the total computed is slightly different from the sum of the individual patch volumes.

This study also showed that it would be better to break the study area into more subregions for a more detailed analysis.  It would be possible to have small patches near the area of interest and larger ones further away.  The size of the patches would also be dependent on the grid size chosen.

## Acknowledgment

The work described herein was funded by the U. S. Army Engineer District, Philadelphia.  Permission of the Chief of Engineers and the Philadelphia District to publish this paper is gratefully acknowledged.

## References

LaGarde, V. E., III and Heltzel, S. B.  1980.  A Data Management System for Finite Element Sediment Transport Models.  Proceedings of the Third International Conference on Finite Elements in Water Resources, the University of Mississippi, University, Miss.

NUMERICAL MODELING OF INLET-ESTUARY SYSTEMS

By H. Lee Butler[1]

ABSTRACT

The Army Corps of Engineers, as part of its mission, has had to
address various problems (hazardous navigation conditions, shoaling and
erosion problems, etc.) associated with tidal inlet hydraulics and to
propose improvement alternatives. This presentation discusses the use
of a two-dimensional numerical model (WIFM) for simulating tidal
hydrodynamics for existing and plan conditions.
To achieve a solution of the governing equations, ADI finite-
difference techniques are employed on a stretched rectilinear grid
system. The model predicts vertically integrated flow patterns as well
as the distribution of water-surface elevations. Code features include
the treatment of regions that are inundated during a part of the
computational cycle, subgrid-scale barrier effects, and a variety of
permissible boundary conditions and external forcing functions.
An application of WIFM is presented to demonstrate model usage in
treating inlet-estuarine system problems. Elements of improvement
plans tested in the various models include installations of jetties
(with and without low weir sections), bulkheads, and training groins;
channel construction and maintenance; and marine developments.

INTRODUCTION

Various mathematical models have been developed to investigate the
hydrodynamic processes of large bodies of water including the design,
operation, and maintenance of various coastal projects. This paper
discusses the development of a two-dimensional finite difference model
(Butler 1980) and the application to a variety of Corps of Engineers
studies.
A two-dimensional model known as the Waterways Experiment Station
(WES) Implicit Flooding Model (WIFM) was first devised for application
in simulating tidal hydrodynamics of Great Egg Harbor and Corson Inlets,
New Jersey (Butler 1978). Program WIFM originally employed an implicit
solution scheme similar to that developed by Leendertse (1970) and has
been applied in numerous studies where tidal, storm surge, and tsunami

---

1.  Research Physicist, Wave Dynamics Division, Hydraulics Laboratory,
    US Army Engineer Waterways Experiment Station, CE, Vicksburg,
    Mississippi 39180.

inundation phenomena were simulated. Basic features of the model include
flood modeling of low-lying terrain, treatment of subgrid barrier effects,
and a variable grid option. Included in the model are actual bathymetry
and topography, time and spatially variable bottom roughness, inertial
forces due to advective and Coriolis acceleration, rainfall, and spatial
and time-dependent wind fields. Horizontal diffusion terms in the
momentum equations are optionally present and can be used, if desired,
for aiding stability of the numerical solution.

GOVERNING EQUATIONS

The basic equations used in modeling hydrodynamics of inland and
coastal waters are derived from the classical Navier-Stokes equations in
a Cartesian coordinate system (Figure 1). By assuming (a) the pressure
varies hydrostatically in the vertical direction; (b) density variations
are negligible except in the buoyancy term; and (c) eddy coefficients
are used to account for turbulent diffusion effects, the equations of
conservation of mass and momentum are:

$$u_x + v_y + w_z = 0 \tag{1}$$

$$u_t = -\frac{1}{\rho} p_x - \left[(u^2)_x + (uv)_y + (uw)_z\right] + fv + (A_H u_x)_x$$
$$+ (A_H u_y)_y + (A_V u_z)_z \tag{2}$$

$$v_t = -\frac{1}{\rho} p_y - \left[(uv)_x + (v^2)_y + (vw)_z\right] - fu + (A_H v_x)_x$$
$$+ (A_H v_y)_y + (A_V v_z)_z \tag{3}$$

$$p_z = -\rho g \tag{4}$$

where u , v , and w are the three-dimensional velocities in the x ,
y , and z directions; t is time; f is the Coriolis parameter; g
is the acceleration due to gravity; p is the pressure; $\rho$ is the
fluid density; $A_H$ and $A_V$ are the horizontal and vertical eddy coefficients. Omitted for brevity are equations for conservation of salinity and temperature, an equation of state, and appropriate boundary
conditions.

If the additional assumption of fluid homogeneity is made and a
depth-averaging process applied, one can derive the usual two-dimensional
form of the governing equations, namely:

$$\eta_t + U_x + V_y = 0 \tag{5}$$

$$U_t = -gd\eta_x - \left[(U^2)_x + (UV)_y\right] + fV + E_H (U_{xx} + U_{yy})$$
$$+ \tau_{xs} - \tau_{xb} = -gd\eta_x + M_x \tag{6}$$

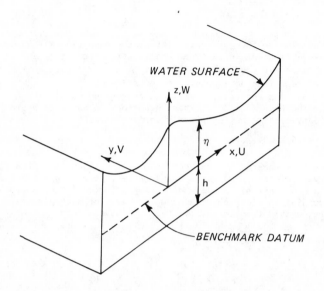

Figure 1.    Cartesian coordinate system.

APPLICATION OF STRETCHED COORDINATES

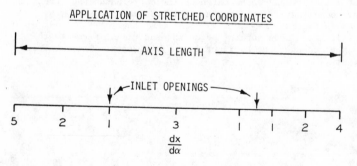

Figure 2.    Illustration of coordinate stretching rates along
a region axis.

$$V_t = -gd\eta_y - \left[(UV)_x + (V^2)_y\right] - fU + E_H (V_{xx} + V_{yy})$$

$$\tau_{ys} - \tau_{yb} = -gd\eta_y + M_y \tag{7}$$

where U and V are the vertically integrated mass fluxes; $\eta$ is the water-surface elevation; d is the local water depth; $\tau_{xs}$ and $\tau_{ys}$ are shear stresses at the free surface; $\tau_{xb}$ and $\tau_{yb}$ are bottom shear stresses; and $E_H$ is a horizontal eddy coefficient.

The discussions that follow will concentrate on solving the vertically integrated Equations 5-7. These equations, along with appropriate boundary conditions, completely define the WIFM model.

## COORDINATE TRANSFORMATION

A major advantage of the subject code is the capability of applying a smoothly varying grid to a given study region, permitting simulation of a complex landscape by locally increasing grid resolution and/or aligning coordinates along physical boundaries. For each direction, a piecewise reversible transformation, which takes the form:

$$x = a_x + b_x \, \alpha^{c_x}$$

$$y = a_y + b_y \, \gamma^{c_y} \tag{8}$$

where coefficients $a_x$, $a_y$, $b_x$, $b_y$, $c_x$, $c_y$ are arbitrary and to be determined, is independently used to map prototype or real space into computational space. Variables $\alpha$ and $\gamma$ are directions in computational space. This procedure treats the computational domain as consisting of a number of regions for which different sets of equations (as in Equation 8 above) apply. The mapping coefficients are determined from an iterative procedure by matching the coordinates and stretching rates, $dx/d\alpha$ or $dy/d\gamma$ at boundaries of adjacent regions (as illustrated in Figure 2).

The stretching does not introduce any additional terms to the equations, but changes the horizontal gradient terms. The resulting equations are:

$$\eta_t + \frac{1}{\mu_x} U_x + \frac{1}{\mu_y} U_y = 0 \tag{9}$$

$$U_t = - \frac{gd}{\mu_x} \eta_x + M_x^\mu \tag{10}$$

$$V_t = - \frac{gd}{\mu_y} \eta_y + M_y^\mu \tag{11}$$

where $\mu_x = dx/d\alpha$ and $\mu_y = dy/d\gamma$ and $M_x^\mu$, $M_y^\mu$ represent transformation effects on remaining components of the momentum equations.

ADI FINITE-DIFFERENCE SCHEME

The differential Equations 9-11 are to be approximated by difference equations. The scheme used in program WIFM is a "leap-frog," three time level scheme which can be derived from the differential equation:

$$\hat{W}_t + A\hat{W}_x + B\hat{W}_y + \hat{M} = 0 \tag{12}$$

where the component variables of $\hat{W}$ are expressed in velocity form and

$$\hat{W} = \begin{pmatrix} \eta \\ u \\ v \end{pmatrix} \quad , \quad A = \begin{pmatrix} o & d/\mu_x & o \\ g/\mu_x & o & o \\ o & o & o \end{pmatrix}$$

$$B = \begin{pmatrix} o & o & d/\mu_x \\ o & o & o \\ g/\mu_x & o & o \end{pmatrix} \quad , \quad \hat{M} = \begin{pmatrix} o \\ M_x \\ M_y \end{pmatrix} \quad ,$$

with $\hat{M}$ containing the velocity form of the nonlinear terms in the governing equations. The approximating difference equation is written as:

$$(1 + 2\lambda_x + 2\lambda_y)\ \hat{W}^{n+1} = (1 - 2\lambda_x - 2\lambda_y)\ \hat{W}^{n-1} - 2\Delta t\ \hat{M}^n \tag{13}$$

where

$$\lambda_x = \frac{1}{2}\frac{\Delta t}{\Delta x} A\ \delta_x \text{ and } \lambda_y = \frac{1}{2}\frac{\Delta t}{\Delta y} B\ \delta_y$$

and $\delta_x$ and $\delta_y$ are central spatial difference operators. By adding the quantity $\lambda_x\lambda_y(\hat{W}^{n+1}-\hat{W}^{n-1})$ to permit factorization, the following relation is obtained:

$$(1 + 2\lambda_x)(1 + 2\lambda_y)\ \hat{W}^{n+1} = (1 - 2\lambda_x)(1 - 2\lambda_y)\ \hat{W}^{n-1} - 2\Delta t\hat{M}^n \tag{14}$$

It can be shown that the addition of the extra term is equivalent to the addition of the truncation error

$$\frac{\Delta t^2}{4} AB\ \frac{\partial^3\hat{W}}{\partial t\partial x\partial y}$$

Thus, the factorized finite-difference Equation 14 is still a second-order approximation to the differential Equation 12. The advantage of using Equation 14 lies in the fact that the solution of the factorized form can be split into two separate one-dimensional operations (the ADI approach).

By factorizing Equation 14 and introducing an intermediate level, $\hat{W}^*$, the structure for WIFM's solution algorithm can be written as:

$$(1 + 2\lambda_x) \; \hat{W}* = (1 - 2\lambda_x - 4\lambda_y) \; \hat{W}^{n-1} - 2\Delta t \; \hat{M}^n \tag{15}$$

$$(1 + 2\lambda_y) \; \hat{W}^{n+1} = \hat{W}* + 2\lambda_y \; \hat{W}^{n-1} \tag{16}$$

The double-sweep solution technique is used to solve each operational step. A functional representation of each sweep can be expressed as:

x-sweep

$$\eta* = F_1 \; [\eta^n, \; \eta^{n-1}, \; u^*, \; u^n, \; u^{n-1}, \; v^{n-1}] \tag{17}$$

$$u^* = F_2 \; [\eta^*, \; \eta^n, \; \eta^{n-1}, \; u^*, \; u^n, \; u^{n-1}, \; v^n, \; v^{n-1}] \tag{18}$$

$$v^* = F_3 \; [\eta^n, \; \eta^{n-1}, \; u^n, \; u^{n-1}, \; v^n, \; v^{n-1}] \tag{19}$$

y-sweep

$$\eta^{n+1} = G_1 \; [\eta^*, \; v^{n+1}, \; v^{n-1}] \tag{20}$$

$$u^{n+1} = G_2 \; [u^*] = u^* \tag{21}$$

$$v^{n+1} = G_3 \; [\eta^{n+1}, \eta^n, \eta^{n-1}, \; u^n, \; u^{n-1}, \; v^{n+1}, \; v^*, \; v^n, \; v^{n-1}] \tag{22}$$

Noting that $v^*$ is an explicit expression, a substitution of $F_3$ (Equation 19) into $G_3$ (Equation 22) is made. Thus, each sweep consists in solving a one-dimensional problem involving $\eta^*$ and $u^*$ in the x-sweep and $\eta^{n+1}$ and $v^{n+1}$ in the y-sweep.

Implicit methods are characterized by a property of unconditional stability in the linear sense. The scheme used in WIFM is limited by a weak condition, namely,

$$\Delta t \leq \min_{x,y} \; [\frac{(\Delta x, \; \Delta y)}{(u^2 + v^2)^{1/2}}] \tag{23}$$

In general, this limitation on $\Delta t$ is at least two orders of magnitude larger than the limit imposed on an explicit scheme by the surface gravity wave, namely,

$$\min_{x,y} \; [\frac{(\Delta x, \; \Delta y)}{(u^2 + v^2)^{1/2}}] \; >> \; \min_{x,y} \; [\frac{\Delta x, \Delta y}{\sqrt{gd}}] \tag{24}$$

APPLICATIONS

Program WIFM has been used successfully in many applications conducted at WES. These include tidal circulation studies for Masonboro, Inlet, North Carolina, and Coos Bay Inlet-South Slough, Oregon;

storm surge applications for Hurricane Eloise, Panama City, Florida, and Hurricane Carla, Galveston, Texas; and tsunami inundation simulations for Crescent City, California, and the Hawaiian Islands. A recent paper (Butler 1980) summarizes these applications.

To exemplify use of the model, a brief description of the application to the Coos Bay Inlet-South Slough, Oregon, area is presented. Coos Bay Inlet routes heavy shipping up the Coos Bay River to North Bend and small-craft traffic (commercial fishing and pleasure craft) to Charleston Harbor just south of the inlet entrance. Continual shoaling problems exist within the entrance channel to Charleston Harbor, and cost of additional maintenance warranted installation of structures that would alleviate the problem as well as provide a reduction in wave damage in the harbor's boat basin. South Slough is a shallow body of water south of Charleston and fed by the Charleston channel. The upper reaches of the slough constitute a National Marine Sanctuary. The major task was to analyze alternate plan conditions as to their impact on the tidal hydraulics of the entire system and of South Slough in particular. Prototype tidal and velocity data were gathered in a field survey conducted by WES. A variable mesh was designed with grid spacing ranging from 150 ft to 900 ft. The finer mesh was focused around the inlet entrance and Charleston Harbor where proposed structures would likely be placed. Figure 3 displays all plan conditions, and some of the major tide and velocity stations used in comparing plan with base conditions. Figures 4 and 5 show typical results obtained for model verification. Figure 6 depicts the comparison between plan and existing conditions (Plans A and E) for currents at Station U located in the entrance to South Slough. Volumetric discharge through various ranges within the system also was computed, and all results indicated that the tidal prism and circulation in South Slough would not be altered by construction of any of the plans tested. Plan A was constructed and has proved successful in allieviating shoaling problems. Postconstruction prototype data indicate that no changes in the hydrodynamic characteristics of the entrance to South Slough occurred after construction.

CONCLUSIONS AND RECOMMENDATIONS

This paper presents details of the development of a two-dimensional finite-difference hydrodynamic model. A coordinate transformation is used to obtain finer resolution in important local areas without sacrificing economical application of the model. References are given for specific model investigations along with a brief description of an inlet-estuary application to the Coos Bay area. The current model is being extended to include modeling baroclinic effects. Additional implementation of boundary-fitted coordinates via the use of elliptic grid generation techniques would improve the model's applicability.

ACKNOWLEDGMENT

The research described and example application presented herein were conducted under various programs of the United States Corps of Engineers by the U. S. Army Engineer Waterways Experiment Station. Permission was granted by the Chief of Engineers to publish this information.

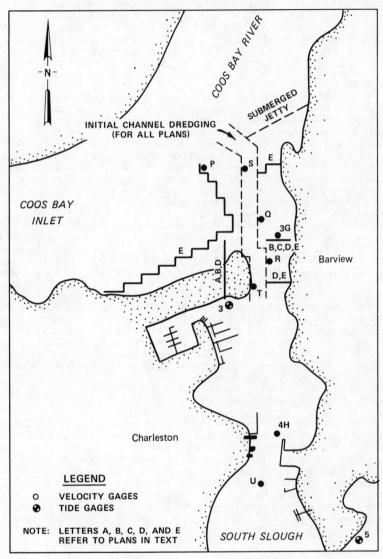

Figure 3.   Display of alternative improvement plans and
            major observation stations.

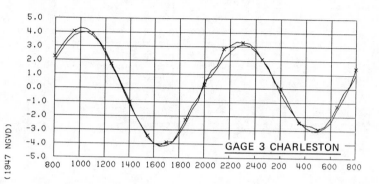

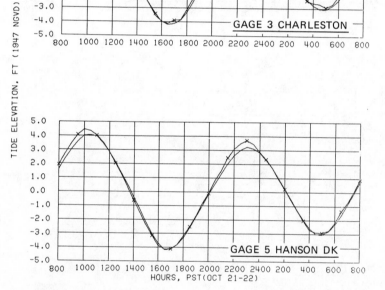

TEST CONDITIONS
OCEAN TIDE RANGE 8.2 FT
21-22 OCT 1976

LEGEND
——— PROTOTYPE
✳————✳ MODEL

Figure 4.  Example computed and measured water level at stations
in Charleston Harbor (3) and South Slough (5).

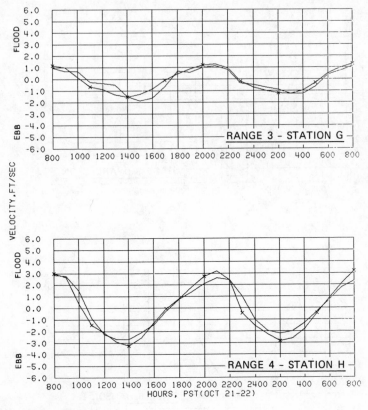

TEST CONDITIONS
OCEAN TIDE RANGE 8.2 FT
21-22 OCT 1976

LEGEND
——— PROTOTYPE
✶——✶ MODEL

Figure 5.   Example computed and measured current velocity at
            stations in the entrance to Charleston Harbor (3G)
            and in the entrance to South Slough (4H).

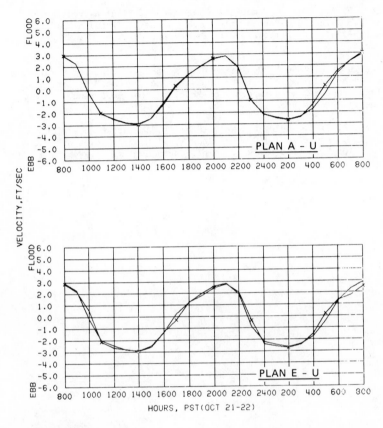

TEST CONDITIONS
OCEAN TIDE RANGE 8.2 FT
21-22 OCT 1976

LEGEND
——— BASE
✗——✗ PLAN

Figure 6. Example model current comparisons of plan with existing (base) conditions at station U for plans A and E.

REFERENCES

1.  Butler, H. L., 1978. "Numerical Simulation of Tidal Hydrodynamics: Great Egg Harbor and Corson Inlets, New Jersey," Technical Report H-78-11, US Army Waterways Experiment Station, CE, Vicksburg, MS, June.

2.  Butler, H. Lee, 1980. "Evolution of a Numerical Model for Simulating Long-Period Wave Behavior in Ocean-Estuarine Systems," Estuarine and Wetland Processes with Emphasis on Modeling, Marine Science Series, Volume 11, Plenum Press, New York.

3.  Leendertse, J. J., 1970. "A Water-Quality Simulation Model for Well-Mixed Estuaries and Coastal Seas, Vol. 1, Principles of Computation," RM-6230-rc, Rand Corp., Santa Monica, CA, February.

# SUBJECT INDEX

Page numbers refer to first page of paper

731